Open School-Conference of NIS Countries - Ultrafine Grained and Nanostructured Materials - 2020 (UFGNM-2020)

IOP Conference Series: Materials Science and Engineering
Volume 1008

Ufa, Russia
5 - 9 October 2020

ISBN: 978-1-7138-4544-7
ISSN: 1757-8981

Printed from e-media with permission by:

Curran Associates, Inc.
57 Morehouse Lane
Red Hook, NY 12571

Some format issues inherent in the e-media version may also appear in this print version.

Institute of Physics
Dirac House, Temple Back
Bristol BS1 6BE UK

Phone: 44 1 17 929 7481
Fax: 44 1 17 920 0979

techtracking@iop.org

Additional copies of this publication are available from:

Curran Associates, Inc.
57 Morehouse Lane
Red Hook, NY 12571 USA
Phone: 845-758-0400
Fax: 845-758-2633
Email: curran@proceedings.com
Web: www.proceedings.com

TABLE OF CONTENTS

IOP Conf. Series: Materials Science and Engineering 1008 (2020) 011001 doi:10.1088/1757-899X/1008/1/011001

PREFACE

Proceedings of the Open School-Conference of NIS Countries «Ultrafine Grained and Nanostructured Materials – 2020» (UFGNM- 2020)

Radik R. Mulyukov, Ayrat A. Nazarov, Renat M. Imayev

The Open School-Conference of NIS Countries «Ultrafine Grained and Nanostructured Materials – 2020» (UFGNM - 2020) was successfully held in Ufa, Bashkortostan Republic, Russia, from October 5 till October 9, 2020. The UFGNM serial conference, which is held biennially since 2008 and has become one of the leading materials science conferences in the territory of the former USSR, aims to address fundamental problems and application prospects of advanced nanostructured and ultrafine grained materials. The conference scope includes but is not limited to various topics of this interdisciplinary field such as nanoparticles, nanowires, nanofilms and coatings, nanostructured composite materials, carbon nanostructures, bulk ultrafine grained metals and alloys, processing methods, structure and properties, interfaces in materials, superplasticity and related phenomena, solid state joining, theory and simulation etc.

UFGNM-2020 was organized by the Institute for Metals Superplasticity Problems of the Russian Academy of Sciences in cooperation with Bashkir State University and was supported by donations from JSC "United Engine Corporation – Aviadvigatel" and LLC "Gazprom Tranzgas Ufa".

Proceedings contain 73 papers presented on the conference both offline and online. All papers have been peer reviewed and passed several stages of edition procedure. We express our sincere gratitude to all people who contributed to this process and did all their best to enhance the scientific merit and quality of the Proceedings. We also are grateful to the authors for their contributions, to the speakers and participants of the school-conference UFGNM-2020. We hope to welcome them along with new participants to the next school-conference UFGNM-2022.

Ufa, Bashkortostan, Russia

IOP Conf. Series: Materials Science and Engineering 1008 (2020) 011002 doi:10.1088/1757-899X/1008/1/011002

COMMITTEES

of the Open School-Conference of NIS Countries «Ultrafine Grained and Nanostructured Materials – 2018» (UFGNM- 2018)

International Organizing Commitee

Chairman

Dr. Sci., Prof., Corresponding Member of RAS R.R. Mulyukov, Institute for Metals Superplasticity Problems of the Russian Academy of Sciences, Ufa, Russia

Co-Chairman

Dr. Sci., Prof. N.D. Morozkin, Bashkir State Univerisity, Ufa, Russia

Committee Members

Dr. Sci., Prof., Corresponding Member of RAS M.I. Alymov, Merzhanov Institute of Structural Macrokinetics and Materials Science of the Russian Academy of Sciences, Chernogolovka, Russia
Dr. Sci., Prof. of the Russian Academy of Sciences J.A. Baimova, Institute for Metals Superplasticity Problems of the Russian Academy of Sciences, Ufa, Russia
Dr. Sci., Prof. A.M. Glezer, Moscow National University of Science and Technology MISIS, Moscow, Russia
Dr. Sci. R.M. Imayev, Institute for Metals Superplasticity Problems of the Russian Academy of Sciences, Ufa, Russia
Dr. Sci., Prof., Corresponding Member of the Russian Academy of Sciences A.A. Inozemstev, JSC "Aviadvigatel", Perm', Russia
Dr. Sci. N.K. Kasmamytov, Institute of Physico-Tehcnical Problems and Materials Science of the National Academy of Sciences of Kyrgyzstan, Bishkek, Kyrgyzstan
Dr. Sci., Prof. N.A. Koneva, Tomsk State University of Architecture and Building, Tomsk, Russia
Dr. Sci. G.F. Korznikova, Institute for Metals Superplasticity Problems of the Russian Academy of Sciences, Ufa, Russia
Dr. Sci., Prof. K.A. Kuterbekov, L.N. Gumilyov Eurasian National University, Astana, Kazakhstan
Dr. Sci. M.V. Markushev, Institute for Metals Superplasticity Problems of the Russian Academy of Sciences, Ufa, Russia
Dr. Sci., Prof. D.L. Merson, Togliatti State University, Togliatti, Russia
Dr. Sci., Prof. M.M. Myshlyaev, Baykov Metallurgy and Materials Institute of the Russian Academy of Sciences, Moscow, Russia
Dr. Sci. A.A. Nazarov, Institute for Metals Superplasticity Problems of the Russian Academy of Sciences, Ufa, Russia
K.S. Nazarov, Institute for Metals Superplasticity Problems of the Russian Academy of Sciences, Ufa, Russia
Dr. Sci., Prof. S.P. Pavlinich, Research Institute of Technology and Manufacture of Engines, United Engine Corporation, Moscow, Russia
Dr. Sci., Prof. V.V. Perevezentsev, Federal Research Center Institute of Applied Physics, Nizhniy Novgorod, Russia
Dr. Sci., Prof., Corresponding Member of the National Academy of Sciences of Belarus V.V. Rubanik, Institute of Technical Acoustics of the National Academy of Sciences of Belarus, Vitebsk, Belarus
Dr. Sci., Prof., Corresponding Member of the Russian Academy of Sciences V.V. Rybin, Peter the Great Saint-Petersburg Polytechnic University, S.-Petersburg, Russia
Dr. Sci., Prof., Corresponding Member of the Russian Academy of Sciences V.V. Sagaradze, M.N. Mikheev Institute of Metals Physics of Ural Branch of the Russian Academy of Sciences, Ekaterinburg, Russia
Dr. Sci., Prof. S.V. Smirnov, Institute of Engineering Science of Ural Branch of the Russian Academy of Sciences, Ekaterinburg, Russia

IOP Conf. Series: Materials Science and Engineering 1008 (2020) 011002 doi:10.1088/1757-899X/1008/1/011002

Dr. Sci., Prof. M.D. Starostenkov, Polzunov Altai State Technical University, Barnaul, Russia
Dr. Sci., Prof., Corresponding Member of the National Academy of Sciences of Ukraine V.N. Varyukhin, Donetsk, Ukraine
Dr. Sci., Prof. R.A. Yakshibaev, Bashkir State Univerisity, Ufa, Russia

Local Organizing Committee

Chairman

Dr. Sci. A.A. Nazarov, Institute for Metals Superplasticity Problems of the Russian Academy of Sciences, Ufa, Russia

Co-Chairman

Dr. Sci., Prof. V.P. Zakharov, Ufa Federal Research Center of the Russian Academy of Sciences, Ufa, Russia

Secretary

Dr. V.M. Valitova, Institute for Metals Superplasticity Problems of the Russian Academy of Sciences, Ufa, Russia

IOP Conf. Series: Materials Science and Engineering 1008 (2020) 011003 doi:10.1088/1757-899X/1008/1/011003

Peer review declaration

All papers published in this volume of *IOP Conference Series: Materials Science and Engineering* have been peer reviewed through processes administered by the Editors. Reviews were conducted by expert referees to the professional and scientific standards expected of a proceedings journal published by IOP Publishing.

- **Type of peer review:** Single-blind

- **Describe criteria used by Reviewers when accepting/declining papers. Was there the opportunity to resubmit articles after revisions?**

The paper is relevant to the topic of the conference and contains new research result(s), is clearly written and structured; figures and tables are informative and correctly organized; references are correct and contain new literature sources. After revision the papers were submitted to the editor; in some cases, two to three revisions were done. This provided the high rate of acceptance.

- **Conference submission management system:**

No special system was used. The papers were submitted to the conference secretariat and then distributed to reviewers by the editors. The reviewers sent their reviews to the editor who then finalised the editing process.

- **Number of submissions received**: 74

- **Number of submissions sent for review:** 74

- **Number of submissions accepted:** 73

- **Acceptance Rate (Number of Submissions Accepted / Number of Submissions Received X 100):** 98.6%

- **Average number of reviews per paper:** 2

- **Total number of reviewers involved:** 9

- **Any additional info on review process (ie plagiarism check system):** No

- **Contact person for queries:** Ayrat A. Nazarov, aanazarov@imsp.ru

IOP Conf. Series: Materials Science and Engineering 1008 (2020) 012001 doi:10.1088/1757-899X/1008/1/012001

Friction: from Leonardo da Vinci to modern nanotechnologies

A V Makarov

M.N. Miheev Institute of Metal Physics, Ural Branch of the Russian Academy of Sciences, 18 S. Kovalevskoy St., Ekaterinburg, 620108, Russia

E-mail: avm@imp.uran.ru

Abstract. The review paper presents an outstanding contribution of Leonardo da Vinci to the science of friction, which is the heart of modern nanotechnologies. It demonstrates a significant role of nanocrystalline friction structures (NCFS) in the formation of the tribological properties of metal alloys and in the implementation of deformation-induced nanostructuring techniques for metal alloy surface processing. Inherent in the very nature of friction, the ability to obtain an NCFS can be effectively used for targeted nanostructuring of product surface layers in order to increase their resistance to wear, corrosion, mechanical loads, and high temperatures. This can be achieved by using simple frictional and ultrasonic impact-frictional surface treatments, as well as combined methods based on the chemical modification of the formed NCFS. It is proposed to use nanostructuring frictional pretreatment and the industrial technology of nanostructuring burnishing in combined methods for surface preparation of structural stainless steels, including those produced by additive laser technology, for applying high-strength nanocomposite thin-film coatings.

1. Introduction

Along with gravitation, friction is the most important process largely determining life on the Earth. A fundamental contribution to the understanding and application of the phenomenon of friction was made by Leonardo da Vinci (1452–1519), a great thinker, scientist, and artist. This became especially evident in 1960s, after two of his unpublished manuscripts, known as the Madrid Codices, had been sensationally found in the Madrid National Library. The great Leonardo "pronounced a sentence" to the perpetual motion machine – many years before the official refusal of the French Academy of Sciences to accept for consideration perpetuum mobile projects as unscientific since frictional losses are inevitable. 200 years before Amonton's experiments and almost three centuries before the publication of Coulomb's studies, Leonardo da Vinci, basing himself on experiments, discovered the basic laws (aspects) of friction, introduced the notion of the friction coefficient into engineering practice, proposed antifriction bronze – the copper and tin alloy similar to the low-melting-point bearing alloy patented by Babbitt in 1839, invented a very complicated globoid gear, discovered anew only two and a half centuries later [1].

In fact, it is friction that underlies effective methods for surface nanostructuring of metals and alloys. Inherent in the very nature of friction is such important condition for deformation-induced nanostructuring as the formation of the stress state implemented during shearing under high pressure. And what about Beilby layers appearing during metal polishing, having high hardness, chemical stability, and adsorptivity? These layers (and G. T. Beilby discovered them as early as in 1921) had for a long time been classed as amorphous materials, and it was not before the development of transmission electron microscopy that researchers managed to reveal superfine (nanocrystalline) structure of the Beilby layers. It may have been these manifestations of the frictional effect (friction) that ushered in the era of nanotechnologies.

This review paper, based on studies by the author and his associates, briefly analyzes one of the acute and practically important problems of physical metallurgy, namely studying and using nanocrystalline friction structures (NCFS).

2. Features of NCFS formation in metals and alloys

Nanocrystalline friction structures form naturally in thin surface layers of metals and layers. They result from a specificity of the stress arising in the friction contact zone. Microvolumes of the metal in the surface layers of the rubbing bodies are affected by external contact shear and compressive stresses. This creates conditions for extremely large values of plastic strain to occur in a 1–10 μm thick layer, which are reachable only under conditions of rotational plasticity, which causes the NCFS formation [2, 3]. NCFSs appear both under sliding friction and under abrasive action in the process of coarse-crystalline (figure 1 a) structure evolution during the sequential implementation of the sliding and twinning mechanisms, the formation of banded (figure 1 b) and cellular (figure 1 c) substructures, ultrafine textured (figure 1 d) and, at the final stage, nanocrystalline (figure 1 e, f) structures with large-angle misorientations of crystallites sized up to 100 nm [4–6]. Effective metal nanostructuring is promoted by a high friction coefficient (to enhance shear strain) and a noncorrosive environment in the friction contact zone (to prevent oxygen-induced embrittlement of the diffusively active nanostructured layer) [3, 5–7].

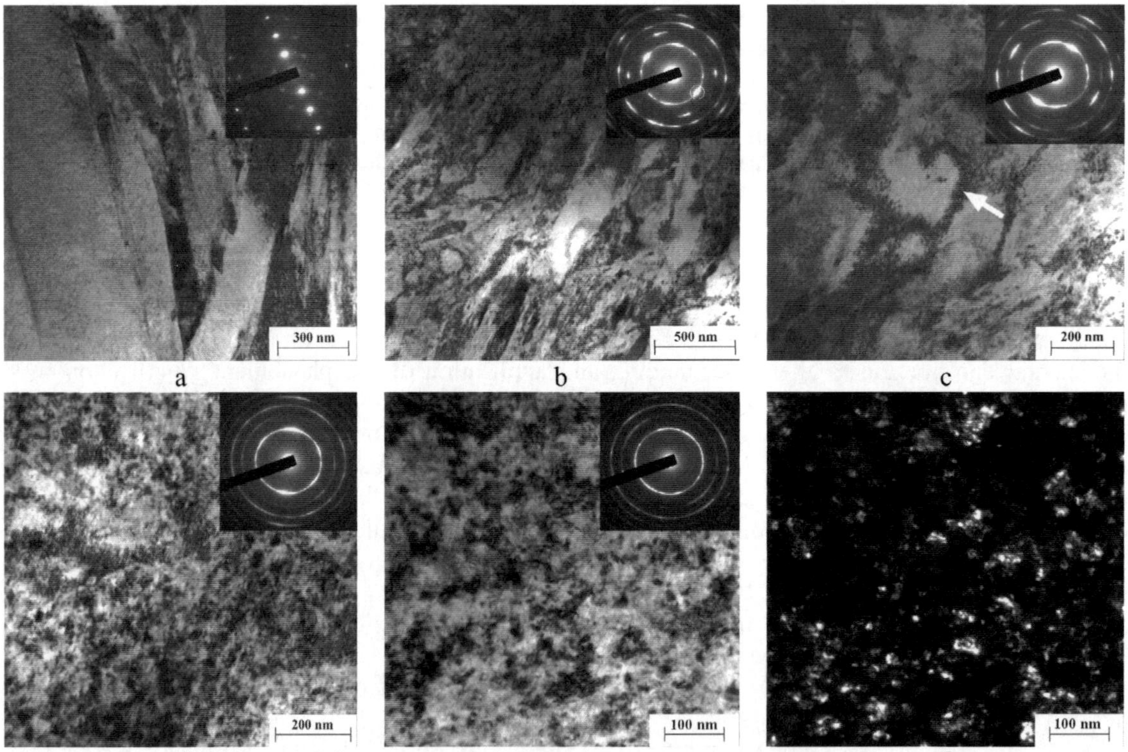

Figure 1. The structure of the 80S4 steel (0.85% C; 4.20% Si) (a) after water quenching from 870°C and cold treatment at −196°C and (b–f) subsequent frictional treatment: (a–e) bright-field images and microdiffractions; (f) dark-field image taken in the (110)α reflection.

3. NCFS contribution to the formation of the tribological properties of metal alloys

The strength characteristics (microhardness, shear strength) and tribological (Latin *tribos* – friction) properties (friction coefficient and wear rate) of NCFS are largely governed by the level of their dispersion, the conditions of dislocation displacement through the bulk of the nanograins, the resistance of NCFS fragments to rotations, the presence of interstitial elements (carbon, nitrogen), the degree of alloying with substitutional elements, the type of the crustal lattice, the presence of dispersed strengthening phases (carbides, nitrides, intermetallics), the magnetic state of austenite, and the effect of the air environment causing NCFS fracture due to oxygen-induced embrittlement [2, 3, 5].

Thus, in low-alloy (<0.1% C) iron alloys, the NCFSs of γ(FCC) austenite and ε(HCP) martensite demonstrate a much higher resistance to adhesive wear than the NCFS of α(BCC) martensite, whose wear resistance increases sharply during alloying with substitutional elements (Si, Ti) enhancing the solid-solution strengthening of martensite. The factors decreasing NCFS viscosity, such as the presence of relatively brittle intermetallic (Ni_3Ti, Ni_3Al) and ε-carbide particles and oxygen saturation of the active surface layer, negatively affect the resistance of steels and alloys to wear. In alloys with a high (~0.8%) carbon or nitrogen content, the maximum strength and resistance to adhesive wear are characteristic of the NCFS of high-carbon α-martensite since the impurity mechanism of dislocation pinning by carbon atoms in the NCFS of α-martensite is more effective than in the NCFS of austenite. Low values of the friction coefficients ($f = 0.25$–0.28) and high resistance to adhesive wear are characteristic of the NCFS of the ε(HCP) phase of high-manganese iron alloys, including steels, and the NCFS of the γ(FCC) phase of nitrogen-containing austenitic steels. This is due to the effect of slight basal slip in the crystals (fragments) of the NCFS ε-phase, isomorphous to α-cobalt, as well as to the degeneration of multiple slip and the intensification of planar slip in NCFS fragments of nitrogen austenitic steels.

4. Using NCFSs in modern deformation nanotechnologies

The indisputable scientific and practical significance of studying NCFSs stems not only from their important contribution to the formation of the tribological properties steels and alloys. The next logical step was intentional formation of a stable nanostructured layer with improved characteristics of wear resistance, heat resistance, corrosion resistance, and mechanical strength on the surfaces of metal products. This is achieved by the application of fairly easy-to-implement frictional treatments (figure 2), enabling even the surfaces of difficult-to-deform steels (laser-quenched, cemented, high-speed) to be nanostructured [5, 6, 8, 9]. For effective implementation of this method of nanostructuring and strengthening, it is necessary to provide strain accumulation in a thicker surface layer without its accelerated destruction (wear) during frictional treatment proper and to decrease the hazard of premature brittle failure of "over-work-hardened" nanostructured layer as early as in the initial operation period.

It is important to note the high potential of practical application of frictional treatments as finishing nanostructuring techniques for producing surfaces with low roughness [10–14]. The metal-physical foundations of frictional treatment find application in the manufacturing technology of nanostructuring burnishing, which is successfully implemented in multifunction turn-milling centers [10, 11]. The proposed ultrasonic impact-frictional treatment (UIFT) is an effective development of ultrasonic treatments [15–19]; as distinct from the standard ultrasonic impact treatment (UIT), it is performed by the method of oblique impact of a pulsing indenter and without using a lubricant (figure 3). This treatment makes it possible to increase the shear component of plastic strain owing to the enhanced friction effect on the metal. This facilitates strain accumulation within a layer and, correspondingly, more effective dispersion of the structure (up to the nanoscale) and steel surface layer strengthening.

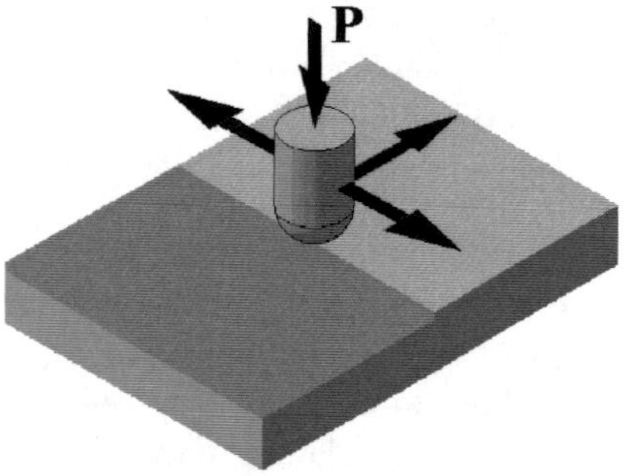

Figure 2. Scheme of frictional treatment.

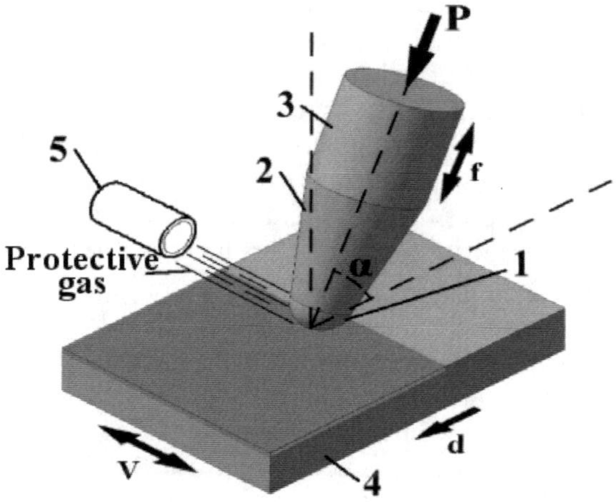

Figure 3. Scheme of UIFT: 1 – tool (indenter); 2 – waveguide; 3 – magnetostrictive or piezoelectric transducer; 4 – specimen; 5 – tube for protective gas feeding.

5. Friction nanostructuring used in combined technologies of modifying metal alloy surfaces

NCFSs formed by frictional treatment are effective in combined deformation-thermal treatments of martensitic [20–22] and austenitic [18, 23] steels, laser-clad Cr-Ni coatings [24], as well as in deformation-thermochemical treatments, particularly, in the intensification of the processes of low-temperature plasma nitriding [25–28]. The accelerated diffusion of elements over the crystal defects NCFSs is provided by their high imperfection, namely long grain and subgrain boundaries, extreme density of dislocations, disclinations, and pin-hole defects.

The application of frictional treatment [6, 18] and the manufacturing technology of nanostructuring burnishing [10, 11] with sliding indenters for the formation of nanocomposite superhard and heat-resistant zirconium-carbonitride- and chromium-aluminum-based coatings on the surface of austenitic metastable [13] and deformation-resistant [29] steels seems to be promising. It is expedient to use nanostructuring burnishing in combinations with plasma nitriding as the postprocessing of stainless steels, produced by the additive laser technology, for subsequent application of nanocomposite thin-film multilayer coatings based on amorphous superhard diamond-like carbon and titanium carbides. The essence of the proposed new approaches lies in overcoming the basic disadvantage of dead-hard and superhard thin-film coatings consisting in their low toughness and high brittleness. This manifests itself in coating cracking and the loss of adhesion to the base when coatings are applied onto products with rather low hardness. Therefore, it is proposed to apply coatings onto a surface pre-strengthened by friction. Such a combination of methods will allow us to obtain a synergic effect by summing and enhancing the advantages of separate methods, namely high surface hardness and a gradient transition from the surface to the base. The significance and relevance of the proposed line of research result from a matchless opportunity to expand the area of application for state-of-the-art nanocomposite coatings on relatively soft structural steels, including those synthesized by the additive technology of selective laser melting (SLM).

6. Conclusion and prospects

Leonardo da Vinci made a fundamental contribution to the understanding and application of the phenomenon of friction. An important manifestation of friction in metal alloys is the formation of nanocrystalline friction structures (NCFS) in their surface layer, which result from the development of rotational plasticity.

The presented analysis of NCFSs, based on publications by the author and his associates, not only promotes the state-of-the-art development of the metal-physical foundations of the friction and wear of metal alloys, but also creates prerequisites for the improvement of already known methods of strain hardening and combined strengthening of metal surfaces with elements of friction-induced nanocrystallization, as well as for the appearance of new ones. The paper notes the high effectiveness of frictional treatment, the proposed ultrasonic impact-frictional treatment, and combined deformation-thermal (thermochemical) treatments aimed at the nanostructuring of steels and alloys and for the improvement of their operationally important properties.

Of particular scientific and practical interest are the prospects of using NCFSs in new combined technologies for the formation of nanocomposite thin-film coatings with improved operational properties on austenitic stainless steel surfaces, as well as in combined post-treatments of maraging and austenitic steel products made by additive laser techniques. The stated problem can be solved by combined treatments including frictional deformation (frictional treatment and nanostructuring burnishing), possible additional plasma nitriding followed by the deposition of nanocomposite superhard, dead-hard, and heat-resistant coatings.

Acknowledgments
The research was carried out within the state assignment of Ministry of Science and Higher Education of the Russian Federation (theme “Structure” No. AAAA-A18-118020190116-6), supported in part by RFBR and Sverdlovsk Oblast (project No. 20-48-660065) and by RFBR and BRFBR (project No. 20-58-00057). The studies were performed at the collaborative access center “Testing center of nanotechnology and advanced materials” of the M.N. Mikheev Institute of Metal Physics of the Ural Branch of the Russian Academy of Sciences (IMP UB RAS).

References

[1] Silin A A, ed. 1983 *Trenie i Ego Rol` v Razvitii Tekhniki* (Moscow: Nauka Publ.) p 176 [in Russian]
[2] Korshunov L G, Makarov A V and Chernenko N L 1994 *Fizika Metallov i Metallovedenie* **78** 128 [in Russian]
[3] Korshunov L G, Makarov A V and Chernenko N L 2000 *Phys. Met. Metallogr.* **90** S48
[4] Makarov A V, Korshunov L G, Schastlivtsev V M, Solodova I L and Yakovleva I L 2004 *Phys. Met. Metallogr.* **98** 428
[5] Makarov A V and Korshunov L G 2004 *Russ. Phys. J.* **47** 857
[6] Makarov A V and Korshunov L G 2019 *Phys. Met. Metallogr.* **120** 303
[7] Makarov A V, Savrai R A, Pozdejeva N A, Smirnov S V, Vichuzhanin D I, Korshunov L G and Malygina I Yu 2010 *Surf. Coat. Technol.* **205** 841
[8] Makarov A V and Korshunov L G 2003 *Trenie i Iznos (Friction and Wear)* **24** 301
[9] Makarov A V, Korshunov L G, Vykhodets V B, Kurennykh T E and Savrai R A 2010 *Phys. Met. Metallogr.* **110** 507
[10] Kuznetsov V P, Makarov A V, Kiryakov A E, Savrai R A and Anikeev A V 2012 *Patent RF* **2458777**
[11] Kuznetsov V P, Makarov A V, Psakhie S G, Savrai R A, Malygina I Yu and Davydova N A 2014 *Phys. Mesomechanics* **17** 250
[12] Makarov A V, Skorynina P A, Osintseva A L, Yurovskikh A S and Savrai R A 2015 *Obrabotka Metallov (Tekhnologia, Oborudovaine, Instrumenty)* **4(69)** 80 [In Russian]
[13] Makarov A V, Skorynina P A, Yurovskikh A S and Osintseva A L 2017 *Phys. Met. Metallogr.* **118** 1225
[14] Savrai R A, Makarov A V, Malygina I Y and Volkova E G 2018 *Mater. Sci. Eng. A* **734** 506
[15] Makarov A V, Malygina I Yu, Burov S V and Savrai R A 2018 *Patent RF* **2643289**
[16] Makarov A V, Savrai R A, Malygina I Yu, Volkova E G and Burov S V 2018 *AIP Conf. Proc.* **2053** 020006 5 p
[17] Lezhnin N V, Makarov A V and Luchko S N 2019 *Letters on Materials* **9** 310
[18] Makarov A V, Savrai R A, Skorynina P A and Volkova E G 2020 *Met. Sci. Heat Treat.* **62** 61

[19] Lezhnin N V, Makarov A V, Luchko S N, Loginov B A and Loginov A B 2020 *Obrabotka Metallov (Tekhnologia, Oborudovaine, Instrumenty)* **22(2)** 16
[20] Makarov A V, Korshunov L G, Malygina I Yu and Solodova I L 2007 *Metal Sci. Heat Treat*. **49** 150
[21] Makarov A V, Savrai R A, Gorkunov E S, Yurovskikh A S, Malygina I Yu and Davydova N A *Phys. Mesomechanics* 2015 **18** 43
[22] Makarov A V, Korshunov L G, Savrai R A, Davydova N A, Malygina I Yu and Chernenko N L 2014 *Phys. Met. Metallogr.* **115** 303
[23] Makarov A V, Skorynina P A, Volkova E G and Osintseva A L 2018 *Phys. Met. Metallogr.* **119** 1196
[24] Makarov A V, Soboleva N N and Malygina I Yu 2017 *AIP Conf. Proc.* **1915** 030012 5 p
[25] Makarov A V, Gavrilov N V, Samoylova G V, Mamaev A S, Osintseva A L and Savrai R A 2017 *Obrabotka Metallov (Tekhnologia, Oborudovaine, Instrumenty)* **75** 55
[26] Makarov A V, Samoilova G V, Gavrilov N V, Mamayev A S, Osintseva A L, Kurennykh T E and Savrai R A 2017 *AIP Conf. Proc.* **1915** 030011 5 p
[27] Lezhnin N V, Makarov A V, Gavrilov N V, Osintseva A L and Savrai R A 2018 *AIP Conf. Proc.* **2053** 040050 5 p
[28] Zhidkov I S, Kukharenko A I, Makarov A V, Savrai R A, Gavrilov N V, Cholakh S O, Kurmaev E Z 2020 *Surf. Coat. Technol.* **386** 125492
[29] Makarov A V, Skorynina P A, Volkova E G and Osintseva A L 2020 *Met. Sci. Heat Treat.* **61** 764

IOP Conf. Series: Materials Science and Engineering 1008 (2020) 012002 doi:10.1088/1757-899X/1008/1/012002

Nano-mechanical testing of Al-Nb metal matrix composites consolidated by high pressure torsion after annealing

R R Mulyukov[1, 2, 3], **K S Nazarov**[1], **R U Shayakhmetov**[1], **G F Korznikova**[1], **E A Korznikova**[1, 4]

[1] Institute for Metals Superplasticity Problems RAS, 39 Khalturin St., 450001 Ufa, Russia
[2] Bashkir State University, 32 Validi st., Ufa, 450076, Russia
[3] Ufa State Petroleum Technological University, 1 Kosmonavtov st., Ufa, 450064, Russia
[4] Ufa State Aviation Technical University, 12 K. Marx St., 450008, Ufa, Russia

E-mail: elena.a.korznikova@gmail.com

Abstract. Among different approaches for obtaining composite structures, one can distinguish the approach of severe plastic deformation that allows obtaining bulk structures from strongly dissimilar metals. The application of large strains is usually associated with grain refinement and, in some cases, with the formation of intermetallic phases, both factors contributing to the strength characteristics of the material. This work presents an investigation of nano-mechanical properties of an Al-Nb composite obtained by high pressure torsion and post deformational annealing. Measurements show that with an increase in the distance from the centre to the edge of the sample, the nanohardness increases from 2.4 to 3 GPa. Young's modulus varies in the range from 182 to 197 GPa, which indicates the non-homogeneity of the composite sample. The Young's modulus of the composite after high pressure torsion was found to be higher than that of pristine Al and Nb, which is probably due to the formation of the intermetallic Al_3Nb phase, having an increased Young's modulus according to ab-initio calculations.

1. Introduction

Following the general trend of design of new materials with enhanced characteristics, several approaches of obtaining composites have been developed, among them one can recall friction stir welding processing [1], plasma sintering [2], stir casting [3,4], spray deposition [5] and diffusion bonding [6]. However, all mentioned methods are inevitably associated with a local temperature increase which, in turn can lead to inhomogeneity and degradation of the properties of the final product. In this respect, consolidation using severe plastic deformation by high pressure torsion (HPT) has several advantages comparing to other methods, namely, it allows reaching large strains, avoiding the fracture of the sample, the control of the strain rate allows preventing extra heating, and the flexible size of anvils provides an opportunity to vary the shape and size in the moderate range [7–10]. Up to date, the successful rapid diffusion bonding of Al-Mg [11], Al-Cu[12–14] Al-Ni [15] has been reported, where the exceptionally high hardness of the final product is usually associated with the strain induced nucleation of intermetallic phases accompanied by a refinement of the structure refinement.

The loading scheme used in HPT devices allows performing the consolidation of powders of different metals, which also results in the formation of metal matrix composites. This approach enables mechanical alloying of strongly dissimilar materials, such as, Al-Ti [16], Al-W[17] and some others, however the purity of the final product in this case is much lower that for those obtained by consolidation of thin discs. A pioneering attempt to perform bonding of two dissimilar metals - Al and Nb by means of disk consolidation was performed by the authors in [18].

The ability of HPT to the synthesis of bulk composites is primarily associated with a significant inflow of strain energy, which is partially accommodated on the formation of new interfaces, and a high concentration of point defects [19], facilitating diffusion in the lattice. Enhanced diffusion, in turn, provides conditions for the mutual solubility of components and the formation of intermetallic phases at temperatures much lower than those in the corresponding equilibrium phase diagram. The phenomenon of point defect assisted mass transfer in nonequilibrium conditions and related phenomena is addressed in [20-23].

The great number of factors affecting the structure and properties of the composite (the initial structure of the material, HPT processing parameters, number of possible intermetallic phases, exc.) require a great diversity of methods of their characterization. Our previous work reported on the microstructural, X-ray and microhardness investigations of the Al-Nb composite obtained by HPT. This work is devoted to the analysis of the nano-hardness and the Young's modulus evolution of the studied composite performed on the ultrafine level that allows to uncover new features of the structure evolution upon severe plastic deformation and subsequent annealing, which, in turn, can provide a new insight on the possibilities to create composites with tailorable properties.

2. Materials and Methods

Aluminum with a purity of 99.5 wt.% and niobium with a purity of 99 wt.% in the form of disks with a thickness of 0.5 mm and a diameter of 12 mm were used as starting materials for the research. Two disks of aluminum were stacked in the form of a sandwich with one disk of niobium located in the middle. Pressure assisted shear deformation was performed at room temperature on a constrained Bridgeman-type HPT installation with an anvil depth of 0.25 mm at a pressure of 5 GPA with a total number of revolutions N=30. Bulk disk-shaped samples of the Al-Nb mixture with a diameter of 12 mm and 0.5 mm thickness were obtained.

Nanomechanical testing has been performed by means of a basic module of the «NanoScan-3D» equipment, which allows performing a comprehensive study of nanomechanical properties in the load range up to 100 mN by indentation and sclerometry methods, and to study surface topography in the semi-contact scanning probe microscopy mode. We have performed the nanoindentation based on the measurement and analysis of the indentation load-displacement data enabling determination of the Young's modulus value and nano hardness of the material by analyzing the load-depth curve (see figure 1) with a load of 30 mN, details of the measurement procedure can be found in [24]. The indentation was performed in three areas (R=0, R=R/2 and R=R, see scheme in table 1) on the cross section of the HPT disc sample. Every value given in table 1 is the result of 5 measurements averaging.

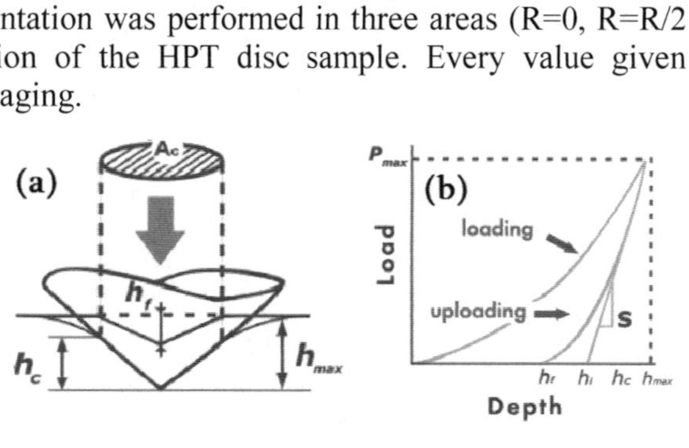

Figure 1. General view of the loading scheme (a), loading curve (b) and contact diagram with the designations of the values used in the method of calculating the modulus of elasticity and hardness. Adopted from [24].

3. Results and discussion

Table 1 presents the results of measurements of nanohardness and Young's modulus in the Al-Nb-Al composite after annealing at 400 and 600 °C. The measurements reveal that after annealing at 400 °C the nanohardness grows with the distance from the center to the edge of the sample, while the Young's modulus reveals the opposite dependence - one can see a decrease from 218 to 118 GPa when moving from the centre to the edge part of the sample. Analysis of nanomechanical characteristics after annealing at 600 °C revealed an overall decrease in all measured characteristics. Thus, an increase in the annealing temperature resulted in the nanohardness decrease from 3.75 to 3.03 GPa and from 3.26 to 2.44 in the vicinity of the sample edge and the sample center, respectively. From the data presented in table 1, one can also conclude that annealing at 600 °C resulted in the alignment of the Young modulus value along the sample profile, while its overall value was found to be approximately the same as after annealing at 400 °C.

Table 1. The Young's modulus E, and nanohardness H of the composite Al-Nb after annealing

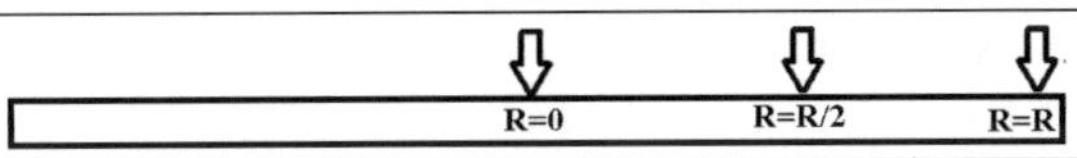	**Al-Nb-Al 400°, 30 min**		**Al-Nb-Al 600°, 30 min**	
zone	**H, GPa**	**E, GPa**	**H, GPa**	**E, GPa**
Sample edge (R=R)	3.75	118.5	3.03	182.45
Middle radius (R=R/2)	2.81	189.75	2.45	196.2
Centre (R=0)	3.26	218.31	2.44	197.74

According to experimental estimations in [25], Young's modulus for Nb in different structural states varies in the range 110-160 GPa, which corresponds to our results. The value of Young's modulus of the composite was higher than that of the original components of the composite. This is probably due to the formation of a large proportion of the dispersed intermetallic phase Al_3Nb, which, according to first-principle calculations has a value of Young's modulus E=260 [25].

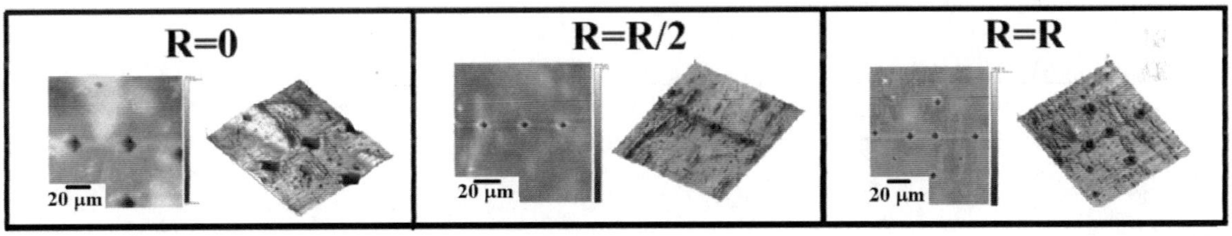

Figure 2. Nanoindentation imprints and corresponding 3D surface profile obtained on the NanoScan-3D for Al-Nb-Al composite annealed at 400 °C in the center of the sample (R=0) in the middle radius area (R=R/2) and on the edge of the sample (R=R) (see scheme in table 1).

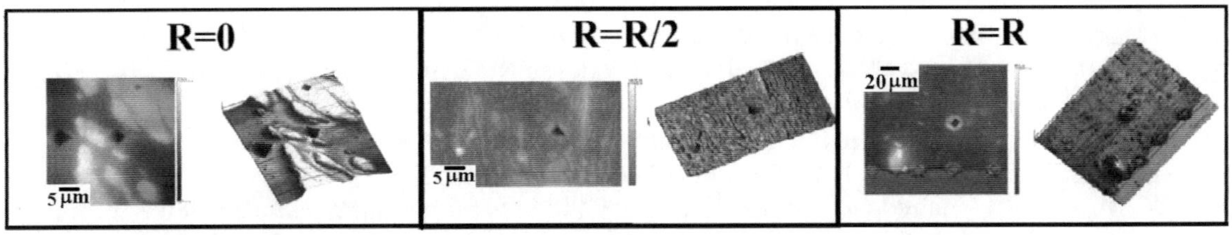

Figure 3. Nanoindentation imprints and corresponding 3D surface profile obtained on the NanoScan-3D for Al-Nb-Al composite annealed at 600 °C in the center of the sample (R=0) in the middle radius area (R=R/2) and on the edge of the sample (R=R) (see scheme in table 1).

Microstructural investigations of the Al- Nb composite revealed the presence of three main zones along the sample radius, namely a coarse fragment mixture of Al and Nb phases in the center of the sample (at R=0), a fine lamellar type structure at the mid radius (R=R/2) and a supersaturated solid solution in the vicinity of the disc edge (R=R) [18]. Post deformational annealing results in a decrease in the microhardness in R=0 and R=R/2 areas. Meanwhile the annealing induced precipitation of Al_3Nb phase contributes to the increase of the sample microhardness in the edge region [18].

The analysis of the nanohardness and Young's modulus of the annealed samples is in general in good coincidence with the investigations performed in [18]. Thus, in the center of the disc sample, at R=0 for both annealing temperatures 400 and 600 °C (figure 2 a and figure 3 a respectively), one can distinguish separate Al and Nb phases with the size of the imprint in Nb being much smaller than that in Al due to the lower strength of the latter. 2D and 3D images of NanoScan-3D imprints in the cross section of R/2 area reveal a laminated structure (figure 2 b and figure 3 b) that correlates well with transmission electron microscopy studies [18]. This structure is stable with respect to annealing at 400 and 600°C, and, as a result one can see that H and E values, as well as the size of the imprint are fairly stable in the studied temperature interval.

The disc edge region (R=R) is characterized by a relatively homogeneous microstructure with the highest values of H and E. This can be due to the high fraction of Al3Nb phase that constitutes ~12% and ~25% after 30 min of annealing at 400 °C and 600 °C, respectively [18]. Since this estimation has been done for the whole sample, the fraction of the intermetallic phase at the periphery of the sample is expected to be even higher, thus considerably contributing to the local mechanical properties of the composite.

4. Conclusions

The study of the nanomechanical properties of the Al-Nb composite after annealing revealed that an increase in the annealing temperature results in an overall decrease of nanohardness, keeping the gradient of its value along the radius of the sample together with alignment of the Young's modulus value along the radius of the sample. The increased E value of the composite comparing to the pristine Al and Nb components is explained by the enhanced Young's modulus of the Al_3Nb phase formed during HPT and subsequent annealing.

Acknowledgements

GFK and RRM gratefully acknowledge the support of the Russian Science Foundation, grant No 18-12-00440. The work was partially supported by the State Assignment of Russian Ministry of Science and Higher Education.

References

[1] Mori K, Bay N, Fratini L, Micari F and Tekkaya A E 2013 *CIRP Annals* **62** 673–94
[2] Ujah C O, Popoola A P I, Popoola O M and Aigbodion V S 2019 *Int. J. Adv. Manuf. Technol.* **101** 2275–82
[3] Mistry J M and Gohil P P 2019 *Composites Part* B **161** 190–204
[4] Bihari B and Singh A K 2017 *Int. J. Eng. Res. Appl.* **7** 42–8
[5] Shockley J M, Strauss H W, Chromik R R, Brodusch N, Gauvin R, Irissou E and Legoux J-G 2013 *Surf. Coat. Technol.* **215** 350–6
[6] Kawasaki M, Han J-K, Lee D-H, Jang J and Langdon T G 2018 *J. Mater. Res.* **33** 2700–10
[7] Zhilyaev A and Langdon T 2008 *Progr. Mater. Sci.* **53** 893–979
[8] Kawasaki M, Figueiredo R B and Zhilyaev A P 2020 *Adv. Eng. Mater.* **22** 1901386
[9] Castro M, Pereira P H, Figueiredo R and Langdon T 2019 *Lett. Mater.* **9** 541–5
[10] Mulyukov Kh Ya, Korznikova G F and Nikitin S A 1996 *J. Appl. Phys.* **79** 8584–7
[11] Kawasaki M, Jung S H, Park J-M, Lee J, Jang J and Han J-K 2020 *Adv. Eng. Mater.* **22** 1900483

IOP Conf. Series: Materials Science and Engineering 1008 (2020) 012002 doi:10.1088/1757-899X/1008/1/012002

[12] Korznikova G, Czeppe T, Khalikova G, Gunderov D, Korznikova E, Litynska-Dobrzynska L and Szlezynger M 2020 *Mater. Charact.* **161** 110122
[13] Bachmaier A, Rathmayr G B, Bartosik M, Apel D, Zhang Z and Pippan R 2014 *Acta Mater.* **69** 301–13
[14] Korznikova G, Kabirov R, Nazarov K, Khisamov R, Shayakhmetov R, Korznikova E, Khalikova G and Mulyukov R 2020 *JOM* **846** 156380
[15] Edalati K, Toh S, Watanabe M and Horita Z 2012 *Scripta Mater.* **66** 386–9
[16] Sun Y, Aindow M, Hebert R J, Langdon T G and Lavernia E J 2017 *J. Mater. Sci.* **52** 12170–84
[17] Edalati K, Toh S, Iwaoka H and Horita Z 2012 *Acta Mater.* **60** 3885–93
[18] Korznikova G, Korznikova E, Nazarov K, Shayakhmetov R, Khisamov R, Khalikova G and Mulyukov R. 2020 *Advanced Engineering Materials* 2000757
[19] Korznikova E A, Mironov S Yu, Korznikov A V, Zhilyaev A P and Langdon T G 2012 *Materials Science and Engineering: A* **556** 437–45
[20] Korznikova E, Sunagatova I, Bayazitov A, Semenov A and Dmitirev S 2019 *Letters on Materials* **9** 386–90
[21] Dmitriev S V, Korznikova E A and Chetverikov A P 2018 *Journal of Experimental and Theoretical Physics* **126** 347–52
[22] Zakharov P V, Korznikova E A , Dmitriev S V, Ekomasov E G, Zhou K 2019 *Surface Science* **679** 1-5
[23] Barani E, Korznikova E A, Chetverikov A P, Zhou K, Dmitriev SV 2017 *Physics Letters, Section A: General, Atomic and Solid State Physics*, **381** 3553-3557
[24] Anon http://nanoscan.info/eng/, accessed 02.09.2020
[25] Nong Z, Zhu J, Yang X, Cao Y, Lai Z and Liu Y 2012 *Physica B: Condensed Matter* **407** 3555–60

Size-dependent phase stability of silver nanoparticles

A M Murzakaev[1,2]

[1]Institute of Electrophysics UB of RAS, 620016, Yekaterinburg, 106 Amundsena St., Russia
[2]Ural Federal University, 620002, Yekaterinburg, 19 Mira St., Russia

E-mail: Amurzak@mail.ru , Aidar@iep.uran.ru

Abstract. A comparative analysis of the structure of silver nanoparticles obtained by chemical synthesis (citrate method) and electrical explosion of wires is carried out. The structure, phase composition, lattice parameters of silver nanoparticles with sizes from 2 to 50 nm were studied by electron diffraction and high-resolution transmission electron microscopy. All silver nanoparticles obtained by the chemical method have only a cubic structure. Silver nanoparticles obtained by the physical method have cubic and hexagonal structures. The existence of the hexagonal phase under normal conditions is explained by the quenching effect. The lattice parameters of the cubic phase of Ag NPs within the experimental determination error (± 0.01 nm), synthesized by the chemical method and by the method electrical explosion of wires do not differ from each other.

1. Introduction

Silver (Ag) nanocrystals have been widely used in several areas (catalysis, biological, antimicrobial, information storage, surface-enhanced Raman spectroscopy, interconnects in ultra large-scale integration circuits etc). It is well known that the physical properties of nanoparticles strongly depend on their microstructure. The crystal structure is a critical parameter for the determination of materials properties of the nanocrystals. Silver in its bulk form possesses a cubic crystal structure (Ag-FCC). The authors of [1] obtained a proof of the presence of a hexagonal phase (Ag-4H) in nuggets of silver. Later it was found that the Ag-4H is a stable phase when the particle size remains smaller than a critical size $d_c < 40$ nm. The stability of the hexagonal polymorph at room temperature and the normal atmospheric pressure has been often discussed in literature as a grain size effect. There are conflicting reports in the literature regarding the phase structure of Ag particles with sizes below 40 nm [2-7]. It is significant that many authors have generally confirmed these results, but the critical size appears to depend strongly on the physical and chemical synthesis techniques.

Changing the size will change the stability range of ordinary bulk materials. This phenomenon has been found in many oxide materials such as ZnO, ZrO_2, Al_2O_3, TiO_2, Fe_2O_3. Various explanations have been proposed for the observed stabilization of the high temperature phase in nanocrystalline oxide particles at room temperature, and controversies still exist in the elucidation of the mechanism of the phase stability. For example, Garvie [8] proposed that the lower surface energy of tetragonal (t) ZrO_2 was the cause for this phase to be present in nanocrystalline form at or below room temperature. He predicted that particles below about 10 nm in diameter are stabilized in the tetragonal form, and those that are above this critical particle size are subject to the monoclinic (m) transformation. Later, in [9] Murase and Kato suggested that water vapor increased the rate of crystallite growth and decreased surface energy having thus significant effect on the t-m transformation. Domain boundaries were also suggested to inhibit the t-m transformation thus leading to tetragonal phase stability [10]. In [11] Srinivasan et al. argued against the concept proposed by Garvie (stabilization due to lower surface energy of the t phase) as they found monoclinic particles with much smaller diameters. They suggested that anionic oxygen vacancies present on the surface control the t-m phase transformation on cooling, and that oxygen adsorption triggers this phase transformation. All the articles cited here, except the one by Garvie [8], have in common that the zirconia nanoparticles were produced by various chemical methods (precipitation and calcination). Chemical factors such as adsorbed atoms and purity of raw materials play an important role in the stability of nanocrystalline zirconia and they may completely alter the original surface energy state. Therefore, none of the above explanations have a general validity for pure nanocrystalline zirconia. One of the most recent explanations is based on an increased effective internal pressure due to the surface curvature and the small particle radius (the Gibbs-Thomson effect): $\Delta p = 2\gamma/r$, where Δp is the difference between the external and the internal pressure, γ is the surface energy, and r is the particle radius [12]. This explanation has been since then quite often used for several nanocrystalline ceramic systems as the stabilization of high-temperature or high-pressure polymorph in nanoparticles appeared to be a more general phenomenon. This size-dependent phenomenon is independent of the fabrication method. One of the main quantitative characteristics of nanoparticles is the Gibbs free surface energy. It differs significantly from the energy for macro objects. It has to be mentioned that other mechanisms were also proposed, e.g. a nanothermodynamic model was also established to reveal the origin of the size effect on the polymorphism (face centered cubic and 4H structures) behavior of Ag nanocrystals with the contributions of surface energy and surface stress to the total Gibbs free energy [7]. It shows that the lower surface energy and surface stress coupled with a higher volume of the Gibbs free energy predominates the thermal stability of 4H-structured Ag in a nanometer scale. A thermodynamic explanation has also been developed to calculate the r_c of Ag nanoparticles at different temperatures [7]. The $d_c(T)$ function of Ag nanocrystals was also obtained with respect to the thermodynamic consideration of $G^{FCC}(r,T) = G^{4H}(r,T)$. Furthermore, several thermodynamic parameters of the Ag-4H are also determined with the developed model.

In this paper, results of detailed studies on the morphology and crystal structure of Ag nanoparticles produced by chemical method and electrical explosion of wire are presented.

2. Experimental

Silver nanoparticles were synthesized by the physical method using electrical explosion of wire (EEW) as described in detail in [13] and the chemical method using a chemical reduction of $AgNO_3$ with N_2H_5OH in the presence of Na-cit as described in detail in [14]. The minimum size of silver nanoparticles was specially obtained by the chemical method. According to [15], the initial stage of reduction yields clusters Ag_2^+, Ag_4^+, Ag_9^+, etc. The nascent Ag_x clusters reacted with citrate, which was followed by their aggregation into coarser particles. Upon reaching the critical size (50 - 100 atoms or 1 - 1.5 nm), the clusters virtually cease to grow by the condensation mechanism. Further clusters growth is highly dependent on the citrate concentration. The growth of NPs at intermediate citrate concentrations, namely, $(1\text{-}5)\cdot 10^{-4}$ mol litre^{-1}, occurred as a result of reduction of silver ions on the cluster surface. Hence, after the cluster nucleation, its further growth proceeded gradually. Silver NPs formed by this mechanism were unsusceptible to aggregation, spherical and had narrow size distribution. At low citrate concentrations, its stabilizing action was insufficient to prevent the aggregation of clusters, which led to NPs of larger diameters. At higher citrate concentrations, the solution ionic strength reached large magnitudes sufficient for the destabilization of the citrate layer, which also resulted in coagulation. Yellow sol was obtained by boiling solution at T = 100 °C.

With the use of x-ray phase analysis (D8 DISCOVER diffractometer, Germany), the phase composition, spacing parameters, and the size of the coherent scattering regions (CSR) were determined. The diffractograms were handled with the use of the TOPAS 3 software. The error of determining the content of phases was about 2 wt%. The particle morphology was studied by transmission electron microscopy (TEM) and scanning electron microscopy (SEM) by microscopes JEM-2100 and LEO-982. Selected-area electron diffraction (SAED) and high-resolution transmission electron microscopy HRTEM (JEM 2100) have been used to determine the morphological and phase features of silver nanoparticles. The calibration of the magnification of the HRTEM and the determination of the device constant for the electron diffraction measurements were performed using a gold foil. This choice is due to the fact that the structure of gold is well known, it rarely has defects, and its interlayer distances are 0.23 nm and 0.20 nm (which corresponded to planes (111) and (200) well resolved by the device); in addition, the electron diffraction pattern of gold can easily be interpreted. The processing and analysis of the images were performed using the 3.9.3 version of the Gatan Digital Micrograph program, which allowed us to calibrate images, to apply masks and filters, to perform fast Fourier-transformations (FFT) both of the entire photograph and of the selected regions, to perform measurements, and to save data in various formats. The interplanar distances of the phases of the samples were taken from the Powder Diffraction File.

3. Results and discussion

3.1. EEW nanopowders

The electric explosion of a wire represents a rapid change in the physical state of a metal as a result of intense heat generation when a high-density pulse current is transmitted. As a result of the passage of a current pulse, the metal overheats above the melting temperature, and the superheated metal is explosively dispersed. The pressure and temperature at the front of the emerging shock wave reach several hundred megapascals and approximately 10^4 K, respectively. As a result of condensation in a stream of rapidly expanding steam, particles of very small sizes are formed. By adjusting the conditions of the explosion, it is possible to obtain powders with particles of different sizes - from 5 to 300 nm (the weighted average particle radius is 70 nm or more). Results of the XDR analysis of the silver nanoparticles showed that more than 97% mass. had a cubic structure ($a = (0.4082 \pm 0.0007)$ nm) with CSR 21 nm, less than 3 % mass. had a hexagonal structure.

3.2. Chemical nanopowders

The EDX analysis was performed on three sample sites. The analysis showed the absence of possible impurities from elements (N, O) in the samples. From these results, it can be concluded that the silver nanoparticles are pure. The results of the XDR analysis of the silver nanoparticles obtained via the chemical method showed only a cubic structure ($a = (0.4085 \pm 0.0007)$ nm) with a CSR of 10 nm. Figure 1a shows TEM images of the synthesized silver nanoparticles. It can be observed that the silver nanoparticles have a spherical shape. Both individual nanoparticles and their agglomerates are present in the figure. Figure 1b shows the distribution of nanoparticles by size in yellow sols. The sol contains particles of different sizes - from 2 to 26 nm (the weighted average particle radius is 7 nm). The structure and phase composition of Ag nanoparticles were determined by the SAED method. The inset in figure 1a shows a SAED pattern of the prepared samples. The SAED pattern of the sample has a point-ring shape, which is indicative of the polycrystalline structure of the samples. All nanoparticles are crystalline, no amorphous nanoparticles were detected. The analysis of SAED patterns has confirmed the crystal structure of particles, which is indexed as a cubic (Fm3m) structure of Ag. An exact determination of the phase composition is possible from the analysis of HRTEM patterns of separate nanoparticles. Figure 1 c, d, e shows three particles of approximately the same size. Figure 1 c shows NPs obtained by the EEW method. The interplanar spacings were obtained from the FFT and they correspond to Ag-4H. In figure 1d,e NPs obtained by the chemical method are shown. The interplanar spacings are obtained from the FFT and correspond to Ag-FCC. After FFT of the whole figure 1e, masking on the FFT, and the reconstructed image by inverting the filtered image is shown in figure 1f. This picture clearly shows Ag clusters ranging in size from 1 to 1.5 nm and NPs up to 5 nm with a cubic structure. The processing of HRTEM images of Ag NPs obtained by the chemical method showed the presence of only the cubic phase. The minimum particle size obtained by the chemical

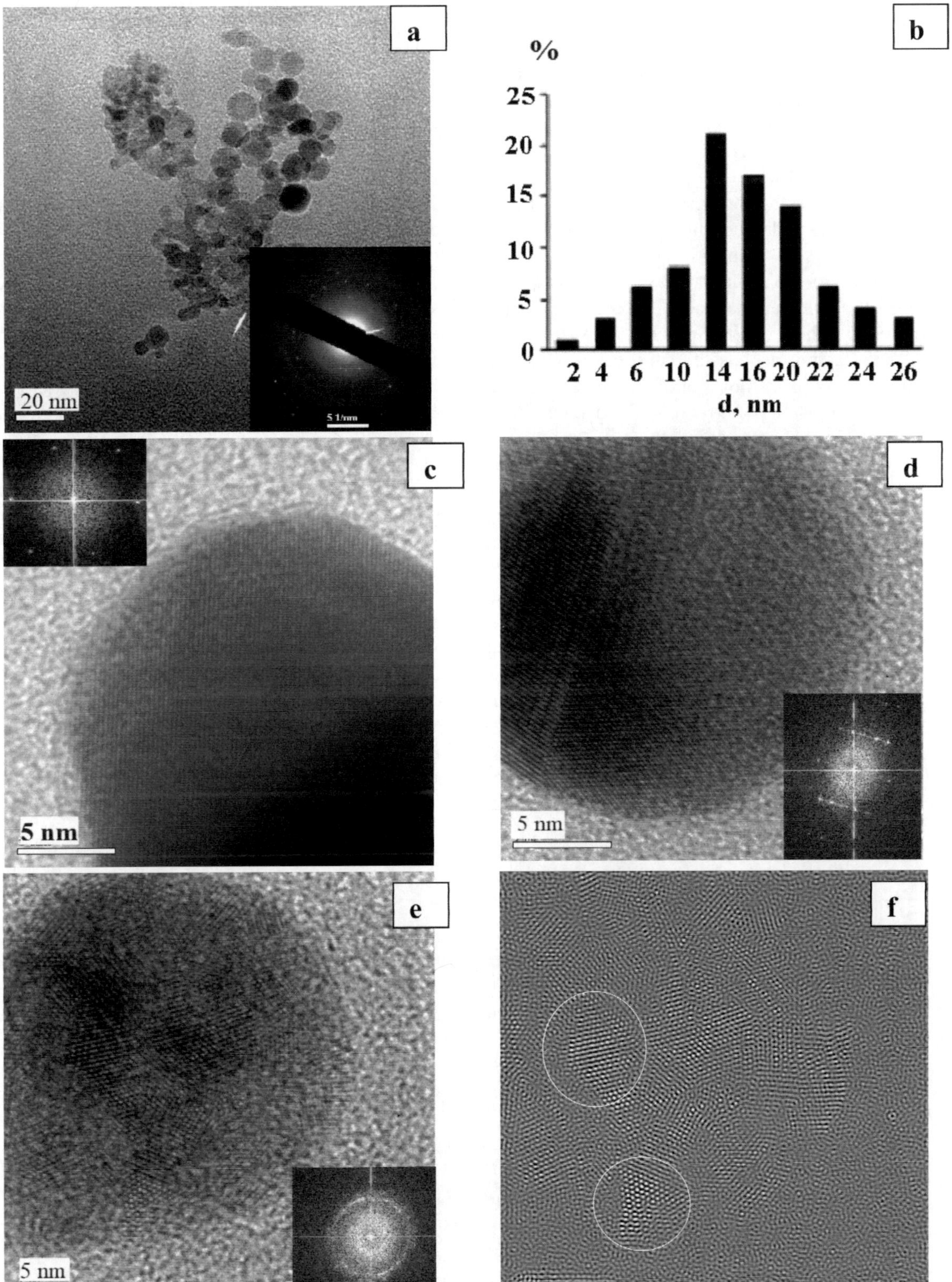
a
20 nm
5 1/nm
b
%
25
20
15
10
5
0
2 4 6 10 14 16 20 22 24 26
d, nm
c
5 nm
d
5 nm
e
5 nm
f

Figure 1.

method is an order of magnitude smaller than the particle size obtained by the EEW method.

Despite this, there are no particles with a hexagonal phase among the particles obtained by the chemical method. The presence of Ag-4H in NPs obtained by the EEW method cannot be explained by the Gibbs-Thomson effect. On the other hand, as shown in [7], the $r_c(T)$ function decreases with increasing T and leads to a higher thermal stability of Ag-4H than Ag-FCC for small sizes of Ag nanocrystals at lower temperatures and vice versa. Thus, this model does not agree with the obtained experimental data.

Next comes the process of formation of the Ag-4H structure. Solidification of the NPs after EEW comes from undercooling of liquid droples. The cooling process is extremely short. The crystalline structure firstly formed in the nanoparticle corresponds to the phase diagram of the material and conforms to the temperature at the point in time of nucleus appearance and the increased surface tension of the nanoparticle. The delay in the transformation into a stable lattice take place due to the extremely short cooling process as well as the perfect structure of nanoparticles (Fig. 1c), in which defects, acting as possible nucleation centres, are rare. After cooling at room temperature, lattice transformation is not possible, since the minimum value of energy required for homogeneous nucleation is an insurmountable barrier. As a result, the nanoparticles produced with EEW are "frozen" in the metastable lattice.

Within the experimental determination error (± 0.01 nm), the lattice periods of Ag NPs synthesized by the chemical method and by the method of wire electric explosion do not differ from each other. It has been established that the unit cell parameter of clusters ranging in size from 1 to 1.5 nm and Ag NPs is not a size-dependent quantity and, within the measurement error, takes on a value characteristic of the bulk state. The results obtained contradict the model concepts put forward in [10].

Conclusion

Studies using a transmission electron microscope have convincingly shown that Ag NPs obtained by the chemical method have only a cubic structure. Ag NPs obtained by the EEW method contain both hexagonal and cubic phases. Based on the analysis of the conditions for obtaining Ag NPs, the mechanism of stabilization of the hexagonal phase under normal conditions is explained by the quenching effect. The lattice periods of the cubic phase of Ag NPs within the experimental determination error (± 0.01 nm), synthesized by the chemical method and by the electrical explosion of wire do not differ from each other.

References

[1] Novgorodova M I, Gorshkov A I and Mokhov A V 1981 *Int. Geol. Rev.* **23(4)** 485
[2] Bublik A I 1954 *Dokl. Academy of Sciences the USSR* **95** 521
[3] Taneja P, Banerjee R, Ayyub P, Chandra R and Dey G K 2001 *Phys. Rev.* B **64** 033405
[4] Liu X H, Luo J and Zhu J 2006 *Nano Lett.* **6** 408
[5] Liang C H, Terabe K, Hasegawa T and Aono M 2006 *Japan. J. Appl. Phys.* **45(7)** 6046-48
[6] Singh A and Ghosh A 2008 *J. Phys. Chem.* **112(10)** 3460
[7] Yang C C and Li S 2008 *J. Phys. Chem.* C **112(42)** 16400
[8] Garvie R C 1965 *J. Phys. Chem.* **69** 1238
[9] Muraze Y and Kato E 1983 *J. Am. Ceram. Soc.* **66** 196
[10] Mitsuhashi T, Ichihata M and Tatsuke U 1974 *J. Am. Ceram. Soc.* **57** 97
[11] Srinivasan R, Rice L, Davis B H 1990 *J. Am. Ceram. Soc.* **73(11)** 3528
[12] Skandan G, Foster C M, Frase H, Ali M N, Parker J C and Hahn H 1992 *Nanostr. Mater.* 1 313
[13] Murzakaev A M 2017 *Phys. Met. Metallogr.* **118(5)** 459
[14] Brainina Kh Z, Galperin L G, Kiryuhina T Yu, Galperin A L, Stozhko N Yu, Murzakaev A M and Timoshenkova O R 2012 *J. Sol. St. Electrochem.* **16** 2365
[15] Krutyakov Yu A, Kudrinskiy A A, Olenin A Yu and Lisichkin G V 2008 *Russ. Chem. Rev.* **77(3)** 233
[16] Qi W H and Wang M P 2005 *J. Nanopart. Res.* **7** 51

Ultrasonic welding of metals: instruments, process parameters, and prospects of welding of ultrafine grained materials

A A Mukhametgalina, M A Murzinova, A A Nazarov, N Yu Parkhimovich, A A Samigullina, E R Shayakhmetova and A P Zhilyaev

Institute for Metals Superplasticity Problems, Russian Academy of Sciences, 39 Khalturin st., Ufa 450001, Russia

E-mail: aanazarov@imsp.ru

Abstract. The results of studies on the development of instruments for ultrasonic welding (USW) of metals, which is the basic process for ultrasonic additive manufacturing, are presented. The influence of process parameters (static loading and welding time) on the microstructure and strength of joints produced by USW of commercially pure copper, nickel and titanium is studied. Preliminary results of the first studies on the USW of metals in an ultrafine grained (UFG) state are presented and the prospects of using UFG metals in the ultrasonic additive manufacturing are evaluated.

1. Introduction

Ultrasonic welding (USW) is a relatively new method of the solid state joining of metals and now is widely used in electrical engineering, electronics, automobile industry etc. for joining foils, plates and wires [1-5]. The interest to USW has increased in the recent years due to an invention of ultrasonic consolidation as one of the methods of additive manufacturing [6-8]. The ultrasonic additive manufacturing is based on successive welding of metallic foils to produce bulk articles. For research needs, this process can also be modeled by means of successive welding of metal sheets using a spot ultrasonic welder. This paper reports the results on the development of instruments for this goal and studies of the structure and properties of joints of three metals, copper, nickel, and titanium, processed by USW with an emphasis on the welding of metals having an ultrafine grained (UFG) microstructure. Finally, the prospects of using metals with UFG structures in ultrasonic additive manufacturing are outlined.

2. Instruments for ultrasonic welding

Ultrasonic horn is a key part of ultrasonic welding equipment, which imparts the ultrasonic energy into the samples of metals under welding. This part must satisfy two main requirements: achieving a sufficient amplitude of vibrations of the welding tip during the welding process, when a significant static load is applied, and the possibility of applying the static load with the least increase of the acoustic resistance. In order to satisfy the former, the horn is made as a kind of ultrasonic concentrator, while the latter condition is satisfied, if the static load is applied at a node. The common way of designing the horns is an analytic calculation (if possible) followed by finite element simulations [9,10].

We have considered three types of horns: 1) a half-wave horn, which has a conical input part and a cylindric output part which is cut from two sides by vertical surfaces having radial transitions (figure 1 a); 2) a half-wave stepped horn with a radial transition (figure 1 b) and 3) one-wave horn with two radially smoothed steps (figure 1 c). All horns have flattened surfaces for applying a static force and / or for spanners. As shown by Merkulov [11], a composite horn consisting of a conical input part and cylindric output part and a stepped horn with an exponential transition section have the highest values of the amplification coefficient. Additional cuts by two vertical surfaces further increases this coefficient for the first type horn. The presence of two stepped transitions provides the third horn also a high amplification coefficient. Khemelev et al. [9] have shown that exponential transition for stepped concentrator can be replaced by a radial one and obtained the correction factors for the lengths of the parts of horns in this case.

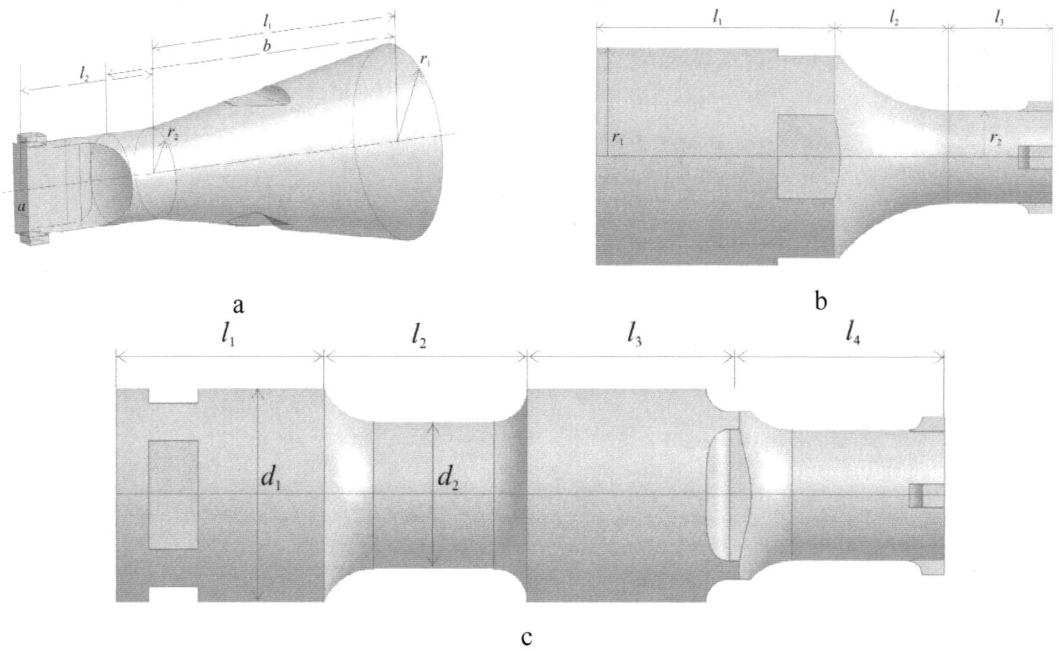

Figure 1. Schemes for design of horns for ultrasonic spot welding: (a) half-wave composite horn consisting of conical input and cylindric output cut by radial surfaces (type 1); (b) half-wave stepped horn with radial transition (type 2); (c) one-wave stepped with radial transitions (type 3).

The analytic calculations do not allow for a sufficiently accurate prediction of the geometrical parameters of horns, since they are valid for the case of lateral dimensions of horns much less than the longitudinal ones. Also, only numerical calculations of real welding horns having weld tips, flattened surfaces for spanners or static load application, is possible.

Starting from the analytically predicted starting configurations obtained for concentrators without weld tips and flattened surfaces, finite element calculations were done for each type horn. The simulations were carried out for Steel 45, which is widely used for making ultrasonic waveguides. The steel bar to fabricate horns had the following properties: density ρ = 7826 kg/m^3, Young's modulus E = 204.25 GPa and Poisson's ratio ν = 0.33. The calculations were done for the target resonance frequency f in the range of 20.0 – 20.1 kHz. The modal analysis of the horn models was performed using ANSYS Workbench Mechanical 19.1.

For the type 1 horn we have considered a horn with the following geometric parameters: r_1 = 31.25 mm, r_2 = 15 mm, l_1 = 82 mm, l_2 = 52.5 mm and different positions of the beginning of radial cuts on the output part of the horn, b. The modal analysis has shown that this type of horn has two neighbor eigenfrequencies, one of which corresponds to torsional vibrations around the waveguide axis and the other to longitudinal oscillations. Figures 2 a,b present the dependences of these frequencies, f, and the amplification coefficient for longitudinal oscillations, M, of the horns on the values of b. As one can see from the figures, the amplification of the waveguide is very high at small values of b and it decreases with b. However, the use of a horn with an elongated narrow output part would result in a worse transmission of vibrations to the welding tip under static load, and it is better to choose a larger value of b at a reasonable amplification. Also, close frequencies of the two vibration modes should be avoided, otherwise torsional vibrations can be excited along with the longitudinal ones leading to lowering of the resulting amplitude. The choice of b = 106 mm can be considered reasonable. Figure 3 represents the displacement map of the horn for this case. One can see that the node of displacements is nicely located on the flattened surface that facilitates applying a static load during USW.

The horn fabricated according to this design, had slightly different sizes (input end diameter 62.5 mm instead of 65 mm, b = 104 mm) and the resonance frequency of 20.20 kHz. Recalculation with these parameters yield the frequency 20.22 kHz, i.e. a value very close to the experimental one.

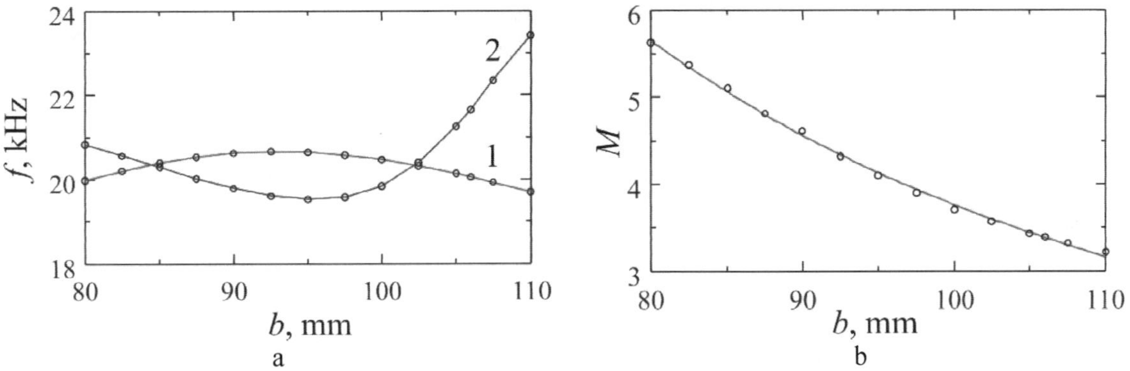

Figure 2. Dependencies of the characteristics of horns of Type 1 on the positions of radial cuts of cylindric parts, b: (a) frequency of longitudinal (1) and neighbor torsional (2) oscillations; (b) amplification ratio for longitudinal vibrations.

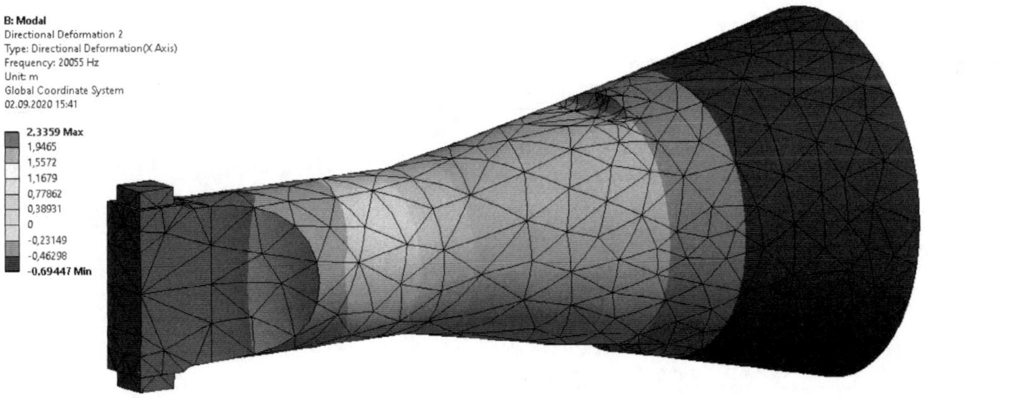

Figure 3. Displacement map of type 1 ultrasonic welding horn with geometrical parameters adjusted for the target frequency.

For the case of type 2 horn, it is important to place the displacement node near the end of the input part, prior to the radial transition. Recently, Khmelev et al. [12] have considered a stepped concentrator with a mounting flange located at the end of the input part and obtained corrected formulae for the lengths of the parts of the concentrator. In the present case, we looked for a configuration, in which the node is located in the middle of flattened surfaces for spanners and the amplification coefficient is reasonably high. The displacement field for the fitted geometrical parameters, l_1 = 73.5 mm, l_2 = 29 mm, l_3 = 37 mm for r_1 = 31.25 mm and r_2 = 14 mm is presented in figure 4. One can see that the resonance frequency is in the target interval and the node is located at a place well suited for applying a static load.

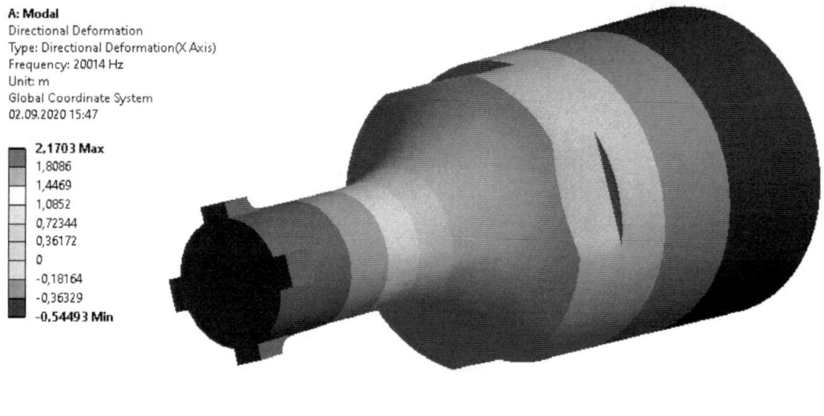

b

Figure 4. Displacement map of type 2 ultrasonic welding horn with geometrical parameters adjusted for the target frequency.

Figure 5 presents the displacement field map for the type 3 horn designed with the geometric parameters $r_1 = d_1/2 = 31.25$ mm, $r_2 = d_2/2 = 25$ mm, $l_1 = l_3 = 64$ mm, $l_2 = l_4 = 62$ mm. Again, one can see that the displacement node is located at the flattened surfaces near the end of the second large-diameter cylinder where the static load can be applied. Despite the smaller ratio between the diameters of the input and output parts, this horn provides the highest amplification coefficient of $M \approx 4.6$.

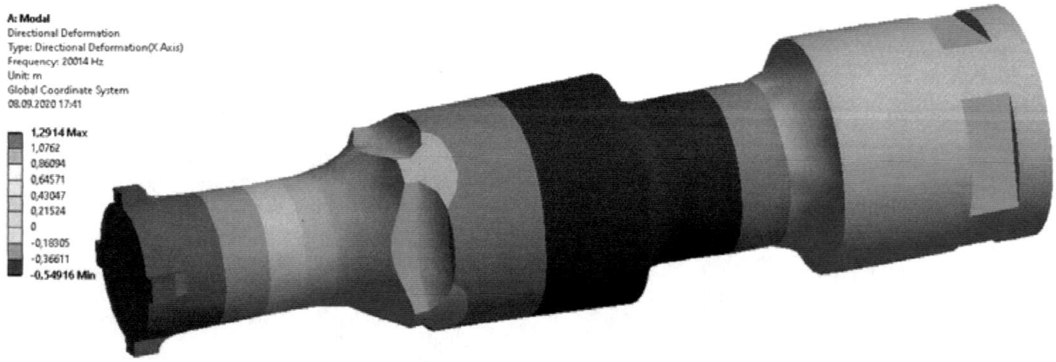

Figure 5. Displacement map of type 3 ultrasonic welding horn with geometrical parameters adjusted for the target frequency.

3. Ultrasonic welding of coarse-grained and ultrafine-grained metals

Samples of metals (commercially pure copper, nickel, and titanium) with the thickness of 0.5 and 0.7 mm were ultrasonically welded using different regimes of welding given by the values of static load and welding time at the amplitude of vibrations of the welding tip of about 20 μm. Two types of joint samples were obtained. By welding of two overlapped sheets, samples for lap shear tests were prepared. By successive welding of sheets, which imitated the process of ultrasonic additive manufacturing, consolidated samples were obtained for a structural characterization. Putting thermal couples between the sheets under welding, the temperature rise during welding was evaluated.

As coarse-grained materials, strips cut from as-received commercial sheets of Cu, Ni and Ti with thickness of 0.5 mm were used. To study the role of UFG structure, samples of UFG copper and nickel were processed by high-pressure torsion (HPT). For this, disks of copper and nickel with diameter of 12 mm and thickness of 10 mm were placed between Bridgman anvils with hollows and subjected to torsion straining up to 5 revolutions under the pressure of 6 GPa [13]. As a result of this treatment, disk samples with diameters 12 mm and thickness of 0.7 mm were obtained. This allows one to obtain UFG structures with grain sizes less than 0.5 μm in copper and nickel [14].

Studies have shown that during the USW process a significant increase of the temperature occurs. The maximum temperature increases with static load *P* and welding time *t* and can amount from 450°C in copper during welding for 1 s in copper up to more than 700°C during welding for 2 s in titanium.

With an increase of *P* and *t* the lap shear strength of USW-processed joints increases. Studies of the microstructure of the joints shows that this is caused by an enhancement of the quality of joints, i.e. the decrease of the density of pores remained in joint interfaces. At the same time, there are optimum values of the static load and above these values the strength of joints decreases due to a significant deformation of the sheets and a change in the failure mode from the interface one to the weld nugget pullout (figure 6).

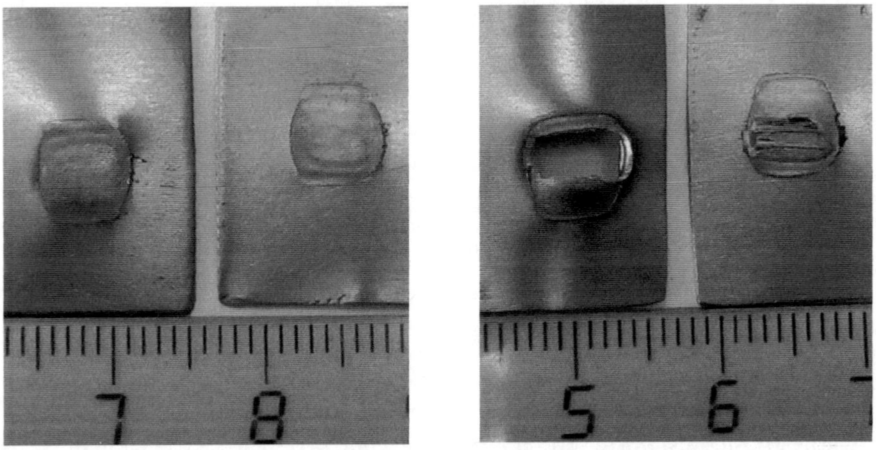

Figure 6. Samples of titanium weld joints processed by USW for 1 s under the static load of 6 (a) and 7 (b) kN after lap shear tests. The left figure indicates the interface failure and the right weld nugget pullout mode of failure.

With an increase of the initial hardness and strength of a material the requirements to the welding parameters providing satisfactory joint quality become more severe: the optimum static load increases and the welding time decreases. This is due to the fact that a higher temperature is required for hard materials to decrease their yield stress and undergo local plastic deformation in the welding zone. For welding of titanium, welding time of 2 s and more was necessary under the load of 5 to 7 kN for the welding tip area of about 30 mm^2, whereas copper and nicked were well welded at lower loads and in time of 1 s.

The most important preliminary result, which was obtained in the studies is that copper and nickel are ultrasonically welded better in the UFG state than in the coarse-grained one. The joints of UFG copper obtained by USW at static load of 3.5 kN had the same strength of about 33 MPa as the joint obtained by welding of commercial copper sheets under the load of 5 kN. The lap shear strength of the joints processed by USW under the load of 6 kN amounted nearly 100 MPa for UFG nickel versus 70 MPa for commercial nickel.

On the other hand, measurements of the microhardness distribution in the cross section of consolidated samples of nickel show that the minimum microhardness is approximately higher in the material consolidated from UFG samples than in the one processed from commercial sheets. This occurs despite the fact that for the regimes used a significant grain growth occurred in the interface regions during welding of UFG materials (figure 7). Nevertheless, the grain size after consolidation is still in the range of a few micrometers that allows the material to retain its high strength and hardness.

Figure 7. Microstructure of the cross section of a joint of initially ultrafine-grained sheets of copper processed by USW. One can see a significant grain growth near the welding interface.

Thus, the preliminary studies demonstrate that by successive welding, i.e. consolidation of sheets or foils with an UFG structure it is potentially possible to obtain bulk materials, which can have higher strength characteristics than the ones obtained by a consolidation from conventional commercial sheets or foils. In order to make better use of this potential, it is important to look further for regimes of USW, which would provide good joint quality at a less significant grain growth.

A considerable grain growth occurs even in the commercial sheets of titanium, although their initial grain size was out of the ultrafine range. The fact that intensive grain growth occurs in UFG copper and nickel and commercial titanium during the very short time intervals during USW is unusual and calls for more careful studies. Most probably, this is due to an effect of not only the enhanced temperature, but also of an intensive cyclic deformation during the process of ultrasonic welding.

4. Conclusions

Modal analysis of three types of horns for ultrasonic welding had been performed using finite-element method based ANSYS software. The results show that in same cases (type 1 horn, for example), torsion vibrations modes with very close frequencies can be excited along with the desired longitudinal vibrations and such modes should be avoided in designing the horns.

Experiments have been carried out on USW of the sheets of commercially pure copper, nickel, and titanium. Copper and nickel with an ultrafine grained microstructure were welded ultrasonically for the first time. These preliminary results demonstrate that, despite a significant grain growth in UFG metals due to the effect of high temperature and intensive oscillatory straining, ultrasonically consolidated samples retain a higher microhardness that the samples consolidated from coarse grained metals. This leads to an important conclusion that ultrasonic additive manufacturing based on USW can have a potential in processing bulk articles with higher strength when using materials with an UFG structure..

Acknowledgments

The present work was supported by the Russian Science Foundation (grant No. 16-19-10126). Electron microscopic studies and mechanical tests were carried out on the facilities of shared services center of IMSP RAS "Structural and Physical-Mechanical Studies of Materials".

References

[1] Silin L L, Balandin G F and Kogan M G 1962 *Ultrasonic Welding* (Moscow: Mashgiz) [in Russian]
[2] Mitskevich A M 1970 Ultrasonic metal welding *Physical Foundations of Ultrasonic Technology* (Nauka: Moscow) [in Russian]
[3] Kholopov Yu V *Ultrasonic Welding of Plastics and Metals* (Leningrad: Mashinostroenie) [in Russian]
[4] Graff K 2005 Ultrasonic metal welding *New Developments in Advanced Welding* ed N Ahmed (Cambridge: Woodhead) p 241-69
[5] Matheny M P and Graff K F 2015 Ultrasonic welding of metals *Power Ultrasonics-Applications of High-intensity Ultrasound* ed J A Gallego-Juárez and K F Graff (Cambridge: Woodhead Publishing, Elsevier Ltd.) chapter 11 p 259–93
[6] White D. Ultrasonic Object Consolidation. US Patent # 6519500, 23 March 2000.
[7] Graff K E 2011 Ultrasonic additive manufacturing *Welding Fundamentals and Processes* ed. T Lienert, T Siewert, S Babu and V Acoff (Russell Township: ASM Int.) p 731-42
[8] Herr A and Norfolk M 2019 *Rapid Prototyp. J.* **26/3** 445-58
[9] Khmelev V N, Khmelev S S, Tsyganok S N and Levin S V 2013 *Sources of Ultrasonic Action. Features of Design and Structures* (AltSTU, Biisk) 196 p [in Russian]
[10] Mukhametgalina A A and Nazarov A A 2019 *Lett. Mater.* **9(4)** 414-8 [in Russian]
[11] Merkulov L G and Kharitonov A V 1959 *Akusticheskii Zhurnal* **5(2)** 183-90 [in Russian]
[12] Khmelev V N, Khmelev S S and Levin S V 2015 *South-Siberian Scientific Bulletin* **2(10)** 17-20 [in Russian]
[13] Khisamov R K, Nazarov K S, Irzhak A V, Shayakhmetov R U, Musabirov I I, Timirayev R R, YumaguzinY M and Mulyukov R R *Lett. Mater.* **9(2)** 212-7
[14] Zhilyaev A P and Langdon T G 2008 *Progr. Mater. Sci.* **58** 893-979

Structural and micromechanical properties of 316L stainless steel produced by selective laser melting

A V Makarov[1], V P Kuznetsov[1,2], V A Sirosh[1], E G Volkova[1], P A Skorynina[3] and A G Merkushev[2]

[1]M .N. Miheev Institute of Metal Physics, Ural Branch of the Russian Academy of Sciences, 18 S. Kovalevskoy St., Ekaterinburg, 620108, Russia
[2]Ural Federal University, 19 Mira St., Ekaterinburg, 620002 Russia
[3]Institute of Engineering Science, Ural Branch of the Russian Academy of Sciences, 34 Komsomolskaya St., Ekaterinburg, 620049, Russia

E-mail: sirosh@imp.uran.ru

Abstract. The structure and micromechanical properties of a 316L stainless steel produced by the selective laser melting (SLM) technique have been studied in its initial state and the following heat treatment: quenching from 1050 ºC and 4-hour annealing at 480 ºC. The heat treatment does not result in changing the steel phase composition, however, it reduces the microhardness by 5%, *HM* hardness (Martens hardness) by 12% and indentation hardness by 16% at H_{IT} maximum load. The heat treatment also increases the E^* contact elasticity modulus by 14% due to the lower porosity of the material. As soon as heat treatment provides no strengthening of the steel synthesized by SLM, deformation processing, chemicothermal treatment and application of thin coatings may become quite promising strengthening methods.

1. Introduction

Additive manufacturing has been gaining a solid ground in modern engineering. This rapidly progressing technology proves to be the most promising for the fabrication of complex parts in the aeronautic, automotive, medical industries, etc. Selective laser melting is one of its applications, which is the layer-after-layer formation of a piece by laser beam scanning of powder layers laid on a substrate. As layers are added, different parts of the piece are subjected to various thermal effects. This leads to residual stresses caused by temperature gradients between the layers. This effect is likely to initiate cracking and delamination of the material [1]. The combination of incomplete fusion [2] and residual stresses brings about high porosity inside the part. Hence, the application of post-processing (deformation processing, heat and chemicothermal treatments, deposition of coating) designed to improve the quality of the material is topical for parts synthesized by the SLM technique.

The purpose of the study was looking into the structural phase states and micromechanical characteristics of stainless chrome-nickel steel samples produced by SLM, including those that were subjected to post-heat treatment.

2. Materials and methods

Stainless austenitic steel samples produced by SLM were studied, their chemical composition being (mass) Cr 15.93%; Ni 10.00%; Mo 2.20%; Mn 1.27%; Si 0.55%; C 0.03%; S 0.02%; P 0.08%, Fe – the rest. An EOS M 280 unit was used to build the samples in gaseous nitrogen employing a 400W laser; the diameter of the laser spot on the surface was 50μm, the thickness of the applied powder layer - 20μm, the rate of the laser beam movement on the powder surface reaching 1000 mm/sec. The structure and micromechanical properties of the steel were investigated in its initial state after forming the samples by the SLM technique and subsequent heat treatment. The aim of heat treatment (HT) was to degass and reduce the amount of vacancies, and to relax residual stresses in the synthesized sample. It was heat treated at 1050 °C for 30 minutes, followed by oil quenching with subsequent 4-hrs ageing at 480 °C and air cooling.

A JEOL JEM-200CX transmission electron microscope was used to study the steel structure. To determine the microhardness, a Shumadzu HMV-G21DT device as well as the residual imprint method were utilized with a load on the Vickers indenter of 0.245 N. An ISO 14577 Fischerscope HM2000 XYm measurement system was applied to determine the microindentation characteristics. The maximum microindentation load on the Vickers indenter was 0.245 N. The micromechanics of the coatings were found with the help of loading and unloading curves gained by the Oliver-Pharr method [3].

3. Experimental results and discussion

The transmission electron microscope survey showed that the initial structure of the 316L steel samples in the planes cut parallel (‖) and perpendicular (⊥) to the Z axis (figure 1), i.e. parallel and perpendicular to the sample building direction during the SLM process, prior to the heat treatment was identical. The structure consists of elongated subgrains with separate dislocations or dislocation cells inside them (figure 2). All the grains were γ-Fe (austenite).

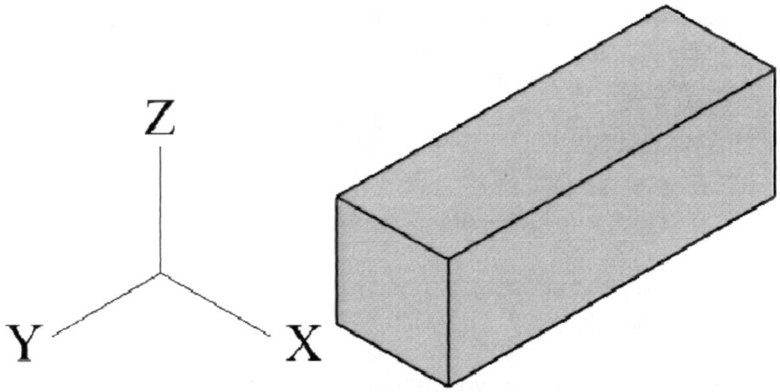

Figure 1. Schematic illustration of the sample synthetization: X,Y - scanning direction, Z - building direction.

The structure of the heat-treated steel in the vertical plane parallel to the Z axis, i.e. in the direction of the sample building (‖), comprises elongated austenitic subgrains with prolonged, straight, band-shaped boundaries (figure 3). The density of dislocations inside the grains is low.

The results of the structural survey fully agree with the published data on the cellular structure in 316L stainless steel formed by SLM [4-6]. The 30 minutes annealing at 1050 °C prior to quenching is expected to make gaseous impurities, which are likely to have evolved during the SLM process, to emit from the γ-solid solution [6]. Uneven distribution of powder in SLM could have led to the formation of the ferrite phase [7], but no α-phase was detected.

The microhardness tests applying the recovered indentation technique revealed that heat treatment results in a slight softening of the steel; the microhardness decreased by 5% from 252±7 HV 0.025 to 240±5 HV 0.025.

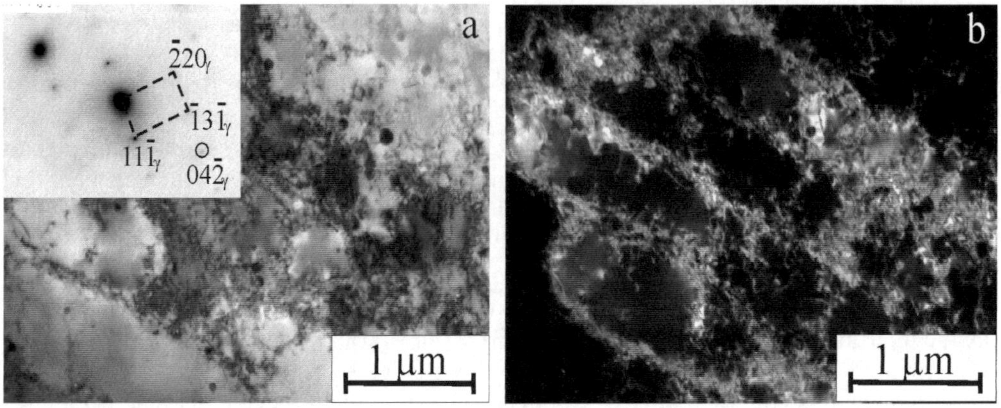

Figure 2. TEM images of the initial 316L (‖) SLM samples structure: bright field and electronic diffraction patterns (a), dark field (b).

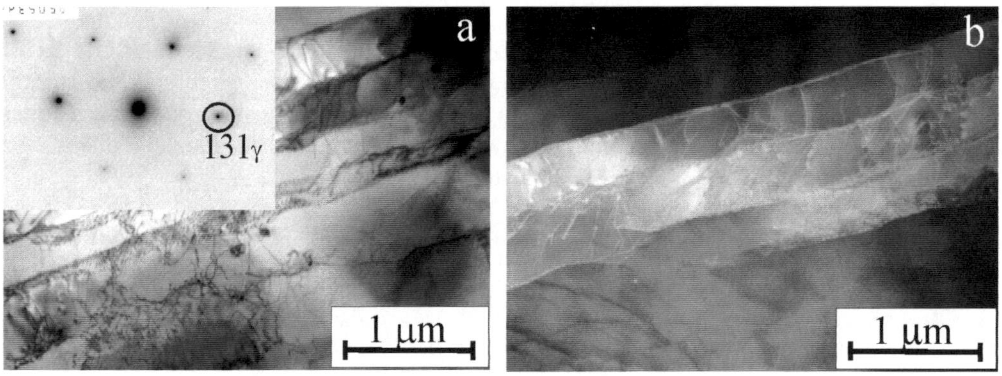

Figure 3. TEM images of the 316L (‖) SLM samples structure after heat treatment: bright field and electronic diffraction patterns (a), dark field (b).

Figure 4 shows the diagrams that describe the loading, namely, the ascending curve, and unloading, represented by the distending curve, at the maximum load F = 0.245 N as determined by the instrumental microindentation method. It is evident that the heat treatment results in shifting the curves towards higher values of the thrust of the indenter h to the right; consequently the values of the maximal (h_{max}) and residual (h_p) indentation depths rise by 7% and 10% respectively, see table 1.The lower strength properties also prove that the material becomes softer (table 1); namely, the Martens hardness HM revealing both plastic and elastic deformations decreases by 12%, from 2.72 ± 0.07 GPa to 2.39 ± 0.05 GPa while the indentation hardness at the maximum load H_{IT} drops by 16% from 3.18 ± 0.09 GPa to 2.67 ± 0.06 GPa. This slight softening of the steel may be attributed to the above-mentioned lower density of dislocations following the heat treatment. It is significant to note that the microindentation data shows higher values of the contact elasticity modulus E^*; it rose by 14% from 171 GPa to 195 GPa (table 1). It reflects a lower porosity (compaction) of the synthesized steel resulting from high temperature heating.

The total mechanical work of indentation W_t determined by the area under the loading curve in figure 4 including the work spent for plastic deformation and elastic recovery, also rises by 8% after heat treatment (table 1), since the less the strength of material, the higher is its deformation under the indenter and consequently the more work is done on deformation. The work of elastic deformation W_e determined by the area under the unloading curve in figure 4, on the contrary, goes down by 16%. The fraction of the elastic deformation in the total deformation decrease; it is also proved by the lower values of the relation H_{IT}/E^* [8] and the elastic recovery $R=((h_{max}-h_p)/h_{max})\times 100\%$ [9] that decrease by 26 and 19%, respectively (table 2). Somehow, the plastic component of the work of indentation calculated using the formula $\delta A=(1-(W_e/W_t))\times 100\%$ [10] grows very insignificantly, by 3.5%, and the resistance to deformation after the transition to plastic flow, which is described by the relation H_{IT}^3/E^{*2} [11], does not change after heat treatment at all (table 2).

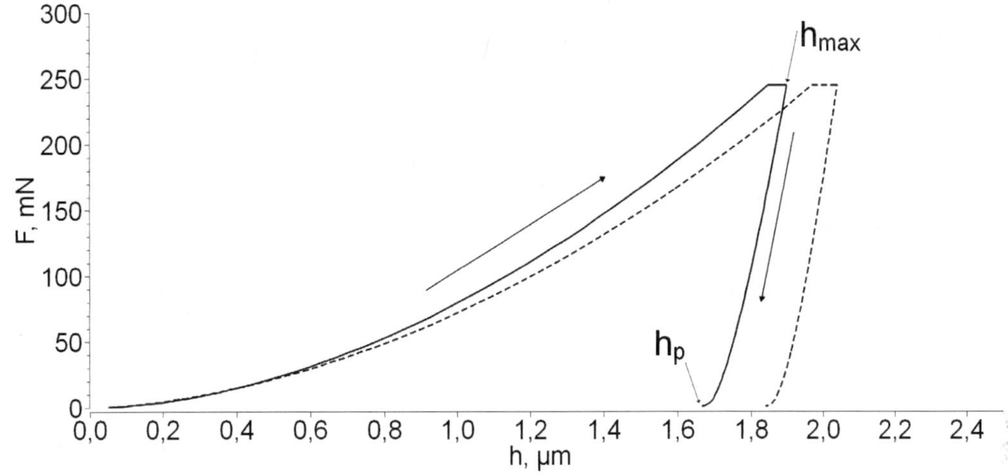

Figure 4. Experimental curves during microindentation of the 316L SLM samples surfaces after (dash-line) and without (solid line) heat treatment.

Table 1. The surfaces microindentation results of 316L SLM samples after and without heat treatment.

Treatment	h_{max} (μm)	h_p (μm)	HM (GPa)	H_{IT} (GPa)	E^* (GPa)	W_t (nJ)	W_e (nJ)
Without HT	1.90 ± 0.03	1.67 ± 0.02	2.72 ± 0.07	3.18 ± 0.09	171 ± 4	159.1± 2.8	22.0± 0.5
After HT	2.04 ± 0.02	1.84 ± 0.02	2.39 ± 0.05	2.67 ± 0.06	195 ± 5	171.8 ± 1.9	18.4 ± 0.4

Table 2. The ability characteristics of 316L SLM samples surface layers to resist mechanical contact action after and without heat treatment.

Treatment	δA (%)	R (%)	H_{IT}/E^*	H_{IT}^3/E^{*2} (GPa)
Without HT	86	12.1	0.019	0.001
After HT	89	9.8	0.014	0.001

The results of the microidentation that have been discussed above support the data on measuring the microhardness applying the recovered indentation technique which shows a slight softening of the SLM steel resulting from the heat treatment. It is shown by lowering the strength characteristics by 5-16%. In addition, it was found that the fraction of the elastic deformation in the total deformation decreases, indicating a faster transition of the heat-treated steel to plastic deformation under contact loading.

Therefore, the heat treatment that has been applied does not result in strengthening of the steel when the mechanism of aging austenite is employed. Deformation strengthening as well as chemicothermal treatment leading to chemical modification of the surface layer might prove to be efficient for strengthening austenitic steel synthesized by SLM.

As the authors of the present paper have shown, efficient strengthening and improved tribological properties of metastable and deformation-stable austenitic chrome-nickel steels is achieved by processing with a sliding indenter [12-16] as well as by combining nanostructuring frictional processing with low-temperature plasma nitriding [17-21]. It seems advantageous to use the described modifying processes for making a gradient-strengthened substrate on the surface of the synthesized austenitic steel for further application of nano-compositional thin-film multilayered coatings on the base of amorphous superhard diamond-like carbon and titanium carbides. We will study these directions in our future research, designed to give scientific credence for the new combined method for post-treatment of SLM samples. This method will make it possible to efficiently use thin-film high-strength coatings in additive manufacturing on parts made from structural steels that are significantly inferior in hardness to these coatings.

4. Conclusion

The austenitic structure with high densities of dislocations is formed by means of synthesizing a 316L steel by the selective laser melting technique. Heat treatment, quenching from 1050 °C and further 4-hour annealing at 480 °C, performed for degassing, reducing the number of vacancies and relaxing residual stresses do not result in altering the phase composition. However, a lower density of dislocations as well as their configurations in austenitic grains is evident. It causes a slight reduction of micromechanical characteristics to occur: the microhardness goes down by 5%, Martens hardness *HM* and indentation hardness at maximum load H_{IT} – by 12 and 16%, respectively. With the help of kinetic microindentation it was determined that the fraction of elastic deformation in the total deformation becomes smaller as caused by heat treatment. A 14% growth of the elasticity modulus E^* was discovered signifying a decrease of the porosity of the steel when heated to a high temperature.

Thus, the studied heat treatment does not provide strengthening through aging austenite in a synthesized steel. To obtain its effective strengthening application of nanostructuring frictional and chemicothermal (plasma nitriding) treatments may turn out to be advantageous as well as additional laying of thin-film coatings on a hardened substrate.

Acknowledgments

The main results were funded by the RFBR and Sverdlovsk Oblast, project number 20-48-660065. The results of the electron diffraction phase analysis were obtained by the M.N. Mikheev Institute of Metal Physics of the Ural Branch of the Russian Academy of Sciences (IMP UB RAS) within the state assignment of Ministry of Science and Higher Education of the Russian Federation (theme “Structure” No. № AAAA-A18-118020190116-6). The studies were performed at the collaborative access center “Testing center of nanotechnology and advanced materials” of IMP UB RAS.

References

[1] Yap C Y, Chual C K, Dong Z L, Liu Z H, Dong Z L, Zhang D Q, Loh L E and Sing S L 2015 *Appl. Phys. Rev.* **2** 041101
[2] Zhou X, Wanga D, Liu X. Zhang D D, Qu S, Ma J, London G, Shen Z and Liu W 2015 *Acta Mater.* **98** 1
[3] Oliver W C and Pharr J M 1992 *J. Mater. Res.* **7** 1564
[4] Amato K N, Gaytan S M, Murr L E, Martienez E, Shindo P V, Hernandez J, Collins S and Medina F 2012 *Acta Mater.* **60** 2229
[5] Pinkerton A J and Li L 2003 *Appl. Surf. Sci.* **208-209** 411
[6] Bazaleeva K O, Tsvetkova E V and Balakirev E V 2016 *Vestn. Mosk. Gos. Tekh. Univ. im. N.E. Baumana, Mashinostr. (Herald of the Bauman Moscow State Tech. Univ., Mech. Eng.)* **5** 117 [in Russian]
[7] Saeidi K, Gao X, Lofajc F, Kvetková L and Shen Z J 2015 *J. Alloys Compd.* **633** 463
[8] Cheng Y T and Cheng C M 1998 *Appl. Phys. Lett.* **73** 614
[9] Petrzhik M I and Levashov E A 2007 *Crystallogr. Rep.* **52** 966
[10] ISO 14577–1:2002
[11] Mayrhofer P H, Mitterer C and Musil J 2003 *Surf. Coat. Technol.* **174-175** 725

IOP Conf. Series: Materials Science and Engineering 1008 (2020) 012005 doi:10.1088/1757-899X/1008/1/012005

[12] Kuznetsov V P, Makarov A V, Osintseva A L, Yurovskikh A S, Savrai R A, Rogovaya S A and Kiryakov A E 2011 *Strengthening Technologies and Coatings* **11** 16 [in Russian]
[13] Makarov A V, Skorynina P A, Osintseva A L, Yurovskikh A S and Savrai R A 2015 *Obrab. Met. (Met. Work. Mat. Sci.)* **4** 80 [in Russian]
[14] Makarov A V, Skorynina P A, Osintseva A L, Yurovskikh A S 2017 *Phys. Metals Metallogr.* **118** 1225
[15] Savrai R A, Makarov A V, Malygina I Yu, Rogovaya S A and Osintseva A L 2017 *Diagnostics, Resource and Mechanics of Materials and Structures* **5** 43 [in Russian]
[16] Makarov A V, Skorynina P A, Volkova E G and Osintseva A L 2020 *Met. Sci. Heat Treat.* **61** 764
[17] Makarov A V, Gavrilov N V, Samoylova G V, Mamaev A S, Osintseva A L and Savrai R A 2017 *Obrab. Met. (Met. Work. Mat. Sci.)* **2** 55 [in Russian]
[18] Makarov A V, Samoylova G V, Gavrilov N V, Mamaev A S, Osintseva A L, Kurennykh T E and Savrai R A 2017 *AIP Conf. Proc.* **1915** 030011
[19] Lezhnin N V, Makarov A V, Gavrilov N V, Osintseva A L and Savrai R A 2018 *AIP Conf. Proc.* **2053** 040050
[20] Makarov A V, Savrai R A, Skorynina P A and Volkova E G 2020 *Met. Sci. Heat Treat.* **62** 61
[21] Zhidkov I S, Kukharenko A I, Makarov A V, Savrai R A, Gavrilov N V, Cholakh S O, and Kurmaev E Z 2020 *Surf. Coat. Technol.* **386** 125492

Effect of electric pulse treatment on the structure and hardness of nickel deformed at room and liquid nitrogen temperatures

R R Ilyasov, A Kh Valeeva, I Sh Valeev, O Sh Sitdikov and M V Markushev

Institute for Metals Superplasticity Problems of Russian Academy of Sciences, 39 Khalturin St., Ufa, Russia

E-mail: ilyasov@imsp.ru

Abstract. The influence of the energy of electro-pulse processing (EPP) on the structure and hardness of pure nickel deformed at room and liquid nitrogen temperatures is studied. The metal was subjected to isothermal rolling with a strain of 90% and subsequent EPP with integral current densities Kj in the range from 0.06 to 0.19×10^5 A^2s/mm^4 corresponding to the calculated temperature range of 130-925°C. It was found that decreasing the rolling temperature of Ni resulted in a more uniform and less misoriented nanocellular structure and a 15-20 HV higher hardness. Subsequent EPP decreased the hardness due to recovery, discontinuous recrystallization with the formation of annealed twins and growth of new grains. It is revealed that a more uniform fine-grained structure with a smaller grain size is formed in the cryorolled metal in a narrower interval and at less EPP energies.

1. Introduction

One of the promising but poorly studied ways of processing the ultrafine-grained (UFG) structure in metals and alloys is the so-called cryogenic deformation, which is carried out at temperatures below 120 K (-153°C) [1-8]. It is considered that low deformation temperatures, suppressing dynamic recovery, promote the accumulation of internal stresses due to the extremely high densities of crystal lattice defects and thus intensify grain refinement under straining and subsequent annealing. However, there are a number of investigations that have shown that deformation even to high strains (with $e > 1$) at cryogenic temperatures does not guarantee the UFG structure processing. For example, severe plastic deformation (SPD) of pure aluminum via cryorolling did not result in the formation of the expected nanocrystalline structure and even did not ensure significant grain refinement [1,7]. The studies of copper have also shown that a decrease in the deformation temperature, suppressing double-cross slip of lattice dislocations, complicates the formation of strain-induced boundaries and thus restricts grain fragmentation (grain subdivision) and the formation of new fine grains via dynamic recrystallization. Nevertheless, in [9-11] it was found that a complex thermomechanical treatment involving cryorolling and subsequent electro-pulse processing (EPP) ensured the formation of the UFG structure. Therewith, the EPP played a dominating role causing processes of static recovery and recrystallization under thermal activation in a highly work-hardened material, allowing extreme grain refinement due to the superfast heating rate and very short time of annealing.

The purpose of the present investigation was to compare the effect of EPP with different energies on the structuring and softening of pure nickel, initially deformed to equal strains at a temperature of liquid nitrogen and ambient temperature.

IOP Conf. Series: Materials Science and Engineering 1008 (2020) 012006 doi:10.1088/1757-899X/1008/1/012006

2. Material and procedure

Pure nickel (designated as NP2 in accordance to the Russian state standard GOST 492-73) was used as the material for the study. The ingots were preliminary subjected to setting at 850°C followed by annealing at 600°C for 2 h. Subsequent multi-pass rolling of mechanically cut out sheet billets was carried out to a total strain of about 90% at 20 and –196°C. Rolling was performed in isothermal conditions using a laboratory six-roll mill with preliminary cooling of removable working rolls and billets in a bathtub with cold water/liquid nitrogen within 60 minutes. The rolling speed did not exceed 100 mm/min with straining per pass of about 5%.

EPP was carried out on flat samples having the dog-bone shape (as samples for tensile tests) with a thickness of 0.4 mm and sizes of the gauge part of 3×4 mm^2. The samples were cut out along the rolling direction and fixed in massive clamping jaws with current connectors. The temperature of heating of the gauge part was calculated using the relationship between the temperature and current density:

$$S(T_c) = \int_{T_1}^{T_2} \rho c \sigma_e dT,\; K_j = \int_0^{\tau} j^2 dt\; S(T_C) = K_j ,$$

where T_C is the calculated temperature, T_l is the reference temperature of the sample, τ is the duration of the electric impulse. For the chosen K_j current parameters (from 0.06×10^5 to 0.19×10^5 A^{2}s/mm^4), the temperature interval of heating the samples was within 130 and 925 °C. With an impulse time of about 10^{-4} s and a current frequency of 10^4 Hz, the depth of its penetration to the sample of nickel was about 0.2 mm [3]. This means that the skin-effect in the samples used could be neglected suggesting a uniform distribution of current on their section.

The microstructure was analyzed in the rolling plane with the use of a "TESCAN MIRA 3 LMH" scanning electron microscope equipped with "HKL Channel 5" software. EBSD maps were obtained using the data from 6 Kikuchi-line scanning with a step of 0.03 and 0.5 μm. The intercrystalline boundary misorientation of 15° was taken as a criterion for dividing into low- and high-angle boundaries (LABs and HABs). The mean size of subgrains and grains (d and D, consequently) were determined by the method of equivalent diameter on the base of at least 300 measurements of crystallites. The average angle of intercrystalline boundary misorientations (Θ) and fractions of HABs and twin boundaries (F_{HAB} and $F_{\Sigma3+\Sigma9}$, consequently) were derived from grain boundary spectra, in which boundaries of less than 1° were not taken into consideration. The size and fraction of recrystallized grains (D_R and F_R, consequently) were also measured using EBSD data and the mentioned software.

Microhardness was determined by the Vickers method with an MVDM 8 "AFFRY" tester at 0.5 N and a loading time of 10 s. To obtain an error not exceeding 5%, more than ten measurements at the point were performed.

3. Results and discussions

The obtained EBSD maps and grain boundary spectra (figure 1) allow one to conclude that the structuring of nickel under room temperature- and cryorolling conditions has a quite similar nature in many respects since it resulted in the formation of the same type of highly work-hardened, partially recrystallized structure. Indeed, in both cases the fragmentation of the initial grains and dynamic recrystallization with the formation of new HABs and fine grains occurred quite inertly and non-homogeneously, especially under cryostraining. Moreover, it could be concluded that the main mechanism of structure transformations under both rolling temperatures was the formation of so called low-energy dislocation structures, which led to the evolution of a predominant and well-developed substructure with nanoscale crystallites of various shapes and sizes divided mostly by LABs. Thus, both deformation structures could be defined, in general, as cellular-type ones, meanwhile in cryorolled condition it was more disperse and less misoriented. As a result, the fractions of HABs and average boundary misorientations were recorded to be a bit higher in the cold-rolled metal (table 1). It should be also noted that the deformed structure is characterized by the absence of deformation twins, which are frequently observed in conventionally cold processed nickel [12].

Judging by the data in figure 2, the higher suppression of dynamic recovery and recrystallization led to a more defected and developed deformation structure, being the main reason for the increased hardness of cryorolled nickel.

From figure 2 also follows that in the low EPP energy range (up to Kj = 0.130×10^5 A^2s/mm^4), the cold-rolled metal continuously softened to some extent. EBSD analysis showed that such behavior was caused by the activation of static recovery and the local occurrence of recrystallization in the work-hardened structure (figure 3, table 1). With a further increase in the EPP energy, the hardness of nickel decreased more extensively and at Kj = 0.170×10^5 A^2s/mm^4, the effect of deformation hardening was completely lost (figure 2). Judging by the maps in figure 3, at such EPP energy conditions the processes of discontinuous static recrystallization were accompanied by abnormal growth of separate new fine grains. As a result, at Kj = 0.152×10^5 A^2s/mm^4, a partially recrystallized structure with an average size of new grains of about 3 μm was initially formed inside the coarse-grained and highly strained structure.

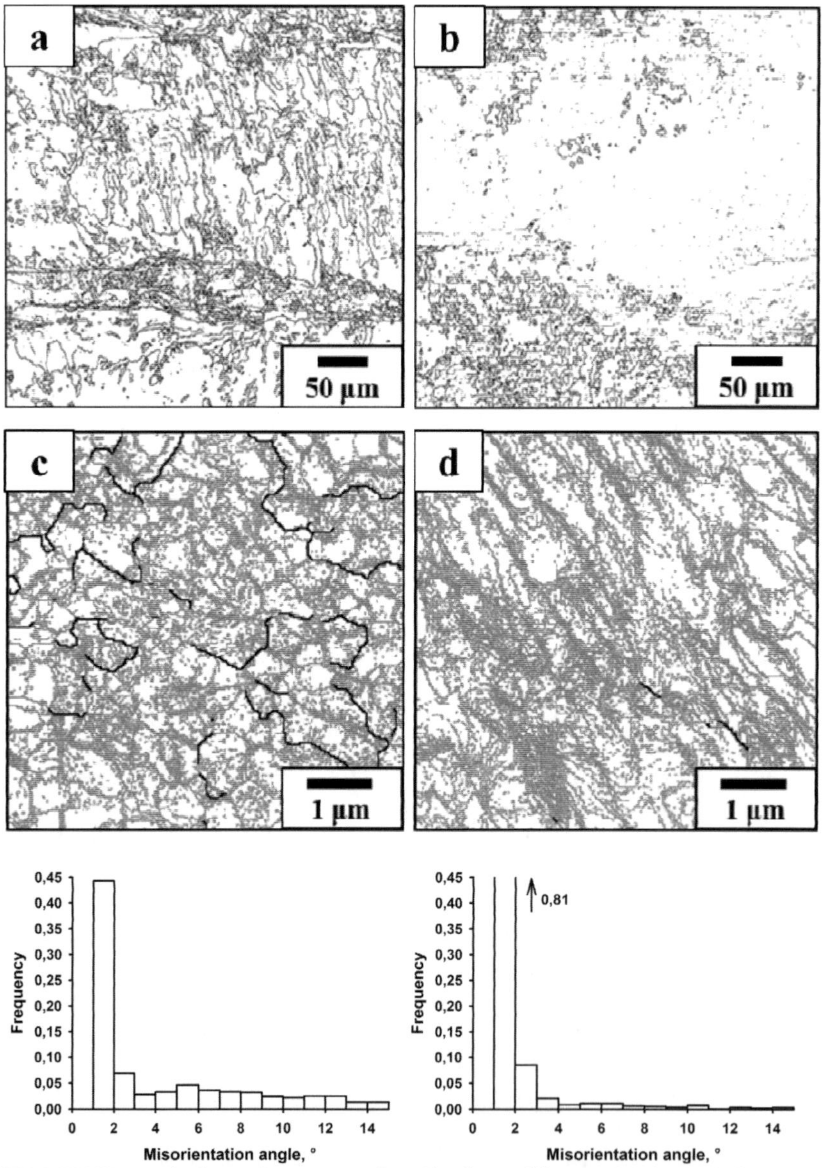

Figure 1. SEM-EBSD grain boundary maps for Ni after rolling at 20 (a,c) and -196°C (b,d) to a total strain of 90% and corresponding LABs spectrums.
(Hereafter, HABs and LABs are indicated by black and red lines, respectively)

Table 1. Parameters of Ni microstructure after rolling and EPP with different energies

Rolling temperature, °C	K_j, $\times10^5$ A^2s/mm^4	d, µm	D_R, µm	F_R, %	Θ, degree	F_{HAB}, %	$F_{\Sigma3+\Sigma9}$, %
20	-	0.08	2	1	6	9	0
	0.152	0.82	3	26	9	14	4
	0.170	2.10	8	79	39	77	27
-196	-	0.07	2	1	5	4	0
	0.106	0.84	2	1	5	4	0
	0.112	1.95	3	90	44	84	39

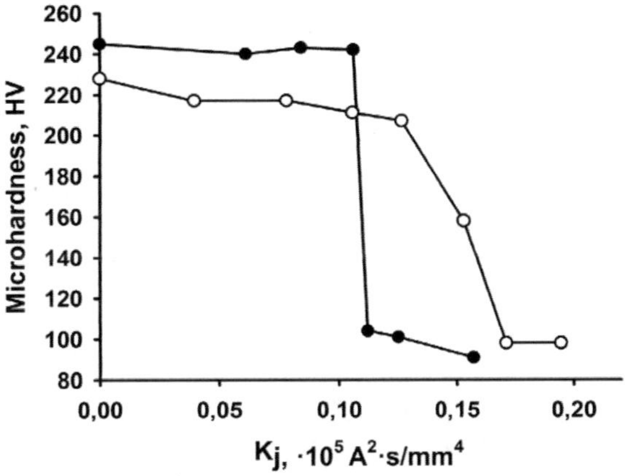

Figure 2. Evolution of microhardness vs the EPP energy for Ni rolled at 20 (○) and -196 °C (•).

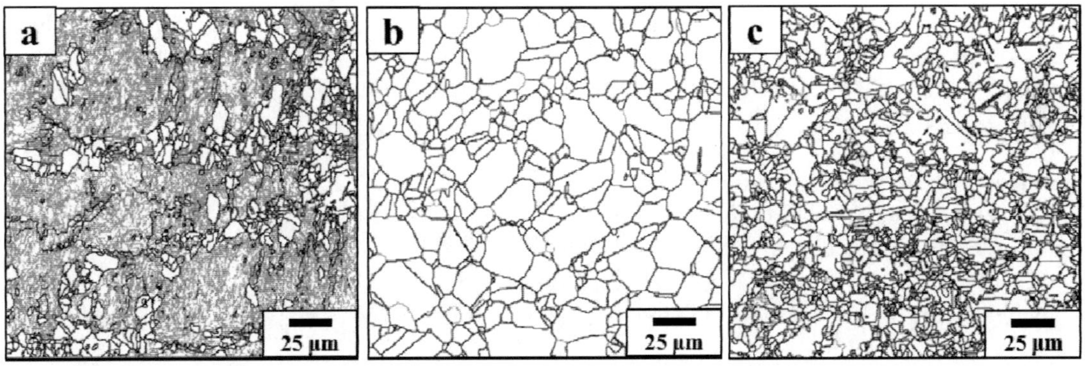

Figure 3. SEM-EBSD grain boundary maps for Ni rolled at 20 (a,b) and -196 °C (c) and further subjected to EPP with K_j= 0.152 (a), 0.170 (b) и 0.112×10^5 A^2s/mm^4 (c). (For better observation LABs of less 2° misorientations are not plotted)

Therewith, these new grains were mainly settled down in the form of chains (bands) with the interspacing close to the initial grain size, demonstrating that sites of their primary nucleation were located in the regions adjacent to the initial boundaries. With increasing the energy of EPP to 0.170×10^5 A^2s/mm^4, recrystallization occupied practically the whole specimen volume and the HABs fraction, the fraction of recrystallized grains and the mean misorientation angle of intercrystallite boundaries reached their maximum values (table 1). However, the intense growth of new grains leading to their noticeable coarsening (figure 3b) was simultaneously observed. It should be noted that the presence of annealed twins, the fraction of which exceeded a quarter (table 1), was a character for the structure processed, confirming the active role of twinning in the transformation of the deformed structure of nickel at EPP.

Unlike cold-rolled nickel, EPP of cryorolled one, performed with small energies (up to 0.106×10^5 A^2s/mm^4 only), did not cause any significant changes in its hardness (figure 2). It revealed a higher stability of the deformation structure, processed by rolling at a lower temperature. Virtually the structure underwent recovery in the form of insignificant transformation (perfection) of dislocation configurations only and even remained highly work-hardened. However, when the current density exceeds $Kj = 0.112\times10^5$ A^2s/mm^4, the stored energy in the cryodeformed nickel rapidly relaxed in a narrow interval of energies (temperatures) of EPP, which was expressed in an intense loss of its strength owing to the formation of an almost completely recrystallized fine-grained structure with a mean grain size about 3 μm (figure 3c). A feature of this recrystallized structure, as well as in the EPP-ed room-temperature rolled Ni, was a quite high fraction of LABs (table 1). Besides, the fraction of twin boundaries increased to a value of about 40%, testifying the even more significant role of twinning and grain growth at EPP of the cryorolled metal. With further increasing the EPP energy, the microhardness of cryorolled nickel decreased to the level of the starting material due to the loss of the Hall-Petch hardening effect because of the coarsening of recrystallized grains.

4. Conclusions

It is established that a decrease in the temperature of rolling of nickel from room temperature to cryogenic leads to the formation of a more uniform and less misoriented nanocellular structure and to an additional increase in its hardness by 15-20 HV. Subsequent impact on the work-hardened metal by powerful impulses of current at EPP resulted in a decrease in its hardness owing to recovery, discontinuous recrystallization with the formation of annealed twins and the growth of new grains. It is revealed that EPP of nickel deformed at a temperature of liquid nitrogen leads to the formation of a completely recrystallized structure at lower energies than that of nickel deformed at room temperature. At the same time, this process proceeds in a narrower interval of current densities of EPP and forms a more uniform fine-grained structure with a smaller grain size.

Acknowledgments

The work was supported by Russian Foundation for Basic Research under grant No. 19-08-00953A. The work was performed using the facilities of the shared services center «Structural and Physical-Mechanical Studies of Materials» at the Institute for Metals Superplasticity Problems of Russian Academy of Sciences.

References

[1] Xiao G H, Tao N R and Lu K 2009 *Mater. Sci. Eng.* A **513-514** 13
[2] Li Y S, Tao N R and Lu K 2008 *Acta Mater.* **56** 230
[3] Zhang Y, Tao N R and Lu K 2008 *Acta Mater.* **56** 2429
[4] Gindin I A, Lazareva M B, Lebedev V P and Starodubov Y D 1967 *Fiz. Met. Metalloved.* **23** 138 (in Russian)
[5] Gindin I A, Lazareva M B, Lebedev V P, Starodubov Y D, Matsevity V M and Hotkevich V I 1967 *Fiz. Met. Metalloved.* **24** 347 (in Russian)
[6] Konkova T, Mironov S, Korznikov A and Semiatin S L 2010 *Acta Mater.* **58** 5262
[7] Huang Y and Prangnell P B 2008 *Acta Mater.* **56** 1619
[8] Voronova L, Degtyarev M, Chashchukhina T, Gapontseva T and Pilyugin V 2018 *Lett. Mater.* **8** 424
[9] Konkova T, Valeev I, Mironov S, Korznikov A, Korznikova G, Myshlyaev M M and Semiatin S L 2016 *J. Alloys Compd.* **659** 184
[10] Konkova T, Valeev I, Mironov S, Myshlyaev M M and Semiatin S L 2014 *J. Mater. Res.* **29** 2727
[11] Valeev I Sh, Valeeva A Kh, Ilyasov R R, Sitdikov O Sh and Markushev M V 2019 *Lett. Mater.* **9** 447
[12] Chang H and Baker I 2007 *Metall. Mater. Trans.* A **38** 2815

IOP Conf. Series: Materials Science and Engineering 1008 (2020) 012007 doi:10.1088/1757-899X/1008/1/012007

Microstructure and properties of solid state joints of titanium sheets produced by ultrasonic welding

A A Mukhametgalina, M A Murzinova and A A Nazarov

Institute for Metals Superplasticity Problems, Russian Academy of Sciences, 39 Khalturin st., 450001 Ufa, Russia

E-mail: a.mukhametgalina@mail.ru

Abstract. The results of the study on the microstructure and properties of solid state joints of titanium sheets obtained by ultrasonic welding (USW) with frequency 20 kHz and vibration amplitude 20 μm during 2 and 3 s under a 6 kN clamping force are presented. The thermo-mechanically affected zone is observed along the defect-free weld line (joint). It was found that the length and width of this zone increased with the welding time. In the thermo-mechanically affected zone, a significant grain growth takes place, a change in the texture is observed and maxima near 60° and 90° in the grain boundary misorientation distributions appear. The maximum lap shear failure load of the joints obtained by USW of 2 and 3 s durations are 1998±122 and 2506±226 N, respectively. There is a change in the failure mode from interface debonding after 2 s USW to nugget pull-out for the welding time of 3 s.

1. Introduction

Ultrasonic welding (USW) is widely used in the electrical, automotive, aerospace, instrument-making and medical industries to obtain joints between thin plates, tapes, foils, and wires. The USW has a number of advantages over other types of welding, the main of which is a low heat generation. The temperature in the welding zone depends on the material used and welding parameters, but, as a rule, it is below the melting temperature of the metals to be joined [1]. In addition, this method of joining metals requires two orders of magnitude less energy in comparison to the resistance spot welding [2]. Due to high-frequency vibrations in the process of USW, the surfaces to be welded are cleaned of dirt and oxide films, therefore, in most cases, preliminary cleaning of the surfaces is not required. Another advantage of the USW is its high productivity: the formation of a joint normally occurs in less than a couple of seconds [1]. As shown in [3], almost any metal and most of their combinations can be welded with ultrasound. Nevertheless, the greatest experience has been accumulated in welding soft metals and alloys with high thermal conductivity based on aluminum or copper [1-3]. The structure and properties of compounds of stronger metals, in particular, titanium and its alloys, obtained by USW, are practically not studied today. To the authors' knowledge, there is only one work in the literature devoted to the USW of titanium [4].

This paper presents the results of studies on the structure and properties of titanium solid state joints of titanium produced by USW.

2. Material and methods

Commercially pure titanium sheet with the thickness of 0.5 mm in the annealed condition was used in this work. Coupons for welding, 20 mm wide and 40 mm long, were cut along the rolling direction (RD) of the sheet, ground and washed with alcohol. USW was carried out with a frequency of 20 kHz and an amplitude of 20 μm. The static load of 6 kN was applied and the duration of ultrasound exposure was equal to 2 and 3 s. The sonotrode tip with dimensions of 5 mm × 6 mm had a serrated surface comprised of seven parallel teeth and its vibrations occurred in the plane of the sheets perpendicular to the rolling direction.

The welded joints were cross sectioned across their center, parallel to the direction of the vibrations. For structural studies, the samples were mechanically ground and polished, and the structure was characterized using optical and scanning electron microscopy on a TESCAN MIRA 3 LMH FEG microscope using orientation analysis (EBSD).

The welded samples were subjected to lap shear tests at room temperature with a crosshead speed of 1 mm / min on an Instron 5982 machine. The results were averaged over three tests with the peak load measured.

3. Results and discussion

3.1 Structure observations

Images of the macrostructure of the resulting joints are shown in figure 1. The microstructure at the periphery and central zones of the welds are shown in figure 2.

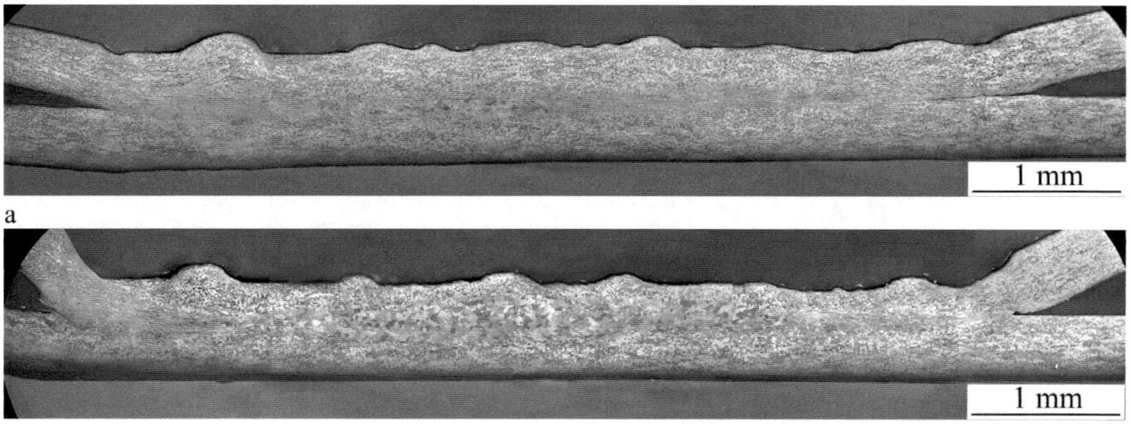

Figure 1. Optical images of the macrostructure of the cross-sections of USW joints produced with welding times 2 s (a) and 3 s (b) (polarized light).

As one can see from the figures, joints of a satisfactory quality were formed under the selected processing conditions. The relative lengths of the defect-free weld joints are respectively 46 and 86% after ultrasonic welding in time intervals 2 and 3 s. Narrow unbonded regions were observed only along the perimeters of the joints. In addition, along the perimeters of the weld nugget, there was a significant reduction in the thickness of the upper sheet. The highest compression deformation was 20% and 35% after USW for 2 and 3 s, respectively.

In these narrow peripheral regions of the welded joints, deformed grains elongated in the direction of vibrations of the weld tip were observed. In areas adjacent to the anvil and the weld tip, the microstructure of the sheets is very close to that of the original sheet. In the central defect-free zone of joints, the microstructure has changed dramatically.

Figures 3 and 4 show the results of EBSD analysis of samples obtained from cross-sections of the initial sheet and from the centers of the weld spots of the welded samples. The microstructure of the alloy in the initial state consisted of equiaxed grains about 7 μm in size (figure 3a). The misorientation spectrum of high-angle boundaries did not have any pronounced maxima (figure 3b). The texture is typical for titanium subjected to rolling and subsequent annealing [5]: the basal planes (0001) are rotated by 20-40° relative to the plane of the sheet around the RD direction, and therefore on the IPF from the sheet plane, an increase in the intensity of reflections in the directions $< 01\bar{1}3 >$ and $< \bar{1}2\bar{1}4 >$ occurred (figure 4a).

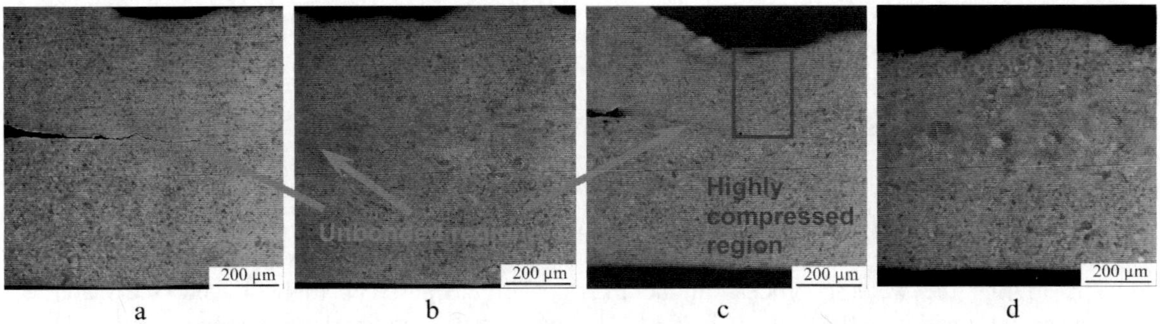

Figure 2. Microstructure of the USW joints produced with welding times 2 s (a,b) and 3 s (c,d); periphery (edge) (a,c) and centre (b,d) of the joints (BSE images).

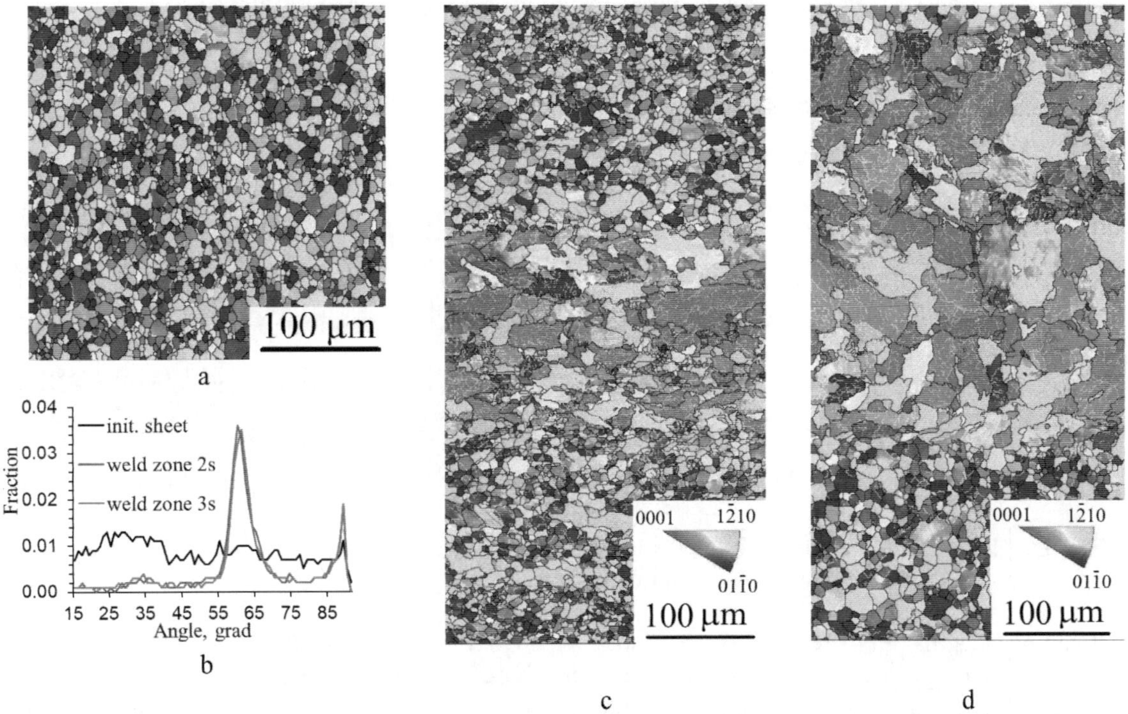

Figure 3. IPF orientation maps of the initial sheet (a) and joints obtained by USW in time intervals 2 s (c) and 3 s (d) and misorientation distribution of high-angle grain boundaries (b).

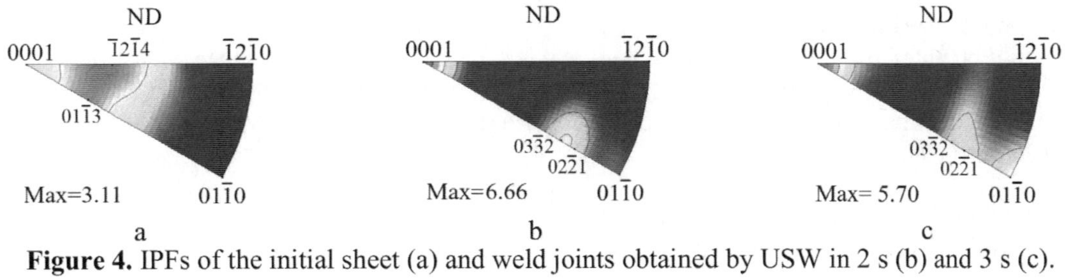

Figure 4. IPFs of the initial sheet (a) and weld joints obtained by USW in 2 s (b) and 3 s (c).

During USW, due to the combined action of high-frequency shear vibrations and static compression force, local heating occurred in the joint zone, and zones of thermo-mechanical influence were formed. In these zones, mainly large grains with a size of 40-80 μm containing a developed substructure were observed (figure 3 b, c). In the misorientation spectra of high-angle boundaries, maxima appeared in the vicinity of the angles 60° and 90° (figure 3d). The texture of the material changed (figure 4 b, c): there is a strong maximum in the <0001> direction on the IPF and less intense maximums were observed in the directions $<03\bar{3}2>$, $<02\bar{2}1>$, and $<10\bar{1}0>$. The texture in the thermo-mechanically affected zone changed little with an increase in the USW time from 2 to 3 s (some enhancement of the component $<10\bar{1}0>$ can be noted). An increase in the duration of USW led to a two-fold increase in the width of the zone of thermo-mechanical influence, which is maximal in the center of the welded joint and decreases towards its periphery.

The changes in the microstructure described above are observed in titanium after β-α transition [6]. Probably, during the welding process at the interface of the surfaces to be joined, the material experiences deformation in the β-phase field and the final structure in the welded zone is produced via β-to-α phase transformation during the cooling cycle. Note that the microstructural changes occurring during USW of titanium differ from the data obtained during similar processing of Al and Cu, where a thin layer with a fine-grained structure was observed in the welded joint zone [2, 7-9].

3.2 Mechanical properties

Typical force vs. crosshead displacement curves obtained during lap shear tests of welded specimens and the appearance of ruptured specimens are shown in figure 5.

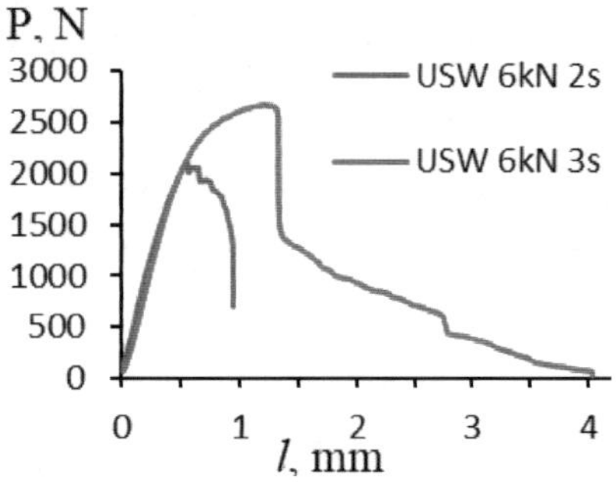

a

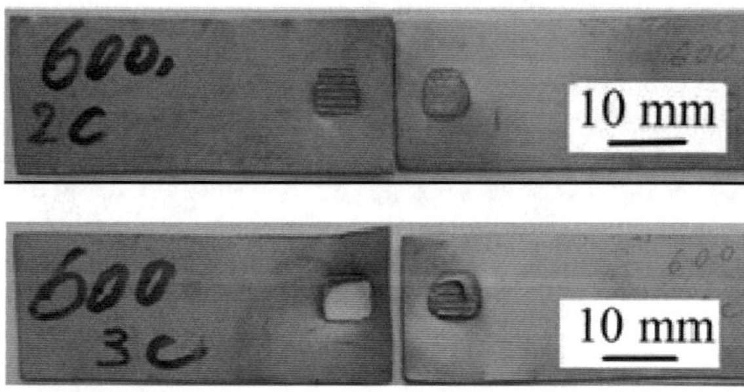

b

Figure 5. Typical lap shear test curves for USW-processed samples (a) and typical appearances of ruptured samples b).

It can be seen that an increase in the duration of the USW from 2 to 3 s (all other conditions being equal) led to an approximately 20% increase in the breaking force (figure 5a), an almost 3 times increase in the fracture work and a change in the fracture mode of the welded joint: fracture occurred at the interface after processing within 2 s, and for longer welding time of 3 s full nugget pull-out took place (figure 5 b). We assume that the main reason for the change in the fracture mode is a significant increase in the area of a defect-free joint and a decrease in the thickness of the top sheet in a narrow region along the perimeter of the welded point.

It should be noted that similar values of the breaking force (maximum lap shear force 1940 N) were obtained during lap shear tests of the joints of sheets of Ti-6Al-4V alloy obtained by roller USW [10]. However, this comparison is not entirely correct, since the true area of the welded joint has not been reliably determined.

4. Conclusions

A comparison of the microstructure and properties of the joints of titanium sheets with a thickness of

0.5 mm obtained by ultrasonic welding with a frequency of 20 kHz and an amplitude of 20 μm under the action of a constant static load of 6 kN for 2 and 3 s has been carried out. It is shown that a defect-free joint is formed in the thermo-mechanically affected zone, the length and width of which increases with the duration of welding. In the thermo-mechanically affected zone, a significant increase in the grain size, a change in texture is observed; in the misorientation spectra of high-angle grain boundaries, maxima appear in the vicinity of angles 60° and 90°. The fracture of the joints obtained with 2 s USW occurs in the interface mode under the action of a force of 1998 ± 122 N. After ultrasonic welding for 3 s, fracture occurs in the weld nugget pull-out mode and the fracture force increases to 2506 ± 226 N.

Acknowledgments
The present work was supported by the Russian Science Foundation (grant No. 16-19-10126). Electron microscopic studies and mechanical tests were carried out on the facilities of shared services center of IMSP RAS "Structural and Physical-Mechanical Studies of Materials".

References
[1] Matheny M P and Graff K F 2015 Ultrasonic welding of metals *Power Ultrasonics-Applications of High-intensity Ultrasound* ed J A Gallego-Juárez and K F Graff (Cambridge: Woodhead Publishing, Elsevier Ltd.) chapter 11 pp 259–93
[2] Bakavos D and Prangnell P B 2010 *Mater. Sci. Eng.* A **527** 6320–34
[3] Graff K F, Devine J F, Keltos J and Zhou N Y 2007 Ultrasonic welding of metals *Welding Handbook, 9th ed.* vol 3, Welding Processes, Part 2, ed A O'Brien (Miami: American Welding Society) pp 263–303
[4] Owa T, Kondo T and Takizawa H. 2010 *Welding Int.* **24** 182-7
[5] Conrad H 1981 Effect of interstitial solutes on the strength and ductility of titanium *Progr. Mater. Sci.* vol 26 iss. 2–4 (Oxford: Pergamon Press Ltd.) 123–403
[6] Gey N and Humbert M 2002 *Acta Mater.* **50** 277–87
[7] Yang J, Cao B and Lu Q 2017 *Materials (Basel)* **10(2)** 193
[8] Haddadi F and Tsivoulas D 2016 *Mater. Charact.* **118** 340–51
[9] Peng H, Chen D and Jiang X. 2017 *Materials (Basel)* **10(5)** 449
[10] Zhu Z, Li M, Su Z, Zhang D and Zhang Y 2018 Study on Microstructure and Mechanical Properties of Ti6Al4V Titanium Alloy Joint with Ultrasonic Consolidation *Transactions on Intelligent Welding Manufacturing* ed S Chen et al (Singapore: Springer) p 120-9

IOP Conf. Series: Materials Science and Engineering 1008 (2020) 012008 doi:10.1088/1757-899X/1008/1/012008

Microstructure and mechanical properties of a nickel-base superalloy heavily alloyed with substitution elements

R I Zainullin, A A Ganeev, R V Shakhov, S K Mukhtarov, V M Imayev and R M Imayev

Institute for Metals Superplasticity Problems of Russian Academy of Sciences, 39 Stepan Khalturin st., Ufa, 450001, Russia

E-mail: shamilm@imsp.ru

Abstract. The present work is devoted to a study of the microstructure and mechanical properties of a new heat resistant nickel-base superalloy Ni-14(Al, Ta)-30.7(Co, Cr, W, Hf) (wt.%) intended for using as a die material. The superalloy contains about 70 vol.% of the $\gamma'(Ni_3Al)$ phase and a small amount (about 3 vol.%) of carbides and topologically close-packed phases. Compression tests were carried for the new superalloy at 1000-1200°C. The superalloy showed the yield stress values at 1100-1200°C comparable with those obtained for the Russian VKNA-superalloys based on the $\gamma'(Ni_3Al)$ phase. Tensile tests were performed for the cast and heat treated and wrought conditions of the superalloy. The tensile properties show that the new superalloy can be considered as an advanced disc material for gas turbine engines as well.

1. Introduction

Critical parts made of nickel-base superalloys for gas turbine engines and produced by casting or powder metallurgy are generally subjected to multi-stage heat treatment and subsequent hot working using hot extrusion, forging or roll-forming [1-4]. The aim of the hot working is to: i) eliminate dendritic segregation, ii) refine the microstructure via recrystallization processes and iii) obtain a shape close to the final part. The solution of these issues is particularly relevant in manufacturing of complex geometry discs made of the most heat resistant nickel-base superalloys (such as Russian superalloys VZh175, SDZhS-15, VV751P et al.) intended for application in gas turbine engines. It is known that the hot working temperatures for such heat resistant superalloys are typically 1100-1200°C. To forge or roll the superalloys at 1100-1200°C, Russian superalloys based on the $\gamma'(Ni_3Al)$ intermetallic phase having a stable structure and not requiring a protective atmosphere during hot working, which are called VKNA alloys, can be used as the die and roller (for the roll-forming mill) materials. However, these superalloys are known to have rather low yield stresses at a temperature of 1000°C and below [5,6].

The present work was aimed to study the microstructure and mechanical properties of an experimental nickel-base superalloy intended first of all for using as a die tool material at temperatures of 1000-1200°C. For the new superalloy, the compression and tensile tests were carried out after different treatments.

2. Material and experimental

The superalloy with the nominal composition Ni-14(Al, Ta)-30.7(Co, Cr, W, Hf) (wt. %) manufactured by ingot metallurgy was taken as a starting material. In addition to these elements, the alloy contained Ce, Zr and B in an amount of 0.06 wt. %. The superalloy composition was found to be very close to its nominal composition. The γ′-phase content in the superalloy was about 70 vol.%. The γ′ solvus temperature (T_s) defined via quenching experiments was $T_s \approx 1260$°C.

The as-cast superalloy was heat treated prior to compression and tensile tests. The heat treatment included homogenization annealing, solid solution heat treatment at 1250°C followed by air cooling and two-stage ageing at 1000 and 880°C. This condition was designated in the text as condition 1. To prepare condition 2, the as-cast superalloy was subjected to homogenization annealing, solid solution heat treatment at 1250°C followed by slow furnace cooling to cause the γ′ phase coarsening and to improve the hot workability of the superalloy. The heat treated material was further subjected to thermomechanical treatment, which included step-by-step forging in a thick-walled can under quasi-isothermal conditions with intermediate recrystallization annealing to a total strain of $e \approx 1.4$. The workpiece preheating temperature was 1210°C, the die temperature was 950°C. The forged workpiece was cooled in air, decanned and subjected to two-stage ageing at 1000 and 880°C. This microstructural condition was designated as condition 2.

Condition 1 was used for compression tests, which were performed at T=1000-1200°C with an initial strain rate of $\dot{\varepsilon} \approx 10^{-3}$ s^{-1} to the engineering strain of ε=10%. To do it, the specimens with a size of Ø8 mm × 12 mm were prepared. The true stress-true strain curves were drawn taking into account a graduate increase of the specimen cross section during compression test. The curves were used to determine the yield stress value ($\sigma_{0.2}$).

Conditions 1 and 2 were taken for tensile tests. Flat samples with a gauge section of 10×3×2 mm^3 were used for tensile testing. The sample surfaces were mechanically polished before testing. Tensile tests were performed at temperatures T=20 and 650-850°C with an initial strain rate of $\dot{\varepsilon} \approx 8.3 \times 10^{-4}$ s^{-1}. The total elongation, δ, the tensile yield stress, $\sigma_{0.2}$, and the ultimate tensile strength, σ_{UTS}, were defined. Microstructure examination was carried out on a scanning electron microscope (SEM) in the back-scattering electron (BSE) mode. The specimen surfaces before studying by SEM were mechanically ground and electropolished.

3. Results and discussion

3.1 Effect of processing on the microstructure

Figure 1 shows BSE images of the superalloy in conditions 1 and 2. In condition 1, the size of γ grains was d_γ=100-500 μm, the size of the primary γ′ phase was $d_{\gamma'}$=1-5 μm, the average size of the secondary γ′ phase precipitated during air cooling and ageing was $d_{\gamma'} \approx 0.2$ μm (figure 1a, b). The volume fraction of the primary γ′ phase was not more than 5%. In the microstructure there were also a small amount of carbides and topologically close-packed (TCP) phases, the volume fraction of which did not exceed 3%.

Thermomechanical treatment led to the occurrence of recrystallization processes and refinement of the initial coarse as-cast structure (figure 1c, d). In condition 2, the size of recrystallized γ grains was determined to be in the range d_γ=2-50 μm. In the microstructure, there were also individual non-recrystallized γ grains with a size of up to 100 μm. The primary γ′ phase not dissolved during thermomechanical treatment occupied about 30 vol.%. The size of the primary γ′ phase was found to amount $d_{\gamma'}$=1-5 μm. The dispersed secondary γ′ phase precipitated during cooling after thermomechanical treatment and subsequent aging. The average size of the secondary γ′ precipitates was $d_{\gamma'} \approx 0.2$ μm. Carbides and TCP phases were also observed in the microstructure. After thermomechanical treatment and ageing their volume fraction slightly increased and amounted to about 3.4%.

3.2 Mechanical behaviour

Figure 2 represents the true stress-true strain curves obtained as a result of compression tests of the superalloy specimens at T=1000-1200°C. One can see that the flow stress decreased with increasing the test temperature. It is worth noting that the flow stress was only slightly changed depending on the strain at all of the test temperatures. This indicates a high stability of microstructure of the superalloy. The compressive yield stress of the superalloy at 1000°C was found to be appreciably higher as compared with that of the VKNA superalloys and the conventional nickel-base superalloy ZhS6U applied as die materials (table 1). The compressive yield stresses of the superalloy at 1100-1200°C were found to be comparable with those of the VKNA and ZhS6U superalloys (table 1) [6-8].

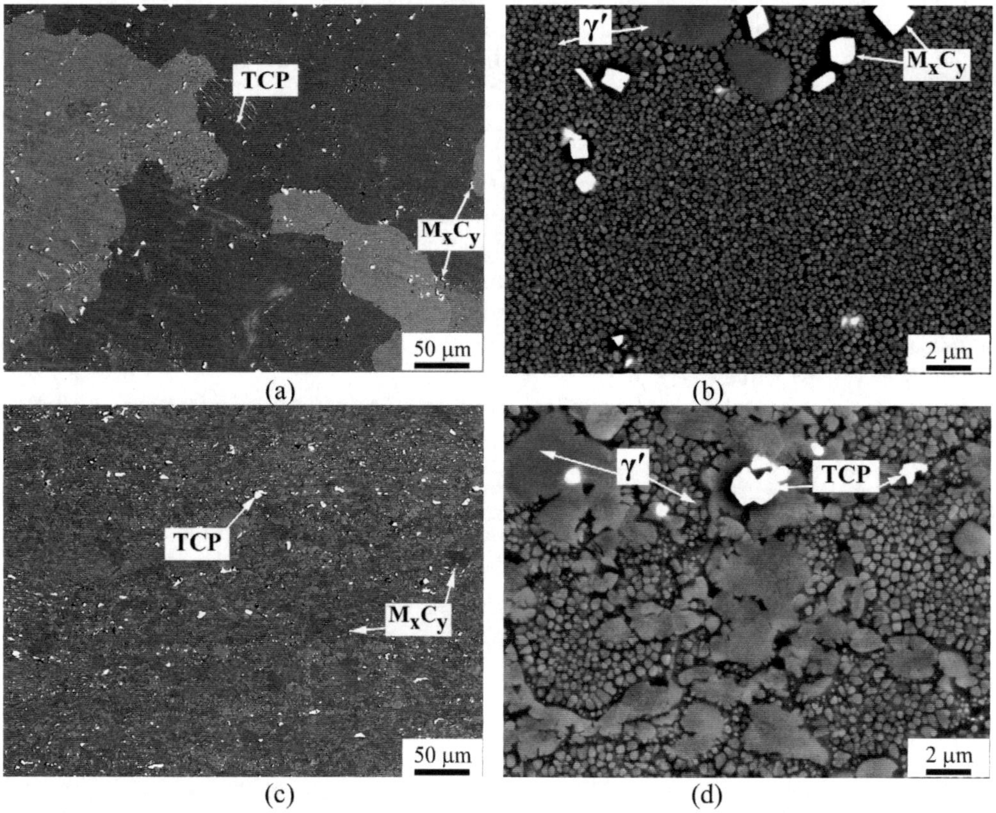

Figure 1. BSE images of the Ni-14(Al, Ta)-30.7(Co, Cr, W, Hf) superalloy: (a, b) condition 1, (c, d) condition 2. The primary and secondary γ' phase, carbides (M_xC_y) and TCP-phases are arrowed.

Table 1. The temperature dependence of the yield stress of the Ni-14(Al, Ta)-30.7(Co, Cr, W, Hf) superalloy in condition 1 as compared with those of other superalloys used as die tool materials

Alloy	Test temperature (°C)			
	1000	1100	1150	1200
Ni-14(Al, Ta)-30.7(Co, Cr, W, Hf)	705	360	234	140
VKNA-4 [6]	290	215	-	115
VKNA-1V [6]	430	350	-	185
VKNA-1VU [6]	575	430	-	145
VKNA-4U [6]	430	370	-	150
VKNA-25 [6]	510	370	-	170
ZhS6U [7]	510	298	-	-

Table 2 represents the results of tensile tests of the superalloy in conditions 1 and 2. The microstructural refinement resulted from thermomechanical treatment led to a significant increase in the strength and ductility of the superalloy at T=20-750°C. At T=850°C, condition 2 showed lower strength properties than condition 1 that can be ascribed to γ grain boundary sliding in condition 2 at this temperature. It should be noted that although the superalloy was designed as a die material and contains unfavorable TCP phases, the tensile properties of the superalloy were comparable with those of some disc nickel-base superalloys. Further research activities can be devoted to study of the effect of solid solution heat treatment on the microstructure and mechanical properties. As expected, the dissolution of the primary γ′ phase and the increase in the volume fraction of the secondary γ′ phase via solid solution treatment will provide further enhancement in strength and creep resistance of the new superalloy.

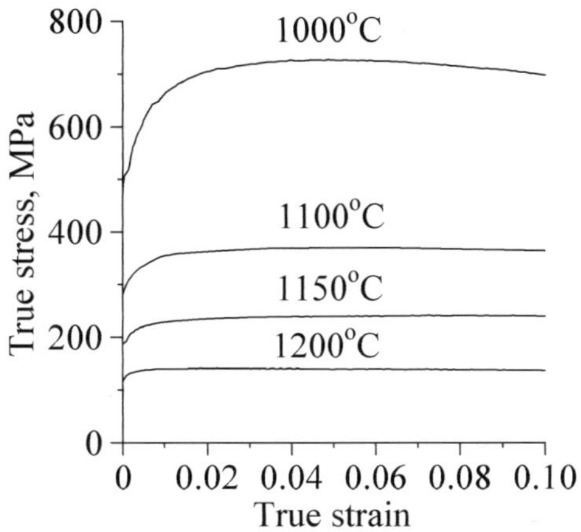

Figure 2. True stress-true strain curves obtained as a result of compression tests of the superalloy specimens (condition 1) at T=1000-1200°C.

Table 2. Tensile properties of the Ni-14(Al, Ta)-30.7(Co, Cr, W, Hf) superalloy in condition 1 / condition 2 (YS – tensile yield stress, UTS - ultimate tensile strength)

T (°C)	YS (MPa)	UTS (MPa)	δ (%)
20	945 / 1003	1048 / 1534	4.1 / 14
650	909 / 1013	1044 / 1240	2.7 / 6.6
750	899 / 1004	1015 / 1102	1.9 / 3.0
850	850 / 700	874 / 787	1.9 / 2.0

4. Conclusion

A new nickel-base superalloy Ni-14(Al, Ta)-30.7(Co, Cr, W, Hf) containing a high volume fraction (about 70%) of the γ′ phase and having a high γ′ solvus temperature (T_s≈1260°C) has been designed. The new superalloy showed appreciably higher yield stress values at temperatures of T=1000-1200°C in comparison with those of conventional nickel-base superalloys and can be used for manufacturing of die tools and rollers for a roll-forming mill for isothermal hot working of the most heat-resistant nickel-base superalloys. Despite the presence of a small amount of TCP phases, the new superalloy also demonstrated quite high tensile properties in the temperature range of 20-850°C.

Acknowledgments

The present work was supported by the Russian Science Foundation (Grant No. 18-19-00594). The work was performed using the facilities of the shared services center "Structural and Physical-Mechanical Studies of Materials" at the Institute for Metals Superplasticity Problems of the Russian Academy of Sciences.

References

[1] Reed R C 2006 *The Superalloys: Fundamentals and Applications* (Cambridge University Press)

[2] Geddes B, Leon H and Huang X 2010 *Superalloys: Alloying and Performance* (ASM International Materials Park)
[3] Sahithya K, Balasundar I, Pant P and Raghu T 2020 *J. Alloys Compd.* **821** 153455
[4] Kaibyshev O A and Utyashev F Z 2005 *Superplasticity: Microstructural Refinement And Superplastic Roll Forming (ISTC Science and Technology)* (Futurepast Inc)
[5] Povarova K B, Bazyleva O A, Drozdov A A, Alad'ev N A and Samsonova M A 2012 *Russian Metallurgy (Metally)* 975
[6] Kablov E N, Ospennikova O G and Bazyleva O A 2011 *Vestnik MGTU im. N.E. Baumana. Ser. Mashinostroenie* **2** 13 (in Russian)
[7] Ospennikova O G 2018 Razrabotka nauchnykh osnov sozdaniya novogo pokoleniya liteinykh jaroprochnykh nanostrukturirovannykh nikelevykh splavov ponijennoi plotnosti s trebuemym kompleksom mekhanicheskikh svoistv: *Dissertacija na soiskanie stepeni doktora tehnicheskih nauk* (Moskva) (in Russian)
[8] Zainullin R I, Ganeev A A, Shakhov R V, Logunov A V, Mukhtarov Sh Kh and Imayev V M 2019 *Letters on Materials* **9** (4) 490

Ultrasonic welding of ultrafine grained nickel

E R Shayakhmetova, A A Nazarov, A A Samigullina and A P Zhilyaev

Institute for Metals Superplasticity Problems, Russian Academy of Sciences, 39 Khalturin st., 450001 Ufa, Russia

E-mail: elvina1408@yandex.ru

Abstract. Solid state joints were obtained for the first time by ultrasonic welding (USW) of ultrafine-grained (UFG) thin disks of Ni processed by high pressure torsion (HPT). Samples of two types were studied: the ones for lap shear mechanical tests were prepared by joining two disks superimposed on each other under static loads of 4.5 and 6.0 kN, and the other type samples for structural characterization were obtained by sequential welding (consolidation) of four disks. The welding time for both type samples was equal to 1 s. The structure of consolidated samples was studied in their cross section by SEM, EBSD analysis, and also their Vickers microhardness was measured.

1. Introduction

Ultrasonic welding (USW) is a relatively new method of joining metallic materials and now is widely used to weld foils, sheets and wires [1-5]. USW is based on the use of the energy of mechanical vibrations generated in materials to be welded by a welding tool of an oscillatory system that performs reciprocating movements with an ultrasonic frequency (19-25 Hz). These vibrations cause the friction of surfaces to be welded, destruction of oxide layers, the plastic deformation and heating of materials in the welding area followed by their seizure. The welding temperature is lower than the melting temperature of the metals, therefore, USW is a method of solid-state joining.

Many studies in the literature have been devoted to elucidate the influence of parameters of the USW process, such as clamping force, vibration amplitude, welding time and welding power, on the quality of joints and lap shear and tensile strengths of welded samples [1-5].

Recently, it has been demonstrated that USW is a promising technology underlying one of the methods of additive manufacturing, ultrasonic consolidation, thus its development is of a considerable interest [6, 7]. The goal of ultrasonic additive technology, like any other metal forming technology, is to obtain parts and structures with high strength characteristics, which depend on both the process parameters and the characteristics of the starting material. From this point of view, it is of great interest to use ultrafine grained (UFG) sheets as a starting material for consolidation, since materials with the UFG structure have increased strength, hardness, and fatigue properties and can have better weldability than coarse-grained (CG) materials.

To date, the literature contains mainly data on the USW of aluminum alloys, copper, copper with aluminum, titanium with aluminum etc. [4,5,8-10]. To the authors' knowledge, there have been practically no works on welding of nickel except for one paper [11]. There is also no data on the welding of metals in the UFG state.

Thus, the main goal of this study is to elucidate the role of the UFG state of nickel in the evolution of its microstructure during USW and to determine the strength of the joints of UFG sheets of nickel obtained by USW.

2. Materials and experimental procedure

Commercially pure nickel with the following composition: Ni ≥ 99.3, Co ≤ 0.2, Si ≤ 0.15, Fe, Mg, Cu, C ≤0.10, Mn ≤ 0.05, Zn ≤ 0.007, S ≤ 0.005, Pb, Ph, Bi, As, Sb, Cd, Sn ≤ 0.002 wt% (grade NP2 according to the Russian classification), was used as a material for the study.

Samples in the form of disks with a thickness of 1.2 mm and a diameter of 12 mm were subjected to high pressure torsion (HPT) under the pressure of 6 GPa on Bridgman anvils at room temperature up to the number of evolutions equal to five. As a result, discs with a thickness of 0.75 mm were obtained. The disks processed by HPT were ultrasonically welded in two ways as follows. Specimens for lap shear tests were obtained by spot welding of two disks superimposed on each other. Ultrasonic consolidation was carried out by sequential welding of four specimens, the lower specimen being welded to a substrate of the same material with the thickness of about 1 mm. The welding sonotrode tip had the shape of a rectangle with sizes 4.5 mm × 6 mm and a serrated surface comprised of seven parallel teeth perpendicular to the axis of the oscillatory system. The amplitude of vibrations of the sonotrode tip was equal to 20 μm, and the static load during USW 4.5 and 6 kN.

To determine the strength of welds made by ultrasonic welding, lap shear tests were carried out. The weld spots on the samples for these tests were located near the middle of their radius, since the microstructure formed during HPT is normally nonhomogeneous and the UFG structure is less developed at the centers of disks [12]. Locating the weld spot near the middle of their radius ensures a reasonable uniformity of the UFG structure. The lap shear tests were carried out at room temperature, with a deformation rate of 0.5 mm/min on an Instron 8862 universal testing machine. The ends of the welded sheets were fixed in grips of the machine maintaining the vertical axis of the entire system. At least three specimens were tested and the average value of lap shear strength was calculated for each type of welding conditions.

The structure of weld interfaces and bulk regions of the layers was characterized by scanning electron microscopy (SEM) on a TESCAN VEGA 3 microscope. The EBSD analysis was performed by a TESCAN MIRA 3 LMH FEG scanning electron microscope equipped with a CHANNEL 5 EBSD analyzer. The scanning was carried out with steps of 100 nm from square grids with side sizes of 2 mm. The specimens for EBSD analysis were prepared by mechanical polishing with a subsequent electromechanical polishing. The obtained EBSD maps were subjected to a clean-up procedure involving a grain tolerance angle of 5° and a minimum grain size of 3 pixels.

The Vickers microhardness of the samples was measured by an AFFRI DM8A testing instrument with a load on the indenter of 10 g and an exposure time of 10 s.

3. Experimental results

The initial structure of UFG nickel processed by HPT in a cross section of the disk sample near the middle of its radius studied by EBSD is shown in figure 1. From the inverse pole figure (IPF) map one can see that in the cross section the grains have an elongated shape, which corresponds to the scheme of deformation of the samples during HPT. The average grain size is about 200 nm along the short axis and 1.5 μm along the long one. An analysis of the misorientation distribution has shown that most of the grain boundaries have high-angle misorientations.

The values of lap shear strength obtained from mechanical tests are presented in table 1. From these data, one can see that for both conditions of welding, the joints of UFG nickel have similar values of strength. A fairly large scatter of the strength values was observed for the load of 6 kN and the failure modes of the samples of this type were different: sample No. 3 failed in the weld nugget pullout mode, while the others in the interface mode like all samples welded under the load of 4.5 kN.

The structural studies have shown that fairly good solid state joints are formed by the USW in both regimes. Only some weld interfaces can be seen at low magnification, most joints are clearly seen only at high magnifications. Typical microstructures of the weld joints formed by USW in the two regimes are presented in figure 2. Figure 2a shows the structure of a joint in the sample processed by USW at a static load of 4.5 kN and the region where EBSD mapping has been done, and figure 2b presents the corresponding map. From this map, one can see that the microstructures of the upper and lower layers are different. In the lower layer, due to the action of deformation at increased temperature during ultrasonic welding, a significant grain growth occurred, while there is a thin layer of UFG structure directly in the joint region. The grain size in the upper layer also increases in comparison to the initial UFG state. To clarify the dependence of the grain size on the layer number, additional studies are required using the EBSD analysis method.

Figure 1. The initial structure of UFG nickel processed by HPT in the cross section of disk samples studied by EBSD analysis.

Table 1. Values of the lap shear strength of samples made by USW of UFG nickel under static loads of 4.5 and 6 kN

Sample number	P=4.5 kN	P=6 kN
1	98.1	81.2
2	91.6	98.3
3	89.5	111.1
Mean	93.1	96.9

The microstructure of a joint of the sample consolidated under the static load of 6 kN is shown in figure 2d, and the location of the area where EBSD mapping has been carried out is shown in figure 2c. It can be seen that in this case the grain size in the layers also increased up to about 4-5 µm, and the grains have an almost equiaxed shape. In the interface area an UFG layer was preserved too.

Presented in table 2 are the results of measurements of the microhardness in specimens consolidated under the loads of 4.5 and 6 kN. Measurements were carried out in the cross-section near the center of weld spot. “Bulk” values correspond to the middles of layers and the interface microhardness to the regions of joints between two layers. The layers are numbered from the bottom to the top. For each bulk and interface region, the results were averaged over measurements at five points. As one can see from the table, both the bulks of the layers and the joint regions are characterized by high values of microhardness (more than 3000 MPa), and in the interface regions the microhardness is noticeably higher. This fact is consistent with the microstructural studies reported above. The values of microhardness in the bulks of layers correspond to the one of nickel polycrystal having the grain size about 1 µm, and in the weld regions the values are slightly lower than the ones corresponding to the UFG structure with the grain size of 200-300 nm. Evidently, the latter fact is due to the presence of micropores in the joints.

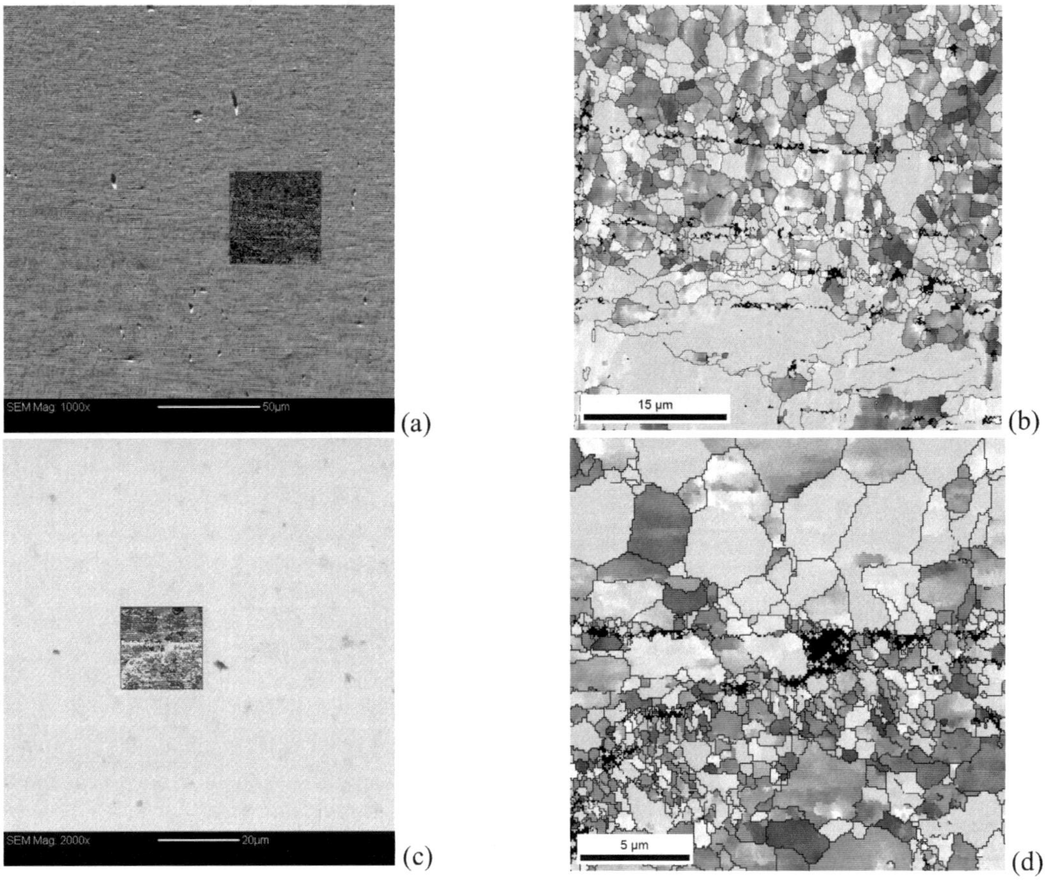

Figure 2. Microstructure of welded joint of UFG nickel at loads of 4.5 and 6.0 kN (b, d). Photos with overlaid EBSD map of the areas from which the survey was made (a, c)

Table 2. Microhardness values (in megapaskals) in the bulk and interface zones of layers in the samples consolidated by USW under the static loads of 4.5 and 6 kN

Layer number		1	2	3	4
P = 4.5 kN	Bulk zone	3052±32	3046±24	3060±20	3050±21
	Interface zone	3156±15	3140±14	3148±18	
P = 6.0 kN	Bulk zone	3132±30	3146±15	3160±20	3150±18
	Interface zone	3266±10	3290±10	3302±12	

Conclusions

In this study, we succeeded to obtain for the first time fairly good solid state joints of nickel sheets with an UFG structure processed by severe plastic deformation using ultrasonic welding. The lap shear strength of the samples obtained ranged from about 90 to about 110 MPa. The scatter of the results of mechanical tests was less significant for the case of static load 4.5 kN and more pronounced for the load of 6.0 kN, and in the latter case one of the specimens failed in the nugget pullout mode. Study of the microstructure of the consolidated samples has shown that during welding process a significant, up to the sizes of 4-5 μm, grain growth in the bulk of the sheets occur and the UFG structure is preserved at interfaces. It has been shown that, nevertheless, the consolidated samples have a fairly high microhardness, which amounted about 3000 MPa in the bulk of layers and 200 MPa higher at interfaces. This demonstrates that the use of UFG samples for consolidation can have a potential in obtaining stronger bulk materials by means of ultrasonic additive manufacturing.

The present work was supported by the Russian Science Foundation (grant No. 16-19-10126). Electron microscopic studies and mechanical tests were carried out on the facilities of shared services center of IMSP RAS "Structural and Physical-Mechanical Studies of Materials".

References

[1] Silin L L, Balandin G F and Kogan M G 1962 *Ultrasonic Welding* (Moscow: Mashgiz) [in Russian]
[2] Mitskevich A M 1970 Ultrasonic metal welding *Physical Foundations of Ultrasonic Technology* (Nauka: Moscow) [in Russian]
[3] Kholopov Yu V *Ultrasonic Welding of Plastics and Metals* (Leningrad: Mashinostroenie) [in Russian]
[4] Graff K 2005 Ultrasonic metal welding *New Developments in Advanced Welding* ed N Ahmed (Cambridge: Woodhead) p 241-269
[5] Matheny M P and Graff K F 2015 Ultrasonic welding of metals *Power Ultrasonics-Applications of High-intensity Ultrasound* ed J A Gallego-Juárez and K F Graff (Cambridge: Woodhead Publishing, Elsevier Ltd.) chapter 11 p 259–293
[6] White D. Ultrasonic Object Consolidation. US Patent # 6519500, 23 March 2000.
[7] Graff K E 2011 Ultrasonic additive manufacturing *Welding Fundamentals and Processes* ed. T Lienert, T Siewert, S Babu and V Acoff (Russell Township: ASM Int.) p 731-42
[8] Peng H, Chen D and Jiang X 2017 *Materials* **10(5)** 449
[9] Mirza F A, Macwan A, Bhole S D and Chen D L 2016 *JOM* **68** 1465-75
[10] Bakavos D and Prangnell P B 2010 *Mater. Sci. Eng.* A **527** 6320–34
[11] Dong S P, Jung H K and Jeong S S 2013 *Int. J. Mining, Metall. Mech. Eng.* **1(2)** 168-72
[12] Markushev M V 2011 *Lett. Mater.* **1(1)** 36-42

Effect of post-forging heat treatment on the microstructure and mechanical properties of a nickel-base superalloy SDZhS-15

K S Mukhtarova[1], R V Shakhov[2], A A Ganeev[2], S K Mukhtarov[2] A V Logunov[3] and V M Imayev[2]

[1] Eötvös Loránd University, Egyetem tér 1-3, Budapest, H-1053, Hungary
[2] Institute for Metals Superplasticity Problems of Russian Academy of Sciences, 39 Stepan Khalturin, Ufa, 450001, Russia
[3] PJSC "UEC-Saturn", 163 Lenin ave., Rybinsk, 152903, Russia

E-mail: shamilm@imsp.ru

Abstract. The newly designed ingot-metallurgy Ni-base superalloy SDZhS-15 intended for disc applications was subjected to multiple heat treatment and forging in the temperature range (T_s-70)-T_s, where T_s is the γ' solvus temperature, followed by solid solution treatment and ageing. A fine-grained microstructure was attained in the forged condition as a result of recrystallization processes occurring during processing. It was revealed that solution treatment at T>(T_s-50) led to a fast γ grain growth, which in turn led to a decrease in strength and ductility of the superalloy. At the same time, the solution treatment at slightly lower temperatures, at (T_s-60)-(T_s-50), allowed to save fine γ grains (d_γ=10-20 μm) and to provide the formation of secondary γ' precipitates with a mean size of about 0.1 μm. The superalloy in the forged and heat treated condition showed superior mechanical properties as compared with those of other disc superalloys.

1. Introduction

Improving the energy efficiency of gas turbine engines (GTE) and similar energy-conversion systems requires the development of new advanced materials with increased strength, heat resistance and/or reduced specific weight. In particular, increasing demands are being placed on nickel-base superalloys, which are widely used for manufacturing of discs for GTE [1]. For this purpose, novel heavily alloyed disc superalloys with a higher content of the γ'(Ni_3Al) forming elements, such as Russian superalloy SDZhS-15, are being designed. In this superalloy, the content of [Al+Ti+Nb+Ta] is 14.8 at.% and the ratio [Ti+Nb+Ta]/Al (in at.%) is 0.86, which indicates a significant solid solution hardening of the γ' phase. The γ matrix chemistry was optimized through a careful balance between the solid solution elements, such as Cr, Co, Mo, W and Re [2,3]. Generally, the goals of alloying were to: i) increase the γ' solvus temperature up to 1200°C and higher; ii) increase the solid solution hardening of the γ matrix and the γ' phase; iii) maintain oxidation and corrosion resistances via significant alloying with Cr; iv) avoid the precipitation of undesirable topologically close-packed (TCP) phases. In spite of heavy alloying, the SDZhS-15 superalloy was characterized by well-balanced alloying not causing formation of harmful phases [2,3]. Multistep heat treatment and hot forging processing have been developed for the superalloy that allowed obtaining sound forgings with a homogeneous recrystallized fine-grained microstructure. It is worth noting that even in the fine-grained condition not subjected to solid solution heat treatment the superalloy showed quite competitive properties in comparison with those of other disc nickel-base superalloys [2]. Nevertheless, to have properly balanced mechanical properties, particularly higher creep resistance, the fine-grained forgings of the SDZhS-15 superalloy should be subjected to post-forging solid solution heat treatment and ageing to obtain a high volume fraction of γ' precipitates. In doing so, the γ grain growth should be as much as possible avoided to retain a small γ grain size. Note that the refined microstructure is very important for GTE discs especially for the hub part, which under service conditions is subjected to high tensile, fatigue and thermocyclic stresses.

The present work was aimed to study the effect of post-forging heat treatment on the microstructure and mechanical properties of the SDZhS-15 superalloy. The heat treatment included different solution heat treatments followed by air cooling and ageing. The microstructure and mechanical properties of the superalloy were compared with the cast and heat treated condition.

2. Material and methods

The SDZhS-15 superalloy with a nominal chemical composition Ni-28(Cr,Co)-12.5(Al,Ti,Nb,Ta)-9(Mo,W,Re)-0.17(C,La,Y,Ce,B) (in wt.%) was manufactured by vacuum induction melting as ingots with a size of about Ø140 mm × 400 mm. The superalloy composition was found to be very close to its nominal composition. The $\gamma'(Ni_3Al)$ solvus temperature was determined as T_s=1220±5°C that is consistent with ref. [2]. The as-cast superalloy was subjected to long-term homogenization annealing at near the γ' solvus temperature, to solid solution heat treatment at 1210°C followed by air cooling with a rate of about 15°C/s and two-stage ageing at 860 (6 h) and 750°C (32 h) (condition 1). The as-cast superalloy was subjected to long-term homogenization annealing at near the γ' solvus temperature followed by slow cooling in order to increase the hot workability of the superalloy. After that, the canned forging with intermediate annealing was conducted at subsolvus temperatures ((T_s-70)-(T_s-40)) followed by air cooling as described elsewhere [2,3]. As a result, sound fine-grained forgings with an approximate size of Ø100 mm × 15 mm were obtained. The post-forging heat treatment included either two-stage ageing at 860 and 750°C (condition 2) or solid solution treatment during 1 h at 1160, 1170, 1180, 1190 and 1200°C followed by air cooling with a rate of about 15°C/s and two-stage ageing at 860 and 750°C. The conditions obtained after solution heat treatment at 1170°C and 1200°C followed by ageing were designated in the text as conditions 3 and 4, accordingly.

The microstructure examination was carried out using scanning electron microscopy (SEM). SEM in the back-scattering electron (BSE) mode was performed in a Mira-3 Tescan microscope, which was equipped with an energy-dispersive X-ray analysis system. The electron backscatter diffraction (EBSD) analysis was performed with a scan-step size of 0.5 μm using the HKL Channel 5 software. The grain boundaries having misorientation angle less than 2° were excluded from the consideration taking into account the measurement accuracy. The grain boundaries having misorientation angle more than 15° were assumed as high-angle ones.

3. Results and discussion

3.1. Effect of processing on the microstructure

Figure 1 shows the microstructure of the superalloy in the cast and heat treated condition (condition 1). Dendritic segregation observed in the as-cast ingot was dissolved during the heat treatment [2,3]. The γ grain size was in the range of 100-500 μm (figure 1a). The primary γ' phase was not observed. The mean size of the secondary γ' precipitates was $d_{\gamma'}$=0.15±0.01 μm (figure 1b). The secondary γ' phase precipitated during air cooling after solution heat treatment and ageing. The carbides having a complex composition were observed, their volume fraction was found to be about 1.5%.

Figure 2 represents the EBSD orientation map obtained from the central part of the forged workpiece (condition 2). One can see that the microstructure was mostly recrystallized and fine-grained. There were also individual large grains with a size of 30-100 μm (figure 2a). The mean size of recrystallized γ grains was defined as d_{γ}=12±1.5 μm. The fraction of high-angle boundaries was equal to 80% and low-angle boundaries were observed only in large γ grains. Primary coarse γ′ precipitates not dissolved at the forging temperature had a size of 3-10 μm and occupied about 15 vol.%. They were mostly located along γ grain boundaries (figure 2b). The secondary γ′ precipitates were formed during air cooling from the forging temperature and ageing. Their average size was $d_{\gamma'}$=0.2±0.02 μm. The carbide particles occupied approximately the same volume fraction as in the condition 1. Unfavorable TCP phases were not detected in condition 2 that is consistent with the earlier performed work [2].

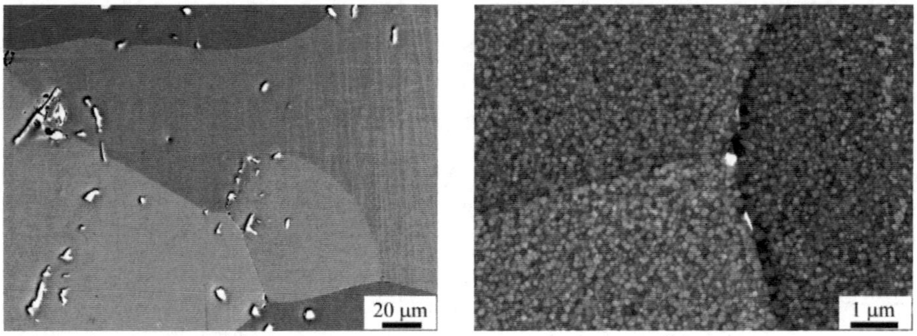

(a) (b)

Figure 1. The BSE images of the cast and heat treated superalloy (condition 1).

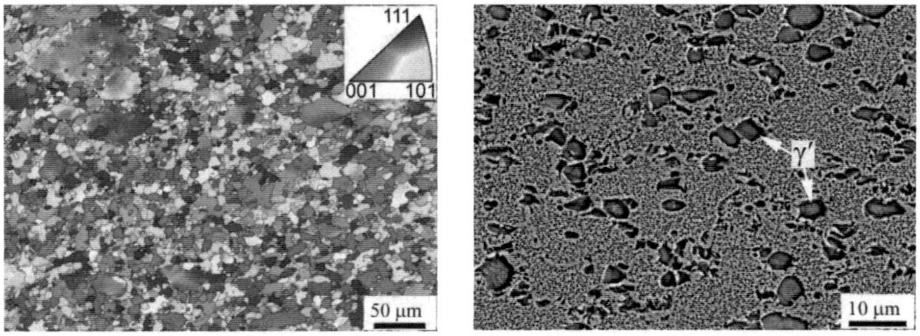

(a) (b)

Figure 2. The microstructure of the superalloy after hot forging and ageing (condition 2): (a) normal-direction EBSD (inverse-pole-figure) map, the forging axis is vertical, high- and low-angle grain boundaries are indicated by black and white lines, respectively; (b) BSE image, the primary γ′ phase is arrowed.

By way of illustration, figure 3 shows the microstructures obtained after hot forging, solid solution heat treatment and ageing. Figure 4 shows the dependencies of the γ grain size and the volume fraction of the primary γ′ phase obtained after solution heat treatment and ageing. After solution treatment at 1160°C and ageing the mean γ grain size was not appreciably changed as compared with the forged condition and averaged d_{γ}=14±1.5 μm. The increase of the solution treatment temperature led to γ grain growth due to static recrystallization and dissolution of the γ′ phase, which restricted the γ grain growth. As the solution treatment temperature increased from 1160 to 1200°C, the γ grain size increased from 14 to about 60 μm and the volume fraction of the primary γ′ phase decreased from 15% to 4.5% (figures 3 and 4). The mean size of the secondary γ′ precipitates after solution heat treatment at 1160-1200°C and ageing was varied within $d_{\gamma'}$=0.1-0.13 μm. The tertiary γ′ precipitates were sometimes observed along γ grain boundaries, their size was smaller than 0.1 μm. Most probably, they were formed during ageing. The volume fraction of carbides after solution heat treatment and ageing was not noticeably changed as compared with conditions 1 and 2.

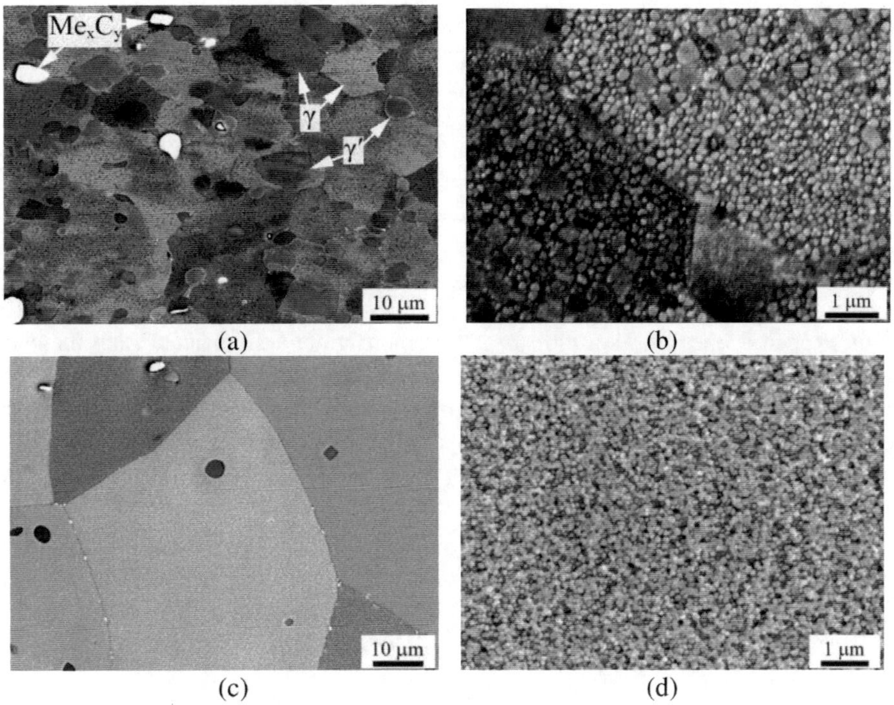

Figure 3. The BSE images of the superalloy after hot forging, solution heat treatment at different temperatures and ageing: (a, b) 1170°C (condition 3), (c, d) 1200°C (condition 4). The arrows show the γ grains, primary γ′ phase and carbides.

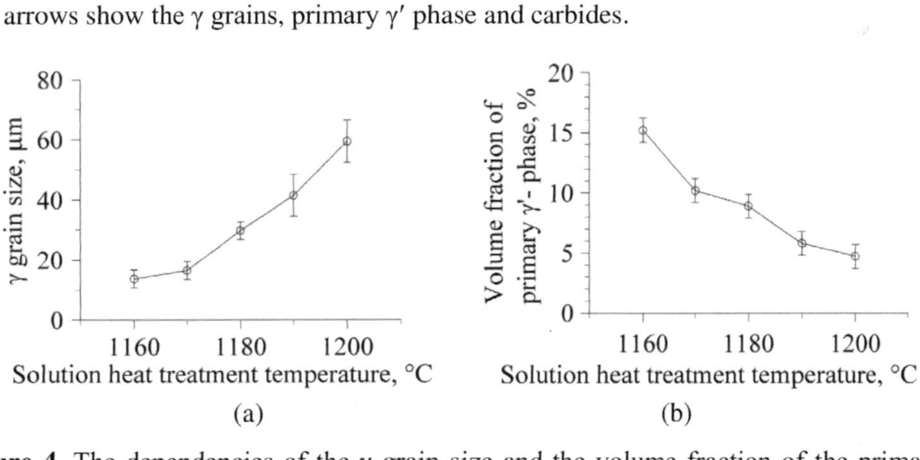

Figure 4. The dependencies of the γ grain size and the volume fraction of the primary γ′ phase obtained after solution heat treatment at different temperatures followed by ageing.

Taking into consideration that the optimal γ grain size lies in the range of 10-20 μm, the temperature range 1160-1170°C was selected as the most appropriate for solid solution heat treatment. After solution heat treatment in this temperature range the γ grain size was defined as d_γ=14-16 μm, the volume fraction of the primary γ′ phase was relatively low (10-15%) and the mean size of the secondary γ′ precipitates was defined as $d_{\gamma'}$=0.1-0.12 μm. Thus, the post-forging solution heat treatment at 1160-1170°C resulted in formation of dispersed γ′ precipitates, while retaining refined γ grains that is a good prerequisite for achieving enhanced mechanical properties.

3.2. Mechanical properties

Figure 5 represents the tensile properties of the superalloy at 20 and 750°C obtained in the conditions 1-4. The forged and heat treated conditions 2-4 demonstrated appreciably higher strength (σ_{UTS}) and ductility as compared with those of the cast and heat treated condition 1. This should be ascribed to finer γ grain size in conditions 2-4. The decrease of strength and ductility in condition 4 in contrast to those in conditions 2 and 3 is associated with coarser γ grain size in condition 4. One can conclude that the increase of the solution treatment temperature led to a decrease of strength and ductility in spite of a decrease of the primary γ′ phase volume fraction.

Figure 6 shows the creep curves obtained for the superalloy in conditions 1-3 at 750°C. One can see that the creep resistance significantly increased at transition from conditions 1 and 2 to condition 3. Thus, as was expected, the solid solution heat treatment gave a significant increment in the creep resistance. That should be ascribed to the smaller size of the secondary γ′ precipitates, the smaller volume fraction of the primary γ′ phase and the optimal γ grain size. One may conclude that condition 3 of the superalloy demonstrated high strength, reasonable ductility and excellent creep resistance. At higher solution heat treatment temperatures (T>1170°C) a fast γ grain growth occurred that led to a decrease of the strength and ductility properties. Therefore, T=1170°C (T_s-50) may be assumed as the optimal solution heat treatment temperature for the SDZhS-15 superalloy.

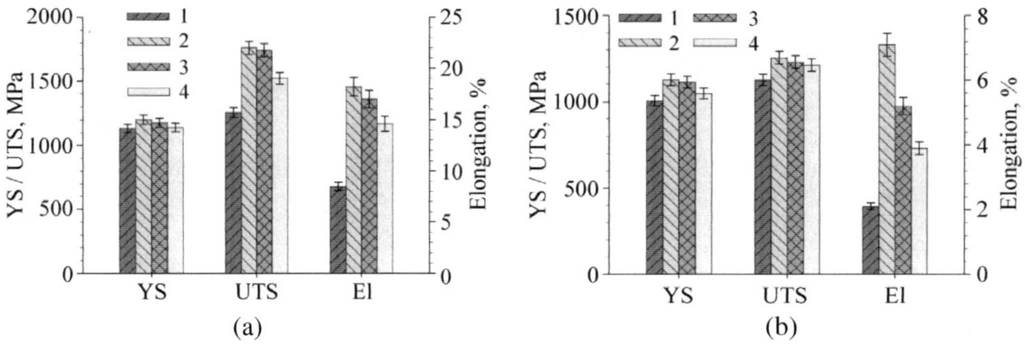

Figure 5. The tensile properties of the superalloy in conditions 1-4: (a) at 20°C, (b) at 750°C.

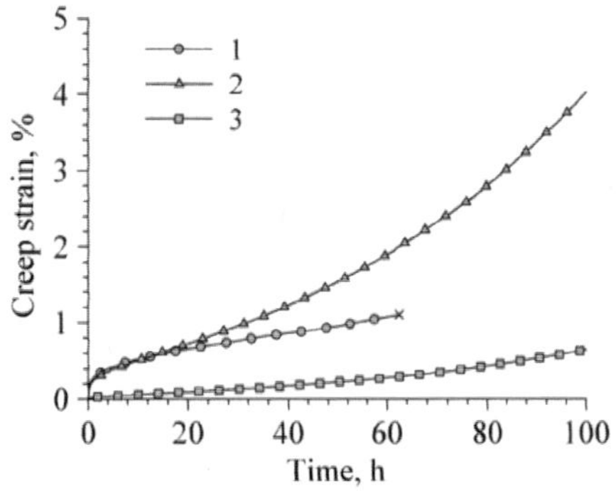

Figure 6. Creep curves obtained for the superalloy in conditions 1, 2 and 3 at 750°C and a loading 700 MPa. In condition 1, the creep failure occurred after testing during about 61 h.

Table 1 shows the tensile properties of the superalloy in condition 3 as compared with those of other nickel-base disc superalloys. Superior tensile properties were mostly obtained for the SDZhS-15 superalloys. It should be emphasized that in contrast to our previous works [2,3], the obtained mechanical properties were reached in larger workpieces and after solution heat treatment, which provided excellent creep resistance. The follow-up task is to realize the obtained mechanical properties in a larger-scale as applied to real discs for GTE.

Table 1. Tensile properties of the SDZhS-15 superalloy in condition 3 as compared with those of other nickel-base disc superalloys (RT – room temperature)

Superalloy	UTS (MPa)	YS (MPa)	δ (%)
SDZhS-15 (present work) RT	1742	1176	17
750°C	1230	1112	5.2
N18 RT [4]	1620	1080	>20
Alloy 10 RT [5]	1650	1210	12-17
AF115 RT [6]	1646	1190	20
TMW-4 RT [7]	1700	1210	9
FGH100 RT [8]	1600	1172	19
705°C	1345	1090	20
LSHR [9]	1700	1210	>15

4. Conclusion

Microstructure and mechanical properties of the newly designed Russian Ni-base superalloy SDZhS-15 intended for disc applications were studied after different treatment. Based on this, effective processing including hot forging, solution heat treatment and ageing was developed for the superalloy. Particularly, it was revealed that the post-forging solution heat treatment at temperatures higher than 1170°C (T_s-50) led to a fast γ grain growth that in turn led to a decrease in strength and ductility. After solution heat treatment at 1170°C the volume fraction of the primary γ′ phase was reduced down to 11% and the size of the secondary γ′ precipitates was about 0.1 μm, while retaining refined γ grain size (d=16 μm). The obtained mechanical properties of the superalloy after hot forging, solution heat treatment at 1170°C and ageing were not inferior or superior to the properties of other Ni-base disc superalloys.

Acknowledgments

The present work was supported by the Russian Science Foundation (Grant No. 18-19-00594). The work was performed using the facilities of the shared services center "Structural and Physical-Mechanical Studies of Materials" at the Institute for Metals Superplasticity Problems of Russian Academy of Sciences.

References

[1] Reed R C 2006 The superalloys: Fundamentals and Applications (Cambridge University Press).
[2] Mukhtarov S K, Imayev V M, Logunov A V, Shmotin Yu N, Mikhailov A M, Gaisin R A, Shakhov R V, Ganeev A A and Imayev R M 2019 *Mater. Sci. Technol.* **35** 1605
[3] Imayev V M, Mukhtarov S K, Logunov A V, Ganeev A A , Shakhov R V and Imayev R M 2019 *Letters on Materials* **9**(2) 249 (in Russian)
[4] Guedou J Y, Lautridou J C and Honnorat Y 1992 *Proc. of the 7th Int. Symp. Superalloys 1992* (Warrendale, Pa.: TMS) p 267
[5] Rice D, Kantzos P, Hann B, Neumann J and Helmink R 2008 *Proc. of the 11th Int. Symp. Superalloys 2008* (Wiley-TMS) p 139
[6] Merrick H, Benn R C and Bhowal P R Patent US 6,468,368 B1 Assignee: Honeywell International, Inc Publ. 22.10.2002
[7] Gu Y F, Cui C, Harada H, Fukuda T, Ping D, Mitsuhashi A, Kato K, Kobayashi T and Fujioka J 2008 *Proc. of the 11th Int. Symp. Superalloys 2008* (Wiley-TMS) p 53
[8] Jia C L, Ge C C and Yan Q Z 2016 *Mater. Sci. Eng. A* **659** 287
[9] Gabb T P, Gayda J, Telesman I and Kantzos P T Patent US 6,974,508 Assignee: US NASA Publ. 13.12.2005

Influence of superplastic deformation on the quality of solid-phase joints obtained by welding of crystalline alloys

R Ya Lutfullin, M Kh Mukhametrahimov, E V Galieva

Institute for Metals Superplasticity Problems, Russian Academy of Sciences, 39 Khalturin st., Ufa 450001, Russia

E-mail: lutfullin.ramil@imsp.ru

Abstract. Pressure welding using the phenomenon of structural superplasticity is a relatively new technological method based on the additive principle of forming parts. The problem of the quality of a solid-phase joint, which is well studied for superplastic titanium alloys, requires a verification of the versatility of the approaches used to solve it for heat-resistant nickel alloys, as well as for welding of dissimilar crystalline materials. The possibility of reducing the temperature of pressure welding by using low-temperature superplasticity of ultrafine-grained (UFG) crystalline alloys is considered.

1. Introduction

It is known that classical diffusion welding of crystalline alloys, which is carried out at temperatures above approximately $0.8T_m$, where T_m is the melting temperature of an alloy, with the welding force limited to values below the yield stress and avoiding a noticeable macroscopic plastic deformation, is a structurally uncontrolled technological process [1]. This circumstance imposes some uncertainties on the possibility of achieving a high quality solid-phase joint after diffusion welding, which is critical for the products of a responsible purpose, in particular in aerospace engineering. Meanwhile, a cardinal solution to the problem of a guaranteed assurance of the quality of solid-phase joints is possible by using the effect of structural superplasticity [2,3]. To date, the determining role of superplastic deformation in achieving the required quality of solid-phase joints at temperatures of about $0.6T_m$ has been experimentally proven for a number of titanium, aluminum, nickel alloys and stainless steels [3-6]. However, the influence of low-temperature superplasticity, which is exhibited by ultrafine-grained (UFG) and nanostructured alloys, on the formation of a reliable solid-phase joints has not yet been fully studied yet and is of scientific and practical interest to researchers and engineers.

2. Solid-phase welding of titanium alloys

One of the first reports on the successful pressure welding of titanium alloys in the state of superplastic deformation was given in an article by Shorshorov et al. [7] published in 1975 in the journal "Welding production". Quite recently, the authors of [6] made an important conclusion on a possibility of achieving a four times reduction in the welding force and a significant, 6 to 30 times, reduction in the duration of the pressure welding process using the superplasticity effect. This result was the basis for further developments of promising pressure welding technologies based on the superplasticuty effect. It is worth to note that the above-mentioned diffusion welding [1], in general, can be considered as a special case of pressure welding, when strict limitations on the macroplastic deformation of the workpieces to be joined are applied.

In [8-9], the role of grain boundary sliding, which is the main mechanism of superplasticity, in the accelerated kinetics of solid-phase joint formation was first established and studied in detail, and a fundamental conclusion was made about the increased solid-state weldability of superplastic titanium alloys. At the same time, it has been shown qualitatively and quantitatively that the determining factor for achieving the strength of the solid-phase joint at the level of the base material is the level of strain in the junction zone [9]. A decrease of the grain size at temperatures of the traditional superplasticity makes it is possible to reduce the necessary strain level to achieve the properties of welded samples approaching the level of mechanical properties of the workpieces being joined. However, given the fact of a structural instability of UFG metals and alloys, in the practice of diffusion welding the average grain size should be taken into account strictly at the moment when the samples to be welded are brought into contact and pressed. For welding massive workpieces made of fine-grained titanium alloy VT6 (Ti-6Al-4V) having the average grain size below 4 μm in the conditions of traditional superplasticity under high vacuum (not worse than $2.0 \cdot 10^{-3}$ Pa), the required strain was below 20% [10]. Spherical pressure vessels, which were fabricated from a package of billets by pressure welding in the conditions of superplasticity followed by superplastic forming and then subjected to full-scale tests applying an excessive internal pressure, failed not along the weld joint but outside it [10]. As has been shown by experiments on pressure welding of titanium alloy VT6 with an UFG structure [11,12], high-quality solid-phase joints with the strength at a level of the base material can obtained at temperatures in the range of 750 to 600°C. The determining factors in this case were shown to be the grain size and strain. Table 1 shows the results of mechanical tensile tests at room temperature of samples obtained by pressure welding at 700 °C from the UFG alloy VT6. A physical model of solid-phase joint formation under conditions of low-temperature superplastic deformation was presented in [13].

Table 1. Mechanical properties of UFG samples of VT6 alloy (average grain size 0.4 μm) joined at a temperature of 700 °C by pressure welding in the state of superplastic deformation ($\dot{\varepsilon} = 7 \cdot 10^{-4}$ s^{-1})

Strain, %	Mechanical properties			
	σ_{02}, MPa	σ_{UTS}, MPa	δ, %	ψ, %
5	1047±35	1054±30	4±3	7±5
10	1046±15	1052±15	17±2	61±2
15	1013±15	1020±15	19±2	65±2

3. Solid-phase welding of nickel-based alloys

One of the most important problems, the solution of which will provide a progress in aircraft engine building, is the development of technological regimes for obtaining a reliable solid-phase joint of heat-resistant nickel-based alloys [14,15]. The main idea of obtaining a reliable solid-phase joint is to implement local superplastic deformation in the junction zone, which was previously successfully implemented for pressure welding of titanium alloys [16-18]. At the same time, the possibility of reducing the pressure welding temperature by using the effect of low-temperature superplasticity was considered as the most important task. Figure 1 shows the result of an original experiment on the pressure welding of a heat-resistant nickel alloy EK61 made at 800°C to obtain a high-quality (virtually pore-free and without additional phases) solid-phase joint. In this experiment, two billets made of coarse-grained heat-resistant nickel alloy EK61 were brought into contact through an UFG superplastic layer made of the same alloy, in which the superplastic deformation necessary to achieve the required quality was localized. It should be emphasized that in this case the coarse-grained samples under welding was preserved almost undeformed. This experiment leads to a new and an important principle for solid-phase welding technology, which tells that in order to obtain a high-quality joint, it is sufficient to provide a superplastic deformation not in both bodies under contact but only in one of them. This principle has been also successfully tested for welding of dissimilar heat-resistant nickel alloys, in particular, the deformable EP 975 alloy with a cast single-crystal alloy of the VKNA type.

Figure 1. Metallography of the cross section of the solid-phase joint of the alloy EK61: upper and lower zones correspond to the coarse-grained structure of workpieces and the middle zone to the UFG structure of the intermediate layer

4. Conclusions

The experimental studies have shown that pressure welding combined with superplastic deformation allows for obtaining pore-free solid-phase joints of coarse grained heat resistant nickel alloy EK61 using ultrafine-grained layer of the same alloy and of the EP975 superalloy having a fine-grained structure with a monocrystalline intermetallic Ni_3Al-based VKNA-type alloy. From these results, one can conclude that to obtain a high quality solid-phase joint between similar or dissimilar materials, it is necessary and sufficient that one of the workpieces should be superplastic. The reduction in the average grain size of EK61 alloy down to the ultrafine scale makes it possible to reduce the temperature of superplastic deformation down to about 800°C. This allows one to achieve pore-free solid-phase joints at lower temperatures.

Acknowledgments

The present work was supported by the state assignment of the Institute for Metals Superplasticity Problems of RAS No. AAAA-A17-117041310221-5. Electron microscopic studies and mechanical tests were carried out on the facilities of shared services center of IMSP RAS "Structural and Physical-Mechanical Studies of Materials".

References

[1] Karakozov E S, Orlova L M, Peshkov V V and Grigorevsky V I 1977 *Diffusion Welding of Titanium* [in Russian] (Moscow: Metallurgy) [in Russian]
[2] Kaibyshev O A 1992 *Superplasticity of Alloys, Intermetallides and Ceramics* (Berlin; New York: Springer-Verlag)
[3] Lutfullin R Ya 2000 *Proceedings of International Conference "Current States of Theory and Practice of Superplasticity in Materials" (Ufa: Gilem)* p 67-74 [in Russian]

[4] Lutfullin R Ya and Kaibyshev O A 1997 *Mater. Sci. Forum* **243-245** 681-6
[5] Wang Z C, Ridley N, Lorimer G W, Knauss D and Briggs G A D 1996 *J. Mater. Sci.* **34** 5199-206
[6] Galieva E V, Lutfullin R Ya, Akhunova A Kh, Valitov V A and Dmitriev S V 2018 *Sci. Technol. Weld. Join.* **27** 612-8
[7] Shorshorov M Kh, Dzneladze E M, Tihonov A S, Karakozov E S and Mattushkin B A 1975 *Welding Production* **11** 20-1 [in Russian]
[8] Kaibyshev O A, Lutfullin R Ya and Berdin V K 1991 *Doklady Akademii Nauk* **319** 615-8 [in Russian].
[9] Kaibyshev O A, Lutfullin R Ya and Berdin V.K 1994 *Acta Metall. Mater.* **42** 2609-15
[10] Lutfullin R Ya, Kruglov A A, Mukhametrakhimov M Kh and Valiakhmetov O R 2004 *Welding Int.* **18** 646-51
[11] Lutfullin R Ya, Kruglov A A, Safiullin R V, Mukhametrahimov M Kh and Rudenko O A 2009 *Mater. Sci. Eng.* A **503** 52-4
[12] Safiullin R V, Mukhametrakhimov M Kh, Safiullin A R, Malysheva S P, Kozlov A N, Berestov A V, Harin S A, Morozov M A 2018 *Lett. Mater.* **8** 329-34 [in Russian]
[13] Lutfullin R Ya and Mukhametrakhimov M Kh 2010 *Rev. Adv. Mater. Sci.* **25** 142-7
[14] Logunov A V and Shmotin U N 2013 *Modern Heat-Resistant Nickel Alloys for Gas Turbine Disks* [in Russian] (Moscow: OOO "Science and Tecknology") [in Russian]
[15] Akhunova A Kh, Valitov V A, Galieva E V 2020 *Lett. Mater.* **10** 328-33 [in Russian]
[16] Lutfullin R Ya and Mukhametrakhimov M Kh 2009 *Advanced Mater.* **7** 189-93 [in Russian]
[17] Lutfullin R Ya 2011 *Lett. Mater.* **1** 59-64 [in Russian]
[18] Kaibyshev O A, Kruglov A A and Lutfullin R Ya 2007 *Method of Manufacture of Article by Superplastic Moulding and Diffusion Welding* Russian Federation Patent # 2291019.

Use of high hydrostatic pressure for food sterilization and seed treatment

A Yu Volkov[1], N A Kruglikov[1], A V Alexandrov[1] and V V Kotkova[2]

[1] M.N. Mikheev Institute of Metal Physics, Ural Branch of RAS, 18 S.Kovalevskaya St., Ekaterinburg, 620108 Russia

[2] Ural State Economic University, 62 March 8 St., Ekaterinburg, 620144 Russia

E-mail: volkov@imp.uran.ru

Abstract. Experiments with high hydrostatic pressure to sterilize meat, sauces, mayonnaise, milk and yoghurts have been conducted. We also used the hydrostatic processing to increase the speed of germination and productivity of various plants (radishes, tomatoes, licorice, etc.). It was found that, after high pressure processing, the microbial contamination of all food products was reduced. It resulted in a significant increase in the shelf life of food (from 3 to 5 times). It was shown that the high pressure processing was a way to significantly increase the crop yield of tomatoes (by ~65%) and the germination of legumes (almost by an order).

1. Introduction

Obtaining bulk samples of fine-grained and nanostructured materials requires applying high pressures and shear stresses to workpieces. This is why modern materials science laboratories are equipped with powerful presses and high-performance hydraulic high-pressure pumps. For example, it was with the use of a hydraulic press with a force capacity of 10 MN that we developed a technology for producing a thin magnesium plate with an ultrafine-grained structure and unusually high plasticity [1]. Hydroextrusion with the use of boosters developing the pressure up to 1.5 GPa made it possible to obtain high-strength binary and ternary composites, which contain up to one million thin Mg-filaments in Cu or Al-matrices [2]. Meanwhile, this powerful equipment is in great demand in areas that are far from the materials science.

For example, increasing the shelf life of food and semi-finished products is an important applied research task. Most often, to achieve this goal, various preservatives are added to food products and modified gas media are used for storage. The most developed methods of physical impact on food products in Russia are ionizing radiation treatment and ultrasonic waves. Meanwhile, a new trend is being dynamically developed, which consists in the processing of food products by means of high hydrostatic pressure (high pressure processing, HPP) [3]. In a number of countries (Germany, USA, Japan), HPP is used in pilot production of "green products". The shelf life of such products is at an acceptable level without the introduction of any preservatives. On the territory of the former USSR, scientists from Donetsk and Lugansk have achieved the most progress in the field of high pressure food processing. In Russia, this field of biotechnology is nearly unknown to commodity producers. The availability of powerful high-pressure equipment gives metal scientists a unique chance to try themselves in a completely new field, i.e. in biophysics.

The purpose of this work was to study the influence of such parameters of HPP as pressure value and exposure time on microbiological, organoleptic and some other parameters of various food products and semi-finished products. In addition, we conducted experiments to find out the effect of high-pressure seed treatment on the speed of their germination and crop yield of plants.

2. Materials and methods

High-pressure devices and equipment designed and manufactured at the Institute of Metal Physics of the Ural Branch of the Russian Academy of Sciences were used for conducting experiments. A laboratory hydrostat was used for pressure processing up to 250 MPa. The capacity of its container is 5 l, high pressure is created in it by an oil pump. If a higher pressure (up to 650 MPa) was required, a 3 l container was used and the pressure in it was created by a plunger connected to a hydraulic press. For experiments under a pressure of up to 1500 MPa, a hydroextrusion unit was used, where a plug was installed instead of a matrix. The diameter of the working part of the container in this unit is 22 mm, the working length of the container is 70 mm. Before conducting the experiments, all samples were placed in a sealed latex shell. When processing seeds, liquids were sometimes added into this shell: water, fertilizer solution, etc. The high pressure processing was carried out at room temperature; the soaking time depended on the processed product and the selected process pressure, but did not exceed 30 minutes.

3. Results and discussion

3.1. HPP of meat products and semi-finished products

The experiments were carried out on meat products (chilled beef) and sausages, and samples of minced meat and chicken were also processed. As an example, Table 1 shows the change in microbiological parameters of sausages Rossiyskie after processing under the pressure of 250 MPa for 15 minutes. As a comparison, parameters of the control group of sausages that were not subjected to HPP are also presented. The experiment was carried out as follows. On the day indicated by the manufacturer as the last day of the shelf life, the packaging was opened and the material of the control and test groups was inoculated to determine mesophyll aerobic and optional-anaerobic microorganisms (QMA&OAMO) and the presence of coliform bacteria. The obtained parameters were taken as a baseline. The subsequent removal of the products for analysis took place after 1, 2 and 3 days of storage beyond the expiry date. In accordance with the storage rules established by the manufacturer for a specific product, both control and test samples were kept in a household refrigerator at a temperature of 4±2°C.

It was found that the sausages Rossiyskie taken for the experiments completely withstood the shelf life specified by the manufacturer: in the baseline parameters, the content of aerobic and anaerobic mesophyll microflora, as well as coliform bacteria, did not exceed the established limits. During further storage, the number of opportunistic pathogenic microorganisms in the control group of sausages begins to significantly exceed the permissible level after two days. In turn, the rate of change in QMA&OAMO in the test group is significantly lower (see table 1). Based on the dynamics of the increase in microbiological points, the processed sausages will still meet the standards for another week of storage beyond the expiry date. Coliform bacteria were not detected in both the test and control samples during the entire experiment.

In the course of the experiments, it was found that HPP increased the shelf life of meat and meat products at least 2.5 times. Figure 1 confirms well the different rates of bacterial reproduction in samples of chilled beef after 10 days of storage in a home refrigerator (4±2°C). No change in the organoleptic, physico-chemical and toxicological properties of the processed meat products was detected.

A certain decrease in the moisture-retaining properties of meat semi-finished products can be indicated as one of the negative consequences of HPP. For example, the mass fraction of moisture in the sausages Rossiyskie, which we studied, is 62.6±0.2%. After HPP under 600 MPa for 1 minute, the mass fraction of moisture in them decreased to 57.0±0.2%. Obviously, this effect should be taken into account in the manufacture of products from processed meat. It was also shown that pressure above 850 MPa led to a protein degradation in meat products. In fact, after such processing, meat products become cooked. Comparative experiments showed that an increase in the pressure of HPP from 250 to 600 MPa causes a slight increase in the shelf life of meat products with a significant increase in the cost of equipment. From this point of view, the 500-600 MPa HPP pressure values specified in some patents seem excessive and are most likely a marketing ploy by equipment manufacturers.

Table 1. Test results of sausages Rossiyskie.

	Control points	Control group	Test group	Norm acc. to Technical Regulation of Customs Union TR TS 034-2016
QMA&OAMO	At the end of the shelf life (baseline)	9.5 x 10^4	1.0 x 10^5	Not exceeding 5.0 x 10^6
	24 hours	4.0 x 10^6	2.5 x 10^5	
	48 hours	2.2 x 10^7	8.0 x 10^5	
	72 hours	Above 3.0 x 10^7	9.0 x 10^5	

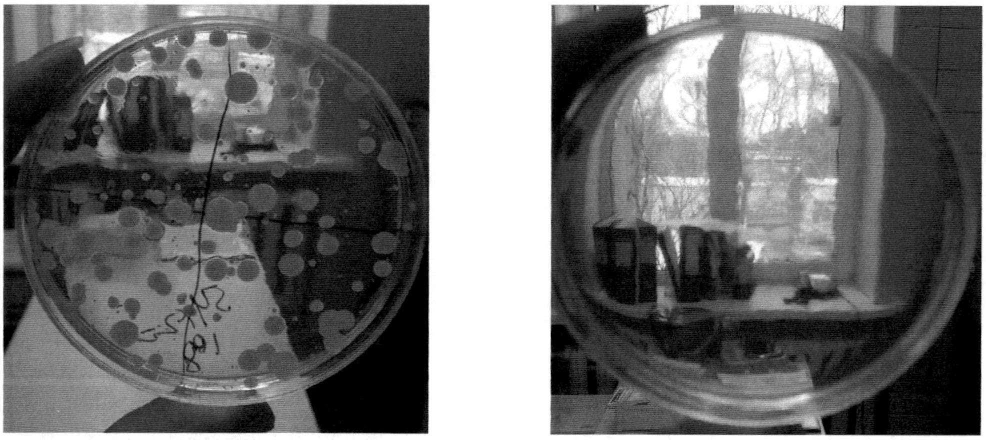

a **b**

Figure 1. Visualization of bacterial and microbial colony growth (QMA&OAMO) in control (a) and test (b) samples of chilled (4±2°C) beef after 10 days of storage. The test beef samples were processed under the pressure of 250 MPa for 30 minutes.

3.2. HPP of oil-fat products and milk

The experiments were carried out on samples of mayonnaise, ketchup, milk and other products. The HPP pressure was 500 MPa. It was found that the organoleptic parameters of all the studied products remain almost unchanged after HPP. However, the microbiological parameters of samples after HPP differ significantly from the control group. For example, as a result of HPP, the total microbial count in mayonnaise samples decreased 100 times; the number of yeast cells – almost 10^3 times. Using the example of a vegetable-fat filled spread, it was found that mold was also sensitive to high-pressure processing. The content of lactobacilli in all samples remained unchanged. Pathogenic microorganisms and a group of indicator bacteria were not detected in both control and test samples.

After HPP, the pasteurized milk with 3.5% fat survived 20 days of storage. It was found that souring is slower compared to the control sample. However, milk after HPP acquires an extraneous taste (presumably due to proteolysis of protein components under high pressure). The resulting taste defect makes it impossible to recommend HPP as a way to increase the shelf life of milk. Experiments with yoghurts showed that excessive pressure over certain values kills lactobacilli.

Based on the results of the experiments, it can be concluded that HPP of mayonnaise, ketchup, milk and various oil and fat products can be used to increase their microbiological parameters. However, in each case, long-term experiments are required to select the optimal processing pressure.

3.3. Effect of high pressure on germination of seeds and crop yield of plants

The use of various physical methods of seed treatment is a well-known way to accelerate their germination and increase of the crop yield of plants. At present, various shock or pulse effects (for example, cryoprocessing, hydraulic shock, etc.) have been tested and used. We carried out experiments of HPP of seeds of different varieties of radishes, tomatoes, potatoes and licorice. It was found that HPP of radish seeds significantly accelerated their germination, but did not lead to an increase in crop yield. At the same time, a significant increase in the volume of tops was noted, which is also important for agricultural producers. In turn, HPP of tomatoes killed some of the seeds, but the crop yield of the surviving seeds increased by more than 60%. It was established that HPP can be used as a method to fight against the seed hardness in legumes: the number of sprouted licorice seeds increased by about an order of magnitude. All the described experiments aimed at an increase in the crop yield were carried out at pressures below 100 MPa. In turn, an increase in the HPP pressure to 500 MPa leads to guaranteed death of all pathogenic microflora on the seeds. It can be used as a method to fight against fungal diseases in cereals.

4. Conclusions

From the results obtained, it follows that short-term exposure of meat and oil-fat products to high pressure does not cause changes in organoleptic parameters, but delays the growth and development of vegetative microflora. Thus, HPP has a positive effect on maintaining the freshness of products for a long time without the use of preservatives. We did not confirm experimentally that the pressure of 500-600 MPa is optimal to HPP. Indeed, HPP that we performed at 250 MPa gave similar results. We can assume with caution that the increased HPP pressure is most likely a marketing ploy. High-pressure processing of legume seeds is an effective measure of fighting against seed hardness. The example of tomatoes showed that, in a number of cases, HPP of seeds led to a significant increase in the crop yield of plants.

Acknowledgments

The work was carried out under the financial support of the Russian Foundation for Basic Research (project № 18-016-00082).

References

[1] Komkova D A, Antonova O V, Petrova V N and Volkov A Yu 2019 *Lett. Mater.* **9** 451
[2] Volkov A Yu, Antonov B D, Patrakov E I, Volkova E G, Komkova D A and Glukhov A V 2020 *Mater. Design* **185** 108276
[3] Cheftel J C and Culioli J 1997 *Meat Science* **46** 211

IOP Conf. Series: Materials Science and Engineering 1008 (2020) 012013 doi:10.1088/1757-899X/1008/1/012013

Accumulative high-pressure torsion of steel 316 and β-Ti alloy

D V Gunderov[1,2], D A Khasanova[2], V V Astanin[2], A A Churakova[1,2], I A Ramazanov[2], S D Prokoshkin[3], V A Sheremetyev[3]

[1] Institute of Molecule and Crystal Physics, Ufa Federal Research Centre RAS, 71 October Avenue, 450054, Ufa, Russia
[2] Ufa State Aviation Technical University, 12 K. Marx st., Ufa, 450008, Russia
[3] National University of Science and Technology "MISiS", 4 Lenin Avenue Moscow, 119049 Russia

E-mail: dimagun@gmail.com

Abstract. As revealed by studies, in the process of high-pressure torsion of hard or work-hardenable metals and alloys, including bulk metallic glasses, the actual produced strain is much smaller than the expected one. The authors proposed a new technique called "accumulative high-pressure torsion" for producing high strains in hard materials, including bulk metallic glasses.

1. Introduction

As revealed by the studies [1–4], in the process of high-pressure torsion (HPT) of hard or work-hardenable metals and alloys including bulk metallic glasses (BMGs), the actually produced strain, is much smaller than the one expected according to the formula $\gamma = 2\pi nR/h$ (1) [1]. This discrepancy can be attributed to the effect of "slippage" of anvils on the sample surface during the HPT processing of hard and work-hardenable materials.

The authors earlier have proposed a new technique, accumulative high-pressure torsion (ACC HPT), for producing high strains in hard materials including BMGs [5]. In the process of accumulative HPT, a disk-shaped sample is subjected to several cycles of deformation according to the following route (figure 1): "HPT processing up to n = 1 or n = 2 anvil revolutions → cutting of the sample into parts → compression of the stacked parts on anvils and subsequent HPT processing up to n = 1 (or another number of anvil revolutions)" [5]. Such cycles can be repeated several times. At the final stage, the stacked segments are subjected to HPT with a large number of revolutions ($n \geq 3$), which consolidates the sample into a monolithic disk (similarly as observed in the HPT-consolidation of amorphous ribbons and powders [1,6]). As a result of ACC HPT, the material experiences a significant total compressional or torsional strain, since in the process of upsetting the sample is known to experience a considerable strain.

In [5] it was shown, that the structure of bulk metallic glasses (BMG) $Zr_{52.5}Cu_{17.9}Ni_{14.6}Al_{10}Ti_5$ (at. %) (Vit105) underwent a more significant transformation during accumulative high-pressure torsion processing than in the case of regular HPT with a similar number of revolutions (n = 5), the X-ray diffraction pattern revealed a larger shear and increment in the broadening of the amorphous halo. It should be noted that the microhardness of initial BMG is very high (Hv>5000 MPa), which causes slippage of samples during regular HPT processing. In contrast, in the process of accumulative HPT at the upsetting stage the material is known to experience strain.

This paper reports the first results on the effect of ACC HPT processing on crystalline materials such as β-Ti alloy Ti-18Zr-15Nb and austenitic steel 316. Each of these alloys contains several alloying elements but have a single-phase composition in the initial as-quenched state. This makes it easier to monitor possible phase transformations during the SPD processing of these materials. It should be noted that there have been many previous studies on the effect of regular HPT processing on steel 316 [7,8], which enables making a comparison between the results on the effect of accumulative HPT processing reported in this study and the previously obtained results.

2. Materials and methods

The starting materials were austenitic steel 316 and β-Ti alloy Ti-18Zr-15Nb. Initial samples of conventional stainless steel 316 (Fe–0.03C–17Cr–0.41Si–1.72Mn–0.01P–0.03S–12.9Ni–2.36Mo, wt%) were received from Prof. T. D. Shen, Yanshan University, Qinhuangdao, China. Experimental Ti-18Zr-15Nb (at.%) alloy is developed in the National University of Science and Technology "MISiS", Moscow, Russia by prof S.D. Prokoshkin's group. The Ti-18Zr-15Nb titanium alloy is the most preferred material for medical implants [9,10]. This shape-memory alloy (SMA) composed of non-toxic components exhibits a high biomechanical compatibility and corrosion resistance. The reversible thermoelastic β↔α" martensitic transformation provides the shape memory effect in this alloy [9]. The ingot was initially produced by vacuum arc remelting, then transformed into a rod by multi-axis forging at a temperature 1050 °C.

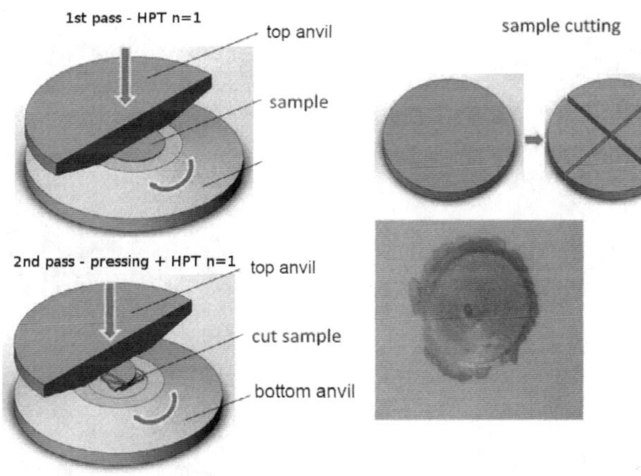

Figure 1. Principle of accumulative HPT and a photo of a sample produced according to this procedure [4].

High-pressure torsion die-set with 20 mm diameter anvils having a 0.4 mm deep groove was used for HPT processing, which was conducted at a pressure of 6 GPa, room temperature, and various numbers of revolutions. The rotation speed of the anvils was 1 rpm. In the "accumulative HPT" procedure, at the first stage, a sample was subjected to HPT up to n = 2 rotations, as shown on figure 1. Obtained disks at the second stage were cut into 4 segments, then these segments were stacked on the HPT anvils on top of each other and HPT up to n = 2 was performed again. The cycles of "accumulative HPT" n=2 were repeated 3 times. At the fourth stage, the procedure was repeated again, but rotation number was n = 4,for both steel 316 and Ti-18Zr-15Nb alloy. As a result, solid HPT disks from all materials were obtained. The total number of revolutions during "accumulative HPT" processing was n=10. It can be asserted that the total strain produced during "accumulative HPT" processing due to the summation of compressive and torsional strains was much greater than that resulting from conventional HPT processing.

The structure and phase composition were studied using Rigaku Ultima IV X-ray diffractometer under Cu-Kα radiation. The microhardness measurements were performed using Durascan-50 at a load of 100 g for 10 s.

3. Results and discussion

To find the actual strain that can be produced by high-pressure torsion in steel 316, two halves of a disk (figure 2, a) were subjected to simultaneous HPT up to the rotation angle of 90° (n = 1/4) according to the principle first presented in [4] (figure 2, b). The samples obtained by this way are shown in figure 1, c. One can see that there was no shear of the lower surface of the half with respect to the upper surface, i.e. the HPT deformation did not occur due to the effect of slippage. Apparently, this is due to the fact that the steel sample was hardened already at the initial stages of deformation. i.e., at the beginning of HPT (up to H_v = 400 after n=1/4). Thus, the experiment on the HPT with n=1/4 for steel 316 halves demonstrates that torsional deformation may be not implemented already at the initial stages of HPT processing by selected modes (figure 1, c). On the other hand, a huge amount of studies indicates a strong refinement of the structure of various materials during HPT [1,2] despite the possible slippage during processing. For example, in steel 316, after HPT n = 10, the grain is refined down to 40 nm [7,8]. Hence, the reasons leading to the structure refinement during HPT, in spite of slippage, require additional studies and analyses.

For comparison, a soft metal, copper, was subjected to HPT for n= 1/4 under a similar scheme. The observed rotation angle of the sample halves corresponds to the rotation angle of the anvils (figure 1, d), i.e. the strain introduced into the sample is consistent with the known formula.

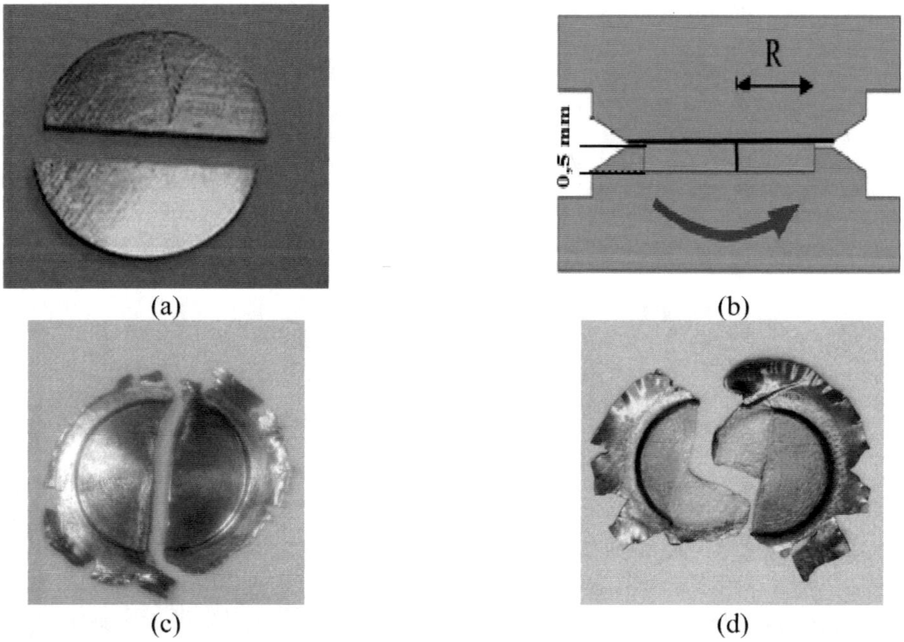

Figure 2. Scheme and view of the samples, subjected to joint HPT: a – two disk halves, b – the principle of joint HPT [4], c – two halves of steel 316 after joint HPT for n=¼, d – two halves of Cu after joint HPT for n=¼.

Therefore, the results presented above demonstrate that the strain actually produced during regular HPT processing for many materials is smaller than the expected one, and the use of accumulative HPT is required to produce really large strains.

An analysis of the XRD data shows that the BCC β-phase is the main phase component in the initial Ti-18Zr-15Nb alloy, both after HPT n=10 and ACC HPT n=10 processing. The X-ray line width of the β-phase increases after HPT n=10. But the largest increase in the X-ray line width is observed after the ACC HPT which produces the largest strain. The increase in the X-ray line indicates an increase in the dislocation density or/and grain refinement. Thus, β-phase in the Ti-18Zr-15Nb alloy turned out to be stable even to very large real deformations.

Table 1. X-ray line width $B_{110\beta}$ and microhardness of the Ti-18Zr-15Nb alloy after the specified HPT processing.

	Initial	HPT, n = 10	ACC HPT, n = 10
X-ray line width $B_{110\beta}$**, deg.	0.31	0.98	1.12
Microhardness, HV (± 8), area ½ R	265	345	330

As a result of HPT processing, the microhardness of Ti-18Zr-15Nb alloy grows noticeably (table 1). After HPT processing for n = 5, HV increases both in the center and at the edge of the disc samples. The increment in the microhardness after HPT processing indicates an increase in strength and yield stress. As the number of revolutions increases above n > 5, HV does not increase further after HPT n = 10 and even after accumulative HPT processing n = 10. This result is unusual and calls for additional investigations.

Analysis of the XRD data shows that the γ phase is the main phase component in the initial steel 316 (figure 3, a). The steel also contained a small amount of the α-phase, although it should be noted that in most other works original steel 316 does not contain the α-phase after correct heat treatment. HPT led to an increase in the α-phase or α'-martensite peaks (figure 3, b). Earlier, the appearance of about 20% α'-martensite in steel 316 after HPT n = 10 was noted in [7]. As a result of ACC HPT processing, phase transformations have become much stronger. The α'-martensite peak after ACC HPT processing became more intense than the γ-phase peak. This indicates that most of the γ-phase has turned into α'-martensite as a result of the very high strain achieved in the ACC HPT.

In addition, after HPT up to n = 10, a very weak line appeared at the left foot of the γ(111) line at 41-41.5°. After ACC HPT with n = 10, this line increased noticeably, and on the right foot of the α'(110) line at 46.5-47° another weak line grew. These peaks are related to the ε-martensite with a hexagonal close packed lattice. The epsilon phase is the "high pressure phase" in the iron, often appearing during severe plastic deformation of the material [11,12].

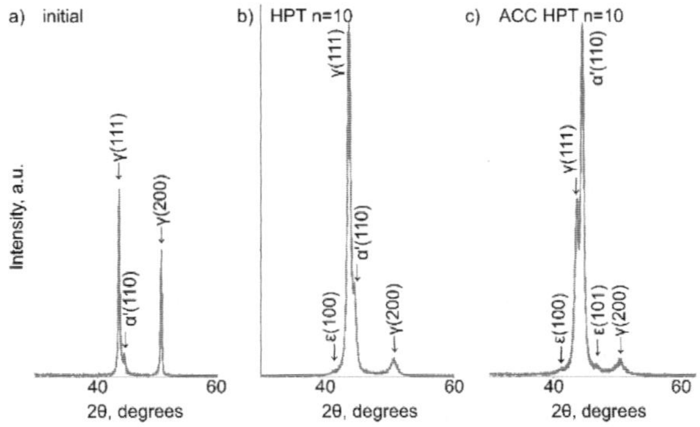

Figure 3. X-ray diffraction profiles of the steel 316 samples: a) initial; b) subjected to HPT n=10; c) subjected to ACC HPT n=10.

As a result of HPT, the H_v of steel 316 also grows significantly (table 2). H_v as a result of ACC HPT increased slightly more than after regular HPT, and an increase in H_v is observed both in the center and at the periphery of the HPT sample-disk. A greater increase in H_v indicates a more efficient refining of the structure of 316 steel after ACC HPT than after regular HPT.

Table 2. Microhardness of 316 steel in different parts of the sample in the initial state and after treatment with the HPT.

Condition	centre	½ R	edge
Initial	170		150
HPT n=10	430	450	465
ACC HPT n=10	475	490	525

4. Conclusions

The studies demonstrate that during HPT of samples of some materials (in particular, austenitic steel) with increasing number of revolutions the strain is not imparted into the sample due to the effect of slippage. The new technique of accumulative high-pressure torsion (ACC HPT) is efficient for producing high strains in metallic materials. The studies show that HPT and accumulative HPT do not lead to a change in the phase composition of Ti-18Zr-15Nb alloy, β-phase remains the main phase even after deformation. In steel, however, HPT n = 10 led to an increase in the content of the α-phase and α'-martensite, but apparently by no more than 20%. At the same time, as a result of ACC HPT, a significant part of the γ-phase in steel 316 has turned into α'-martensite and high-pressure ε-martensite. Hence, the ACC HPT can more efficiently transform the structure of alloys than regular HPT.

Acknowledgments

This work was supported by the Russian Foundation for Basic Research under grant 19-58-80018 and the Russian Science Foundation under project 20-69-47029.

References

[1] Valiev R Z, Zhilyaev A P and Langdon T G 2013 *Bulk Nanostructured Materials* (Hoboken, NJ: John Wiley & Sons)
[2] Zhilyaev A and Langdon T 2008 *Prog. Mater. Sci.* **53** 893–979
[3] Kovács Z, Schafler E, Szommer P and Révész Á 2014 *J. Alloys Compd.* **593** 207–12
[4] Gunderov D V, Churakova A A, Boltynjuk E V, Ubyivovk E V, Astanin V V, Asfandiyarov R N, Valiev R Z, Xioang W and Wang J T 2019 *J. Alloys Compd.* **800** 58–63
[5] Gunderov D V, Churakova A A, Astanin V V, Asfandiyarov R N, Hahn H and Valiev R Z 2020 *Mater. Lett.* **261** 127000
[6] Glezer A M, Sundeev R V. and Shalimova A V. 2011 *Dokl. Phys.* **56** 476–8
[7] Etienne A, Radiguet B, Genevois C, Le Breton J-M, Valiev R and Pareige P 2010 *Mater. Sci. Eng.* A **527** 5805–10
[8] Dobatkin S V, Rybalchenko O V, Enikeev N A, Tokar A A and Abramova M M 2016 *Mater. Lett.* **166** 276–9
[9] Prokoshkin S, Brailovski V, Dubinskiy S, Zhukova Y, Sheremetyev V, Konopatsky A and Inaekyan K 2016 *Shape Mem. Superelasticity* **2** 130–44
[10] Kudryashova A, Sheremetyev V, Lukashevich K, Cheverikin V, Inaekyan K, Galkin S, Prokoshkin S and Brailovski V 2020 *J. Alloys Compd.* **843** 156066
[11] Tavares S S M, Gunderov D, Stolyarov V and Neto J M 2003 *Mater. Sci. Eng.* A **358** 32–6
[12] Huang C X, Yang G, Gao Y L, Wu S D and Li S X 2007 *J. Mater. Res.* **22** 724–9

IOP Conf. Series: Materials Science and Engineering 1008 (2020) 012014 doi:10.1088/1757-899X/1008/1/012014

Computer simulation of friction stir processing of plate from 2024 aluminum alloy

A Kh Akhunova, M F Imayev and A Kh Valeeva

Institute for Metals Superplasticity Problems, Russian Academy of Sciences, 39 Khalturin st., Ufa 450001, Russia

E-mail: akhunova.a.a@gmail.com

Abstract. In the present work, three-dimensional finite element modeling of the friction stir processing of a 3-mm-thick 2024 aluminum alloy plates using DEFORM-3D software product has been carried out. The optimal pin length was determined based on the calculation of the distribution of effective deformation, temperature, and displacement of material points. Numerical modeling was done on pins in the form of a cylinder with a diameter of 2 mm and a length of l = 1.6, 1.8, 2.0, and 2.2 mm. Friction stir processing was carried out at a traversing rate of 2 mm/s, a tool rotation rate of 1000 rpm, and an axial force of 20 kN. The material behavior of the sheet was described using the Johnson-Cook model included in the standard DEFORM-3D library. The results showed that the most preferable under the considered processing conditions is a pin with a length of 2.0 mm because it allows one to obtain the widest and most symmetrical regions of intense heating and deformation.

1. Introduction

Aluminum alloys are widely used in many industries - automotive, shipbuilding, aerospace, etc. However, these alloys are poorly exposed to traditional welding methods due to hydrogen dissolution and oxidation, so structural parts are often assembled using permanent riveted joints. This method has its drawbacks such as waviness of the material, radial cracks around the holes, additional weight, etc. One of the methods of joining structural parts, which allows avoiding the listed disadvantages, is friction stir welding (FSW) [1].

According to this technology, a rotating tool consisting of a shoulder and a pin moves along the contact surface of two butted and rigidly pressed workpieces. The shoulder is in a firm contact with the surface of the workpieces and the pin is immersed into the material. The rotation and movement of the tool along the contact surfaces increase the temperature and softens the metal in the adjacent areas, which facilitates mixing of the material.

In the course of such welding, the material undergoes very large and high strain rate deformation, in which the true strain can reach the values of about 50, and the strain rate about 100 s^{-1}. A side effect of the welding process is the formation of a fine-grained microstructure in the weld zone. Thus, this method can also be used to obtain metal semifinished products with a modified microstructure. This technology is more commonly referred to as friction stir processing (FSP).

In the FSP process, frictional mixing of the metal being processed in the solid-phase state is done by a rotating tool, which is similar to the tool in the FSW. The difference between the FSW and FSP is that in the latter case the parts are not joined but a monolithic workpiece is processed.

The geometry of the tool pin, in particular its length, plays an important role in both the FSW and FSP, since affects the heat generation, material flow, forces, and other process output characteristics [2]. To select the effective parameters of the FSW/FSP process, it is advisable to use computer simulation, which can significantly reduce the cost of manufacturing an expensive special tool [3-5].

In this work, a three-dimensional finite element modelling of the friction stir processing of a sheet of 2024 aluminum alloy using DEFORM-3D software product has been carried out to select the optimal pin length based on calculating the distribution of effective deformation, temperature, and displacement of material points.

2. Computer simulation

Finite element modeling was carried out in order to select the optimal tool pin length in friction stir processing using DEFORM-3D software. To reduce the computation time and avoid the instability of the solution, the 2024 aluminum alloy plates were modeled as a single body 3 mm thick, 40, and 30 mm long and wide, respectively (figure 1).

Friction stir processing was carried out at a traversing rate of 0.5 mm/s, a tool rotation rate of 1000 rpm, and an axial force of 2 kN.

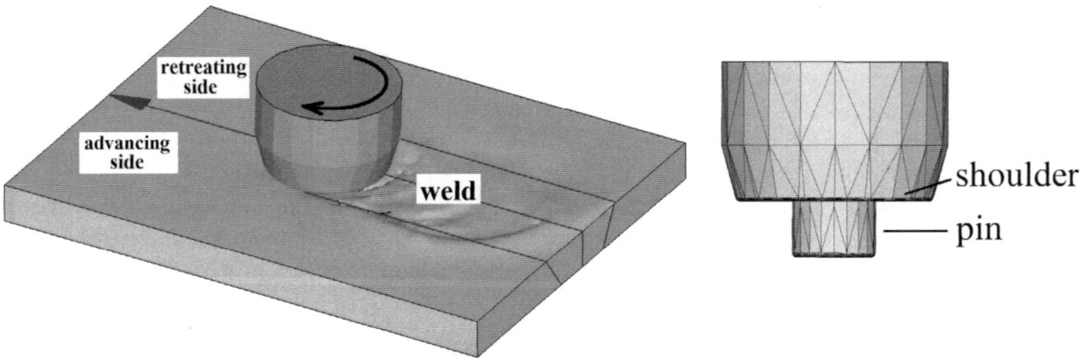

Figure 1. Schematic of the friction stir processing.

Pins in the form of a cylinder with a diameter of 2 mm and lengths of l = 1.6, 1.8, 2.0, and 2.2 mm were considered. A general view of a pin is shown in figure 2.

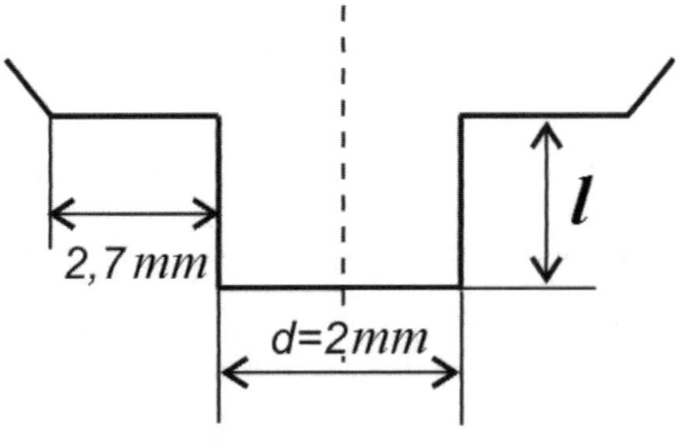

Figure 2. Pin profile.

The behavior of the workpiece material was described using the Johnson-Cook model included in the standard DEFORM-3D library. The following physical constants of the alloy were assumed: density - 2640 kg/m^3; thermal conductivity - 122 W/(m×K); heat capacity at constant pressure - 0.922 kJ/(kg×K). During friction stir processing of the workpiece, the condition of heat exchange with the environment was set. The initial temperature of the workpiece, tool, and environment was assigned the value of 20 °C. Convection with a heat transfer coefficient of 11 W/(m^2×K) was set on all external surfaces. The tool was taken as a perfectly rigid body since the yield stress of its material is much higher than that of the workpiece. AISI-D2 tool steel was chosen as the material for the tool. The friction coefficient was taken equal to 0.5.

3. Results and discussion

The FSP process is characterized by very high temperature and strain gradients across the thickness of the workpiece, which leads to an inhomogeneity of the microstructure formed in the treatment zone. One way to avoid this kind of inhomogeneity is to use the optimal pin geometry.

Figure 3 shows typical diagrams of the distributions of temperature, effective strain and displacement of material points in the cross-section of the workpiece under processing for different values of the length of the tip. The cross-section runs along a plane perpendicular to the seam through the center of rotation of the tool.

Based on the simulation results, it can be concluded that in the FSP process, nonuniform temperature distribution is formed in the treatment area in the transverse direction of the workpiece (figure 3 a), while the highest temperature value is observed on the advancing side. It should be noted that as the length of the pin increases, the maximum heating temperature of the workpiece also increases. At l = 1.8 mm, the temperature at the point of contact with the shoulder reaches about 300°C, and the area of its distribution is small. At l = 2 mm an extensive and fairly uniform region is formed with a temperature in the range of 383-455 °C. At l = 2.2 mm, the area heated to T = 383-455 °C decreases and shifts towards the advancing side closer to the joint root.

The results of study show that the workpiece area being worked out becomes more extended as the pin length increases, while the area of high values of effective strain (43.8-50 mm/mm) increases as the value of l increases and shifts to the root of the joint (figure 3 b). The widest and most symmetrical area of effective strain at the level of 43.8-50 mm/mm is observed at l = 2 mm.

According to [6-7], the highest quality treatment of the workpiece material occurs at temperatures of about 0.8 T_m, since heating to such temperatures provides high plasticity and good mixing of the material. For 2024 aluminum alloy the value of 0.8 T_m depends on variations in composition and is 385-425 °C.

Analysis of the simulation results allows us to conclude that the material in the lower region of the workpiece is practically not subject to mixing, and the shorter the pin length, the larger is this area. The areas in the middle and upper parts of the workpiece are mainly mixed, the length of the areas being worked increases with the length of the pin. From the diagrams of the distribution of the displacement of material points (figure 3 c), it follows that the workpiece material from the advancing side is less involved in mixing, remaining mainly in its area, while the material from the retreating side actively moves to the advancing side and mixes with the material from this area. This pattern is observed for all pin lengths.

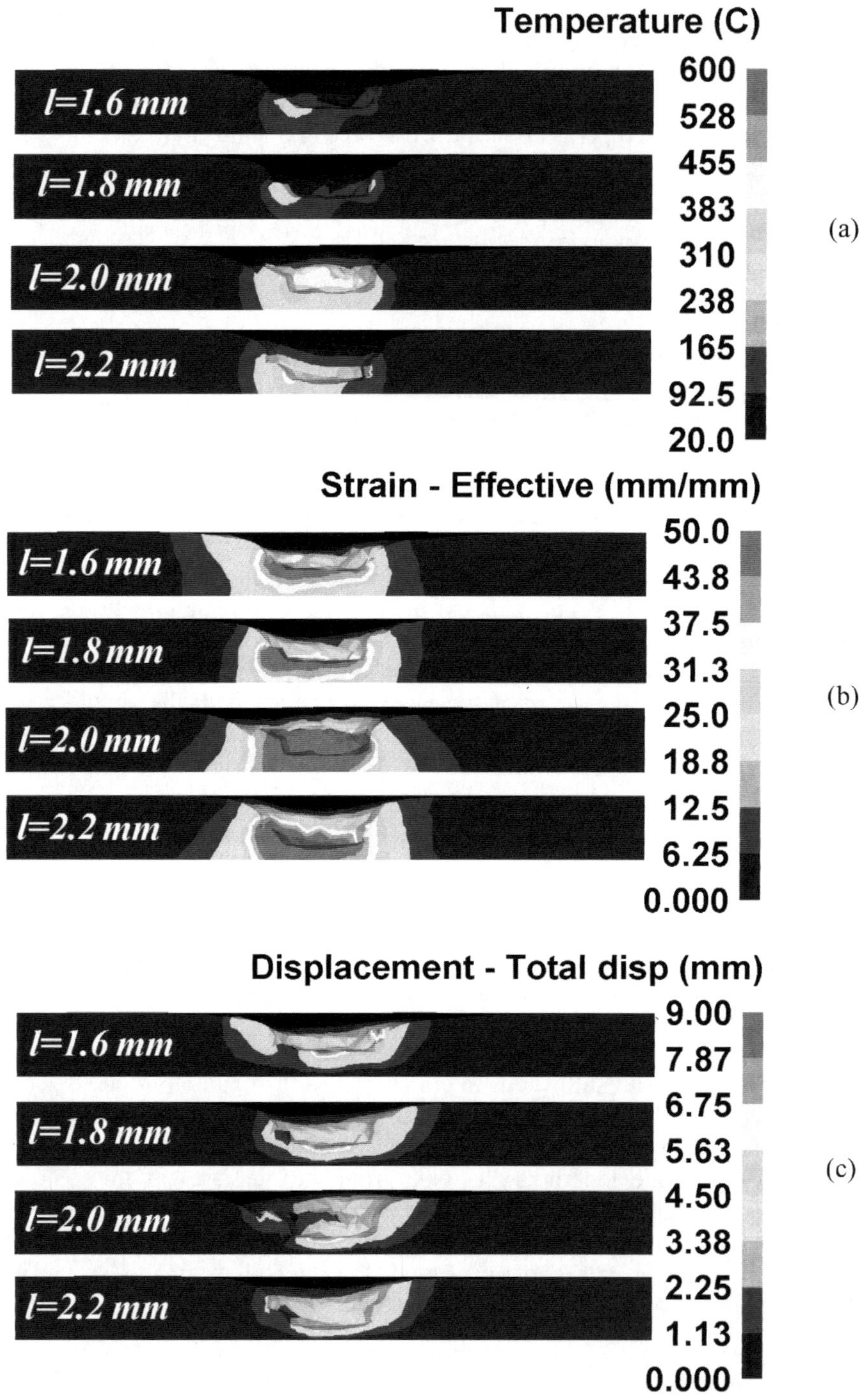

Figure 3. Distribution of temperature (a), effective strain (b), and displacement of material points (c) depending on the length of the pin in the cross-section of the workpiece during friction stir processing. It was found that a 2.0-mm-length pin is preferable for friction stir processing under these conditions from all the considered pin lengths.

Conclusions

Thus, by means of computer simulation, it has been established that of all the considered pins, a pin with a length of 2.0 mm is preferable for FSP / FSW under these conditions. When processing with such a pin, wide and symmetrical areas are formed with a temperature of about 0.8 T_m favorable for the formation of a high-quality weld and the largest homogeneous area of high effective strain values.

Acknowledgments

The present work was accomplished according to the state assignment of IMSP RAS.

References

[1] Grujicic M, Arakere G, Pandurangan B, Hariharan A, Yen C-F and Cheeseman B 2011 *J. Mater. Eng. Perform.* **20** 11–23
[2] Ramachandran K K, Murugan N and Kumar S S 2015 *Mater. Sci. Eng.* A **639** 219–33
[3] He X, Gu F and A Ball 2014 *Progr. Mater. Sci.* **65** 1–66
[4] Jain R, Pal S K and Singh S B 2018 *Int. J. Adv. Manuf. Technol.* **94** 1781–97
[5] Akhunova A K, Imayev M F and Valeeva A K 2019 *Lett. Mater.* **9(4)** 456-9 (in Russian)
[6] Mubiayi M P, Akinlabi E T and Makhatha M E 2019 *Current Trends in Friction Stir Welding (FSW) and Friction Stir Spot Welding (FSSW) An Overview and Case Studi* (Switzerland: Springer International Publishing) p 206
[7] Mishra R S, De P. S and Kumar N 2014 *Friction Stir Welding and Processing: Science and Engineering* (Switzerland: Springer International Publishing) p 338

Finite-element modeling of the stress-strain state in disk-shaft type parts from dissimilar Ni-based alloys

A Kh Akhunova, V A Valitov and E V Galieva

Institute for Metals Superplasticity Problems, Russian Academy of Sciences, 39 Khalturin st., Ufa 450001, Russia

E-mail: akhunova.a.a@gmail.com

Abstract. Finite element modeling of the stress-strain state of disk-shaft type parts of gas turbine engines during pressure welding has been made. Pressure welding was carried out by three schemes of deformation of a disk and/or a shaft. In Scheme 1, the shaft was inserted into the disk by setting the displacement of the upper die, in Scheme 2 simultaneous action of a die on the upper end surface of the disk and shaft was set, and in Scheme 3 the shaft was inserted into the disk by setting the displacement of the die, then the action of the die on the upper end surface of the disk was set. Computer simulation is performed using the DEFORM-2D software package. The deformable nickel base alloys EP975 and EK79 were taken as materials for a disk and a shaft, respectively. The results of computer simulation have shown that for increasing quality of the welded joint between the disk and the shaft it is preferable to use pressure welding which is carried out with a combination of sequence actions on the shaft and disk, since in this case an effective summarized compressive stresses and radial strain were achieved which leads to an increase of the joint quality.

1. Introduction

Small gas turbine engines (GTEs) are widely used not only in aviation, but also in the field of small power engineering [1]. The rotor of a small GTE is a specially profiled disc, on the rim of which the blades are attached with the lock. Torque is transmitted to the disk from the shaft due to their connection to each other. Because of their relatively small dimensions, the disk and the rotor shaft of a small GTE can be manufactured as a whole by hot deformation methods followed by turning processing.

However, the significant labor intensity of this technology makes it preferable to manufacture such parts separately with their subsequent connection. The most promising method of joining individual elements of such structures is pressure welding under conditions of superplasticity [2-3]. One of the main advantages of pressure welding is the ability to preserve the microstructure of the workpieces being welded, which is important for products operated at high temperatures and made, respectively, from heat-resistant nickel alloys [4]. An improvement in the quality of a solid-phase joint can be favored by a scheme of deformation, in which a shear strain components takes place in the plane of the joint [3-5].

In this paper, methods for producing a welded joint by pressure welding carried out by the action of a deforming tool on the disk and/or the shaft in contact with the disk are considered. Deformable nickel alloys EP975 and EK79, respectively, were chosen as the material for the disk and shaft.

IOP Conf. Series: Materials Science and Engineering 1008 (2020) 012015 doi:10.1088/1757-899X/1008/1/012015

2. Computer simulation

Computer simulation is performed using the DEFORM-2D software package. The shaft is a cylinder with a diameter of 12 mm and a height of 12 mm coupled with a truncated cone with a height of 20 mm having the angle 3°. The hole in the disk narrows with the same cone angle as that of the shaft. The total height of the disk is 14 mm and its diameter 28 mm. The disk and shaft are assumed to be made of ultrafine-grained heat-resistant nickel base alloys EP975 and EK79, respectively. The material properties were determined using experimental stress-strain curves obtained under uniaxial compression of the alloys at the welding temperature, which was equal to 1100°C. The value of 0.5 was assumed for the friction coefficient.

To determine the influence of the pressure welding scheme of the samples on the plastic deformation in the zone of their connection, three schemes were considered, set by means of the upper die. In Scheme 1, the shaft was inserted into the disk by setting the displacement of the upper die by 2 mm with a movement speed of 1.9 mm/min, which corresponds to the initial strain rate of the cylindrical part of the shaft 10^{-3} s^{-1}; in Scheme 2, simultaneous action of a die on the upper end surfaces of the disk and shaft was set. The die displacement was 2 mm with a movement speed of 1.9 mm/min. Finally, in Scheme 3 the shaft was inserted into the disk by setting the displacement of the die by 2 mm with a movement speed of 1.9 mm/min. Then, the action of the die on the upper end surface of the disk was set by setting the offset of the die by 2 mm with a movement speed of 1.9 mm/min.

Figure 1 shows the starting positions of the disk and the shaft for corresponding pressure welding schemes. Deforming tools (die) were assumed to be absolutely rigid, deformable bodies (disk and shaft) had elastic-plastic properties.

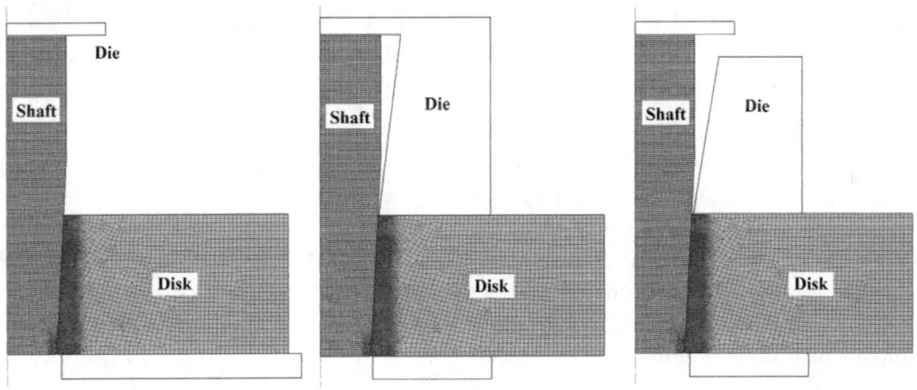

(a) (b) (c)

Figure 1. The finite element model of disk-shaft for welding by Scheme 1 (a), Scheme 2 (b) and Scheme 3 (c).

3. Results and discussion

The results of simulation have shown that during the welding following Scheme 1, a defect typical for the processes of expansion appears, namely, a small radial groove between the shaft and he disk at the upper surface of the latter is formed. This is completely contrary to the requirements put to the pressure welding. It is possible to prevent the formation of this groove by clamping the disk between two technological plates, or, alternatively, pressing the disk between dies as during welding by Schemes 2 and 3 and simultaneously deforming the disk. The material pressed out into the gap between the dies and the shaft produces pressure on the shaft surface. A compression force arises, which is directed counter to the force applied during the expansion of the disk, and the welding zone is influenced by compression, which is necessary to create physical contact and activate the surfaces to be welded. In addition, when the disk material is pressed out into the gap between the dies and the shaft, all pores that can form during the creation of physical contact at the point of junction from the shaft diameter to the disk diameter are healed.

The degree of contact of the inner wall of the disk in the shaft can be estimated from the radial components of strain and stress in the contact zone of the billets.

From the distribution of the radial components of strain ε_R (figure 2), it can be concluded that when welding according to Scheme 1, strain of only one sign "-" are observed in the area of contact between the disk and the shaft; radial strain in the workpieces is unidirectional, while they are very small (figure 2 a).

When welding according to Scheme 2 (figure 2 b) and Scheme 3 (figure 2 c), the picture changes. Strains of different signs are observed on the contact surfaces of the shaft and the disk, which means that the flow of the material of the shaft and the disk goes towards each other. When the die is pressed into the disk under the action of compression, the disk expands in width, exerting significant pressure on the shaft, which promotes the formation of physical contact between the disk and the shaft and improves the quality of the joint. The general nature of the distribution of strain is the same for both schemes. In this case, compared with Scheme 1, the $\varepsilon_R$values in the near-contact areas of the shaft and disk increase by one and two orders of magnitude, respectively. It can also be noted that with pressure welding according to Scheme 3, the radial strains in the shaft are distributed more evenly.

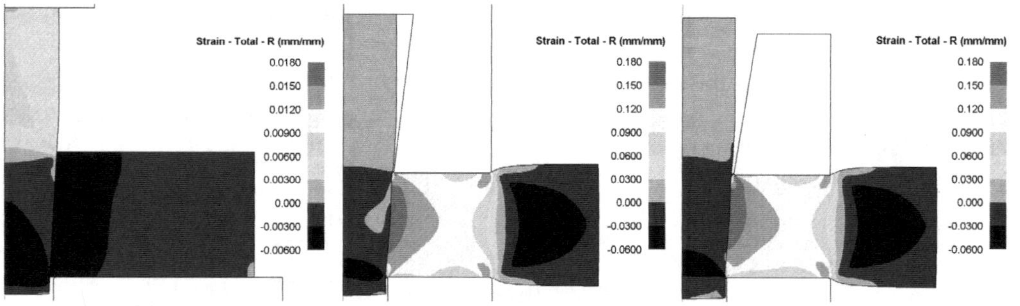

(a) (b) (c)

Figure 2. The distribution of radial strain ε_R during pressure welding by Scheme 1 (a), Scheme 2 (b) and Scheme 3 (c).

The distribution of the radial stress components σ_R (figure 3) showed that in all cases of welding, compressive stresses are observed on the contact surfaces of the shaft and disk. At the same time, when welding according to Schemes 2 and 3, the values of σ_R in the joint zone are 3 times greater than when welding according to Scheme 1.

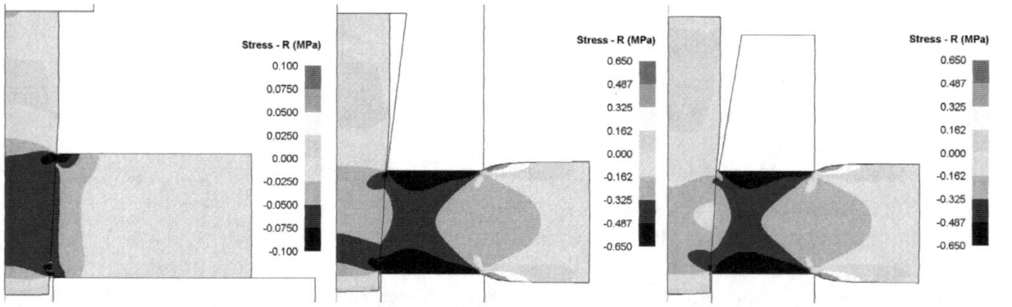

(a) (b) (c)

Figure 3. The distribution of radial stress σ_R during pressure welding by Scheme 1 (a), Scheme 2 (b) and Scheme 3 (c).

From the distribution of the circumferential components of strain ε_θ, it can be concluded that when welding according to Scheme 1 (figure 4 a), strain with the "+" sign are observed on the contact surface of the disk, which arise due to the fact that the disk material is stretched around the shaft due to an increase in the diameter of the disk hole during distribution. When welding according to Schemes 2 (figure 4 b) and 3 (figure 4 c) due to the fact that the disk is compressed between the dies, the material flow is directed to the surface of the shaft and compression strains are formed in the contact area. It can also be noted that with pressure welding according to Scheme 3, circumferential deformations in the shaft are distributed more evenly than during pressure welding according to Scheme 2.

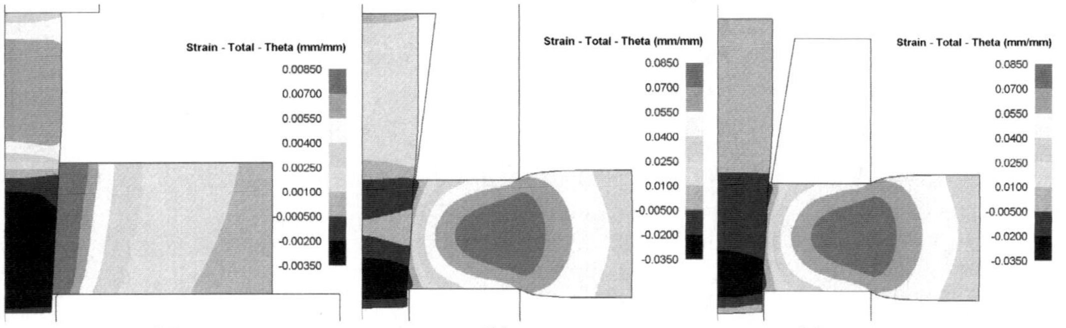

Figure 4. The distribution of circumferential strain ε_θ during pressure welding by Scheme 1 (a), Scheme 2 (b) and Scheme 3 (c).

From the distribution of circumferential stress components σ_θ it can be seen that when welding according to Scheme 1 (figure 5 a), tensile stresses are localized in the near-contact region of the disk, while during welding according to Schemes 2 (figure 5 b) and 3 (figure 5 c), this area is dominated by compressive stresses. In this case, the maximum value of σ_θ when welding according to Scheme 2 is 2 times greater than when welding according to Scheme 3.

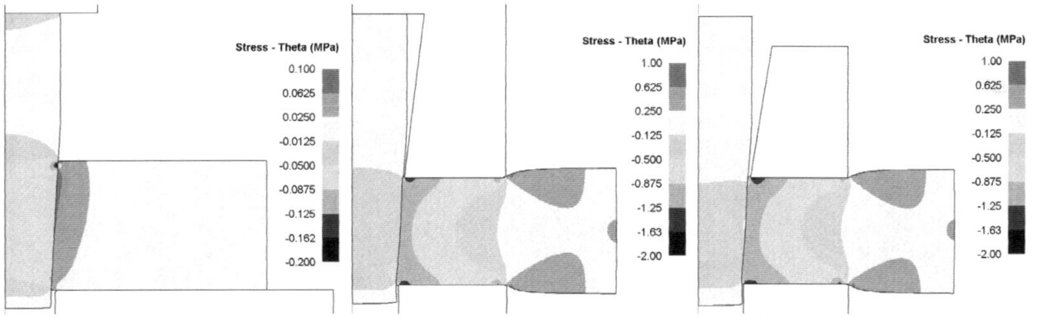

Figure 5. The distribution of circumferential stress σ_θ during pressure welding by Scheme 1 (a), Scheme 2 (b) and Scheme 3 (c).

4. Conclusions

Thus, the results of the studies made have shown the following:

Pressure welding, carried out with a combination of actions on the shaft and the disk, changes the nature of the distribution of circumferential and radial stresses in the contact zone compared to pressure welding carried out only by the introduction of the shaft into the disk.

To create a permanent connection between the shaft and the disk, it is preferable to use pressure welding, carried out using a combination of actions on the shaft and the disk, since in this case there is an increase in radial compressive stresses and the appearance of circumferential compressive stresses on the contact surface of the disk. Under the action of compressive stress, the disk material exerts a significant pressure on the surface of the shaft, which promotes the formation of a physical contact between the disk and the shaft, and, as a result, improves the quality of the joint.

When welding using a combination of actions on the shaft and disk, it is necessary to use a sequence of operations, since in this case a more uniform distribution of the radial and circumferential strain components in the shaft is achieved.

Acknowledgments
The computer modeling was supported by the Russian Science Foundation (Grant No. 18-19-00685). The part devoted to mechanical properties of EK79 and EP975 was supported by the IMSP RAS State assignment No. AAAA-A17-117041310215-4.

References

[1] Logunov A V, Shmotin Yu N 2013 *Modern Heat-Resistant Nickel-Base Alloys for Gas Turbine Disks* (Moscow: Nauka v Tekhnologii) p 256 (in Russian)
[2] Valitov V A, Akhunova A Kh, Galieva E V, Dmitriev S V, Lutfullin R Ya and Zhigalova M Yu 2017 *Lett. Mater.* **7(2)** 180-5 (in Russian)
[3] Akhunova A Kh, Valitova E V, Dmitriev S V, Valitov V A and Lutfullin R Ya *2016 Welding Int.* **30(6)** 488-91
[4] Galieva E V, Lutfullin R Ya, Akhunova A Kh, Valitov V A and Dmitriev S V 2018 *Sci. Technol. Weld. Join.* **23(7)** 612-8
[5] Akhunova A Kh, Galieva E V, Drozdov A A, Arginbaeva E G, Dmitriev S V and Lutfullin R Ya 2016 *Lett. Mater.* **6(3)** 211-6 (in Russian)

Protective PEO-coatings on titanium shape memory alloy for medical implants

S R Shechtman[1], R G Farrakhov[1], I A Ramazanov[1], V A Sheremetyev[2] and E V Parfenov[1]

[1]Ufa State Aviation Technical University, 12 Karl Marx Street, Ufa, 450008, Russia
[2]National University of Science and Technology "MISiS", 4 Leninsky pr., Moscow, 119049, Russia

E-mail: shex@inbox.ru

Abstract. The use of titanium shape memory alloy Ti-18Zr-15Nb with promising coatings for the manufacture of medical implants is considered. In this paper, we examined plasma electrolytic oxidation (PEO) in a pulsed bipolar mode with different frequencies to protect the surface from the aggressive environment of the human body.. Two frequencies of 300 and 1000 Hz were used. The results of SEM, EDS, XRD analysis, and electrochemical tests of the coatings were discussed. As a result of comparison, it was shown that carrying out the PEO process at 1000 Hz provides the formation of a more uniform coating with higher quality than the coating obtained at 300 Hz.

1. Introduction

The development of the processes of transplantation of biomaterials into the human body has revealed the problem of the lack of technologies for the creation of bone implants that are biologically compatible with bone tissue. The practice of using titanium implants in surgical traumatology for bone reconstruction in bone tumors, revision arthroplasty does not always guarantee the absence of inflammation or necrosis of the tissues surrounding the material and poses the problem of developing a technology for creating bone implants with a high level of biocompatibility [1].

An example of a new structural biomaterial for permanent bone implants is a nickel-free shape memory alloy (SMA) based on the Ti-Zr-Nb system, containing only biocompatible (non-toxic) components and having a low Young's modulus close to that of bone tissue (up to 30 GPa). Due to the unique combination of biochemical and biomechanical compatibility, this alloy is the most promising among metallic biomaterials for permanent bone implants [2]. To have a smooth transition of the physicochemical properties of the implant material to the human bone, the surface of the device must be modified with a protective functional coating. Currently, one of the promising technologies to improve the corrosion, wear resistance and biocompatibility of permanent bone implants made of titanium and titanium alloys is plasma electrolytic oxidation (PEO) [3, 4].

The PEO of β-titanium alloys Ti-Zr-Nb and Ti-Zr-Nb-Ta was studied using direct current (DC) and pulsed DC regimes, and the improvement in the biocompatibility was shown [5-7]. However, compared to these regimes, pulsed bipolar PEO appears to be an advanced modification of the plasma electrolytic oxidation which previously showed promising results in the development of biocompatible coatings on Ti [8-10]. Therefore, the aim of this study is an investigation of the PEO coatings produced at different frequencies in pulsed bipolar regime on a novel Ti-18Zr-15Nb shape memory alloy.

2. Experimental

The studies were carried out on samples in the shape of disks with a diameter of 9 mm cut from the β-titanium alloy Ti-18Zr-15Nb developed at the National Research Technological University MISiS. Plasma electrolytic oxidation was carried out in a pulsed bipolar mode at frequencies of 300 and 1000 Hz. An aqueous solution of 20 g / L $Na_3PO_4 \cdot 12H_2O$ was used as the electrolyte; its temperature was kept constant at 20 ± 1 ° C; the electrolyte volume was 5 L. Other parameters of the PEO process are shown in Table 1.

Table 1. Parameters of plasma electrolytic oxidation of Ti-18Zr-15Nb alloy.

Sample code	Frequency (Hz)	Voltage pulse amplitude (V)	Positive pulse duty cycle (%)	Negative pulse duty cycle (%)	Time (min)
TZN_300	300	470	51	26	5
TZN_1000	1000	430	26	26	2.5

The top view of PEO coating was studied using the JEOL JSM-6490LV scanning electron microscope with INCAX attachment for the EDS analysis. The distribution of the points for measuring the elemental composition was chosen in such a way as to estimate the characteristic areas on the surface of the sample and to be able to construct a map of the distribution of various elements in the studied PEO coatings. The phase composition of the coating was investigated using X-ray diffractometer (XRD) Ultima IV (Rigaku, Japan) equipped with a Cu-Kα X-ray source. The scanning was performed from $2\theta = 20°$ to $2\theta = 80°$ with the step of 0.02° and the measurement time at each step amounted 1 s. The phase composition of the surface was studied for both the uncoated alloy and all coatings obtained. Phase identification and semiquantitative analysis were performed using Philips X'Pert HighScore Plus software with the PDF2 databank. The coating porosity was calculated from the SEM images using ImageJ software. The surface roughness Ra was measured with a TR 220 profilometer. Coating thickness h was measured with a Defelsko Positector 6000 eddy current probe with an accuracy of +/– 0.1 μm. Electrochemical studies were carried out using an Elins P-5X potentiostat-galvanostat-impedance meter (Russia) in Ringer's solution in a three-electrode cell with a volume of 100 ml with a silver chloride reference electrode and a graphite electrode. The corrosion potential and current density as well as the polarization resistance were evaluated.

3. Results and discussion

The resulting PEO coatings for two processing modes at 300 and 1000 Hz appear as dense gray films. At the same time, the coatings obtained in the 300 Hz mode have a lighter white-gray tint, while when formed at a frequency of 1000 Hz, a uniform gray coating is observed.

SEM images presented in figure 1 show the top views of the PEO coatings on Ti-Zr-Nb alloy for the two processing modes at 300 and 1000 Hz. An analysis of the images obtained at the magnification of 500 shows that the protective layers formed at 1000 Hz (figure 1, b) have a more uniform distribution of pores compared to that at 300 Hz (figure 1, a). The surface layer at 300 Hz has predominantly round pores, while at 1000 Hz elongated pores and grooves appear. The pore sizes vary from 0.5 to 10 μm, while the grooves have the widths of 3-5 μm and lengths up to 30 μm. The oxide films obtained have the thickness of 21.2 ± 1.2 μm for the frequency of 300 Hz and 17.3 ± 5.3 μm for 1000 Hz.

Also, as a result of the PEO process mechanism, rough surfaces with a significant number of open pores were formed, both in the case of TZN_300 and TZN_1000. Application of the frequency of 1000 Hz allows obtaining a more uniform coating without surface defects, which appear at 300 Hz in the form of areas with very small and very large pores. The porosity of the coatings obtained is 8.0 - 10.7% at 300 Hz and 12.5 -15.5% at 1000 Hz. The results of the study of the surface roughness showed that the coating obtained at 1000 Hz had a lower roughness in comparison with the PEO coating processed at a frequency of 300 Hz (figure 1).

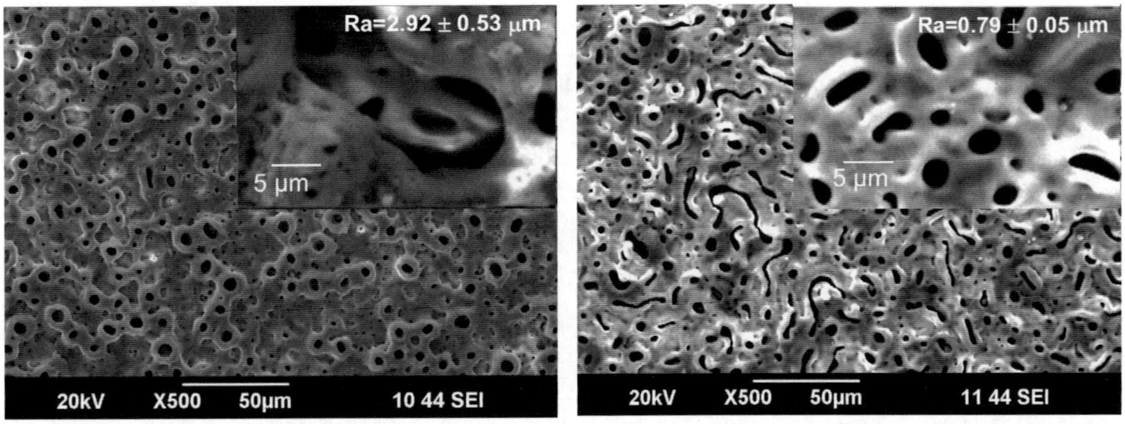

Figure 1. SEM images of the PEO coatings at different frequencies: 300 Hz (a) and 1000 Hz (b).

A detailed analysis of micrographs of the surface of PEO coatings shows that there is a change in the surface morphology. Also, a partial sealing of the pores is observed at 300 Hz. The presence of pores on the surface of samples with PEO-coatings promotes better adhesion of osteogenic cells and adsorption of proteins. Moreover, this also increases the contact area during the bone integration and decreases the level of mechanical stress [11-12].

As follows from the EDS analysis (table 2), the PEO coating includes not only the constituent elements of the substrate but also electrolyte component elements, in this case, phosphates which participate in the formation of the coating and promote the coating growth.

The qualitative composition of the investigated coatings containing Ti, Zr, Nb, O, P is very similar. Analysis of the EDS mapping shows that the coating contains oxides of Ti, Zr, and Nb. Titanium is evenly distributed over the surface, while Zr, Nb, and O appear less in the pores of the coating. For the TZN_300 sample, the surface defects contain a significant amount of P, while for the TZN_1000, phosphorus was not detected in the phase composition. This may indicate that longer PEO pulses favor the deposition of the electrolyte component into the coating.

Figure 2 shows X-ray diffraction patterns of samples made of Ti-Zr-Nb alloy in the initial state and with the PEO coatings obtained at different frequencies. A comparison of the samples with the coatings and in the initial state shows that the three peaks coincide and correspond to the β-phase of the titanium alloy. Analyzing the PEO coatings obtained at different frequencies, one can assume the presence of TiO_2 (anatase and rutile) in the surface layer. Also, other oxides based on Nb (Nb_2O_5, NbO_2) and Zr (ZrO_2) appear in the surface layer of the TZN_300 sample, while TZN_1000 contains only traces of NbO_2. This can be caused by the effect of the higher power of the microdischarges appearing due to the longer voltage pulses. Titanium has the lowest melting temperature (T_m = 1670°C) among the substrate elements and to oxidize zirconium (T_m = 1852°C) and niobium (T_m = 2468°C), higher microdischarge energy is required.

Analysis of the corrosion properties of PEO coatings obtained at different frequencies for uncoated and PEO coated samples is shown in figure 3.

The study of the corrosion potential (E_{corr}) shows (figure 3a) that TZN_300 treatment provides surface passivation in comparison with the substrate material TZN, while the TZN_1000 has increased surface activity. Probably, higher porosity for this sample facilitates the penetration of corrosive media, which reduces the surface passivation. Corrosion current, i_{corr}, correlates with the porosity of the coating and its thickness (figure 3b). Uncoated material has the highest value of i_{corr}, while TZN_1000 shows the lowest corrosion current. The polarization resistance R_p (figure 3c) for TZN_1000, accordingly, increases. Therefore, TZN_1000 shows better corrosion performance.

Table 2. Elemental composition of the surface of the Ti-18Zr-15Nb alloy with the PEO coating.

Sample code	Point	*O*	*Ti*	*Nb*	*Zr*	*P*	Others
	1	48.38	19.89	8.17	20.10	3.45	-
TZN_300	2	32.11	27.08	9.65	25.17	5.99	-
	3	20.80	33.90	12.62	27.65	5.03	-
	4 (pore)	0.00	78.90	3.90	13.03	4.17	-
	1	30.93	23.28	1.80	42.10	-	1.89
TZN_1000	2	35.14	23.07	3.19	37.26	-	1.34
	3 (pore)	11.51	49.50	4.39	33.41	-	1.19

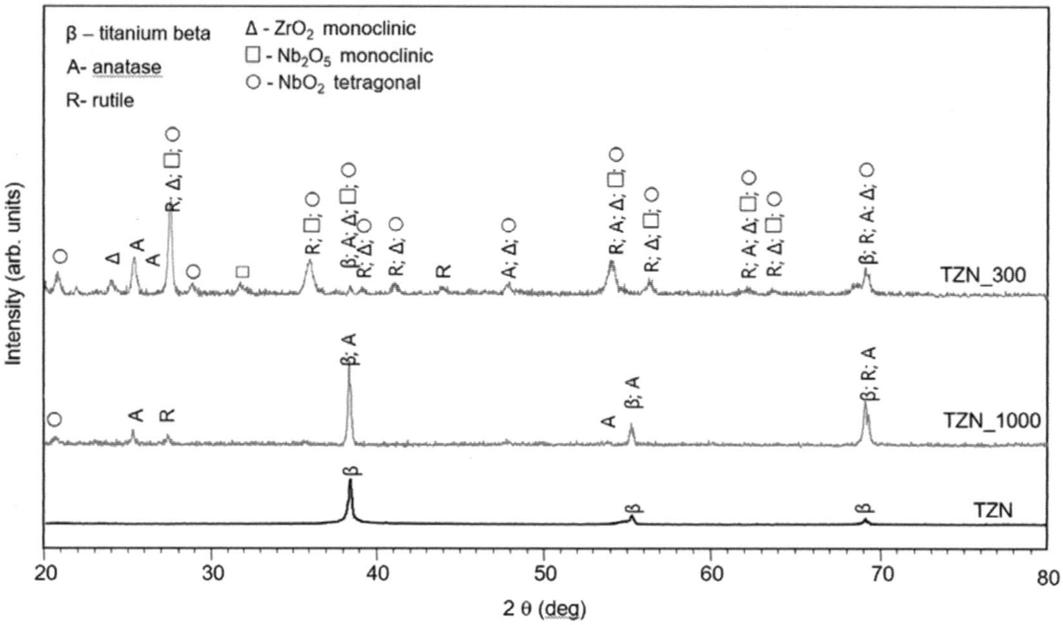

Figure 2. X-ray diffractograms of the Ti-18Zr-15Nb samples with the PEO coating obtained at the frequencies 300 and 1000 Hz.

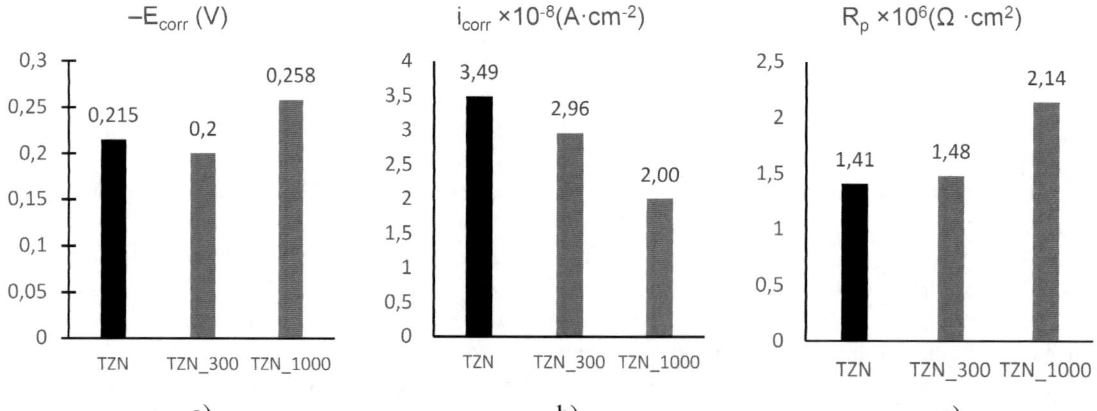

Figure 3. Results of electrochemical tests of Ti-Zr-Nb alloy samples without and with PEO coating in different modes: corrosion potential $-E_{corr}$ (a); corrosion current density i_{corr} (b); polarization resistance R_p (c).

4. Conclusion

The study of samples with PEO coatings on the Ti-18Zr-15Nb shape memory alloy showed the possibility of obtaining continuous coatings on this alloy. Plasma electrolytic oxidation of Ti-Zr-Nb alloy at a frequency of 1000 Hz provides a more uniform surface layer with a better quality than when processing at 300 Hz. It is expected that the observed porous morphology of the coating will facilitate integration with the bone tissue, and the presence of an inert oxide will reduce the release of metal ions into body fluids.

Acknowledgments
The authors are grateful to the Russian Science Foundation (grant No. 20-69-47029) for the support of interdisciplinary scientific cooperation with NITU MISIS which allowed us to carry out this work.

References

[1] Niinomi M 2003 *Science and Technology of Advanced Materials*. **4** p 446
[2] Sheremetyev V, Kudryashova A, Cheverikin V, Korotitskiy A, Galkin S. Prokoshkin S. and Brailovski V 2019 *Journal of Alloys and Compounds* **800** 320
[3] Rafieerad A R, Ashra M R, Mahmoodian R and Bushroa A R 2015 *Materials Science and Engineering* C **57** 397
[4] Santos-Coquillat A, Mohedano M, Martinez-Campos, Arrabal E R., Pardo A and Matykina E. 2019 *Materials Science and Engineering* C **97** 738 A.
[5] Sowa M, Parafiniuk M, Mouzelo M, Kazek-Kęsik A and Zhidkov I S, 2019 *Electrochimica Acta* **302** 10
[6] Cordeiro J, Nagay B, Ribeiro A 2019 *Journal of Alloys and Compounds* **770** 1038
[7] Kazek-Kesik A, Lastowka D, Donesz-Sikorska A Smieszek A and Grzesiak J. 2015 *Journal of Electrochemical Society* **162** D589
[8] Gowtham S, Arunnellaiappan T and Rameshbabu N 2016 *Surface and Coatings Technology* **301** 63.
[9] Yerokhin A, Parfenov E V and Matthews A 2016 *Surface and Coatings Technology* **301** 54
[10] Parfenov E V, Mukaeva V R, Farrakhov R G, Parfenova L V, Lukina E S, Dyakonov G S, Valiev R Z and Danilko K V 2019 *Surface and Coatings Technology* **357** 669
[11] Santos-Coquillat A, Martínez-Campos E, Martínez-Corriá R, Ramos V, Mohedano M, Arrabal R and Matykina E *Surface and Coatings Technology* **347** 358
[12] Yeung W K, Reilly G C, Matthews A and Yerokhin A. 2013 *Journal of Biomedical Materials Research* B **101** 939

Models and parameters of Cosserat hexagonal lattices with chiral microstructure

A A Vasiliev[1], **I S Pavlov**[2]

[1] Tver State University, Sadoviy per. 35, 170102 Tver, Russia
[2] Mechanical Engineering Research Institute of the Russian Academy of Sciences, 85 Belinsky Street, 603024 Nizhny Novgorod, Russia

E-mail: alvasiliev@yandex.ru

Abstract The article develops a structural approach to construction of Cosserat lattice models with chiral microstructure. A variant of the phenomenological potential of connecting cites (particles) of the lattice has been introduced. The parameters of this potential have been found for special cases of complex connections of finite-sized particles and complex beam compounds. Based on the potential, a discrete model and a long-wavelength continuum chiral model have been elaborated. Expression of macroparameters, in particular, the Poisson's ratio, through the parameters of the phenomenological potential has been found. The dependence of macroparameters on the parameters of the microstructure has been obtained for chiral hexagonal lattices composed of finite-sized particles with complex connections and composite beam compounds.

1. Introduction

Elaboration of models and investigation of the special properties of chiral lattices are of prime interest. In particular, phenomenological theories and models of specific structures are developed, auxetic properties of lattices for static problems and the property of a frequency filter in dynamics problems [1-6] are studied.

Two approaches can be distinguished in the development of the theory and models of generalized mechanics: phenomenological and structural. The phenomenological approach is quite general, but it requires a theoretical justification for correctness, a study of the limits of variation of parameters, as well as experimental research for practical application. The structural approach is less general, but historically it played a key role in the assessment, development and application of various models and theories of generalized mechanics. One of the advantages of structural models is their ability to establish a relationship between the generalized structural macroparameters and microstructure parameters. In particular, it can be important for designing materials with desired properties.

This article summarizes and develops some results of the paper [5], where the auxetic properties of a particular case of chiral hexagonal lattices composed of finite-sized particles were studied, as well as the papers [6-8], in which the structural approach was applied for non-chiral lattices.

2. The model of the complex connection of finite-sized particles

We consider a complex connection of particles with spring-type connections. Fragments of particles are shown in Fig. 2a. The centers of particles located at points 1 and 2, as well as the attachment points of the connections have been marked. The figure also shows the rigidities of the connections, e_k. The

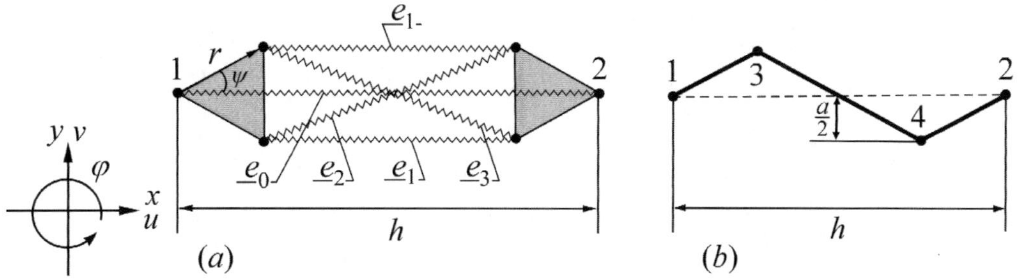

Fig. 1. The complex connection of finite-sized particles (a). The composite beam compound (b)

rigidities of the diagonal particle connections can be the same, $e_2 = e_3$. In this case, there are a symmetrical connection and a lattice with non-chiral microstructure. If the rigidities are different, i.e. $e_2 \neq e_3$, we have a lattice with chiral microstructure.

The potential energy of the springs is proportional to the square of the relative elongation:

$$P_i = e_i \varepsilon_i^2 / 2, \quad \varepsilon_i = \left(L_i - L_{0,i}\right) / L_{0,i},$$

where L_i and $L_{0,i}$ are the lengths of the connections before and after deformation. In order to construct linear models, we linearize elongation $\Delta L = L_i - L_{0,i}$ with respect to small displacements u and v and rotations φ. For compact recording, let us introduce notation $E_i = e_i / L_{0,i}^2$.

The potential of a complex connection is obtained by summarizing the potentials of five springs that make up the connection. After transformation it takes on the form

$$P = \frac{1}{2} K_{11} \Delta u^2 + \frac{1}{2} K_{22} \gamma^2 + \frac{1}{2} K_{33} \Delta\varphi^2 + K_{12} \gamma \Delta u. \tag{1}$$

Here the following designations are used:

$$\Delta u = u_2 - u_1, \; \gamma = v_2 - v_1 - h\left(\varphi_2 + \varphi_1\right)/2, \; \Delta\varphi = \varphi_2 - \varphi_1.$$

Coefficients K_{ij} have the form:

$$K_{11} = E_0 + 2E_1 + \left(\frac{b}{d}\right)^2 \left(E_2 + E_3\right), \; K_{22} = \left(\frac{a}{d}\right)^2 \left(E_2 + E_3\right), \; K_{33} = \frac{1}{2} E_1 a^2, \; K_{12} = \left(\frac{a}{d}\right)\left(\frac{b}{d}\right)\left(E_2 - E_3\right), \tag{2}$$

where $a = 2r\sin(\psi)$, $b = h - 2r\cos(\psi)$, $d = \sqrt{a^2 + b^2}$.

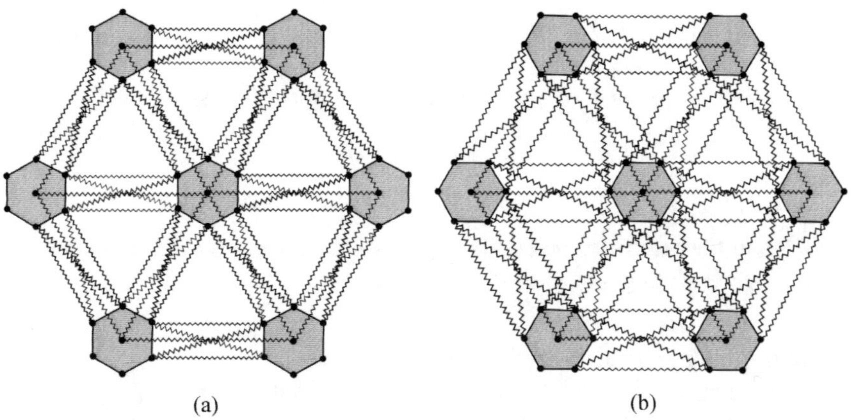

Fig. 2. Two types of symmetrical hexagonal cells composed of finite-sized particles

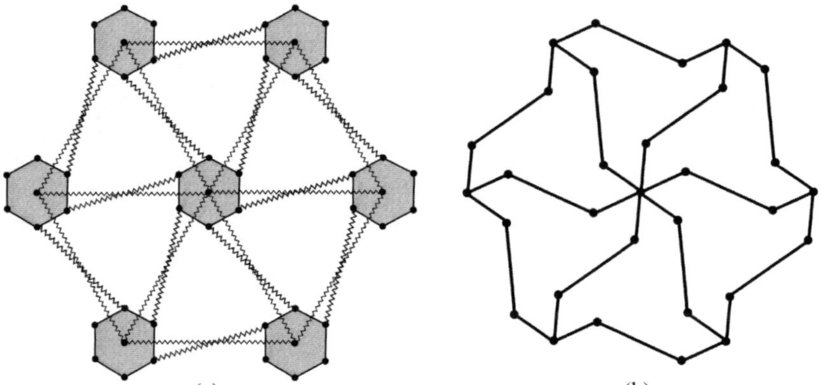

(a) (b)

Fig. 3. A chiral hexagonal cell composed of finite-sized particles, (a), and a chiral cell with composite beam compounds (b)

Potential (1) with coefficients (2) can be used to construct lattice models of two types for hexagonal lattices (Fig. 2). The angles ψ are equal to $\pi/6$ and $\pi/3$ for the cells shown in Fig 2a and Fig 2b, respectively. An example of a cell of a chiral lattice with $E_2 \neq E_3$, $E_3 = 0$, and $E_1 = 0$ is presented in Fig. 3a.

3. The model of the composite beam compound

Let us consider a beam connection composed of three beams (Fig. 1b). For $a=0$ there are a simple beam connection and an ordinary beam hexagonal lattice. And in the case, when $a\neq 0$, there is a composite beam compound that is used for construction of lattices with chiral microstructure.

By eliminating the degrees of freedom of the internal cites 3 and 4, one can construct a connection potential. Vectors $\bar{U}_{1,2} = [u_1,\ v_1,\ \varphi_1,\ u_2,\ v_2,\ \varphi_2]^t$ and $\bar{U}_{3,4} = [u_3,\ v_3,\ \varphi_3,\ u_4,\ v_4,\ \varphi_4]^t$ are introduced, which components are displacements u_k, v_k and rotations φ_k in cites 1-4. The system of static equations is written as $C_{11}\bar{U}_{1,2} + C_{12}\bar{U}_{3,4} = 0,\ C_{21}\bar{U}_{1,2} + C_{22}\bar{U}_{3,4} = 0$. Components $\bar{U}_{3,4} = -C_{12}^{-1}C_{11}\bar{U}_{1,2}$ are found from the first equations of the system. Substitution into the second half of the system of equations gives the rigidity matrix of the connection $\left[C_{21} - C_{22}C_{12}^{-1}C_{11}\right]\bar{U}_{1.2} = 0.$

The potential of the complex compound has the form: $P = (1/2)\bar{U}_{1,2}^t\left[C_{21} - C_{22}C_{12}^{-1}C_{11}\right]\bar{U}_{1,2}$. After transformations, it can be reduced to the form of Eq. (1) with coefficients K_{ij}

$$K_{11} = 4\left[48a^2J + Ah^2\left(h^2 + 4a^2\right)\right]k,\quad K_{22} = 4\left[12h^2J + Aa^2\left(h^2 + 4a^2\right)\right]k,$$

$$K_{33} = J/\sqrt{h^2 + 4a^2},\quad K_{12} = -6ahA\left(h^2 + 4a^2\right)k,$$

$$k = E\frac{48JA\sqrt{h^2 + 4a^2}}{7h^2a^2\left(h^2 + 4a^2\right)^2A^2 + 192\left(h^4 + 4a^4\right)\left(h^2 + 4a^2\right)AJ + 9216a^2h^2J^2}$$

and standard notation for beams is used: *A* is a cross-sectional area, *J* is the second moment of inertia of the cross-section, *E* is Young's modulus.

4. The continuum model and the lattice macroparameters

We construct lattice models on the base of potential (1). The discrete equations are derived for the middle particle (middle cite) of the hexagonal cell, but they are bulky. These equations are not given here, since the main task of our work is finding expressions of macroparameters in terms of structural parameters. Using the expansion into Taylor series and taking into account the second-order derivatives, it is possible to obtain the long-wavelength continuum equations:

$$c_1 u_{xx} + c_2 u_{yy} + c_3 v_{xy} + c_4 \varphi_y - c_5 \left(2u_{xy} - v_{xx} + v_{yy} + 2\varphi_x\right) = 0,$$

$$c_2 v_{xx} + c_1 v_{yy} + c_3 u_{xy} - c_4 \varphi_x + c_5 \left(u_{xx} - u_{yy} + 2v_{xy} - 2\varphi_y\right) = 0,$$

$$c_6 \left(\varphi_{xx} + \varphi_{yy}\right) - c_4 \left(u_y - v_x + 2\varphi\right) + 2c_5 \left(u_x + v_y\right) = 0. \quad (3)$$

The parameters c_k are expressed in terms of the connection potential coefficients by the formulas

$$c_1 = \frac{\sqrt{3}}{4}(3K_{11} + K_{22}),\; c_2 = \frac{\sqrt{3}}{4}(K_{11} + 3K_{22}),\; c_3 = \frac{\sqrt{3}}{2}(K_{11} - K_{22}),$$

$$c_4 = \sqrt{3}K_{22},\; c_5 = \frac{\sqrt{3}}{2}K_{12},\; c_6 = \frac{\sqrt{3}}{4}\left(4K_{33} - h^2 K_{22}\right).$$

These parameters are averaged over the area of the cells $S = \sqrt{3}h^2/2$.

Account of higher-order derivatives in the expansions enables one obtaining equations of the higher-order gradient micropolar theory and expressing the parameters of this model in terms of the coefficients of potential (1).

Comparing equations (3) with the micropolar theory equations

$$(\lambda + 2\mu)(u_{xx} + v_{xy}) + (\mu + \alpha)\left(u_{yy} - v_{xy}\right) + 2\alpha\varphi_y - A\left(2u_{xy} - v_{xx} + v_{yy} + 2\varphi_x\right) = 0,$$

$$(\lambda + 2\mu)(u_{xy} + v_{yy}) + (\mu + \alpha)\left(v_{xx} - u_{xy}\right) - 2\alpha\varphi_x + A\left(u_{xx} - u_{yy} + 2v_{xy} - 2\varphi_y\right) = 0,$$

$$4B(\varphi_{xx} + \varphi_{yy}) - 2\alpha\left(u_y - v_x + 2\varphi\right) + 2A\left(u_x + v_y\right) = 0,$$

one can find the relationships between the parameters of these. As a result, expressions of the micropolar parameters through the parameters of potential (1) yield:

$$\lambda = \frac{\sqrt{3}}{4}\left(K_{11} - K_{22}\right),\; \mu = \frac{\sqrt{3}}{4}\left(K_{11} + K_{22}\right),\; \alpha = \frac{\sqrt{3}}{2}K_{22},\; B = \frac{\sqrt{3}}{16}\left(4K_{33} - hK_{22}\right),\; A = \frac{\sqrt{3}}{2}K_{12}.$$

Due to these equalities it is possible [5] to express the Poisson's ratio in terms of the connection potential parameters (1):

$$\nu = \frac{K_{11}K_{22} - K_{22}^2 - 2K_{12}^2}{3K_{11}K_{22} + K_{22}^2 - 2K_{12}^2}.$$

Acknowledgements
The research was carried out under the financial support of the Russian Foundation for Basic Research (projects №№ 19-08-00965-a, 18-29-10073-mk).

References

[1] Spadoni A and Ruzzene M 2012 *J. Mech. Phys. Solids* **60**(1) 156
[2] Liu X N, Huang G L and Hu G K 2012 *J. Mech. Phys. Solids* **60**, 1907
[3] Smith C W, Grima J N and Evans K E 2000 *Acta mater.* **48** (17) 4349
[4] Bacigalupo A and Gambarotta L 2016 *Int. J. Solids Struct.* **83**, 126
[5] Vasiliev A A and Pavlov I S 2020 *Phys. Status Solidi B* **257**, 1900389
[6] Erofeev V I, Pavlov I S and Leontiev N V 2013 *Composites: Mechanics, Computations, Applications* **4** (4) 239
[7] Potapov A I, Pavlov I S and Lisina S A 2009 *J. Sound Vibration* **322** (3) 564
[8] Pavlov I S, Vasiliev A A, Porubov A V 2016 *J. Sound Vibration* **384** 163

Evolution of particles of secondary phases in alloys of the Cu-Cr-Zr system in the state of a supersaturated solid solution in the process of deformation-heat treatment

R G Chembarisova and A V Galaktionova

Department of Physics, Ufa State Aviation Technical University, Karl Marx str. 12, 450008 Ufa, Russia

E-mail: chroza@yandex.ru

Abstract. In this work, the evolution of particles of secondary phases in the process of equal-channel angular pressing and subsequent heat treatment of the Cu-1.0 wt.% Cr-0.7 wt.% Zr-0.2 wt.% Al alloy, preliminarily subjected to quenching from the pre-melting temperature, is studied. It is shown that in the process of deformation of the samples, partial dissolution of small particles of the secondary phases occurs. As a result of the subsequent aging, the Cr and Al atoms that have passed into the solution are again separated into small particles. Additionally, there was a partial separation of Cr atoms from the solid solution into small particles of chromium.

1. Introduction

Copper alloys are widely used for the manufacture of electrical conductors, which requires them to have high electrical conductivity. It is known that particles of secondary phases in the alloys affect the electrical conductivity and strength of conductors [1]. In this regard, studies of their evolution in the process of deformation-heat treatment of alloys are relevant [2, 3].

The aim of this work was to analyze the evolution of particles of secondary phases in the process of equal-channel angular pressing (ECAP) and subsequent aging of the Cu-1.0 wt.% Cr-0.7 wt.% Zr-0.2 wt.% Al alloy preliminarily subjected to high-temperature heat treatment with subsequent quenching basing on the known experimental data on the character of the change in its specific electrical conductivity [3].

2. Experimental and theoretical approaches

A detailed description of the experiments, the results of which were used in analytical modeling, are presented in [3]. Samples of the Cu-1.0 wt.% Cr-0.7 wt.% Zr-0.2 wt.% Al alloy in the initial states (IS) with an average grain size equal to d=90±4 μm were obtained by annealing at a temperature of 1050°C for 5 h with subsequent quenching in water to form a supersaturated solid solution (SS). Samples of coarse-grained (CG) alloy obtained in this way were deformed using 8 passes of ECAP on equipment with a channel intersection angle of 90° via route Bc at room temperature. As a result, an ultrafine-grained (UFG) structure (8ECAP state) was formed in the alloy. Samples of the UFG alloy were subjected to subsequent aging at a temperature of 450 °C for 0.5 h, which led to the formation of the 8ECAP+A state.

The microstructure was studied by optical metallography, scanning and transmission electron microscopy. The conductivity measurements were performed using an eddy current device at room temperature.

During the preliminary heat treatment of the alloy, an increase in the size of particles that did not dissolve at a temperature of 1050 °C was observed. According to [3], these particles are most likely intermetallic compounds, which were formed during casting of the alloy. The phase analysis of the microstructure of samples of a Cu-Cr-Zr alloy containing Al atoms subjected to thermomechanical treatment showed that they could contain dispersed particles of secondary phases of different sizes and compositions, such as Al_3Zr_5, AlCrZr, $ZrAl_2$, $ZrAl_3$ and Zr_2Al_3 [2], the size and density of which change depending on the external influence. In the IS, against the background of the obtained α - solid solution, large inclusions with the size D=4.0±0.4 μm were observed. The distance between them L_l was established based on the available images of the microstructure and turned out to be equal to 8.3 μm. Small particles were also detected. Their size distribution is shown in figure 1a. The average distance between them L_s amounted to 300.0±7.0 nm.

After 8 ECAP passes, the grain size decreased to 200.0±5.0 nm. The average distance between small particles increased to 340.0±8.0 nm. The fraction of small particles with a size of ≈10 nm has increased (figure 1b) [3].

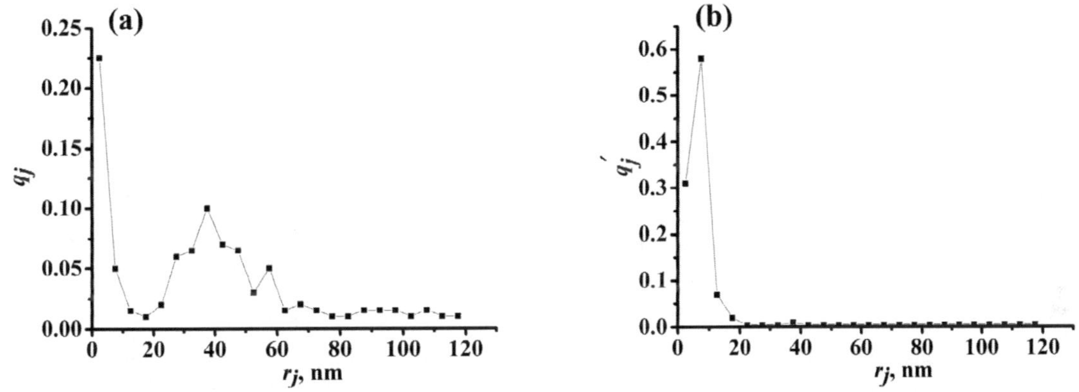

Figure 1. Distribution of particle sizes in the initial CG state, q_j – the fraction of particles of radius r_j (a); distribution of particle size in the 8ECAP state, q_j' – the fraction of particles of radius r_j (b).

Subsequent aging resulted in the decomposition of the SS. The most probable size of small particles lays in the range of 15–20 nm; the distance between the particles decreased to 155.0±3.5 nm [3].

The specific electrical resistance of the studied alloy in its various structural states is due to the scattering of electrons on various defects of the crystal lattice, alloying atoms, thermal vibrations of the lattice and grain boundaries [4 - 8].

Alloying atoms a dissolved in the matrix contributes to the resistivity of the alloy, which is directly proportional to their atomic fraction C_a in solution [5]:

$$\rho_a = \frac{k^3 \hbar l_a^4 n_a C_a}{9e^2 \pi n_e}\left[1 - \frac{\sin 2kl}{2kl}\right], \quad (1)$$

where n_a is the volume concentration of the matrix atoms; $k=2\pi/\lambda$ - wave number, $\lambda=h/m^* u_F$ - electron wavelength, m^* - effective electron mass, u_F - Fermi velocity; $\hbar$ is Planck's constant; e - electron charge value; n_e - density of free electrons; l - size of the region of interaction of an electron with an alien atom, $l_a \approx 1.35 b_B$, $b_B \approx a\sqrt{2}/2$, a - lattice parameter.

The contribution of lattice dislocations ρ_{disl} is directly proportional to their density q_c [6, 5]:

$$\rho_{disl} = \frac{\hbar k^2 l_c^2 q_c}{9 n_e e^2}\left(1 - \frac{\sin 2kl}{2kl}\right),\ l_c \approx 1.55 b_B. \quad (2)$$

To determine the value of q_c, the dislocation density in grain boundaries (GBs) was preliminarily calculated. It was assumed that GBs are formed by one set of edge dislocations spaced apart from each other at a distance $H = b/2\sin(\Theta/2)$. Then the dislocation density in them will be equal to $q_w = 1/Hd$. The average density of dislocations in high-angle GBs of the alloy takes the value

$$q_{wh} = \frac{8}{\left(\Theta_s - 2\Theta^*\right)} \int_{\Theta^*/2}^{\Theta_s/4} \frac{1}{d}\frac{\sin Z}{b} dZ = \frac{1.054 \cdot 10^9}{d}\ \text{m}^{-2}, \quad (3)$$

where $\Theta_s = \pi/2$, $\Theta^* = 16°$; the value of the Burgers vector in the case of high misorientation angles Θ is equal to $b = 2h_c \sin(\Theta^*/2)$, $h_c = 7b_B$ [7]. Then the density of lattice dislocations q_c, taking into account their average value q and the volume fraction of GBs $f = 0.03$ [8], takes the value

$$q_c = (q - f q_{wh})/(1 - f). \quad (4)$$

Knowing the resistivity of linear defects ρ_w, which make up GBs, per unit of their density q_w and the distance between them H, it is possible to calculate the specific electrical resistance of GBs per unit of their area S in a unit volume Ω: $\rho_{gb}/N_{gb} = \rho_w / q_w H$, $N_{gb} = S/\Omega \approx 1/d$. High-angle GBs of CG samples increase the resistivity by $\rho_{gb} \approx 2.67 \cdot 10^{-16} N_{gb}$ $\Omega \cdot$m [8]. The contribution of high-angle GBs of UFG alloy samples was determined on the basis of experimental data [9, 8] and amounted to $\rho_{gb}/N_{gb} \approx 5.60 \cdot 10^{-16}$ $\Omega \cdot \text{m}^2$.

An electron moving in the lattice with an average velocity υ collides with the particles of the secondary phases, approaching the center of the particle at a distance of its radius r. Consequently, the mean free path of an electron is equal to $\lambda = 1/n_{ps}\pi r^2$, where n_{ps} is the number of particles per unit volume. According to Ohm's law, electrical resistivity is

$$\rho = \frac{m_e \upsilon}{n_e e^2 \lambda}, \quad (5)$$

where m_e, n_e are the mass and density of electrons, respectively. Taking into account that electrons obey the Fermi-Dirac statistics, we will take the average velocity equal to the Fermi velocity: $\upsilon = u_F$. In this case, the electron scattering cross section is associated with the geometric cross section of the particle, since its dimensions are much larger than the atomic dimensions. Small and large particles of secondary phases will increase the resistivity of the alloy by ρ_{secs} and ρ_{secl} [4], respectively:

$$\rho_{secs} \approx \frac{m_e u_F}{n_e e^2} \sum_j f_{secsj} \frac{\pi r^2(j)}{V_{psj}},\ \rho_{secl} \approx \frac{m_e u_F}{n_e e^2} f_{secl} \frac{\pi R^2}{V_{pl}}, \quad (6)$$

where f_{secsj}, V_{psj} - volume fraction and volume of small particles of radius $r(j)$, respectively; f_{secl}, V_{pl} is the volume fraction and volume of large particles; R is the size of the region of interaction of an electron with a large particle, which is equal to its radius $D/2$.

It was assumed that the processes of electron scattering by various defects of the crystal lattice are independent. Therefore, the resistivity of the alloy was determined according to the Matthiessen rule as the sum of their contributions.

3. Results and discussion

Electrical resistivity is a structurally sensitive quantity that can be easily measured. Therefore, it has been used to detect the character of changes in the particles of secondary phases as a result of ECAP and subsequent thermal effect on the alloy.

In the cast state, the structure of alloys of the Cu-Cr-Zr system consists of dendrites of a solid solution based on copper with phase particles distributed along the boundaries of the dendrites in accordance with the phase diagram. In alloys containing Cr, chromium-based particles with an average size of ≈1.5 μm, which are identified as Cr particles, precipitate from the SS. Alloys containing Zr atoms have large (≈3.5 μm) particles of the intermetallic compound Cu_5Zr. Subsequent hot forging and quenching lead to the formation of a more uniform structure and to the partial dissolution of the second phases [10]. The solubility of Cr in Cu at 1050 °C is 0.6 wt.% (0.73 at.%), the solubility of Zr in Cu is 0.15 wt.% (0.10 at.%). The solubility of Al in Cu at 1050 °C is 6.9 wt% (14.8 at.%). It decreases with increasing temperature [1]

Since the content of alloying elements exceeds their solubility in Cu at the quenching temperature, particles of undissolved phases appear in the structure of the alloys. It was assumed that, as a result of the growth of intermetallic compounds, the Al and Zr atoms were completely separated from the solution, and also the Cr atoms were partially separated. Cr atoms are included in only one intermetallic compound AlCrZr (table 1). Therefore, it was assumed that the solution contains not 0.6 wt% Cr, but 0.4 wt% Cr, that corresponds to the experimental value of the specific electrical conductivity of the alloy.

Large rounded inclusions were identified as Cr particles and Cu_5Zr compounds [10]. The number of small particles is γ^3 times greater than that of large ones: $N_{ps}=\gamma^3 N_{pl}$, where N_{ps} and N_{pl} are the total number of small and large particles, respectively, $\gamma = L_l/L_s$. In this case, the probabilities of detecting small particles of different phases, as well as the probabilities of detecting large particles of different chemical compositions, were considered the same, equal to 1/g and 1/p, respectively. Then the mass fraction of Cr atoms precipitated into particles of secondary phases can be calculated according to the formula:

$$W_{\text{secCr}} = \frac{\sum_j \sum_i q_j N_{ps} N_{\text{pCrs}i}(j)\mu_{\text{Cr}} / g + N_{pl} N_{\text{pCrl}} \mu_{\text{Cr}} / p}{\left(n_{\text{Cr}}\mu_{\text{Cr}} / \mu_{\text{Cu}} + n_{\text{Zr}}\mu_{\text{Zr}} / \mu_{\text{Cu}} + n_{\text{Al}}\mu_{\text{Al}} / \mu_{\text{Cu}} + 1\right) N_{\text{Cu}}\mu_{\text{Cu}}}, \quad (7)$$

where g and p are the number of phases of small and large particles, respectively. g=5 (table 1), p=2. $N_{\text{pCrs}i}(j)$ – the number of Cr atoms in one small particle of the secondary phase i and radius r_j (table 1); N_{pCrl} – number of Cr atoms in one large particle of the secondary phase, $N_{\text{pCrl}}=4\pi(D/2)^3/3\Omega_{\text{Cr}}$. The volumes of Al, Cr, and Zr atoms are equal, respectively: $\Omega_{\text{Al}}=\mu_{\text{Al}}m_0/\rho_{\text{densAl}} \approx 16.60\cdot 10^{-30}$ m^3, $\Omega_{\text{Cr}}=\mu_{\text{Cr}}m_0/\rho_{\text{densCr}}\approx 12.00\cdot 10^{-30}$ m^3, $\Omega_{\text{Zr}}=\mu_{\text{Zr}}m_0/\rho_{\text{densZr}}\approx 23.28\cdot 10^{-30}$ m^3, where ρ_{densAl} – Al density, ρ_{densCr} – Cr density, ρ_{densZr} – Zr density; m_0 – atomic mass unit; μ_{Al}, μ_{Cr}, μ_{Zr} – relative atomic masses of Al, Cr and Zr, respectively. From expression (7) it follows that the number of large particles in the sample is

$$N_{pl} = \frac{gpW_{\text{secCr}}\alpha N_{\text{Cu}}\mu_{\text{Cu}}}{\mu_{\text{Cr}}\left(p\sum_j \sum_i q_j \gamma^3 N_{\text{pCrs}i}(j) + gN_{\text{pCrl}}\right)}, \quad \alpha = n_{\text{Cr}}\mu_{\text{Cr}} / \mu_{\text{Cu}} + n_{\text{Zr}}\mu_{\text{Zr}} / \mu_{\text{Cu}} + n_{\text{Al}}\mu_{\text{Al}} / \mu_{\text{Cu}} + 1 \quad (8)$$

Table 1. The number of atoms of aluminum $N_{\text{pAls}i}$, chromium $N_{\text{pCrs}i}$ and zirconium $N_{\text{pZrs}i}$ in one small particle of the secondary phase i.

	$ZrAl_3$	Zr_2Al_3	$ZrAl_2$	AlCrZr	Al_3Zr_5
$N_{\text{pZrs}i}$	$\frac{V_{\text{ps}j}}{\Omega_{\text{Zr}}+3\Omega_{\text{Al}}}$	$\frac{2V_{\text{ps}j}}{2\Omega_{\text{Zr}}+3\Omega_{\text{Al}}}$	$\frac{V_{\text{ps}j}}{\Omega_{\text{Zr}}+2\Omega_{\text{Al}}}$	$\frac{V_{\text{ps}j}}{\Omega_{\text{Al}}+\Omega_{\text{Cr}}+\Omega_{\text{Zr}}}$	$\frac{5V_{\text{ps}j}}{5\Omega_{\text{Zr}}+3\Omega_{\text{Al}}}$
$N_{\text{pAls}i}$	$\frac{3V_{\text{ps}j}}{\Omega_{\text{Zr}}+3\Omega_{\text{Al}}}$	$\frac{3V_{\text{ps}j}}{2\Omega_{\text{Zr}}+3\Omega_{\text{Al}}}$	$\frac{2V_{\text{ps}j}}{\Omega_{\text{Zr}}+2\Omega_{\text{Al}}}$	$\frac{V_{\text{ps}j}}{\Omega_{\text{Al}}+\Omega_{\text{Cr}}+\Omega_{\text{Zr}}}$	$\frac{3V_{\text{ps}j}}{5\Omega_{\text{Zr}}+3\Omega_{\text{Al}}}$
$N_{\text{pCrs}i}$	-	-	-	$\frac{V_{\text{ps}j}}{\Omega_{\text{Al}}+\Omega_{\text{Cr}}+\Omega_{\text{Zr}}}$	-
i	1	2	3	4	5

The volume fraction of small particles of radius r_j, the fraction of which in the total number of small particles is equal to q_j (figure 1a), was calculated based on its definition:

$$f_{\mathrm{sec}\,sj} = \frac{V_{\mathrm{ps}j}N_{\mathrm{ps}j}}{V_{\mathrm{s}}} = \frac{V_{\mathrm{ps}j}q_j\gamma^3 N_{\mathrm{pl}}}{V_{\mathrm{s}}} = \frac{gpV_{\mathrm{ps}j}q_j\gamma^3 W_{\mathrm{sec\,Cr}}\alpha\mu_{\mathrm{Cu}}}{(n_{\mathrm{Cr}}\Omega_{\mathrm{Cr}} + n_{\mathrm{Zr}}\Omega_{\mathrm{Zr}} + n_{\mathrm{Al}}\Omega_{\mathrm{Al}} + \Omega_{\mathrm{Cu}})\mu_{\mathrm{Cr}}(p\sum_j q_j\gamma^3 N_{\mathrm{pCrs}i}(j) + gN_{\mathrm{pCrl}})}, \quad (9)$$

where $N_{\mathrm{ps}j}$ – number of small particles of radius r_j; V_{s} – volume of sample, which was considered equal to the sum of the volumes of its constituent atoms; Ω_{Cu} – volume of Cu atom, $\Omega_{\mathrm{Cu}}=\mu_{\mathrm{Cu}}m_0/\rho_{\mathrm{densCu}}=11.82\cdot10^{-30}$, ρ_{densCu} - Cu density. According to the calculations $f_{\mathrm{secs}} = \sum_j f_{\mathrm{secs}j} = 5.87\cdot10^{-3}$. The volume fraction of large particles was determined by the expression

$$f_{\mathrm{sec\,l}} = \frac{V_{\mathrm{pl}}N_{\mathrm{pl}}}{V_{\mathrm{s}}} = \frac{gpV_{\mathrm{pl}}W_{\mathrm{sec\,Cr}}\alpha\mu_{\mathrm{Cu}}}{(n_{\mathrm{Cr}}\Omega_{\mathrm{Cr}} + n_{\mathrm{Zr}}\Omega_{\mathrm{Zr}} + n_{\mathrm{Al}}\Omega_{\mathrm{Al}} + \Omega_{\mathrm{Cu}})\mu_{\mathrm{Cr}}(p\sum_j q_j\gamma^3 N_{\mathrm{pCrs}i}(j) + gN_{\mathrm{pCrl}})}, \quad (10)$$

where $n_k=W_k\mu_{\mathrm{Cu}}/\mu_k(1-W_{\mathrm{Zr}}-W_{\mathrm{Cr}}-W_{\mathrm{Al}})$, k=Cr, Zr, Al; W_{Cr}, W_{Zr}, W_{Al} - mass fractions of Zr, Cr, Al atoms in the alloy. According to the calculations $f_{\mathrm{secl}}=1.42\cdot10^{-2}$.

The electrical resistivity of the alloy in the IS is due to thermal vibrations of atoms in the nodes of the crystal lattice and defects in its structure. When calculating the electrical resistivity ρ_{T} due to thermal vibrations of the crystal lattice, the contribution of the regions occupied by large Cr particles was taken into account: $\rho_{\mathrm{T}}=(1-f_{\mathrm{secCrl}})\ \rho_{\mathrm{TCu}}+f_{\mathrm{secCrl}}\rho_{\mathrm{TCr}}$, where $\rho_{\mathrm{TCu}}=1.72\cdot10^{-8}$ Ω·m is the electrical resistivity of pure CG copper; $\rho_{\mathrm{TCr}}=2.70\cdot10^{-8}$ Ω·m is the electrical resistivity of chromium at RT. It was assumed that the resistivity of the particles of the Cu_5Zr compound has a value close to that for the matrix, which somewhat underestimates the value of ρ_{T}. The contribution of the regions occupied by fine particles was neglected. The density of lattice dislocations (4), taking into account the density of dislocations in GBs (3) $q_{\mathrm{wh}}\approx1.17\cdot10^{13}$ m^{-2} and the average density of dislocations equal to $q\approx3.00\cdot10^{12}$ m^{-2} [11], takes the value $q_c\approx2.70\cdot10^{12}$ m^{-2}. The contribution of factors that increase the electrical resistivity of the alloy in the IS is shown in table 2.

Table 2. Specific electrical resistivity ρ, calculated δ and experimental δ_{exp} values of specific electrical conductivity, contribution of thermal vibrations of atoms of crystal lattice and defects in the crystal structure of Cu to the specific electrical resistivity of the Cu-1.0 wt.% Cr-0.7 wt.% Zr-0.2 wt.% Al alloy in its various structural states.

	Initial	8ECAP	8ECAP+A
ρ_{secs} (Ω·m)	$3.86\cdot10^{-11}$	$8.07\cdot10^{-12}$	$5.78\cdot10^{-10}$
ρ_{secl} (Ω·m)	$1.76\cdot10^{-12}$	$1.76\cdot10^{-12}$	$1.76\cdot10^{-12}$
ρ_{T} (Ω·m)	$1.73\cdot10^{-8}$	$1.73\cdot10^{-8}$	$1.74\cdot10^{-8}$
ρ_{Cr} (Ω·m)	$2.51\cdot10^{-8}$	$2.62\cdot10^{-8}$	$4.81\cdot10^{-9}$
ρ_{Al} (Ω·m)	-	$0.80\cdot10^{-8}$	-
ρ_{disl} (Ω·m)	$4.66\cdot10^{-13}$	$1.48\cdot10^{-10}$	$2.51\ 10^{-11}$
ρ_{gb} (Ω·m)	$2.97\cdot10^{-12}$	$0.28\cdot10^{-8}$	$0.24\ 10^{-8}$
ρ (Ω·m)	$4.25\cdot10^{-8}$	$5.45\cdot10^{-8}$	$2.52\cdot10^{-8}$
δ (% IACS)	40.6	31.6	68.4
δ_{exp} (% IACS)	42±1	32±1	70±1

After 8 passes of ECAP, the distance between small particles increased and the fraction of particles with a size of ≈10 nm increased [3]. Consequently, their fragmentation took place. At the same time, atoms of alloying elements were released into solution from small particles of secondary phases as a result of their partial dissolution. The deformation by the ECAP method created favorable conditions for this. If we assume that the distance between micrometric particles has not changed significantly, then the number of small particles will be $\gamma' = L_l/L'_s$ times the number of large particles: $N'_{ps}=\gamma'^3 N_{pl}$. Let us assume that Al atoms are released into the solution. The number of Al atoms in small particles changed by ΔN_{Al}: $\Delta N_{Al} = N_{pl}\left[\sum_j\sum_i \gamma^3 q_j N_{pAlsi}(j) - \sum_j\sum_i \gamma'^3 q'_j N'_{pAlsi}(j)\right]/g$. The number of Cr atoms released from small particles was $\Delta N_{Cr} = N_{pl}\left[\sum_j \gamma^3 q_j N_{pCrs4}(j) - \sum_j \gamma'^3 q'_j N'_{pCrs4}(j)\right]/g$. According to calculations, the atomic fraction of the released elements is equal to $\Delta C_{Al} = \Delta N_{Al}/(n_{Cr}+n_{Zr}+n_{Al}+1)N_{Cu} \approx 1.56\cdot10^{-3}$, $\Delta C_{Cr} = \Delta N_{Cr}/(n_{Cr}+n_{Zr}+n_{Al}+1)N_{Cu} \approx 2.07\cdot10^{-4}$, where N_{Cu} – the number of Cu atoms in the alloy. N'_{pAlsi}, N'_{pCrs4} is the number of Al and Cr atoms in one small particle of the secondary phase i and radius r_j in the 8ECAP state, respectively (table 1). After ECAP, the fraction of Cr atoms in the solution reached the value $C_{Cr} \approx 5.08\cdot10^{-3}$, the fraction of Al atoms – $C_{Al} \approx \Delta C_{Al}$. It was also assumed that the Zr atoms were not released into the solution. Zr atoms are strongly associated with crystal lattice defects. The volume fractions of small and large particles, according to the calculations, are equal to $f_{secs}=1.36\cdot10^{-3}$ and $f_{secl}=1.42\cdot10^{-2}$, respectively. The dislocation density after 8 ECAP passes reached the value $q \approx 1.00\cdot10^{15}$ m^{-2} [11]. The average dislocation density in high-angle GBs, according to (3), was equal to $q_{wh} \approx 1.054\cdot10^{9}/d = 5.27\cdot10^{15}$ m^{-2}. The average density of lattice dislocations is $q_c=0.87\cdot10^{15}$ m^{-2}. The electrical resistivity in this state was determined mainly by the Cr and Al atoms dissolved in the matrix, the thermal vibrations of atoms, and GBs (table 2).

Subsequent heat treatment led to the decomposition of the SS [3]. In the modeling, it was assumed that the atoms of Al and Cr, which passed into the solution during partial dissolution of small particles during ECAP, newly allocated in the secondary phase particles. Accordingly, the volume fraction of small particles gained a value equal to that one in the IS. But at the same time, Cr atoms were additionally separated from the solution into small particles of chromium. Based on the known value of the specific electrical conductivity of the alloy after aging, it can be assumed that the fraction of additionally precipitated Cr atoms was $\Delta C_{Cr}=3.94\cdot10^{-3}$. The volume fraction of fine particles of precipitated Cr, calculated according to its definition (9) by the formula $f_{secCrs} = \Delta C_{Cr}\left(n_{Cr}+n_{Zr}+n_{Al}+1\right)m_0\mu_{Cr}/\left(n_{Cr}\Omega_{Cr}+n_{Zr}\Omega_{Zr}+n_{Al}\Omega_{Al}+\Omega_{Cu}\right)\rho_{densCr}$, in which the volume of the particles was expressed in terms of the density of chromium, the number of particles through an atomic fraction ΔC_{Cr}, amounted $f_{secCrs}=3.98\cdot10^{-3}$. In the 8ECAP+A state, the fraction of Cr atoms in the solution decreased to $C_{Cr} \approx 0.93\cdot10^{-3}$.

The average size of small particles slightly increased and became equal to 17.5 nm. It was assumed that the distances between micrometric particles and their sizes did not change significantly. As studies of the thermal stability of the UFG structure of Cu-Cr-Zr bronzes have shown [11], annealing for 1 h at a temperature of 300-500 °C does not lead to any significant changes in the microstructures formed during ECAP. Therefore, the average grain size after annealing was taken equal to 230 nm. The average density of dislocations took the value $q \approx 2.80\cdot10^{14}$ m^{-2} [11]. The average density of lattice dislocations, calculated taking into account the density of dislocations in GBs of the UFG material (3) equal to $q_{wh} \approx 4.60\cdot10^{15}$ m^{-2}, made up $q_c=1.50\cdot10^{14}$ m^{-2}. The specific electrical resistivity of the alloy in this state in addition to thermal vibrations of atoms, is due to the contribution of GBs, as well as to the alloying Cr atoms. The contribution of small particles increased by two orders of magnitude compared to that in the 8ECAP state (table 2).

4. Conclusions

According to the modeling results, the experimentally observed character of the change in the specific electrical conductivity of the Cu-1.0Cr-0.7Zr-0.2Al alloy during thermo-mechanical treatment is possible if all Zr and Al atoms, as well as some Cr atoms, were in the particles of the secondary phases in the IS. As a result of ECAP, Cr and Al atoms passed from small particles into solution. Their partial dissolution took place. As a result of the subsequent aging, the Cr and Al atoms, which passed into the solution during ECAP, again precipitated into the particles of the secondary phases. At the same time, additional precipitation of fine Cr particles from solution into fine Cr particles occurred.

References

[1] Osintsev O E and Fedorov V N 2004 *Copper and Copper Alloys. Domestic and Foreign Brands: Reference Book* (Moscow: Mechanical Engineering) p 336 [in Russian]
[2] Faizov I A, Raab G I, Faizova S N, Zaripov N G and Aksenov D A 2016 *Lett. Mater.* **6** (2) 132–7 [in Russian]
[3] Faizov I A, Raab G I, Faizova S N, Aksenov D A, Zaripov N G, Gunderov D V and Golubev O V 2016 *Lett. Tambov University. Ser. Natur. and Tech. Sci. Phys.* **21**(3) 1387–91 [in Russian]
[4] Chembarisova R G, Aleksandrov I V and Yamileva A M 2019 *Tech. Phys.* **64** 162–70
[5] Chembarisova R G 2018 *IOP Conf. Ser.: Mater. Sci. and Eng.* **447** 012085
[6] Harrison W A 1958 *J. Phys. Chem. Sol.* **5** 44–6
[7] Karolik A S and Luhvich A A 1994 *J. Phys: Condens. Matter* **6** 873–86
[8] Chembarisova R G 2020 *J. Tech. Phys.* **90**(4) 618–26
[9] Pękala K and Pękala M 1995 *NanoStr. Mater.* **6** 819–22
[10] Shangina D V 2018 *Diss. for the Degree of Cand. Tech. Sci.* (Moscow: IMET) p 142 [in Russian]
[11] Morozova A I, Mishnev R V, Belyakov A N and Kaibyshev R O 2018 *Rev. Adv. Mater. Sci.* **54**(1) 56–92

Synthesis and Characterization of $(Ag_2S)_x(ZnS)$ Heteronanostructures

Stanislav I. Sadovnikov[1], **Eugeny Yu. Gerasimov**[2]

[1] Institute of Solid State Chemistry, Ural Branch of the Russian Academy of Sciences, 620990 Ekaterinburg, 91 Pervomaiskaya st., Russia
[2] Boreskov Institute of Catalysis, Siberian Branch of the Russian Academy of Sciences, 630090 Novosibirsk, 5 Lavrentiev st., Russia

E-mail: sadovnikov@ihim.uran.ru

Abstract. $(Ag_2S)_x(ZnS)$ heteronanostructures were synthesized by two-stage chemical co-deposition of silver and zinc sulfides using aqueous solutions of zinc nitrate, silver nitrate, sodium sulfide, trisodium citrate and Trilon B. Prepared heteronanostructures were characterized for phase composition and nanoparticles size using X-ray diffraction and the high-resolution transmission electron microscopy. The size of Ag_2S nanoparticles in the obtained heteronanostructures depends on the ratios of reagent concentrations and ranges from 9 to 30 nm, the size of ZnS nanoparticles is equal to 4-5 nm.

1. Introduction

Nanostructured semiconductor sulfides are a new class of materials, having dimensions in the 2-30 nm range, which provide one of the greatest potentials for improving performance and extended capabilities of products in a number of industrial sectors especially in electronics. Semiconductor silver sulfide Ag_2S and zinc sulfide ZnS are widely used as materials for modern electronics devices. The development of semiconductor nanocomposites consisting of Ag_2S and ZnS nanoparticles allows adjusting the band gap of such heteronanostructures.

Production of heteronanostructures based on Ag_2S and ZnS allows one to improve laser optics devices, expand the spectral sensitivity of semiconductors to the visible long-wavelength region, and to obtain new nanomaterials for direct conversion of solar energy into electricity, for solid-state ultraviolet (UV) lasers, beta-voltaics and high-speed resistance switches [1-6]. In [7], matchstick-shaped Ag_2S-ZnS heteronanostructures were synthesized. These heteronanostructures in the form of ZnS quantum rods with Ag_2S quantum dots on the top of the rod were obtained by thermal co-decomposition of such single-source precursors as silver diethyldithiocarbamate $AgS_2CN(C_2H_5)_2$ and zinc diethyldithiocarbamate $Zn[S_2CN(C_2H_5)_2]_2$. The matchstick-shaped Ag_2S-ZnS heteronanostructures have been shown to display photoluminescence in both UV/blue and near-infrared (NIR) regions. It is known than doping of zinc sulfide by silver sulfide leads to a large decrease in the band gap. For example, an increase in the Ag_2S content in ZnS-Ag_2S heteronanostructures up to 10 mol.% is accompanied by an appreciable decrease in the optical band gap E_g from ~3.6 to ~2.1 eV, and also by an increase in the intensity of the absorption band in the infrared region [1]. Thus, change of the content of silver sulfide in heteronanostructures allows one to change the band gap and control optical properties in a wide range.

However, the problem of simple synthesizing heteronanostructures based on nanostructured zinc and silver sulfides has not been solved yet.

In the present work, a study of chemical deposition of $(Ag_2S)_x(ZnS)$ heteronanostructures is carried out for the first time. A simple method for synthesizing the heteronanostructures based on zinc sulfide doped with silver sulfide is proposed.

2. Materials and Methods

Trisodium citrate, Trilon B and sodium sulfide Na_2S from Sigma Aldrich, zinc nitrate and silver nitrate from domestic market were used. All of the chemicals were of analytical grade (A.C.S) and used without further purification. Aqueous solutions of reagents for synthesis of heteronanostructures based on Ag_2S and ZnS sulfides were prepared with the use of high-purity deionized water as a solvent. High-purity deionized water was produced using the Milli-Q Reference (Merck, Millipore) water preparation system. The specific resistivity of water is equal 18.2 MΩ·cm at 298 K (electrical conductivity is below $5.5 \cdot 10^{-6}$ S·m^{-1}), the total content of organic carbon did not exceed 5-10 ppb (5-10 μg·L^{-1}).

The solubility products K_{sp} of sulfides ZnS ($K_{sp} = 2.5 \cdot 10^{-22}$) and Ag_2S ($K_{sp} = 6.3 \cdot 10^{-50}$) [8] differ very strongly. Therefore, $(Ag_2S)_x(ZnS)$ heteronanostructures have been synthesized in two stages. First, silver sulfide Ag_2S was synthesized by chemical deposition from aqueous solutions of silver nitrate $AgNO_3$ and sodium sulfide Na_2S in the presence of sodium citrate Na_3Cit. To obtain colloidal solutions of silver sulfide without Ag impurities and without a citrate shell, reaction mixtures with a small relative excess of sodium sulfide Na_2S and a minimum concentration of Na_3Cit were used. The synthesis of a colloidal Ag_2S solution was carried out in the dark in a neutral medium at pH $\approx$ 7. Next, Trilon B was added to an aqueous zinc nitrate solution under continuous stirring. Then, the solution prepared was mixed simultaneously with a sodium sulfide solution and synthesized colloidal silver sulfide solution. The use of Trilon B promoted the formation of nanostructured zinc sulfide on a surface of silver sulfide particles, to occurrence of ZnS thin lamelliform nanoparticles, and to formation of $(Ag_2S)_x(ZnS)$ heteronanostructures at a final stage of synthesis. The compositions of reaction mixtures are given in Table 1. Conditions of synthesis of $(Ag_2S)_x(ZnS)$ heteronanostructures are described in detail in study [9].

The synthesized powders were washed and decantated. The synthesized samples were dried by sublimation methods in an Alpha 1-2 LDplus freeze dryer (Martin Christ) at an ice condenser temperature of -55°C (218 K).

Table 1. Composition of reaction mixtures, lattice constant a_{ZnS}, average size D of Ag_2S nanoparticles and average size h of ZnS nanoparticles in synthesized $(Ag_2S)_x(ZnS)$ heteronanostructures

Sample	Composition	Reagent concentration in the reaction mixtures (mmol·L^{-1})						a_{ZnS} (nm)	D±0.5 (nm)	h±1 (nm)
		ZnS synthesis			Ag_2S synthesis					
		$Zn(NO_3)_2$	Na_2S	Trilon B	$Ag(NO_3)_2$	Na_2S	Na_3Cit			
1	$(Ag_2S)_{0.025}(ZnS)$	50	50	3.125	2.5	1.25	1	0.53755	10	4
2	$(Ag_2S)_{0.10}(ZnS)$	50	50	25	10	5	1	0.53974	9	4
3	$(Ag_2S)_{0.25}(ZnS)$	50	50	-	25	12.5	5	0.53664	28	5
4	$(Ag_2S)_{0.50}(ZnS)$	50	50	-	50	25	10	0.53631	30	5

Deposited dried sulfide nanopowders were examined by X-ray diffraction (XRD) method on a Shimadzu XRD-7000 diffractometer in Cu$K\alpha_{1,2}$ radiation. The determination of the crystal lattice parameters and final refinement of the structure of the synthesized sulfide nanopowders were carried out with the use of the X'Pert Plus software suite [10].

The microstructure, particle size and element chemical composition of powders of $(Ag_2S)_x(ZnS)$ heteronanostructures were studied by the scanning electron microscopy (SEM) method on a JEOL-JSM LA 6390 microscope coupled with a JED 2300 Energy Dispersive X-ray Analyzer and by the high-resolution transmission electron microscope (HRTEM) method on a JEOL JEM-2010 microscope.

In addition, the average particle size D (more precisely, the average size of coherent scattering regions (CSR)) in synthesized nanopowders of $(Ag_2S)_x(ZnS)$ heteronanostructures was estimated by XRD method from the diffraction reflection broadening using the dependence of reduced reflection broadening $\beta^*(2\theta) = [\beta(2\theta)\cos\theta]/\lambda$ on the magnitude of scattering vector $s = (2\sin\theta)/\lambda$ [11, 12].

3. Results and Discussion

The XRD patterns of $(Ag_2S)_x(ZnS)$ heteronanostructures are shown in Figure 1. For comparison, the XRD patterns of nanostructured Ag_2S with a particle size ~40 nm and nanostructured ZnS with a particle size ~4 nm are shown.

A quantitative analysis of XRD patterns of $(Ag_2S)_x(ZnS)$ heteronanostructures and comparison with data [13, 14] has shown the presence of diffraction reflections of cubic ZnS and monoclinic Ag_2S. Diffraction reflections of Ag_2S and, especially, ZnS in XRD patterns of heteronanostructures are strongly broadened. It testifies to the small size of sulfide particles of which $(Ag_2S)_x(ZnS)$ heteronanostructures consist.

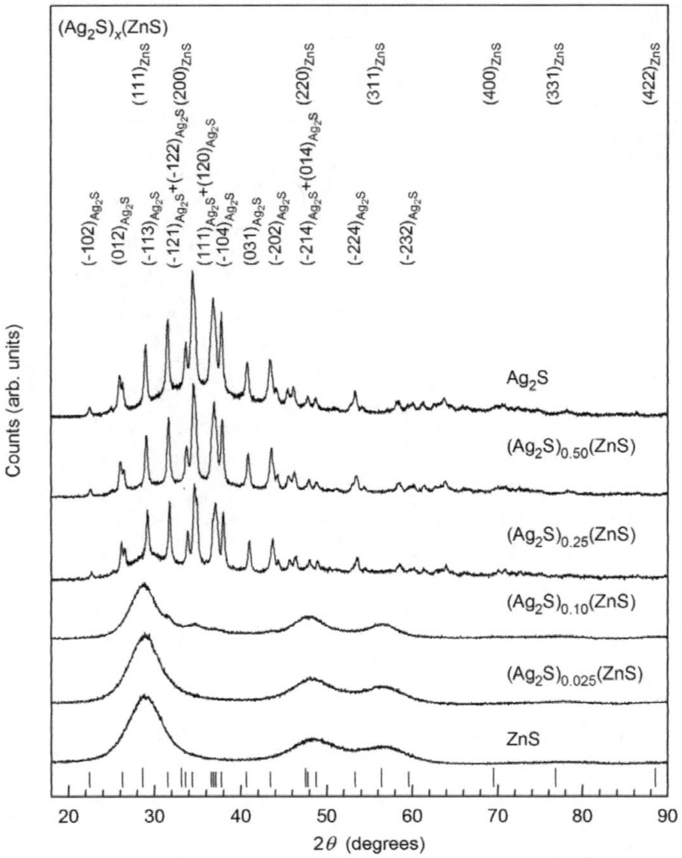

Figure 1. XRD patterns of $(Ag_2S)_x(ZnS)$ heteronanostructures and nanostructured Ag_2S and ZnS.

The observed set of diffraction reflections corresponds to cubic (space group *F*-43*m*) zinc sulfide with structure of the sphalerite (zinc blende) type and to silver sulfide with monoclinic (space group $P2_1/c$) α-Ag_2S acanthite structure. Crystal lattice constant a_{ZnS} of ZnS is 0.5344-0.5398 nm.

According to the results of an energy-dispersive X-ray analysis, the contents of Zn, Ag and S in heteronanosrtructure 1 are equal to 63.1±0.3, 5.2±0.1 and 31.7±0.2 wt.%, and in heteronanostructure 2 to 53.4±0.3, 17.6±0.2 and 28.7±0.2 wt.%, respectively. The contents of Zn, Ag and S in heteronanosrtructure 3 are 40.9±0.3, 33.9±0.2 and 25.1±0.2 wt.%, in heteronanostructure 4 are equal 29.6±0.2, 48.6±0.3 and 21.8±0.2 wt.%, respectively. Considering the contents of elements determined and the ratios of reagent concentrations in reaction mixtures 1-4 (*see* Table 1), prepared heteronanostructures 1, 2, 3, and 4 have approximate compositions $(Ag_2S)_{0.025}(ZnS)$, $(Ag_2S)_{0.10}(ZnS)$, $(Ag_2S)_{0.25}(ZnS)$, and $(Ag_2S)_{0.50}(ZnS)$.

The formation of the heteronanostructure is confirmed by the HRTEM image of a nanoparticle (Fig. 2) obtained from the reaction mixture 2. It is observed that the central part of the nanoparticle is formed by silver sulfide, the interplanar distance of ~0.246 nm coincides with the distance between the atomic planes (111) of silver sulfide with a monoclinic (space group $P2_1/c$) α-Ag_2S acanthite structure. The surface of the nanoparticle is partially covered with a layer of cubic zinc sulfide: the observed interplanar distance of ~0.311-0.312 nm corresponds to the distance between the atomic planes (111) of cubic (space group *F*-43*m*) ZnS (Fig. 2). The size of Ag_2S particle is 15 nm, the size of ZnS particles is about 4 nm and consistent with the size estimated from XRD data (*see* Table. 1).

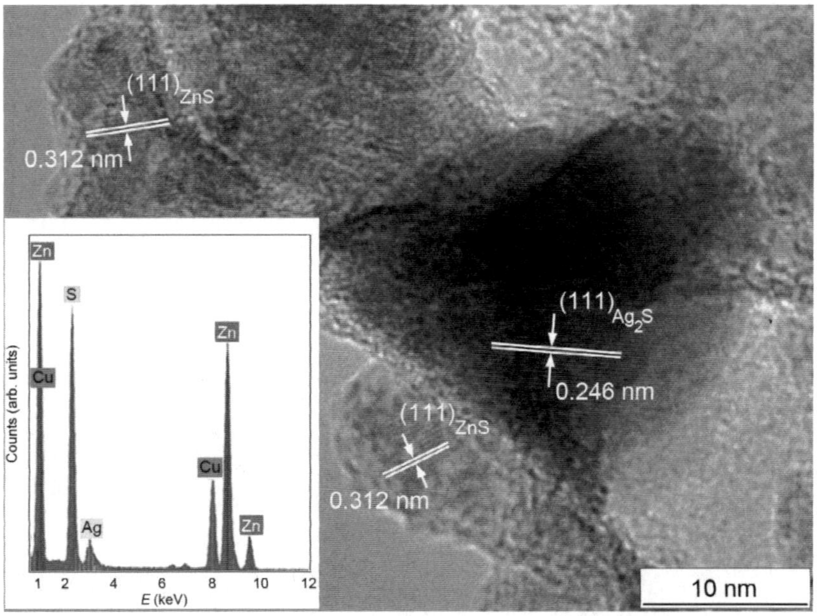

Figure 2. HRTEM image of a heteronanoparticle obtained from the reaction mixture 2. The central part of the nanoparticle is formed by Ag_2S, and the surface of the nanoparticle is partially covered with a layer of cubic ZnS. The inset presents the EDX spectrum of $(Ag_2S)_{0.10}(ZnS)$ heteronanostructure.

Some particles of the heteronanostructures synthesized represent plate-like nanoparticles of silver sulfide with the sizes up to 10 nm and more, which consist from disoriented crystal blocks with the sizes from 2 to 5 nm. Particles of silver sulfide are covered partially by more thin lamellar nanoparticles of cubic zinc sulfide.

Thus, the proposed two-stage chemical co-deposition of Ag_2S and ZnS sulfides allows synthesizing $(Ag_2S)_x(ZnS)$ heteronanostructures. Apparently, such synthesis procedure is also applicable to the preparation of heteronanostructures of other metal sulfides having different solubility.

4. Conclusions

A comparative study of chemical deposition of $(Ag_2S)_x(ZnS)$ heteronanostructures is carried out for the first time. The heteronanostructures were synthesized by chemical co-deposition of silver and zinc sulfides from aqueous solutions. Nanopowders of $(Ag_2S)_x(ZnS)$ sulfide heteronanostructures with different nanoparticle sizes and different phase compositions were obtained.

A change in the ratios of reagent concentrations allows one to control the size of Ag_2S nanoparticles in the prepared heteronanostructures in the range from 9 to 30 nm, and the size of the ZnS nanoparticles from 4 to 5 nm.

Acknowledgement

This study was supported by the Russian Science Foundation (grant no. 19-79-10101) through the Institute of Solid State Chemistry of the Ural Branch of the RAS.

References

[1] Fang X, Zhai T, Gautam U K, Li L, Wu L, Bando Y and Golberg D 2011 *Progr. Mater. Sci.* **56** 175
[2] Wang X, Huang H, Liang B, Liu Z, Chen D and Shen G 2013 *Crit. Rev. Solid State Mater. Sci.* **38** 57
[3] Cui C, Li X, Liu J, Hou Y, Zhao Y and Zhong G 2015 *Nanoscale Res. Lett.* **10** 431
[4] Sadovnikov S I and Gusev A I 2017 *J. Mater. Chem. A* **5** 17676
[5] Sadovnikov S I, Rempel A A and Gusev A I 2018 *Russ. Chem. Rev.* **87** 303
[6] Sadovnikov S I 2019 *Russ. Chem. Rev.* **88** 571
[7] Shen S, Zhang Y, Peng L, Du Y, Wang Q 2011 *Angew. Chem. Int. Ed.* **50** 7115
[8] Patnaik P 2004 *Dean's Analytical Chemistry Handbook* (New York: McGraw-Hill)
[9] Sadovnikov S I, Ishchenko A V and Weinstein I A 2020 *J. Alloys Comp.* **831** 154846
[10] X'Pert HighScore Plus. Version 2.2e (2.2.5). © 2009 PANalytical B. V. Almedo, the Netherlands.
[11] Williamson G K and Hall W H 1953 *Acta Metall.* **1** 22
[12] Gusev A I and Rempel A A 2004 *Nanocrystalline Materials* (Cambridge: Cambridge Intern. Science Publ.)
[13] Swanson H E and Fyatt R K 1953 *NBS Circular 539*, **2** 16
[14] Sadovnikov S I, Gusev A I and Rempel A A 2015 *Phys. Chem. Chem. Phys.* **17** 12466

Simulations of ZnS deposition on Ag_2S surface and formation of Ag_2S/ZnS heteronanostructure

Ilya A Balyakin[1,2,3], Stanislav I Sadovnikov[1]

[1] Institute of Solid State Chemistry, Ural Branch of the Russian Academy of Sciences, 620990 Ekaterinburg, Pervomaiskaya st. 91, Russia
[2] Institute of Metallurgy, Ural Branch of the Russian Academy of Sciences, 620016 Ekaterinburg, Amundsen st. 101, Russia
[3] Ural Federal University named after the first President of Russia B. N. Yeltsin , NANOTECH Centre, 620002 Ekaterinburg, Mira st. 19, Russia

i.a.balyakin@gmail.com

Abstract. Simulation of ZnS deposition from aqueous solution on a surface of crystalline Ag_2S has been performed to determine the features of formation of Ag_2S/ZnS heteronanostructures. The classical molecular dynamics and density functional theory have been used to study the features of the initial stages of ZnS growth on Ag_2S [001] surface. By classical molecular dynamics it was established that sulfur atoms initially are adsorbed on the Ag_2S. Zinc atoms were shown to be adsorbed after sulfur atoms. However, the final location of the first adsorbed Zn layer was closer to the Ag_2S surface than location of the first adsorbed S layer. Density functional theory calculation showed that Zn atoms were indeed closer to Ag_2S surface and confirmed classical molecular dynamics results. However, density functional theory results are more precise since they made possible to account for Ag_2S surface reconstruction.

1. Introduction

Semiconductor silver sulfide Ag_2S and zinc sulfide ZnS are demanded materials of modern electronics [1-4]. Modifying of electronic and optical properties of sulfides and composites of different sulfides can be reached by a change of band gap. The band gap of sulfides varies with a decrease in the size of particles and with the creation of sulfide heteronanostructures (nanocomposites) of two different metals [5]. It is known than doping of zinc sulfide by silver sulfide leads to a large decrease in the band gap. The creation of semiconductor nanocomposites consisting of Ag_2S and ZnS nanoparticles allows adjusting the band gap of Ag_2S/ZnS heteronanostructures. One of the methods for the production of heteronanostructures based on nanostructured zinc and silver sulfides is the co-deposition from aqueous colloidal solutions [6, 7].

Experimental synthesis of Ag_2S/ZnS heteronanostructures in aqueous solutions has some problems. In particular, two types of particles of Ag_2S/ZnS heteronanostructure were observed. Some particles of the heteronanostructures synthesized represent plate-like nanoparticles of silver sulfide, which consist from disoriented crystal blocks with the sizes from 2 to 5 nm. Particles of silver sulfide are covered partially by thinner lamellar nanoparticles of cubic zinc sulfide. Ag_2S/ZnS heteronanoparticles of the second type are core-shell type nanoparticles. The core of these nanoparticles consists of silver sulfide and has a size 10-30 nm. Ag_2S core is covered by ZnS shell with a thickness about 4-5 nm.

IOP Conf. Series: Materials Science and Engineering 1008 (2020) 012020 doi:10.1088/1757-899X/1008/1/012020

The sequence of deposition of atoms of a different type (Zn or S) on silver sulfide is not quite clear. Possibility of deposition of silver sulfide on ZnS is not clear too.

The features of the formation of the Ag_2S/ZnS heteronanostructure in an aqueous solution can be found by simulations.

The main task of mathematical modeling of nanostructures is to predict their properties. There is still no single method for modeling nanostructures. This is primarily due to the limited capabilities of computer technology because the most accurate first-principle calculations based on quantum mechanics [8] do not allow modeling large arrays of atoms. Therefore, the classical molecular dynamics (CMD) method based on Newtonian mechanics is used for this purpose [9, 10].

In the present work, the process of capture of ions of zinc and sulfur by Ag_2S surface has been simulated using the molecular dynamics. Since in CMD simulation Ag_2S layers were frozen, the reconstruction of its surface wasn't accounted for. For considering the surface reconstruction, density functional theory simulation was performed. The main objective of simulation was to determine the features of formation of the first ZnS layer on Ag_2S [001] surface.

2. Simulations Details

Classical molecular dynamics simulation was performed by means of LAMMPS software [11] (http://lammps.sandia.gov) using a model with implicit solvent. It means that solvent molecules are not simulated, and their effect is accounted for by applying random force to dissolved atoms. Thus, in the simulated system only four types of atoms are presented: Ag and S from Ag_2S and dissolved Zn and S. Figure 1 represents the initial configuration of the simulation. For Ag and S atoms of Ag_2S surface the equations of motion are not solved, thus, these atoms don't change their positions with time, thus Ag_2S surface is "frozen". For Zn and S atoms from the solution the equations of motion are presented by Langevin equation [12]:

$$m\boldsymbol{a} = -\frac{\boldsymbol{v}}{b} - \nabla U + \boldsymbol{F}_L, \quad (1)$$

where m is the mass of the particle; **a** and **v** are the acceleration and speed of a particle; b is mobility of the particle; U is the potential energy of the particle due to the interaction with other particles and external field; $\mathbf{F_L}$ is Langevin force (which replaces the effect of solvent molecules).

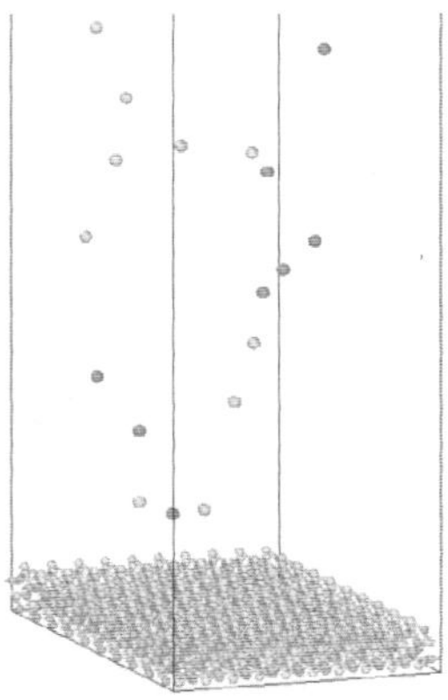

Figure 1. Initial configuration of CMD simulation

Types of interaction for CMD simulations between different types of atoms are presented in the Table 1. There is no interaction between pairs of frozen atoms. The detailed description of Tersoff potential can be found elsewhere [13]. For all Lennard-Jones interactions the cut-off distance is equal to corresponding σ parameter, thus, only repulsion is described by Lennard-Jones interactions. For fixing of adsorbed sulfur atom three Ag-S bonds are required, thus, 3 different sets of parameters are presented in Table 1 for the pair Ag-S_{sol}. These parameters were found from the "energy-volume" dependence of Ag_2S unit cell obtained by density functional theory calculation.

Table 1. Interaction between different atom types in the system

Pair of atoms	Type of interaction	Parameters
S_{Ag2S}-Zn S_{sol} -Zn	Tersoff potential	A=2475.8 eV B=225.5 eV $\gamma = 1$ λ_1 =3.0099 Å^{-1} λ_2 = 1.7322 Å^{-1} λ_3 =0 $\beta = 1.1 \cdot 10^{-6}$ m = 3 n = 0.78734 c = 100390 d = 16.217 $\cos\theta_0$ = -0.59825 R = 3.059 Å D = 0.15 Å
S_{Ag2S}-S_{sol} S_{sol}-S_{sol}	Lennard-Jones potential*	ε = 0.01538 eV σ = 3.85 Å
Ag-Zn	Lennard-Jones potential	ε = 0.01295 eV σ = 3.11 Å
Ag-S_{sol}	Parabolic function**	r_{01} = 2.593 Å k_1 = 6.432 eV·Å^{-2} r_{02} = 2.664 Å k_2 = 6.901 eV·Å^{-2} r_{03} = 3.047 Å k_3 = 4.658 eV·Å^{-2}

* $U(r) = 4\varepsilon[(\sigma/r)^{12} - (\sigma/r)^6]$,
** $E = (k/2)(r - r_0)^2$.

For accounting of features of Ag_2S surface reconstruction and more physical description of ZnS unit deposition on Ag_2S surface density functional theory (DFT) simulations were performed using VASP software [14]. The cut-off energy for plane wave basis set was chosen to be 450 eV. For bulk calculations 7x7x7 Monkhorst-Pack [15] grid was used, and for [001] surface calculation 7x7x1 grid was used. Perdew-Burke-Ernzerhof [16] pseudopotentials were used for all calculations.

First, the energy of 1x1x3 unreconstructed slab was computed (E_{slab}^{ur}). Then, the atoms in two neighboring cells were allowed to change their positions until each component of force acting on them become lower than 10^{-2} eV/Å. The minimization of forces was performed using conjugate gradient algorithm [17]. Obtained reconstructed structure had the energy E_{slab}^{r}. Surface reconstruction energy was computed as

$$E_r^{001} = (E_{\text{slab}}^r - E_{\text{slab}}^{\text{ur}})/ab, \tag{2}$$

where a and b are lattice parameters of Ag_2S (4.2264 Å and 6.9282 Å respectively). It should be mentioned that the absence of factor 2 in denominator of (2) is connected with the fact that only one of two surfaces of a slab was allowed to relax.

For obtaining S and Zn positions the following steps were performed. First, a sulfur atom was placed at not reconstructed Ag_2S surface at its "ideal" position (the position which would this sulfur atom occupy in perfect infinite crystal). After that, structural relaxation was performed keeping atoms in the bottom unit cell frozen. Then, Zn atom was added to this relaxed 3 Ag_2S+S system. The initial position of Zn atom was calculated as follows

$$\boldsymbol{r}_{\text{Zn}} = \frac{\boldsymbol{r}_{\text{S1}}+\boldsymbol{r}_{\text{S2}}}{2} + \frac{[(\boldsymbol{r}_{\text{S2}}-\boldsymbol{r}_{\text{S1}})\times[(\boldsymbol{r}_{\text{S2}}-\boldsymbol{r}_{\text{S1}})\times\boldsymbol{k}]]}{|[(\boldsymbol{r}_{\text{S2}}-\boldsymbol{r}_{\text{S1}})\times[(\boldsymbol{r}_{\text{S2}}-\boldsymbol{r}_{\text{S1}})\times\boldsymbol{k}]]|} \cdot \sqrt{d_{\text{ZnS}}^2 - (\boldsymbol{r}_{\text{S2}} - \boldsymbol{r}_{\text{S1}})^2/4}, \tag{3}$$

where $\mathbf{r}_{\text{S1,2}}$ are the positions of two sulfur atoms closest to the surface, **k** is {001} vector, d_{ZnS} is the smallest distance between zinc and sulfur in cubic ZnS, [**a**×**b**] means cross product of **a** and **b**.

For computing adsorption energy of ZnS unit, energies of single Zn and single S atoms were calculated. Then, the adsorption energy was computed as follows

$$E_{\text{ads}} = E_{\text{Ag2S+ZnS}} - (E_S + E_{\text{Zn}} + E_{\text{slab}}^r), \tag{4}$$

where $E_{\text{Ag2S+ZnS}}$ is the energy of relaxed 1x1x3 Ag_2S + ZnS structure.

3. Results and Discussion

3.1 Classical MD simulation

A model with the following approximations was used for the final calculation of ZnS deposition on the Ag_2S surface: (1) Ag_2S is represented by a plate having a thickness of one unit cell and containing 15×15 surface cells of silver sulfide; (2) above each Ag_2S unit cell, zinc and sulfur atoms are located one at a time randomly in planes which are parallel to the Ag_2S surface, and are at a distance of 10 Å from each other; (3) the initial speed of zinc and sulfur atoms was the same, directed perpendicular to the Ag_2S surface and equal to $4.1\cdot10^{-3}$ Å·fs^{-1}; (4) every 10000 machine steps or every 10 ps, forces directed perpendicularly downwards to the Ag_2S surface and equal to 0.043 and 0.086 eV·Å^{-1}, were applied to the sulfur and zinc atoms, respectively; (5) to eliminate the high deposition speed, which can be caused by the periodic application of these forces, the Nose-Hoover algorithm was used for constant temperature 300 K molecular dynamics simulations. All these approximations are aimed to imitate Ag_2S surface in very diluted ZnS solution.

In this calculation, it is important to find out what positions are occupied by the atoms of sulfur and zinc on the surface of Ag_2S.

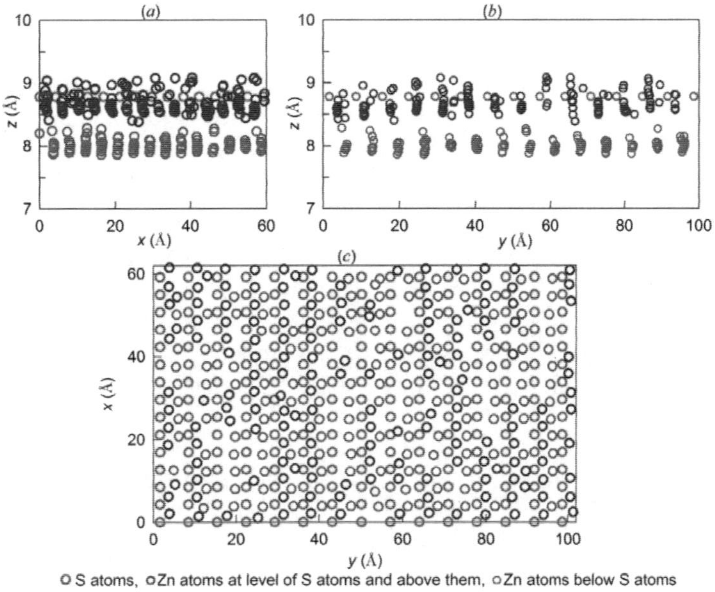

Figure 2. Projections of the positions of Zn and S atoms that are situated above the Ag_2S surface on the (*a*) [*x*0*z*], (*b*) [0*yz*], and (*c*) [*xy*0] planes.

The positions of S atoms on the surface of Ag_2S are indirectly set in advance because Ag-S bonds make the position of S atoms unambiguous. The calculated projections of the positions of Zn atoms that are situated over the Ag_2S surface and averaged over time are shown in Figure 2. The projections of the positions of surface S atoms on the [*xy*0] plane are presented in Fig. 2*c*. In fact, these are the positions around which the deposited sulfur atoms should oscillate.

For CMD calculations mean distance between atoms in the closest tothe surface Zn layer (green circles on Fig. 2) and sulfur atoms (red circles on Fig. 2) is 3.07 Å which significantly differs from the smallest Zn-S distance in cubic ZnS (d_{ZnS} = 2.34 Å).

3.2 Density functional theory calculation

The energy of 1x1x3 unreconstructed slab obtained by DFT calculations is equal to E_{slab}^{nr} = -113.059 eV, and the energy of 1x1x3 reconstructed slab to E_{slab}^{r} = -114.163 eV. Thus, reconstruction energy was estimated according to (2) to be -0.038 eV/Å^2 or -0.603 J·m^{-2}. Energy of final relaxed 1x1x3 Ag_2S + ZnS structure is $E_{Ag2S+ZnS}$ = -120.389 eV, and energies of single Zn and S atoms are E_{Zn} = -0.011 eV and E_S = -0.875 eV, respectively. According to (4), the adsorption energy of ZnS unit on [001] Ag_2S surface is 5.34 eV.

The structural changes of Ag_2S surface + ZnS are presented in Figure 3.

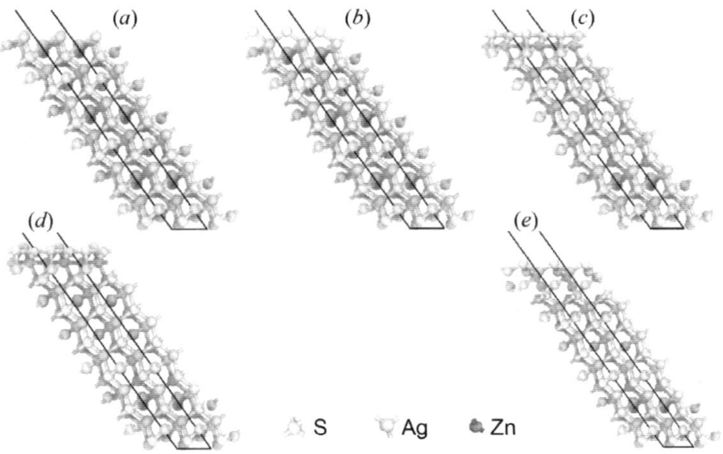

Figure 3. Structure of Ag_2S [001] surface with adsorbed ZnS: (*a*) pure unreconstructed Ag_2S surface; (*b*) unreconstructed Ag_2S surface with additional sulfur atom, i. e. Ag_2S+S system; (*c*) relaxed Ag_2S+S system; (*d*) relaxed Ag_2S+S system with Zn atom placed according eq.(4); (*e*) relaxed Ag_2S+S+Zn system.

From Fig. 3 we observe that Zn atoms are indeed closer to Ag_2S surface than adsorbed S atoms, which agrees with CMD simulations. But the distance between Zn and S atoms in this case equals 2.16 Å, which is much closer to Zn-S distance in cubic ZnS comparing to CMD study. Moreover, downward difference is reasonable, as far as there are no attractive forces acting atoms from vacuum, which makes the bond lengths shorter comparing to bond lengths in bulk material.

4. Conclusions

CMD simulations of ZnS deposition on Ag_2S surface demonstrated, that sulfur atoms are first deposited from the solution. However, in spite of the fact that Zn atoms are deposited after S atoms, positions of Zn atoms are closer to the Ag_2S surface.

Density functional theory simulation demonstrated severe relaxation of Ag_2S surface which is characterized by relatively high relaxation energy. However, structural relaxation of ZnS unit on the [001] Ag_2S surface demonstrated the same tendency as CMD simulation – Zn atom was located closer to the Ag_2S surface then S atom. Zn-S distance in CMD simulations significantly differs from this one in cubic ZnS, while for DFT simulations reasonable small downward difference is observed. This indicates, that CMD simulation could be used for ZnS deposition on Ag_2S only quantitatively but not qualitatively. However, DFT simulations has low performance and using DFT for direct modeling of described deposition is demanding to computer recourses and time. Thus, future work will be aimed for constructing of classical interatomic potential for the system Ag-S-Zn based on *ab initio* data, as far as such potential would allow to perform accurate simulation of dynamics of Zn-S deposition on Ag_2S surface due to combining accuracy of DFT and performance of CMD.

Acknowledgement
This study was supported by the Russian Science Foundation (grant no. 19-79-10101) through the Institute of Solid State Chemistry of the Ural Branch of the RAS.

References
[1] Fang X, Zhai T, Gautam U K, Li L, Wu L, Bando Y and Golberg D 2011 *Progr. Mater. Sci.* **56** 175
[2] Sadovnikov S I and Gusev A I 2017 *J. Mater. Chem.* A **5** 17676
[3] Sadovnikov S I, Rempel A A and Gusev A I 2018 *Russ. Chem. Rev.* **87** 303
[4] Sadovnikov S I 2019 *Russ. Chem. Rev.* **88** 571
[5] Sadovnikov S I, Rempel A A and Gusev A I 2018 *Nanostructured Lead, Cadmium and Silver Sulfides: Structure, Nonstoichiometry and Properties* (Cham - Heidelberg: Springer Intern. Publ. AG)
[6] Zhang X, Liu X, Zhang L, Li D and Liu S 2015 *J. Alloys Comp.* **655** 38
[7] Sadovnikov S I, Ishchenko A V and Weinstein I A 2020 *J. Alloys Comp.* **831** 154846
[8] Kohn W 1999 *Rev. Mod. Phys.* **71** 1253
[9] Frenkel D and Smit B 2002 *Understanding Molecular Simulation. From Algorithms to Applications* (New York: Academic Press)
[10] Field M J 2007 *A Practical Introduction to the Simulation of Molecular Systems* (Cambridge: Cambridge Univ. Press)
[11] Plimpton S 1995 *J. Comput. Phys.* **117** 1
[12] Langevin P 1908 *C. R. Acad. Sci. (Paris)* **146** 530
[13] Tersoff J 1988 *Phys. Rev.* B **37** 6991
[14] Kresse G and Hafner J 1993 *Phys. Rev.* B **47** 558
[15] Monkhorst H J and Pack J D 1976 *Phys. Rev.* B **13** 5188
[16] Perdew J P, Burke K and Ernzerhof M 1996 *Phys. Rev. Lett.* **77** 3865
[17] Press W H, Teukolsky S A, Vetterling W T and Flannery B P 2007 *Numerical Recipes* (New York: Cambridge University Press)

Optimization of high pressure torsion processing for fabrication of the Al-Nb hybrid system

G R Khalikova[1,2], G F Korznikova[1], K S Nazarov[1], R Kh Khisamov[1], S N Sergeev[1], R U Shayakhmetov[1], and R R Mulyukov[1,2]

[1]Institute for Metals Superplasticity Problems, Russian Academy of Sciences, 39 Khalturin st., Ufa 450001, Russia
[2]Ufa State Petroleum Technological University, 1 Kosmonavtov st., Ufa 450062, Russia

gulnara.r.khalikova@gmail.com

Abstract. A hybrid composite material fabricated from the Al-Nb system using high-pressure torsion (HPT) up to 30 turns has been studied in the present work. To fabricate the composite, deformation of a three-layer Al-Nb-Al package was carried out at room temperature on Bridgman anvils with grooves under a pressure of 5 GPa at N=10, 25 and 30 revolutions, at a strain rate of ω=1 and 2 rpm. Initial disc diameters from pure metals and HPT conditions were experimentally optimized to obtain monolithic and defect-free composite samples. The most intensive fragmentation and stirring of niobium in the aluminium matrix was observed if diameter of aluminium discs was 10 mm and deformation conditions N=25 and 30 revolutions and ω=2 rpm were applied. Three microstructural zones were observed after HPT under optimal conditions: the central zone with wide curved layers of niobium in aluminium, the mid-radius zone with finely dispersed layered structure, and periphery with a uniform distribution of niobium in the aluminium matrix. It was shown that HPT led to occurrence of strain-induced ageing resulting in formation of the intermetallic Al_3Nb phase. The microhardness measured along the diameter of the obtained composite materials changed nonmonotonically depending on the produced structure (microstructural zone).

1. Introduction

To create aluminum-matrix composite, from standpoint of approximation of the interaction system of the reinforcing phase and the Al matrix to a thermodynamically stable one, various methods based on the in-situ principle are actively used [1-3]. In this case, hardening intermetallic phases playing the role of reinforcements are formed in the Al matrix due to their nucleation and growth. One of the methods is severe plastic deformation by high-pressure torsion (HPT) [4]. Different layered metal-matrix composites have been recently fabricated via HPT of a stacking of discs of two dissimilar metals such as Al-Mg [5], Al-Cu [6], Ag-Cu [7]. In this case, hybrid materials having a unique multilevel structure resulted from the occurrence of diffusion processes are formed. An aluminum-matrix composite based on the Al-Nb system is of great interest to create the hybrid composite material, because the melting point of niobium is 2.5 times higher than that of pure Al and the solid state solubility of Nb in Al is near zero. This will expectedly lead to the greatest contribution of the hardening intermetallic phase(s) synthesized during HPT to the strength and other mechanical properties.

The aim of the present work was to access the feasibility of fabrication in-situ of the hybrid composite materials based on the Al-Nb system via HPT.

2. Materials and methods

The initial materials were Al (with purity 99.5%) and Nb (with purity 99%). The diameters of the Al discs were 6, 7, 8, 10, and 12 mm, and the Nb discs were 12 mm in diameter. The thickness of the Al discs was in the range of 0.3-1.2 mm and the Nb discs were 0.5 mm in thickness. The aluminium and niobium discs were laid in the three-layer Al-Nb-Al package.

HPT of the Al-Nb-Al package was carried out at the room temperature (RT) on Bridgman anvils under a pressure of 5 GPa at N=10, 25, and 30 revolutions, at a rotation speed of ω=1 and 2 rpm. The final size of the aluminum-matrix composite samples obtained via HPT was Ø12 mm × 0.6 mm.

The microstructure examination and microhardness measurements were carried out from the cross sections of the composite samples. Microstructural observations were conducted using a Tescan Mira 3LMH scanning electron microscope in the backscattering electron (BSE) mode. Transmission electron microscopic (TEM) studies were performed on a JEM 2100 Plus microscope at 200 kV accelerating voltage. The selected area diffraction pattern was obtained from a 2 μm^2 area. The X-ray diffraction analysis was performed for the cross section of the disc using a high-resolution Rigaku Ultima IV X-Ray diffraction system and Cu-K_α radiation. Microhardness was measured by the Vickers method on an AFFRI DM8A microhardness tester with 10 g indentation load.

3. Results and discussion

As has been shown elsewhere [6], the formation of monolithic aluminum-matrix composite of three-layer Al-Cu-Al packets by HPT took place at an optimal pressure of about 5 GPa, ω=1 rpm and N=10. In this case, the diameter of the Al and Cu discs was 12 mm, the initial thickness was about 0.5 mm. However, the use of the same diameters of the Al and Nb discs and the same HPT conditions, intensive stirring of the layers did not occur (figure 1). The sample was retained layered and only some fragmentation of layers was detected at periphery of the sample (figure 1). Therefore, the varied initial diameters of Al discs (6, 7, 8, and 10 mm) were used for HPT of the Al-Nb-Al packages. In these experiments, the geometry of the Nb discs remained unchangeable. SEM observations of the composite samples produced by HPT (N=10, ω=1 rpm) showed that the most intensive fragmentation and stirring of niobium in the aluminum matrix were achieved with an Al disc diameter of 10 mm (figure 1). So, the next variations of HPT conditions were fulfilled using the Al discs of 10 mm in diameter.

Increased deformation rate due to an increase in the number of sliding systems involved in deformation [8], led to a more significant fragmentation and a better niobium distribution within the sample. The Al-Nb-Al samples were deformed to 25 and 30 revolutions for further microstructure refinement and mixing of the Al and Nb layers. In doing so, monolithic deformed samples free of any visible cracks and chips were obtained. In the samples deformed to 25 and 30 revolutions the intensive fragmentation and mixing of the Al and Nb layers was attained practically in the entire volume of the deformed samples (figure 1) that is consistent with ref. [9].

The HPT of the Al-Nb-Al samples at N=25 and 30 revolutions led to the formation of vortex structures as a result of laminar flow of the composite material. Note that similar observations were earlier made in ref. [10]. In the center of samples wide curved layers of the starting materials are observed (figure 2a). At the mid-radius area of samples, the initial materials were transformed into a dispersed layered structure with a thickness of the observed layers not more than 100 nm (figure 2b). At the periphery of the samples, some mixed structure consisting of separate areas of dispersed interlayers of the components was formed (figure 2c). In this case, the increase of the revolution number during HPT from 25 to 30 expanded significantly the regions of a homogeneous structure while maintaining small individual zones with a layered structure (figure 1). In addition, during HPT at room temperature and N=25 and 30, the aluminum based solid solution supersaturated with niobium was formed, followed by strain-induced ageing leading to synthesis of the dispersed particles of the Al_3Nb intermetallic phase. This conclusion is based on the results of X-ray diffraction analysis and TEM observations (figure 3).

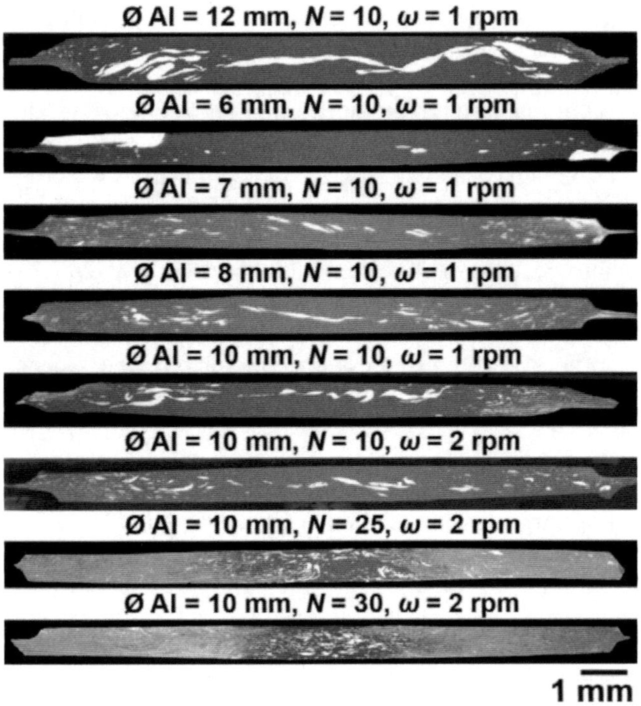

Figure 1. BSE images of polished cross sections of Al-Nb composites obtained using different initial diameters of Al discs and HPT conditions (*N* and *ω*). Dark contrast corresponds to pure aluminum and bright contrast corresponds to pure niobium.

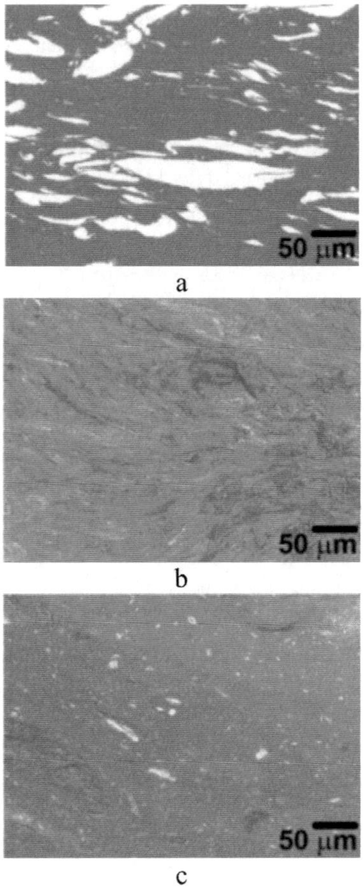

Figure 2. The typical BSE images of the the Al-Nb composites produced by HPT at N=30 and ω=2 rpm: in the center (a), in the mid-radius (b), at the periphery (c) of the sample.

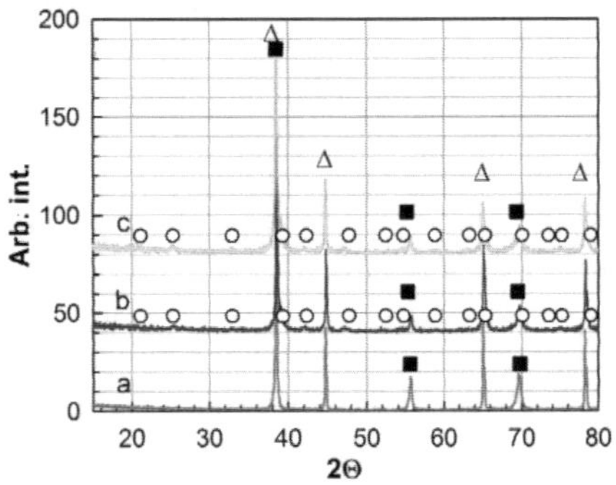

Figure 3. X-Ray diffraction pattern obtained for the Al-Nb-Al samples using Al disc diameter of 10 mm, processed by HPT at ω=2 rpm and N=10 (a); N=25 (b); N=30 (c).

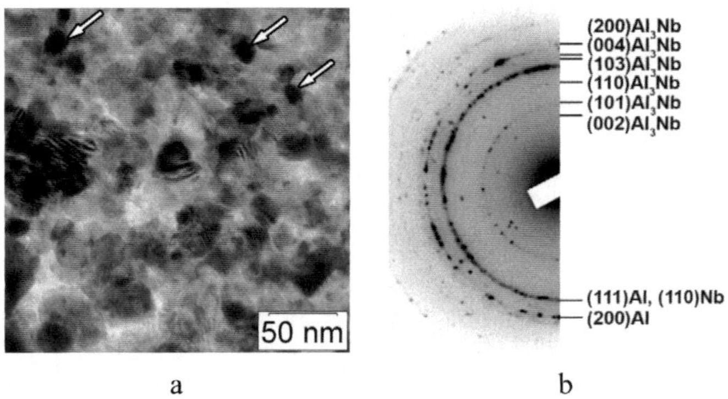

Figure 4. TEM bright-field image and the selected area diffraction pattern obtained from the mid-radius area of the Al-Nb-Al sample using Al disc diameter of 10 mm, processed by HPT at N=30 and ω=2 rpm. Some of the Al_3Nb intermetallic precipitations are arrowed (a). Selected area diffraction pattern corresponds to the mid-radius area (b).

Figure 5 represents the results of the microhardness measurements of different samples subjected to HPT with ω=2 rpm. So, at N=10 in the center and the mid-radius of samples, in which large inclusions and layers of Nb in Al were inhomogeneously disposed, the microhardness was comparable with that of the deformed Al sample and amounted to about 75 HV. At the periphery of the sample, in which dispersed inclusions of Nb in Al were allocated, the microhardness increased and reached about 110 HV.

The increase of the revolution number generally led to the increase of the microhardness. In this case, the changes of the microhardness along the diameter of the samples subjected to HPT at N=25 and 30 revolutions were similar. In the center of samples, where the microstructure was refined in the least extent and distinct dispersed layers and inclusions of Nb in Al were observed, the microhardness was minimal and amounted to about 100 HV in the both states (at N=25 and 30). The increase of the strain value from the center to the periphery of samples led to the increase of microhardness to the maximum values about 280 and 300 HV (for N=25 and N=30, respectively) in the mid-radius. This resulted from the laminated structure with enhanced area of interlayers and the higher density of defects. At the periphery of the samples the maximum value of shear deformation resulted in the formation of the strain-induced solid solution, which in turn resulted in the decrease of the interface hardening component, which was probably not compensated for by the solid solution hardening. Therefore, at the periphery of the samples the microhardness decreased to values about 130 HV (at N=25 and 30).

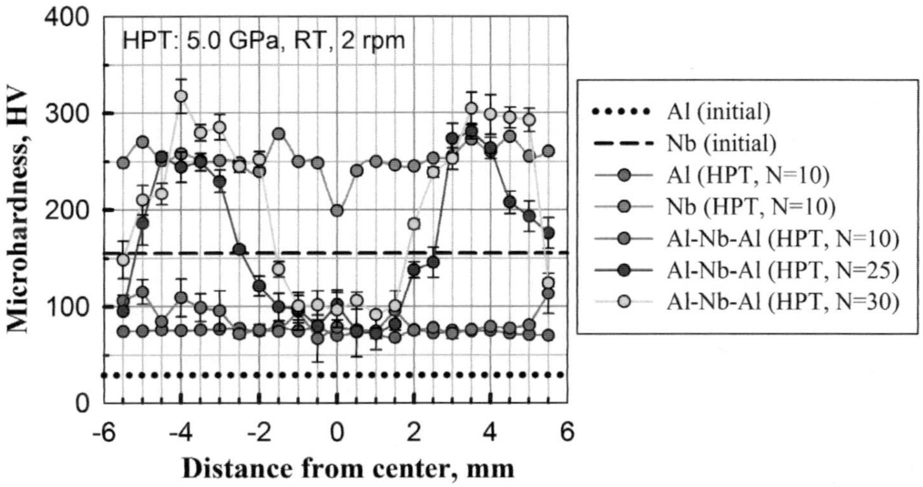

Figure 5. The microhardness measured along the diameter of the cross sections of the Al-Nb-Al composite samples obtained by HPT (with initial Al disc diameter of 10 mm) and the pure Al and Nb discs in the initial state and after HPT.

4. Summary

Monolithic composite samples based on the Al-Nb hybrid system free of any cracks were successfully produced by HPT using different processing conditions. Optimal combination of the initial disc diameters of Al and Nb was experimentally chosen. It was 10 and 12 mm for Al and Nb discs, respectively. HPT of the discs of the chosen diameters provided the most intensive fragmentation and stirring of niobium in the aluminium matrix. It was revealed that the best fragmentation and the most uniform distribution of Nb in the Al matrix were attained at a rotation rate of 2 rpm. The expansion of the significant structural refinement area and the formation of the metal-matrix structure took place at the revolution number of 25 and 30. The formation of the hardening intermetallic Al_3Nb phase due to the occurrence of the strain-induced ageing has been proven by means of X-ray diffraction analysis and TEM observations. The microhardness measurements and microstructure observations showed that the changes in the microhardness along the diameter of the composite samples were consistent with the microstructures formed in the samples after HPT. The microhardness values in the center of the samples were minimum and amounted to about 100 HV, in the mid-radius they were maximum and amounted to about 280 and 300 HV for N=25 and 30, respectively. At the periphery of the samples the microhardness was slightly higher than in the center and amounted to about 130 HV.

Acknowledgments

The present work was accomplished according to the state assignment of IMSP RAS and supported by the Russian Science Foundation (Grant No. 18-12-00440). Experimental studies were carried out using the facilities of the shared services center "Structural and Physical-Mechanical Studies of Materials" at the Institute for Metals Superplasticity Problems of Russian Academy of Sciences.

References

[1] Suryanarayana C 2019 *Madridge J. Nanotechnol. Nanosci.* **4** 127
[2] Srivatsan T S, Sudarshan T and Lavernia E 1995 *Prog. Mater. Sci.* **39** 317
[3] Zeidabadi S R H and Daneshmanesh H 2017 *Mater. Sci. Eng.* A. **702** 189
[4] Viswanathan V, Laha T, Balani K, Agarwal A and Seal S 2006 *Mater. Sci. Eng. R* **54** 121
[5] Han J-K, Liss K-D, Langdon T G and Kawasaki M 2019 *Sci. Rep.* **9** 17186
[6] Korznikova G F, Nazarov K S, Khisamov R Kh, Sergeev S N, Shayachmetov R U, Khalikova G R, Baimova J A, Glezer A M and Mulyukov R R 2019 *Mater. Lett.* **253** 412
[7] Pouryazdan M, Kaus B J P, Rack A, Ershov A and Hahn H 2017 *Nat. Commun.* **8** 1611
[8] Valiev R Z, Zhilyaev A P and Langdon T G 2013 *Bulk Nanostructured Materials: Fundamentals and Applications* (Wiley-Blackwell) 440 p
[9] Hernández-Escobar D, Rahman Z U, Yilmazer H, Kawasaki M and Boehlert C J 2019 *Phil. Mag.* **99** 557
[10] Kulagin R, Beygelzimer Y, Ivanisenko Yu, Mazilkin A, Straumal B and Hahn H 2018 *Mater. Lett.* **222** 172

IOP Conf. Series: Materials Science and Engineering 1008 (2020) 012022 doi:10.1088/1757-899X/1008/1/012022

Effect of nanostructuring on the thermal expansion and saturation magnetization of Fe - 36% Ni and Fe - 50% Ni alloys

K A Krylova[1,2], I Kh Bitkulov[1] and R R Mulyukov[1,2]

[1] Institute for Metals Superplasticity Problems of RAS, 39 Stepan Khalturin St., Ufa 450001, Russia
[2] Bashkir State University, 32 Zaki Validi St., Ufa 450076, Russia

E-mail: bukreevakarina@gmail.com

Abstract. Nanostructured samples of Fe - 36% Ni (Invar alloy) and Fe - 50% Ni alloys with a fragment size of about 100 nm were obtained by high pressure torsion using Bridgman anvils. The formation of the nanostructure leads to a decrease in the coefficient of thermal expansion of the Fe - 36% Ni and Fe - 50% Ni alloys by 2.2 and 1.2 times, respectively. Annealing of the Fe - 36% Ni alloy after nanostructuring leads to the formation of a dispersed bcc phase, which is not observed in the coarse-grained Invar alloy. The release of this phase affects the anomalous growth of thermal expansion (above the values typical for a coarse-grained Invar alloy) and the retention of saturation magnetization at temperatures above the Curie temperature of 260 °C. It is found that heating the nanostructured Fe - 50% Ni alloy above the Curie temperature (460 °C) saves the saturation magnetization. By analogy with the results for the Fe - 36% Ni, it can be assumed that this is due to the appearance of a dispersed bcc ferromagnetic phase. Detailed structural analysis will be carried out in future works.

1. Introduction

Alloys of the Fe-Ni system are distinctive materials due to the fact that the variation in the amount of dissolved Ni atoms in Fe has a significant effect on the physical properties of an alloy. A unique alloy of this system is Invar [1], which contains 36% Ni and has an abnormally low value of the coefficient of thermal expansion (CTE) equal to $\alpha \approx 1.6 \cdot 10^{-6} C^{-1}$ in a certain temperature range. Due to this, it has found wide application in modern technology and instrument making. However, a change in the nickel content by only ±5% leads to a sharp increase in the CTE. For example, for the Fe-50%Ni alloy, it has a value as high as $\alpha \approx 8.9 \cdot 10^{-6} C^{-1}$ [1].

An important disadvantage of Fe-Ni alloys, limiting their application is their low strength and hardness. For example, the tensile strength of the alloy Fe - 36% Ni is less than 500 MPa [2,3], and the tensile strength of a strip made of Fe - 50% Ni alloy is about 450 MPa. To date, the method of alloying with Be [4], Mo or W [5] atoms is used to harden the Fe - 36% Ni alloy, which most often leads to an increase in the CTE. Another effective method for increasing the strength properties of alloys is deformation nanostructuring. It was established in [6] that the microhardness of the Fe - 36% Ni alloy in the nanostructured (NS) state was 4250 MPa, although in the coarse-grained (CG) state it was 1300 MPa [3]. Nanostructured materials are not only characterized by high strength but also have unique physical properties that are not typical for materials in the CG state [7, 8, 9, 10, 11]. Therefore, a detailed study of the effect of severe plastic deformation on the properties of Fe-Ni alloys in different structural states is of fundamental and practical interest.

The present work is devoted to the study of the effect of deformation nanostructuring on the CTE and saturation magnetization of alloys of the Fe-Ni system: Fe - 36% Ni and Fe - 50% Ni. This work is a continuation of the previous works of the authors on the investigation of the phase composition and thermal expansion of Invar alloy in different structural states, from nanocrystalline to microcrystalline ones [7, 9,10].

2. Experimental

Nanostructured samples of the alloys Fe - 36% Ni (Invar) and Fe - 50% Ni were processed by severe plastic deformation using high pressure torsion on Bridgman anvils. The deformation was carried out up to the true logarithmic strain value of $e = 7$. The structures of Fe - 36% Ni and Fe - 50% Ni alloys after nanostructuring are shown in figures 1, a and b, respectively. It can be seen, that the alloys have a structure consisting of equiaxed fragments with a size of about 100-200 nm. Fragment boundaries are relatively thin. Reflexes on electron diffraction patterns obtained from several fragments are arranged in a circle. To obtain a different structural state, the samples after severe plastic deformation were annealed in a vacuum for 45 min and then cooled with the oven. Annealing of alloys after nanostructuring leads to the formation of equiaxed grains, and their average size increases monotonically with the annealing temperature T_{an}.

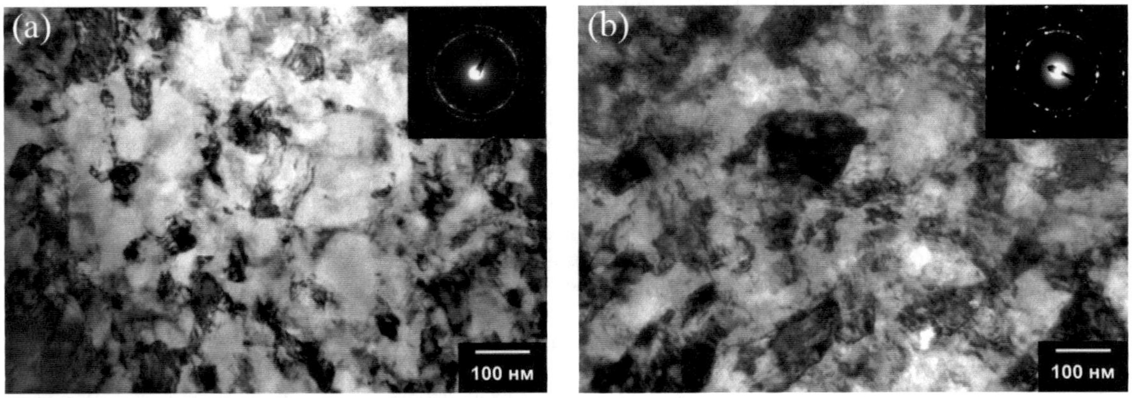

Figure 1. TEM images and diffraction patterns of NS Fe - 36% Ni (a) and Fe - 50% Ni (b) alloys obtained by severe plastic deformation.

The coefficient of thermal expansion $\alpha(T)$ is measured on a Dh 1500 RHP ULVAC SINKU-RIKO dilatometer. The NS samples are heated in the process of measuring $\alpha(T)$ at a constant rate of 2 °C/min and then cooled with the oven. The microstructure is studied using transmission electron microscopy (TEM) on a JEM-2000EX operated at 120 kV. Changes in the dependence of the saturation magnetization on the temperature and annealing time are carried out by automatic magnetic vacuum microbalance [12]. The magnetic field, which is enough to achieve saturation, is 240 kA/m.

3. Results and discussion

Figure 2 shows the CTEs at 20 °C of the Fe - 36% Ni and Fe - 50% Ni alloys after deformation nanostructuring as functions of the annealing temperature. It can be seen, that severe plastic deformation leads to a decrease in the coefficient $\alpha_{20\,°C}(T)$ by 2.2 and 1.2 times for Fe - 36% Ni (a) and Fe - 50% Ni (b) alloys, respectively. In [13], a decrease in the CTE of Invar alloy (Fe - 35% Ni - 0.49% Mn) was also observed after the plastic deformation by upsetting with the strain $e = 1.1$, which was explained by the arising distortion of the crystal lattice. This results in the weakening of the interatomic interactions and contributes to the appearance of negative bulk magnetostriction.

Annealing leads to an increase in the thermal expansion of the alloys, due to a decrease in the defects density, which reduces the distortion of the crystal lattice. After annealing at 500 °C, the value of $\alpha_{20\,°C}$ of both alloys approaches the one typical for the CG state. However, on the $\alpha_{20\,°C}(T_{an})$ curve for the Fe - 36% Ni alloy, an anomalous increase in the CTE is observed in the interval T_{an} = 350 – 420 °C. This is due to the precipitation of a less close-packed bcc phase in the Fe - 36% Ni alloy, which was previously discovered by the authors in the NS Invar alloy using X-ray diffraction analysis [10]. The deterioration of the Invar properties during the separation of the bcc phase in the Invar alloy obtained by electrodeposition was observed in [14]. At the annealing temperature above 420 °C, the ferromagnetic bcc phase dissolves, which decreases $\alpha_{20\,°C}$ to the values characteristic of CG Invar. The abnormal formation of the bcc phase after annealing of the NS Fe - 36% Ni alloy became possible due to a sharp increase in the diffusion capacity of alloys after deformation nanostructuring associated with nonequilibrium grain boundaries [15, 16, 17].

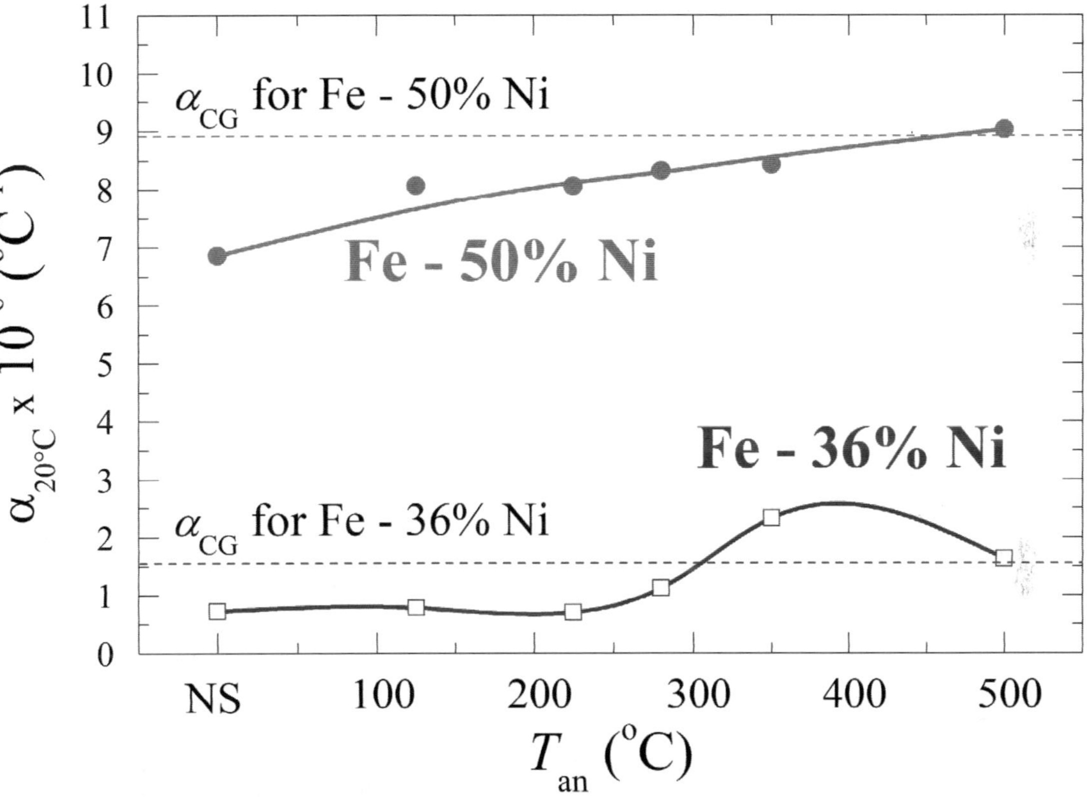

Figure 2. Thermal expansion coefficient at 20 °C $\alpha_{20\,°C}$ as the function of annealing temperature T_{an} for the NS Fe - 36% Ni (squares) and Fe - 50% Ni alloys (circles). The dashed line shows the CTE values corresponding to these alloys in the CG state.

Figure 3 shows the changes in the saturation magnetization of the Fe - 36% Ni (a) and Fe - 50% Ni (b) alloys during heating after deformation nanostructuring. One can see that the $\sigma(T)/\sigma_0$ curves show the preservation of the saturation magnetization of the studied alloys at T above the Curie temperature (T_c = 260 and 460 °C for Fe - 36% Ni and Fe - 50% Ni, respectively). In figure 3, the Curie temperature is shown with a vertical dashed line. It should be noted that upon cooling of NS alloys after annealing at 500 °C (blue curves in figure 3 (a) and (b)), it appears only at T_c, which is typical for alloys in the CG state. The retention of the saturation magnetization at $T > T_c$ for the Fe - 36% Ni alloy is associated with the formation of the ferromagnetic bcc phase, which dissolves during annealing at 500 °C [10, 18]. It was found in [18], that as a result of the annealing of the NS Invar alloy, about 10%

of the bcc phase was formed. Presumably, the formation of the bcc phase also occurs in the Fe - 50% Ni alloy, which leads to the observed magnetization of the alloy above the Curie temperature.

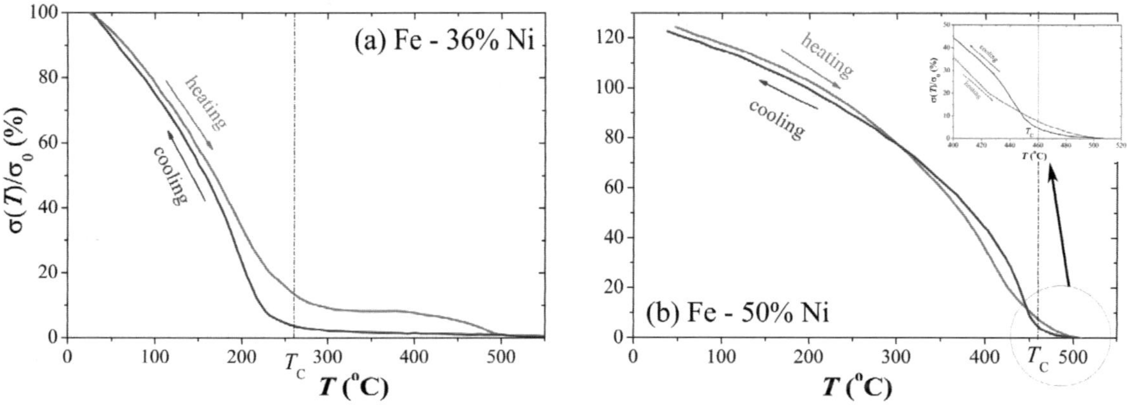

Figure 3. Saturation magnetization of the Fe - 36% Ni (a) and Fe - 50% Ni (b) alloys after nanostructuring as the function of temperature. The saturation magnetization, σ_0, of NS alloys at room temperature after annealing at 500 °C.

4. Conclusions

Using severe plastic deformations by high pressure torsion on Fe - 36% Ni and Fe - 50% Ni alloys, a NS with fragment sizes of about 100-200 nm is obtained. It is found that the formation of the NS state leads to a decrease in the CTE of the Fe - 36% Ni and Fe - 50% Ni alloys by 2.2 and 1.2 times, respectively, in comparison with the CG state. This is due to the effect of nonequilibrium grain boundaries on the magnetic subsystem of alloys and interatomic interaction, leading to an increase in negative magnetostriction [13].

After annealing of NS Invar alloy above 300 °C, an anomalous increase in thermal expansion above the CTE of the alloy in the CG state is observed. This is due to the formation of a less densely packed bcc phase [10, 18], which is not typical for the cast Fe - 36% Ni alloy. The formation of the bcc phase occurs due to an increase in the diffusion capacity of NS alloy [15, 16].

In the NS Fe - 36% Ni and Fe - 50% Ni alloys, anomalous retention of saturation magnetization is observed at temperatures above the Curie temperature (260 and 460 °C, respectively), which occurs due to the release of the ferromagnetic bcc phase, which dissolves after annealing at 500 °C.

The data obtained show that by varying the annealing temperature after deformation nanostructuring of Fe - 36% Ni and Fe - 50% Ni, it is possible to achieve the required thermal expansion value of alloys under study while maintaining their magnetic properties in a wide temperature range. Also, deformation nanostructuring leads to an increase in the mechanical properties of alloys [6], which is an important parameter for expanding the scope of their application.

Acknowledgments
The work is supported by the State Assignment of IMSP RAS.

References

[1] Callister W D 2007 *Materials Science and Engineering: An Introduction* 7th edition (NY: John Wiley & Sons) p 729
[2] Tu Y, Xu J, Zhang J, Zhang B and Zhao D 2007 *Univ. Sci. Technol. Beijing* **14** 46
[3] Vinogradov A, Hashimoto S, and Kopylov V I 2003 *Mater. Sci. Eng.* A **355** 277-85
[4] Gulyaev A A and Svistunova E L 1995 *Scripta Metall. Mater.* **33** 1497-1503
[5] Wang C, and Yuan S Q, Yao C G and Feng ZP 2013 *Adv. Mater. Res.* **690-693** 290-93
[6] Mulyukov R R, Sharipov I Z, Bukreeva K A, Bitkulov I Kh 2010 *Phys. Metals Metallogr.* **109(3)** 234

[7] Mulyukov R R, Kazantsev V A, Mulyukov Kh Ya, Burkhanov A M, Safarov I M and Bitkulov I Kh 2006 *Rev. Adv. Mater. Sci.* **11** 116
[8] Khisamov R, Timiryaev R, Safarov I and Mulyukov R 2020 *Lett. Mater.* **10** 223-6
[9] Bitkulov I Kh, Burkhanov A M, Kazantsev V A, Mulyukov R R, Mulyukov Kh Ya and Safarov I M 2006 *Phys. Metals Metallogr.* **102** 91-6
[10] Krylova K A, Bitkulov I K and Mulyukov R R 2018 *IOP Conf. Ser.: Mater. Sci. Eng.* **447** 012015
[11] Mulyukov R R, Pshenichnuk A I and Baimova J A 2015 *Lett. Mater.* **5** 485–90
[12] Mulyukov K Y, Shakirov I Z and Absalyamov S 1998 Instr. Exper. Techn. **41(3)** 433-4
[13] Nadutov V, Vashchuk D, Volosevych P, Svystunov Y, Beloshenko V, Spuskanyuk V and Davidenko A 2012 *Metallofizika Novejshie Technologii* **34** 395–414 [in Russian]
[14] Nagayama T, Yamamoto T and Nakamura T 2016 *Electrochim. Acta* **205** 178–87
[15] Nazarov A A and Mulyukov R R 2002 Nanostructured Materials. *Handbook of Nanoscience, Engineering, and Technology* (CRC Press, USA), chapter 22 p 22-1
[16] Straumal B, Kilmametov A, Ivanisenko Y, Mazilkin A, Valiev R, Afonikova N, Gornakova A and Hahn H 2018 *J. All. Compd.* **735** 2281–86
[17] Wurschum R, Kubler A and Gruss S 1996 *Annalles de Chimie* **21** 471
[18] Mulyukov R R, Bitkulov I K and Sharipov I Z 2014 *Lett. Mater.* **4** 11–4

Structure, thermal stability and mechanical properties of composite wires made of conducting microalloyed aluminum alloys

A V Nokhrin[1], I S Shadrina[1], V N Chuvil'deev[1], A A Bobrov[1], V I Kopylov[1,2], N N Berendeev[1] and A V Piskunov[1]

[1] Lobachevsky State University of Nizhny Novgorod, 23 Gagarina ave., Nizhny Novgorod, 603950, Russian Federation
[2] Physical and Technical Institute, National Academy of Sciences of Belarus, 10 Kuprevich st., Minsk, 220141, Belarus

E-mail: nokhrin@nifti.unn.ru

Abstract. Thermal stability of the structure and mechanical properties of composite wires made of microalloyed aluminum alloys were studied. The wire was produced by co-rolling of the copper-coated aluminum alloy. It has been demonstrated that annealing at 200 °C for 200 hours leads to the onset of recrystallization, reduced hardness and strength of the wire as per the Hall–Petch equation. The grain size in the annealed wire was less than 1-2 μm, manifesting the stabilized effect of precipitates of $Al_3(Sc,Zr)$ phases. At that the rate of diffusion of copper into the aluminum was quite low. The formation of a uniform fine-grained structure and zero copper-induced embrittlement resulted in increased by up to ~10% plasticity of the wire.

1. Introduction

Small-sized (with a diameter of less than 0.5 mm) composite wires are one of the key elements of modern aircraft and spacecraft electric systems, they are thereby subject to stricter requirements concerning electric conductivity, strength, and thermal stability [1-5].

Today, the Russian aircraft industry uses small-sized composite wires made of 01417 aluminum alloy (developed by All-Russian Institute of Light Alloys) with the addition of 7-9% cerium, lanthanum, and praseodymium, the surface of which has a protective copper coating that ensures better corrosion resistance and facilitates sealing of the wire into connectors. Working on similar tasks, the French manufacturer Filatex uses 1%Mg alloys (similar to the Russian-produced AMg1 alloy) [6]. Low strength of 01417 coarse-grained aluminum alloy, low resistance to wire twisting, characterized by fatigue resistance of the material, modest electric conductivity, and low thermal stability pose the key scientific and technical challenges in the development of new ultralight fine composite "coated aluminum" wires.

One of the biggest challenges of modern electrical engineering is to replace copper wires with aluminum ones while ensuring the same level of thermal stability, high electrical conductance and strength at elevated temperatures. It is particularly challenging to ensure this behavior in small composite "coated aluminum" wires, where prolonged exposure to elevated temperatures leads to an intensive diffusion of copper or silver from the coating into the aluminum wire, rendering it brittle. In some cases, in order to prevent the diffusion of copper or silver atoms into the aluminum wire, various buffer subcoatings are required.

The purpose of this study is to investigate the thermal stability of fine-grained microalloyed aluminum alloys and small-sized composite wires made of such alloys.

2. Methods and materials

The research focused on the structure of small-sized composite "coated aluminum" wires with a diameter of 0.2-0.5 mm, produced by co-rolling in grooves at room temperature. Microalloyed magnesium-free aluminum alloys with a varying content of scandium, zirconium, hafnium, and ytterbium were used as base materials. Samples of alloys measuring 22×22×200 mm were produced by induction casting, using an INDUTHERM VTC-200 casting machine. High-purity copper was used for coating. Small composite wires were produced by co-drawing at room temperature.

Industrially produced by FSP-One Inc. (France), a small aluminum copper-coated composite wire (with a diameter of 0.126-0.129 mm) with a silver subcoating was used for benchmark studies. This wire is used in electric systems of modern European civil aircraft (A380, etc.). As per the supplier certificate, this wire has a silver coating (over 2 μm thick) and a copper buffer subcoating, high electrical conductance (of at least 60% IACS, which corresponds to 2.676 μΩ·cm), and satisfactory strength (tensile strength of 138 MPa, elongation of at least 6%). There are no thermostability requirements. Coatings are electrodeposited while wires are drawn.

A Lloyd Instruments LR5KPlus universal materials testing machine was used to study rupture strength of the wire. Special grips were used to stretch-test the wire. NEXYGEN MT software was used to control the test machine and to process test results. All tests were performed at a crosshead speed of 10 mm/min. For the purposes of statistical validity, at least three samples in each batch were tested. Microhardness measurements were taken with an HVS-1000 microhardness tester. Microstructure was analyzed using a Jeol JSM-6490 scanning electron microscope (SEM) with an Oxford Instruments INCA 350 EDS microanalyzer. Specific electrical resistivity (SER) was measured using the eddy current method with a SIGMATEST 2.069 instrument. The wires were annealed in an air furnace.

In order to study the thermal stability of the structure and mechanical properties of the wires as well as the kinetics of the copper diffusion into the surface of the fine-grained aluminum wire, the samples were subjected to annealing in sealed glass capsules.

3. Experimental results

It is found that, in its initial state, the commercial wire manifests intense interdiffusion of copper into the silver buffer subcoating as well as interdiffusion of aluminum into the silver subcoating, and silver diffusion into the surface of the aluminum wire. The wire has a fine-grained homogeneous microstructure with 2-3 μm grain size in the central part of the wire. Microhardness of the surface layer was 20-25% higher than that of the central part (see figure 1c). Besides, energy-dispersive spectroscopy detected intermetallic particles of Al-Cu and Al-Fe in the aluminum matrix. All these factors, in our opinion, suggest that the wire was manufactured by hot deformation.

It has been demonstrated that long-time annealing at 200 °C leads to the diffusion of copper and silver into the surface of the commercial aluminum wire. The impact of the annealing temperature and time on its microhardness, strength, and plasticity was as follows. It has been found that annealing for up to 500 hours leads to a sharp decrease in the composite wire plasticity with an increase in tensile strength and a decrease in microhardness in the central part of the wire by 15-20%. The results of energy-dispersive spectroscopy suggest that such a behavior was caused by the intense diffusion of copper and silver into the aluminum wire surface.

Induction casting was used to obtain aluminum alloys with a total content of rare earth elements (REE) and transient metals (TM) of less than 0.4 wt%. Al-0.4%Zr alloy, where the zirconium part (0.1, 0.2, 0.3%Zr) was being progressively substituted by 0.1%Sc, 0.1%Hf, and 0.1%Yb microadditives, was used as the base material. The selection of the casting modes and the chemical composition of the alloy was determined in a way which minimizes the volume of Al_3(Zr,REE,TM) primary phases, which could cause the wire to break during cold rolling. In order to define the maximum acceptable value, the concentration of the selected alloying agent in pure aluminum was varied in increments of 0.03-0.05 wt% under the selected casting modes, while the presence and size of particles was recorded for three sections throughout the height of the bar – bottom, central, and upper part (near the head). Wires made of A99 aluminum were also used for comparative purposes.

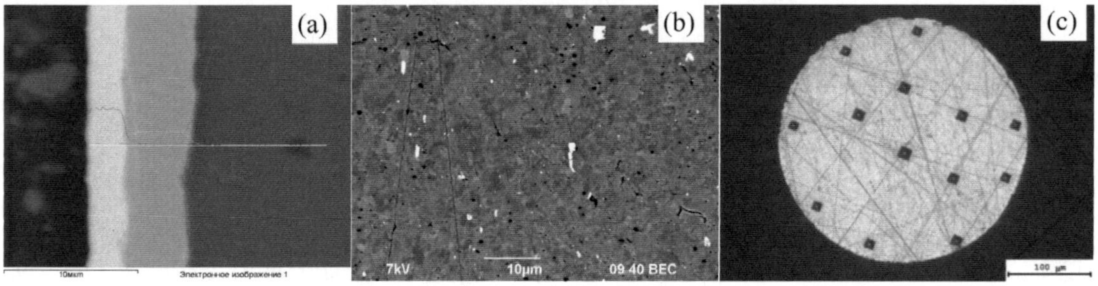

Figure 1. EDS results (a), microstructure image (b) and microhardness measurements (c) of the surface layers of the composite wire produced by FSP-One Inc. (France)

The aggregated findings suggest that for most alloys the presence of primary phases is a consequence of partial decomposition of a supersaturated solid solution under crystallization. This leads to a decrease in the specific electrical resistivity (SER) of the alloy – the SER value appears to be below the theoretical value (ρ_{th}), calculated using the additivity rule.

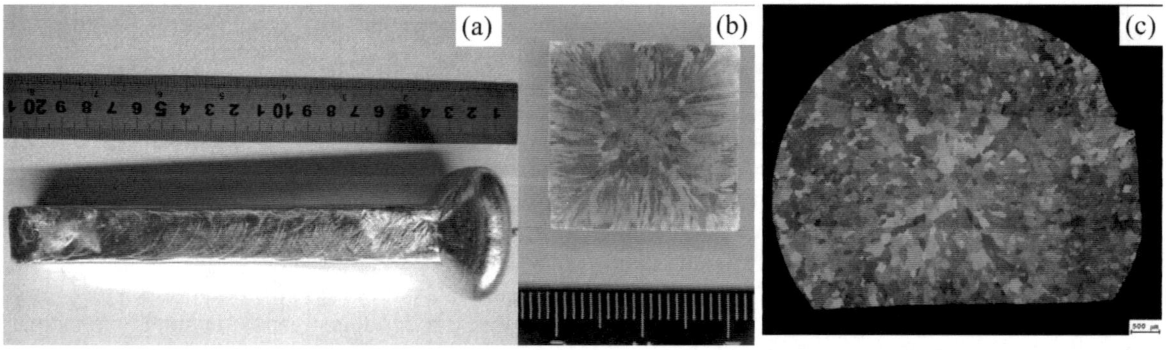

Figure 2. Photo of an Al-0.3Zr-0.1Yb bar (a), typical macrostructure of the cross-section after annealing (b), and after cogging with subsequent annealing (c)

The bars were subjected to severe plastic deformation followed by recrystallization annealing in order to eliminate the heterogeneity of the chemical composition and microstructure, resulting in a decrease in the technological plasticity of alloys (figure 2c). The annealing conditions were selected individually for each alloy to ensure a homogeneous recrystallized structure with a specific electrical resistivity setpoint (no more than 2.8-2.9 μΩ·cm). The dependencies presented in figure 3 demonstrate that annealing at 320 °C leads to a fairly noticeable increase in electrical conductance for most of alloys. This goes to prove that the solid solution decomposition begins, and Al_3(REE,TM) the precipitation starts. Most significant changes in electrical conductivity are observed after annealing cast scandium-containing alloys: μAl-0.3Zr-0.1Sc, Al-0.2Zr-0.1Sc-0.1Yb, Al-0.2Zr-0.1Sc-0.1Hf, and Al-0.1Zr-0.1Sc-0.1Hf-0.1Yb. Similar electrical conductance behavior at this stage for all the alloys with an addition of 0.1 wt.% Sc allows an assumption that during low-temperature annealing (320 °C, 2 h), Al_3Sc precipitates are formed. This assumption is further corroborated by a decrease in electrical conductance after heating these alloys to temperatures above 450°C. In our opinion, the lower electrical conductance can be caused by the partial dissolution of Al_3Sc particles at higher heating temperatures. This increases the concentration of scandium in the solid solution and reduces specific the electrical resistivity as a result. It should also be noted that the solid solution decomposition in scandium-containing alloys begins at a lower temperature (275 °C) than in alloys with a higher zirconium content. In our opinion, this result shows that Al_3Sc and Al_3(Zr,REE,TM) precipitates are formed independently in the alloys under consideration.

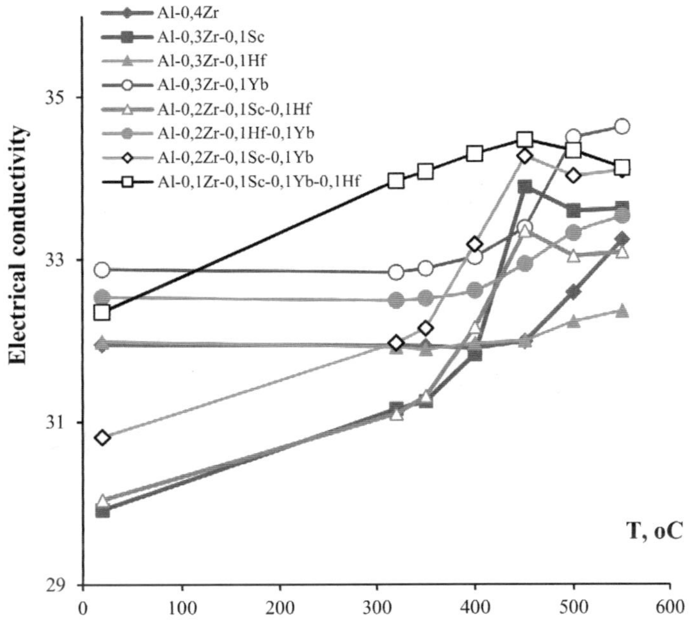

Figure 3. Dependence between the electrical conductivity of the bars made of new conducting alloys and the temperature of two-hour-long annealing

The workpieces were then scalped to remove oxides and subjected to multistage cold rolling in grooves. No intermediate annealing was applied in order to prevent copper diffusion into the surface of the aluminum wire. Rolling was performed on copper-coated workpieces. As a result, samples of composite aluminum wires with a diameter of 0.26 mm and a copper coating approx. 10 μm thick were produced (figure 4). Rolling to a diameter under 0.5 mm caused isolated wire breaks due to single micropores and submicron Al_3(Zr,REE,TM) phases in the central part of the bar.

Figure 4. Prototypes of small-sized composite aluminum wires obtained from the new conducting alloys

Mechanical tensile tests of the wires at room temperature demonstrate that they are quite strong – after drawing, the tensile strength of the wires made of the new alloys exceeds 500 MPa. The hardness of the wires made of Al-0.3Zr-0.1Sc and Al-0.2Zr-0.1Sc-0.1Hf alloys runs to 480-491 MPa. There is virtually no steady plastic flow in stress-deformation diagrams, which indicates low plasticity of the work-hardened metal. After long-term heat resistance tests at 200 °C (200 h), the wire strength reduces significantly, demonstrating, however, very high plastic performance, which implies that copper diffusion from the buffer layer causes only insignificant embrittlement (figure 5a). The hardness of the aluminum wire drops slightly during annealing – for example, during annealing at 200 °C (200 h), the hardness of the wire made of pure A99 aluminum is reduced from 226 MPs to 190 MPa while for the hardness of the wire made of Al-0.3Zr-0.1Sc alloy goes from 480 MPa down to 436 MPa. As figure 5b shows, post-annealing hardness distribution in the cross-section of the wire remains uneven, with hardness near the surface layer 50-70 MPa higher than in the central part.

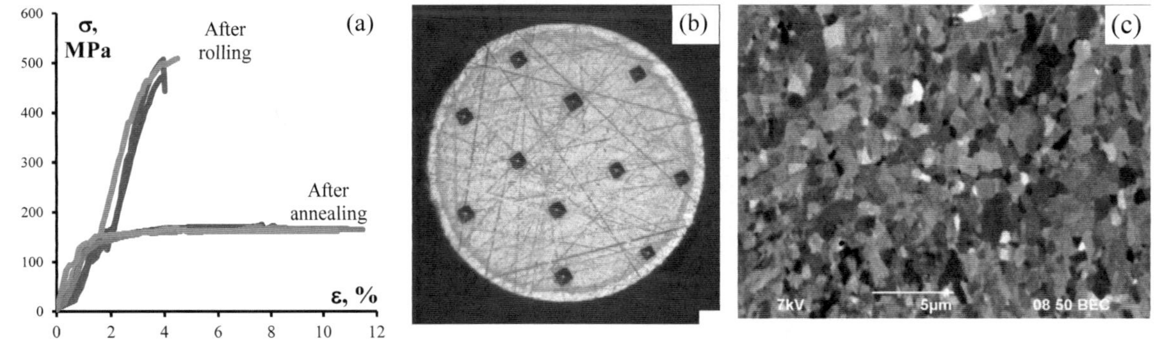

Figure 5. Tensile stress-deformation diagrams of the small wire samples (a), microhardness distribution (b) and microstructure (c) over a cross section of small wire prototypes

The tensile strength of the wire made of Al-0.3Zr-0.1Sc alloy is reduced to 160-170 MPa after annealing at 200 °C (200 h), whereas the percentage elongation to fracture increases by up to ~10%.

The fracture analysis has shown that the annealed samples after tensile tests demonstrate ductile failure. No detachment of the copper buffer layer from the aluminum wire surface was observed. The results of energy-dispersive spectroscopy showed that the depth of copper diffusion into the surface of the aluminum alloy never exceeded 10 μm.

The microstructural analysis demonstrated that prolonged annealing at 200 °C leads to secondary recrystallization, which in turn causes an increase in the average grain size in the aluminum wire (see figure 5c). The average grain size in the wire made of Al-0.3Zr-0.1Sc alloy increases from 0.2-0.3 μm in the initial state up to ~1-2 μm after annealing at 200 °C, 200 hours (see figure 5c). After similar annealing, the average grain size in the wire made of pure A99 aluminum exceeds 200 μm. This result suggests that during prolonged annealing, partial decomposition of the solid solution and precipitation of Al3Sc and Al3Zr particles of nanoscale and submicron sizes R occur [3]. In accordance with the Zener equation ($d_R = \alpha \cdot R/f_v$, where d_R is the stable grain size, f_v is the volume ratio of precipitated particles, α is the numerical factor), this stabilizes the fine-grained structure of the aluminum alloy.

In our opinion, the formation of a homogeneous recrystallized fine-grained microstructure in the wire contributes to an increase in the plastic flow stability of the aluminum wire during tensile tests, even if accompanied by a noticeable decrease in strength (see figure 5a). In our opinion, this marked decrease in strength can be due to the simultaneous onset of recrystallization processes, which leads to a lower tensile strength in accordance with the Hall–Petch equation: $\sigma_b = \sigma_0 + K \cdot d^{-1/2}$, where σ_0 is the yield stress, K = 0.68 MPa·m$^{1/2}$ [3] is the grain boundary strengthening coefficient (Hall–Petch coefficient). Inserting the experimental values of grain sizes (d_1 = 0.2-0.3 μm, d_2 = 1-2 μm) into the Hall–Petch equation, it can be demonstrated that at σ_0 = const, the recrystallization process should lead to a decrease in the tensile strength from ~ 500 MPa to 190-270 MPa. The estimates are quite close to the experimentally measured value σ_b = 160-170 MPa (see figure 5a). In our opinion, the lower experimental values of tensile strength (160-170 MPa) compared to the calculated value (190-270 MPa) can be due to partial coagulation of $Al_3(Sc,Zr)$ particles during prolonged annealing. In accordance with the Orowan equation [6], this leads to a decreased contribution of the second phase particles to the aluminum alloy strength.

4. Conclusions

The thermal stability of the structure and mechanical properties of small composite copper-coated wires made of microalloyed fine-grained aluminum alloys have been studied. It has been shown that the alloys have high thermal stability at long-term isothermal exposure within the temperature range of 180-200 °C. It has been confirmed that annealing is accompanied by recrystallization processes that impair the mechanical properties of the wire. The rate of copper diffusion into the surface layers of the aluminum wire is negligible.

Acknowledgments

The study was performed with the support of the Russian Science Foundation (grant No. 20-19-00672).

The authors wish to thank V.V. Zakharov (All-Russian Institute of Light Alloys, Moscow) for his recommendations on the composition of aluminum alloys and their casting modes.

References

[1] Yang C, Masquellier N, Gandiolle C and Sauvage X 2020 Multifunctional properties of composition graded Al wires *Scripta Materialia* **189** 21-4.

[2] Medvedev A, Arutyunyan A, Lomakin I, Bondarenko A, Kazykhanov V, Enikeev N, Raab G and Murashkin M 2018 Fatigue properties of ultra-fine grained Al-Mg-Si wires with enhanced mechanical strength and electrical conductivity *Materials* **8** 1034.

[3] Nokhrin A, Shadrina I, Chuvil'deev V and Kopylov V 2019 Study of structure and mechanical properties of fine-grained aluminum alloys Al-0.6wt.%Mg-Zr-Sc with ratio Zr:Sc = 1.5 obtained by cold drawing *Materials* **12** 316.

[4] Moisy F, Gueydan A, Sauvage X, Keller C, Guillet A, Nguen N, Martinez M and Hug E 2018 Elaboration of architectured copper clad aluminum composites by a multi-step drawing Process *Materials Science Forum* **941** 1914-9.

[5] Matveyev Y A, Gavrilova V P and Baranov V V 2006 Light conducting materials for aircraft wires *Cables and Wires* **5** 22-3 (in Russian).

[6] Martin J W 1980 *Micromechanisms in Particle-Hardened Alloys* (Cambridge, UK: Cambridge University Press) 167 p.

Corrosion resistance of ultrafine-grained pseudo-α titanium alloy PT-3V

A V Nokhrin[1], **V I Kopylov**[1,2], **P V Andreev**[1], **V N Chuvil'deev**[1], **C V Likhnitsckii**[1], **M K Chegurov**[1], **A A Murashov**[1] and **D A Gudz'**[1]

[1] Lobachevsky State University of Nizhny Novgorod, 23 Gagarina ave., Nizhny Novgorod, 603950, Russian Federation
[2] Physical and Technical Institute, National Academy of Sciences of Belarus, 10 Kuprevich st., Minsk, 220141, Belarus

E-mail: nokhrin@nifti.unn.ru

Abstract. Ultrafine-grained (UFG) samples of PT-3V pseudo-α titanium alloy were subjected to diffusion welding through spark plasma sintering. It has been shown that the destruction of welded UFG samples under hot salt corrosion (HSC) is two-staged: first, intercrystalline corrosion (ICC) is manifested, which then graduates to pitting corrosion. It has been established that the corrosion resistance of the weld joints is determined by the concentration of vanadium at the grain boundaries, the size and volume ratio of the β-phase particles, and the presence of pores in the weld joints. It has been shown that welded UFG samples possess a higher hardness and resistance to ICC than coarse-grained samples.

1. Introduction

Today, titanium pseudo-α Ti-Al-V alloys are widely used in shipbuilding, nuclear power engineering, etc. [1-2] Forming an ultrafine-grained (UFG) structure in titanium alloys by severe plastic deformation represents one of the promising ways to improve their performance. UFG titanium alloys are characterized by increased strength and better creep resistance and fatigue strength [3, 4]. One of the industrial application challenges for UFG titanium alloys is their welding behavior. The commonly used argon-arc or electron-beam welding will not preserve the UFG structure.

One of the worst hazards for titanium alloys in nuclear power engineering and naval aviation is hot salt corrosion (HSC) [4, 5]. This type of corrosion occurs when salt deposits are formed on the surface of equipment operating at elevated temperatures. HSC in titanium alloys is a multi-stage deterioration process under exposure to elevated temperatures, corrosive mediums (salts), and oxygen [4, 5].

The purpose of this work is to study the deterioration behavior of weld joints in samples made of PT-3V UFG alloy when exposed to HSC and, in particular, to study the impact that microstructure parameters of welded samples have on the HSC resistance.

2. Methods and materials

The object of the study was the pseudo-α titanium alloy PT-3V (Ti-4.73%Al-1.88V). The UFG structure was processed by Equal Channel Angular Pressing (ECAP) (T_{ECAP} = 450 °C, N_{ECAP} = 4, B_c mode).

Spark plasma sintering (SPS) [6, 7] is one of the latest techniques for solid state welding of structural materials, including titanium alloys [8, 9]. An important advantage of SPS is the ability to implement high heating rates (up to 2500 °C/min), which will preserve the UFG structure with high mechanical properties in titanium alloys. A Dr. Sinter SPS-625 spark plasma sinterer was used for diffusion welding of the samples. Heating rate (V_h) varied from 10 to 350 °C/min, applied pressure (P) – from 50 to 100 MPa, welding time – from 0 to 90 min. Welding was performed in the α-phase field (600 °C), near the α-β-phase boundary (700 °C), at the beginning of the two-phase (α+β)-field (800 °C), and in the β-phase field (1030, 1140 °C). Welding was performed in a vacuum, without molds, which caused tangent stresses, perpendicular to the uniaxial pressure axis.

A Jeol JSM-6490 scanning electron microscope and a Jeol JEM-2100 transmission electron microscope were used for microstructural studies. A Shimadzu XRD-7000 diffractometer was used for X-ray diffraction analysis (XRD). A Duramin Struers 5 durometer with a load of 2 kg was used for microhardness (H_v) measurements. HSC tests were performed in a 300:1 mixture of NaCl and KBr crystal salts at 250 °C during 500 hours with oxygen access. A Leica DM IRM metallographic microscope was used to assess corrosive damage to the surface of samples.

3. Experimental results

Macrostructure studies show incomplete fusions of ~50-70 μm at the edges of coarse-grained (CG) samples welded under low pressures, at low temperatures, or high heating rates. At the top of the macrodefects, large pores are observed, the volume fraction of which decreases farther from the edge of the sample (away from the areas of the maximum tensile stresses). The remaining larger area of the weld joint has micropores, the volume fraction of which depends on the welding mode. It should be noted that low volume fractions of macro- and micropores are observed in the case of diffusion welding of CG samples over an interval of 700-800 °C and under low stresses (50 MPa).

There are virtually no macrodefects in the weld seams of UFG alloys, the volume fraction of macropores and micropores is small and is observed chiefly after welding at low temperatures (600 °C) or under high pressures (70-100 MPa). Varying basic SPS parameters has little effect on the average grain size in CG samples (Table 1) with an α/α'-structure near the weld joints, where β-phase particles vigorously precipitate along the boundaries of elongated grains of the α'-phase.

Table 1. Parameters of microstructure of the PT-3V alloy specimens after the diffusion welding

Welding regime				Coarse-grained alloy					UFG alloy				
T, °C	V_h, °C/min	*t*, min	*P*, MPa	d_α, μm	d_β, μm	P(*)	H_v, GPa Weld	H_v, GPa Alloy	d_α, μm	d_β, μm	P	H_v, GPa Weld	H_v, GPa Alloy
Initial state				10-100	3-15	-	-	2.0-2.1	0.5	-	-	-	3.1-3.2
600				16	3.2	H	2.5	2.4-2.5	4.8	2.4	H	3.0	3.0-3.1
700				17	3.0	M	2.5	2.5-2.6	5.9	3.0	L	2.8	2.8-2.9
800	100	10	50	18-19	5.1	M	2.3	2.3-2.4	6.9	3.3	M	3.2	3.2
1030				12	-	L	2.5	2.4	-	-	-	-	-
1142				-	-	-	-	-	5.2		L	2.6	2.6-2.7
	10			16-18	5.0	L	2.6	2.5	6.8	3.1	H	2.5	2.5-2.6
700	50	10	50	17-18	3.5	M	2.6	2.4-2.3	5.5	3.3	M	2.7	2.8-2.9
	100			18-19	3.0	M	2.4	2.5	5.9	3.0	L	2.7	2.7-2.9
	350			18	3.0	H	2.4	2.5	3.8	3.3	L	2.9	3.0-3.1
			50	18-19	3.5	L	2.5	2.4-2.6	5.9	3.3	L	2.7	2.8-2.9
700	100	10	70	15	3.2	H	2.7	2.6-2.7	5.3	2.6	H	2.7	2.9-3.0
			100	10	2.9	H	2.7	2.5-2.6	4.9	2.5	H	2.7	3.0-3.1
		0		18-19	3.5	H	2.5	2.5-2.6	4.7	2.0	H	2.9	2.9-3.1
700	100	10	50	18-19	3.0	H	2.4	2.4-2.5	5.9	3.3	M	2.7	2.8-2.9
		50		16-17	2.0	H	2.3	2.3-2.6	9.1	3.5	L	2.6	2.8
		90		12-13	2.0	W	2.4	2.4-2.6	9.3	3.5	L	2.5	2.5-2.6

Note: The volume fraction of macropores and micropores: H – High, M – medium, L – low

Table 2. Results of the corrosion testing of the welded joints of the PT-3V alloy

Welding regime				Coarse-grained alloy				UFG alloy			
T, °C	*V_h*, °C/min	*t*, min	*P*, MPa	HSC character Zone I	HSC character Zone II	h_{max}, μm	h_{cp}, μm	HSC character Zone I	HSC character Zone II	h_{max}, μm	h_{cp}, μm
State before SPS				ICC		~600	~400	ICC		~150	~100
600				C	P, IGC	223	162 ± 43	C	P, IGC	188	124 ± 34
700				C	P, IGC	235	231 ± 39	-	P, IGC	420	273 ± 62
800	100	10	50	C	P	280	265 ± 48	C	P, IGC	400	285 ± 59
1030				-	P	358	184 ± 35	-	-	-	-
1142				-		-	-	C	P, IGC	557	211 ± 45
	10			C	P, IGC	238	182 ± 45	-	P	171	132 ± 24
700	50	10	50	C	P, IGC	235	231 ± 39	-	P, IGC	442	304 ± 51
	100			C	P, IGC	358	184 ± 75	-	P, IGC	420	273 ± 42
	350			C	P, IGC	444	220 ± 92	-	P, IGC	451	189 ± 36
			50	C	P, IGC	290	215 ± 49	-	P, IGC	130	110 ± 30
700	100	10	70	C	P, IGC	132	108 ± 21	-	P	138	113 ± 28
			100	C	P, IGC	246	178 ± 77	-	P	163	93 ± 40
		0		C	P, IGC	390	211 ± 82	C	P, IGC	385	249 ± 58
700	100	10	50	C	P, IGC	358	184 ± 75	-	P, IGC	358	184 ± 55
		50		C	P, IGC	354	191 ± 67	-	P, IGC	495	237 ± 40
		90		C	P, IGC	163	125 ± 62	-	P, IGC	512	297 ± 47

Note: P – pitting corrosion, IGC – intergranular corrosion, C – crevice corrosion

Varying the basic welding parameters has little effect on the average grain size in CG samples. The structure of the alloy structure near the weld joints has a coarse-grained α/α'-structure. The most vigorous precipitation of the β-phase is observed along the boundaries of α'-phase elongated grains, with the volume fraction of β-phase particles in the weld joint being greater than in the initial CG alloy (Fig 1). Increased heating rate, pressure, or time of exposure reduce the average size of β-phase particles in CG alloys.

It should be noted that the grain size in the UFG samples is 3-4 times less than in the CG samples, while occasional β-phase particles of 2-3 μm are observed at grain boundaries. Increasing welding temperature from 600 °C to 800 °C raises the average grain size in UFG alloys from 4.8 to 6.9 μm. Higher welding temperatures and longer times increase the size of β-phase particles and grain size. Heating rates have no marked impact on the size of β-phases but lead to smaller UFG grains.

Salt deposits on the surface of the samples after HSC tests are mixtures of NaCl, titanium oxides (TiO_2, TiO), and alumina, as well as vanadium-based constituents ($VO_{1.15}$, $V_2Ti_3O_9$) and Al_3V. Corrosion products (salt deposits) on the samples produced by SPS at temperatures above 1000 °C also consisted of NaCl, TiO_2, TiO, and aluminum oxide, with aluminum constituents ($TiAl_2Cl_8$, $AlTi_3$) and traces of VO_2 detected instead of vanadium-based constituents ($VO_{1.15}$,, $V_2Ti_3O_9$,, Al_3V). It is important to note that XRD detected vanadium oxide peaks in corrosion products of CG samples while vanadium oxide was not found in the corrosion products of the UFG samples.

The macrostructure studies demonstrate the difference in corrosion mechanisms in the weld joints and outside. In the area of CG alloy joints, there is a combination of pitting and crevice corrosion, and the depth of crevice corrosion along the weld joint is determined by porosity of the seam. The corrosion depth in highly porous seams produced under suboptimal welding modes can exceed 300 μm. Farther away from weld joints in CG alloys, there is a combination of intercrystalline corrosion (ICC) and pitting corrosion. The largest corrosion pits were evenly distributed over the sample area. The worst corrosion failures are observed in the α'-phase area with an increased volume fraction of the β-phase particles. This can explain a heterogeneous distribution of corrosion pits across the surface. ICC can be observed under large corrosion pits. Consequently, ICC sets in fine-grained α'-phase fields during the first stage of HSC, and pitting corrosion develops next. The pressure rates and SPS time have the highest impact on corrosion depth in weld joints of CG samples (Table 2). It should be noted that the size of corrosive defects in CG samples is smaller than in alloy in its initial state (see [2]).

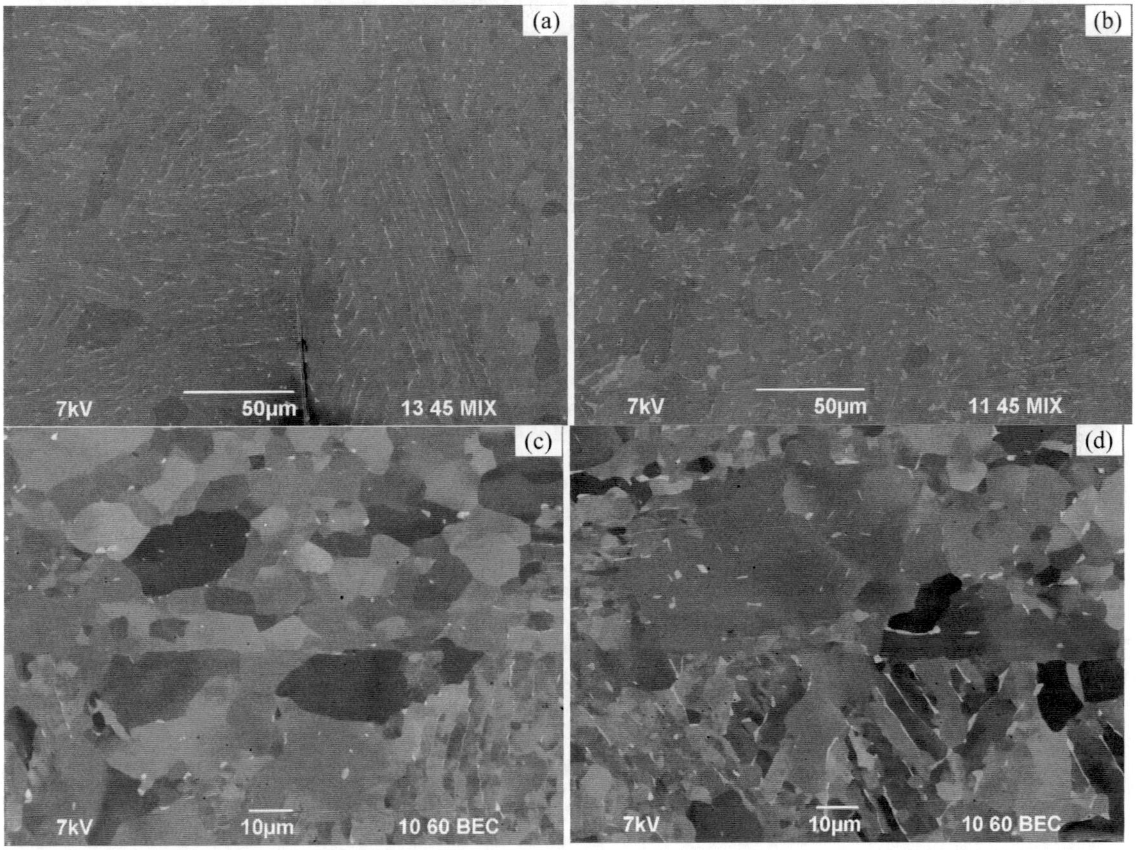

Figure 1. Microstructure of the CG alloy after welding at P = 50 MPa for t = 10 min and heating rate to temperatures of: (a) 50 °C/min, 600 °C; (b) 50 °C/min, 800 °C; (c) 100 °C/min, 700 °C; (d) 350 °C/min, 700 °C

UFG samples manifested crevice corrosion only in weld joints with higher porosity. Farther away from the welding line, a combination of ICC and pitting corrosion is observed, as in the case of CG samples. An increase in SPS time and temperature leads to an increased corrosion depth in UFG alloys weld joints (Table 2). For lower welding temperatures, heating rates, or pressure, corrosion was quite shallow without exceeding the ICC depth in post-ECAP UFG alloys (see [8]). Welding at 700 °C, near the (α + β)-phase boundary under low pressure (50 MPa) and a heating rate of 10 °C/min proves to be optimal – with an average corrosion depth of 110-130 μm and no crevice corrosion in weld seams. It should be emphasized that for these SPS modes, an average pitting depth in UFG samples was 1.5-2 times less than that in CG samples. Hence, it can be concluded that UFG samples have a higher corrosion resistance as compared to CG samples.

The HSC tests on UFG samples demonstrate that an increase in the SPS temperatures from 600 to 800 °C leads to an increase in the average depth of corrosion pits from 124±34 μm up to 285±59 μm, with the maximum corrosion depth growing from 188 up to 400-420 μm. An increase in the heating rates from 50 to 350 °C/min reduces the average corrosion depth from 304±51 down to 189±36 μm. The applied pressure and time of exposure had low-to-no impact on the corrosion depth in the UFG samples.

Figure 2. The nonuniform distribution of large corrosion pits at different sides of the joint line of the coarse-grained titanium alloy Ti-5Al-2V after HSC testing

Hardness studies have suggested that SPS modes have little impact on H_v of seams and the parent metal in CG samples. The hardness of CG samples is 2.4-2.6 GPa. The higher hardness of the samples versus the PT-3V alloy in its initial state (~2.1-2.2 GPa) may be due to its deformation hardening. This finding is indirectly confirmed by seam H_v increasing to 2.7 GPa when welded under a pressure of 100 MPa (Table 1). The hardness of the seams and parent metal in UFG samples is higher than that of CG samples by ~0.4–0.8 GPa. The highest seam hardness (~3.0–3.2 GPa) is observed in UFG samples obtained at low temperatures, high heating rates, and fast times of exposure. Increasing heating rates to 350 °C/min can form a high-density seam with a UFG structure and high hardness (2.9–3.1 GPa).

A model has been offered to describe the impact of microstructure parameters on the susceptibility of UFG titanium alloys to HSC. It has been shown that the following has the biggest impact on the alloy susceptibility to HSC: (i) the concentration of corrosive alloying elements (AE) – vanadium, most of all – at grain boundaries, (ii) the presence of β-phase particles at grain boundaries, (iii) pores in weld joints. According to [8], the formation of grain-boundary segregations of corrosive AE and the presence of β-phase particles with a high concentration of β-stabilizing elements (vanadium) can cause microgalvanic couples that accelerate electrochemical corrosion and can lead to an accelerated chemical corrosion of grain boundaries. The negative impact of pores is primarily manifested through an increase in the available surface area involved in chemical or electrochemical reactions.

4. Conclusions

UFG samples have been shown to have better weldability than CG ones, the weld joints of which manifested incomplete fusions that cause crevice corrosion. There were virtually no seam macrodefects in UFG samples while seam porosity was very low, which ensures high resistance of the weld joints to crevice corrosion of the UFG alloy.

The corrosion resistance of welded CG samples is higher than that of PT-3V alloy in its initial state. Corrosive defects on the surface of CG samples are primarily concentrated in the field of the fine-grained α'-phase, the grain boundaries of which manifest a higher volume fraction of the β-phase.

Destruction of CG samples under HSC is two-staged: first, ICC is manifested, which then graduates to pitting corrosion.

Corrosion resistance of UFG samples depends mainly on the concentration of vanadium at the grain boundaries, the presence of β-phase particles at the grain boundaries and pores in the weld joints. It has been established that increases in the concentration of vanadium at migrating grain boundaries as SPS temperature and time grow, adversely affect the resistance to HSC. It has been shown that diffusion welding in the high-speed heating mode (350 °C/min) up to a temperature close to the phase boundary (700 °C), under a pressure of 50 MPa without holding facilitates the formation of a fine-grained structure with increased hardness and better corrosion resistance.

Acknowledgments

The study was performed with the support of the Russian Science Foundation (grant No. 19-73-00295).

References

[1] Leonov V P, Gorynin I V, Kudryavtsev A S, Ivanova L A, Travin V V and Lysenko L V 2015 Titanium alloys in steam turbine construction *Inorg. Mater. Appl. Res.* **6** 580-90.

[2] Oryshchenko A S, Gorynin I V, Leonov V P, Kudryavtsev A S, Mikhailov V I and Chudakov E V 2015 Marine titanium alloys: Present and future *Inorg. Mater. Appl. Res.* **6** 571-9.

[3] Zherebtsov S, Kudryavtsev E, Kostjuchenko S, Malysheva S and Salishchev G A 2012 Strength and ductility-related properties of ultrafine grained two-phase titanium alloy produced by warm multiaxial forging *Mater. Sci. Eng. A* **536** 190-6.

[4] Chuvil'deev V N, Kopylov V I, Nokhrin A V, Tryaev P V, Kozlova N A, Tabachkova N Yu, Lopatin Yu G, Ershova A V, Mikhaylov A S, Gryaznov M Yu and Chegurov M K 2017 Study of mechanical properties and corrosive resistance of ultrafine-grained α-titanium alloy Ti-5Al-2V *J. All. Compd.* **723** 354-67.

[5] Chevrot Th 1994 *Pressure Effects on the Holt-Salt Stress-Corrosion Cracking of Titanium Alloys* (Cranfield University. School of Industrial and Manufacturing Science) 215 p.

[6] Chuvildeev V N, Panov D V, Boldin M S, Nokhrin A V, Blagoveshensky Yu V, Sakharov N V, Shotin S V and Kotkov D N 2015 Structure and properties of advanced materials obtained by Spark Plasma Sintering *Acta Astron.* **109** 172-6.

[7] Tokita M 2013 *Spark Plasma Sintering (SPS) Method, Systems, and Applications* (Academic Press) pp 1149-77.

[8] Chuvil'deev V N, Nokhrin A V, Kopylov V I, Boldin M S, Vostokov M M, Gryaznov M Yu, Tabachkova N Y and Tryaev P 2019 Spark plasma sintering for high-speed diffusion bonding of the ultrafine-grained near-α Ti-5Al-2V alloy with high strength and corrosion resistance for nuclear engineering *J. Mater. Sci.*. **54** 14926-49.

[9] Yang J, Wang G F, Jiao X, Gu Y, Liu Q and Li Y 2018 Current-assisted diffusion bonding of extruded Ti-22Al-25Nb alloy by Spark Plasma Sintering: Interfacial microstructure and mechanical properties *J. Mater. Engin. Performance* **27** 3035-43.

Effect of friction stir processing on the strengthening of the 2024 aluminum alloy

A Kh Valeeva, A Kh Akhunova, M F Imayev and R F Fazlyakhmetov
Institute for Metals Superplasticity Problems, Russian Academy of Sciences,
39 Khalturin st., 450001 Ufa, Russia

E-mail: valeevs@mail.ru

Abstract. A study of the influence of the tool pin length on the strengthening of the 2024 aluminum alloy under friction stir processing (FSP) was carried out. According to the results of real and model experiments, it was found that for a workpiece with a thickness of 3.0 mm, the necessary and sufficient pin length is 2.0 mm, because in this case, mixing throughout the thickness of the workpiece is provided. It was found that during the FSP of the 2024 alloy, solid-solutional hardening of the processed area occurred due to the release of particles of intermetallic phases.

1. Introduction

Friction Stir Processing (FSP) modifies the surface of the metals and provides considerable changes in microstructure and other properties [1-3]. FSP induces the processes of thermomechanical activation in selective areas on the surface and changes properties of specimens of metal. FSP is based on the same concepts as Friction Stir Welding. The only difference is that FSP doesn't join the metals together, the workpiece is initially monolithic. As a result of FSP, the metallic material attains a fine grained structure and increased mechanical properties [3, 4].

The result of the FSP depends, in particular, on the design of the tool, which affects heat generation, material flow, forces, and other output characteristics [5-7].

This work is aimed to determine the dependence of the microstructure and hardness of the 2024 aluminum alloy on the immersion depth of the tool into the workpiece, as well as to simulate the temperature distribution in the processed area depending on the pin length.

2. Experiment Techniques

The industrial deformable heat-strengthened 2024 aluminum alloy in the form of sheets with a thickness of 3 mm was selected as the material for study. The model of a tool used to conduct FSP is shown in figure 1. The tool with an extendable pin with a left-hand thread and the shoulder with two spiral ledges was used. The pin length ranged from 1.6 mm to 2.2 mm. Treatment at a tool moving speed of v = 3 cm/min, a rotation frequency of 1000 rpm, and an axial force of P = 2 kN was carried out.

Microstructure formed at the processed zone during the FSP was tested using a light microscope and scanning electron microscopy (SEM). Measurements were performed on specimens with a width of 22 and a thickness of 3 mm cut from the friction stir processed zone.

Microhardness was determined using Vickers digital microhardness tester MVDM 8 "AFFRY" at a load of 0.5 N and a duration of 10 seconds. Measurements at the intersection points of imaginary lines in the specimen scheme shown in figure 2 were carried out, with an error not exceeding 5 %.

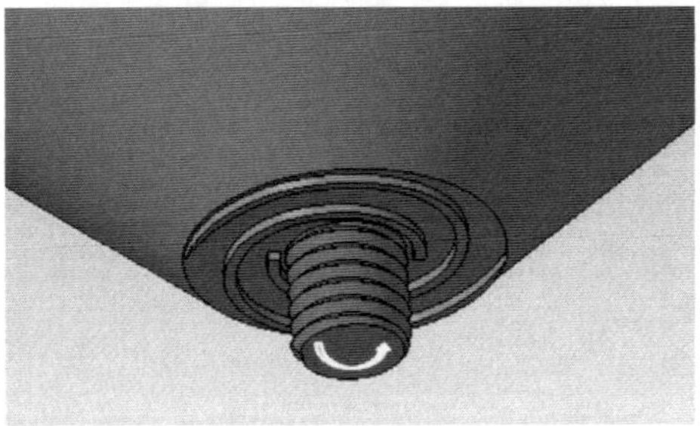

Figure 1. Model of the tool for FSP. By the arrow, the pin rotation direction is shown.

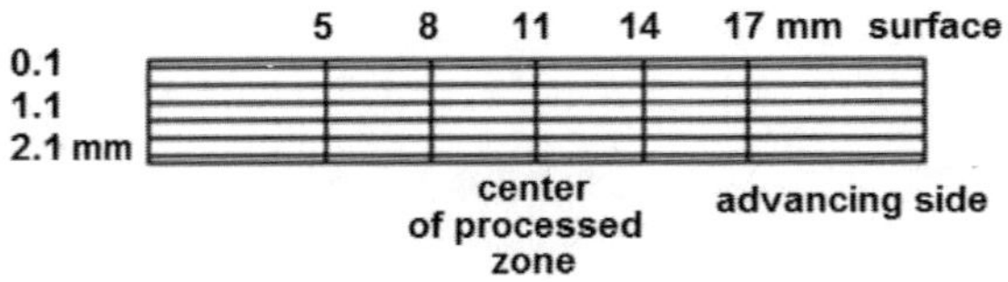

Figure 2. The scheme of the specimen with points in which microhardness was measured.

The modeling of the heat dissipation processes in the specimens during FSP was performed using the DEFORM-3D software application package. The numbers of elements of the workpiece and the tool were equal to 60000 and 32000, respectively. The behavior of the workpiece material was described using the Johnson-Cook model included in the standard DEFORM-3D library. To reduce the calculation error, the thermophysical properties of the 2024 alloy were taken to be constant and equal to the values acquired by the material at a temperature of 600 °C. When processing the workpiece, the conditions of heat exchange with the environment were set. The tool was taken as a rigid body. AISI-D2 tool steel was chosen as the material for the tool. We examined the pins in the form of a cylinder with different lengths: 1.6, 1.8, 2.0, 2.2 mm. The diameter of the lower part of the pin in all cases was equal to 2 mm. The friction coefficient was taken equal to 0.5.

3. Results and discussions

The results of microhardness measurements are shown in figure 3. The processing of the material by stirring led to a significant increase in microhardness in the center of the processed zone and on the advancing side from 53±4 HV to 110±8 HV.

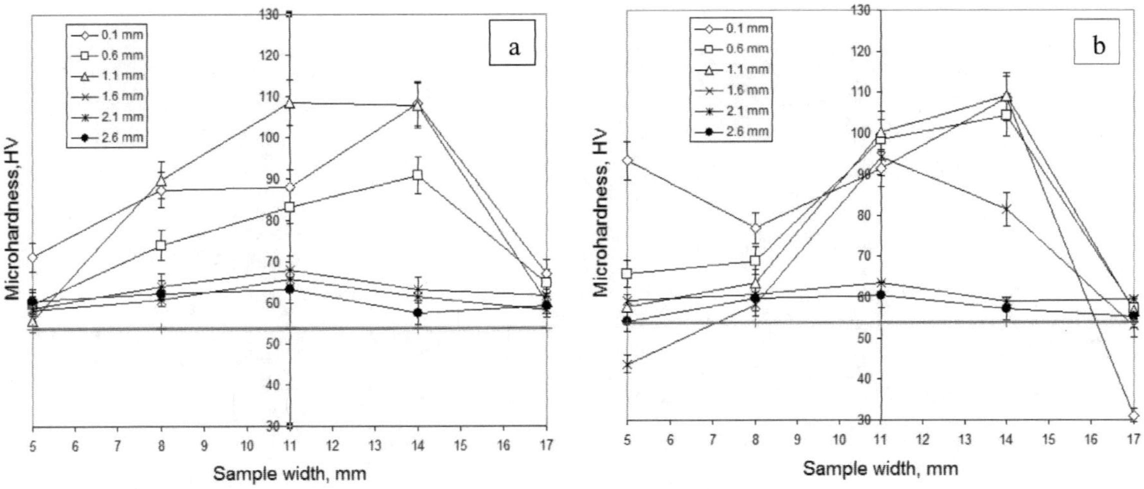

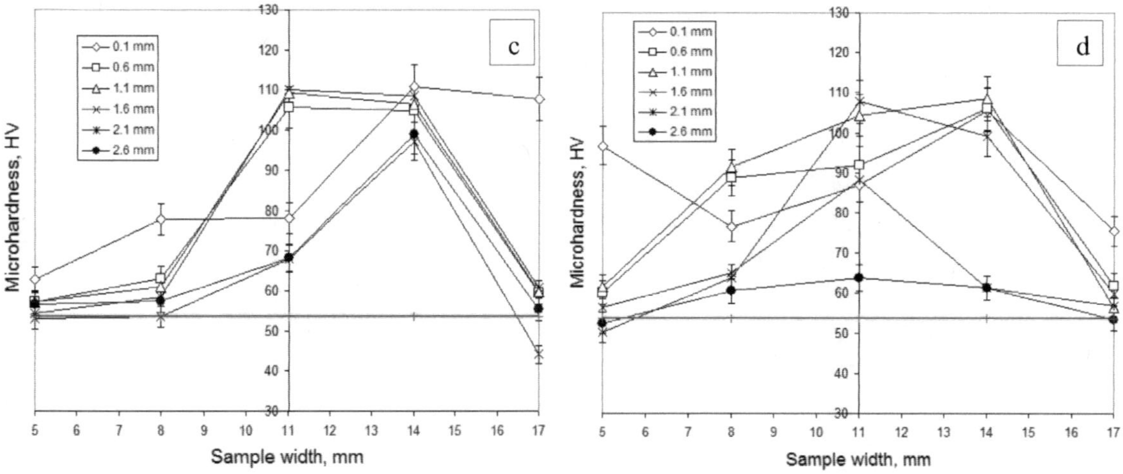

Figure 3. Microhardness of specimens treated with a pin of different lengths: 1.6 mm (a), 1.8 mm (b), 2.0 mm (c), 2.2 mm (d). The horizontal line shows the initial microhardness.

It can be seen that with a pin length of 1.6 mm, a change in microhardness occurs only in the upper part of the specimen, to a depth of 1.6 mm. With an increase in the pin length to 1.8 mm, changes in microhardness are observed in deeper layers of the specimen (up to 2.1 mm) (figure 3b). When processing with a 2.0 mm pin, strengthening occurs in all layers of the specimen (figure 3c). It should be noted a fairly strong (two-fold) hardening of the advancing side, while the highest values of microhardness practically do not depend on the length of the pin. A further increase to 2.2 mm, similarly increases the microhardness of all layers of the specimen and provides a more uniform hardening of both sides of the processed zone.

The study of the microstructure of the processed zone by light microscopy method made it possible to see pronounced bands on the etched surface, the so-called "onion rings", as well as the zone of thermomechanical impact (figure 4). The presence of the so-called onion-ring structure was also reported in other FSP studies [8-10]. As the stirring pin rotates rapidly during FSP, the cylindrical sheets of processed metal are extruded in each rotation.

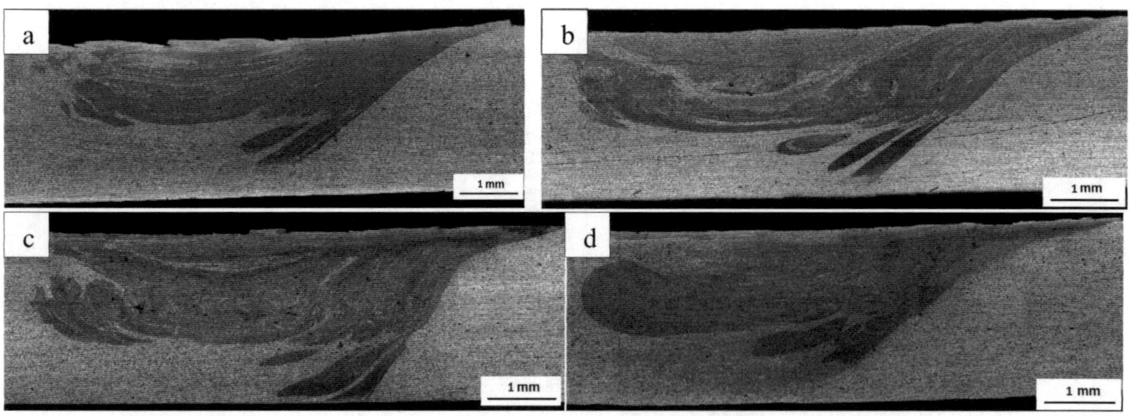

Figure 4. Composed macroscopic image of specimens treated with a pin of different lengths: 1.6 mm (a), 1.8 mm (b), 2.0 mm (c), 2.2 mm (d).

Scanning electron microscopy studies show the following. In the initial state particles of S(Al_2CuMg) and θ(Al_2Cu) excess phases exist in equilibrium with the α-solid solution of copper and magnesium in aluminum matrix [11]. After FSP, particles of two excess phases with different morphologies are observed. Large complex (skeletal) particles are mainly located on the advancing side. Smaller compact particles are present throughout the zone of thermomechanical impact. A study of the chemical composition of particles, performed by energy dispersive analysis, allows us to attribute them to the strengthening $Al_{15}Si_2(CuFeMn)_3$ and θ(Al_2Cu) phases, respectively.

The formation and coagulation of the stable θ(Al_2Cu) phase, as well as the appearance of the $Al_{15}Si_2(CuFeMn)_3$ phase, are usually observed after heating to 300–400 °C [12]. Apparently, during the FSP, the material is heated to such temperatures and the cooling time of the sample to room temperature is quite long. This time was sufficient for particles that were formed and the hardness of the processed area of the 2024 alloy increased.

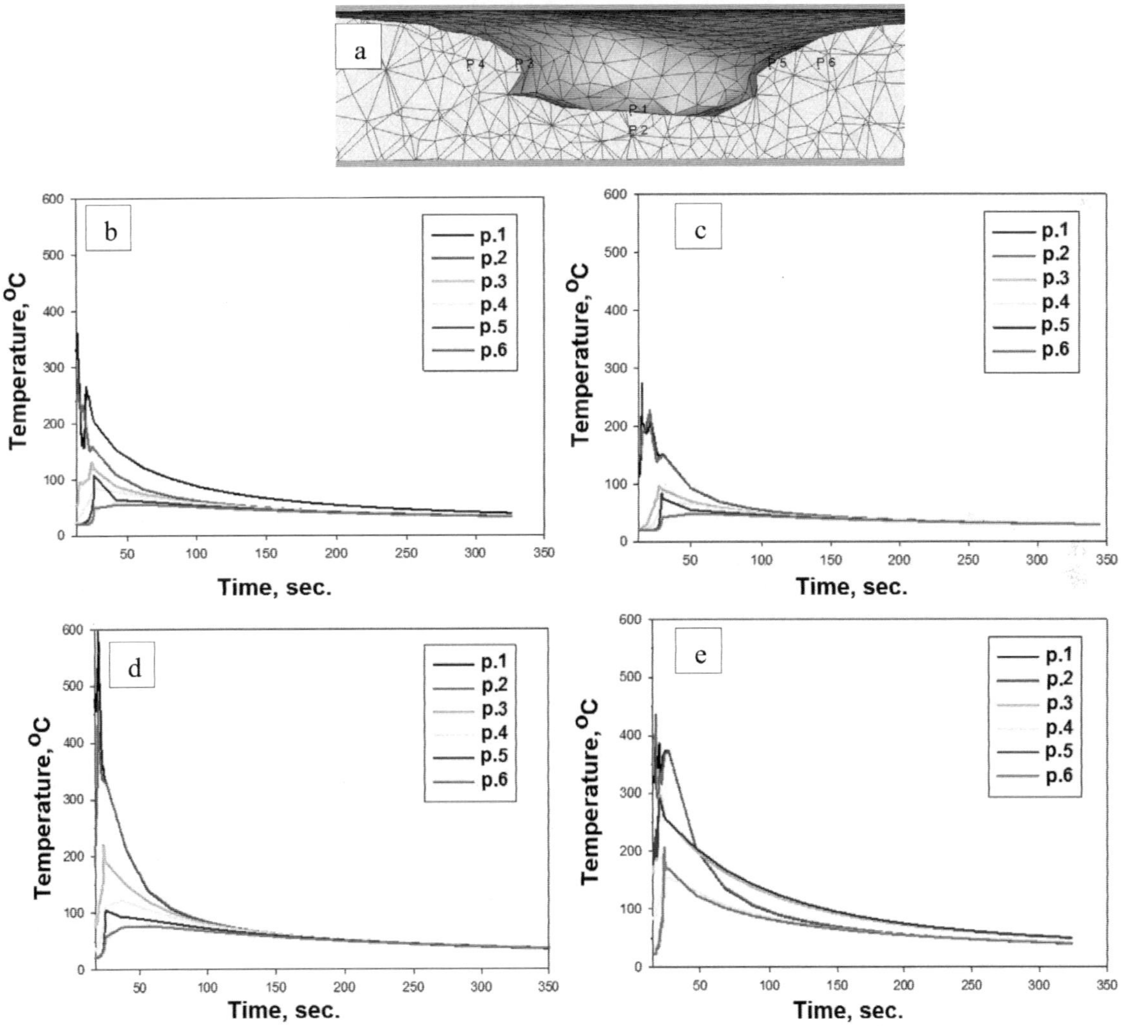

Figure 5. The arrangement of the controlled points in the specimen (a) and the cooling dependence for different pin lengths: 1.6 mm (b), 1.8 mm (c), 2.0 mm (d), 2.2 mm (e).

These results are partially confirmed by the data of the temperature distribution in the zone of thermomechanical effects obtained by finite element modeling. In Figure 5 the location of the temperature monitored points, as well as the graphs of the specimens cooling after the end of the FSP, carried out by a tool with a different pin length.

It can be seen that the temperature for all pin lengths, at least below the pin, reaches or exceeds 300 °C, which is consistent with the observation of the θ(Al_2Cu) phase and the $Al_{15}Si_2(CuFeMn)_3$ phase. Also, processing a workpiece with a 2.0 mm long pin provides more uniform heating in the entire zone of thermomechanical action (Fig. 5d). Thus, based on the results of computer simulation, it can be concluded that for the most noticeable modification of properties in the zone of thermomechanical action, the required and sufficient length of the pin is 2.0 mm.

Conclusions

1. According to the results of real and model experiments, it was found that for specimens from the 2024 aluminum alloy with a thickness of 3.0 mm, a necessary and sufficient pin length at friction stir processing is 2.0 mm, because mixing throughout the thickness is observed in this case.
2. During friction processing with stirring of a 2024 aluminum alloy at the treated area solid-solutional hardening occurs with the formation of intermetallic $Al_{15}Si_2(CuFeMn)_3$ and Al_2Cu phases.

Acknowledgments

The present work was accomplished according to the state assignment of IMSP RAS.

References

[1] Liu H J, Shen J J, Huang Y X, Kuang L Y, Liu C and Li C 2009 *Sci. Technol. Weld. Join.* **14** 577.
[2] Sarvghad Moghaddam M, Parvizi R, Haddad-Sabzevar M and Davoodi A 2011 *Mater. Des.* **32** 2749.
[3] Xue P, Xie G M, Xiao B L, Ma Z Y, Geng L 2010 *Metall. Mat. Trans.* A **41** 2010.
[4] Vinothkumar H, Saravanakumar S, Ramesh C, Prakash P, Ragul Vignesh A and Naveen S 2020 *Mater. Today: Proc.* In press.
[5] Elyasi, Derazkola H, Hosseinzadeh M 2016 *J. Eng. Man.* **230** 1234.
[6] Assidi M, Fourment L, Guerdoux S, Nelson T 2010 *Int. J. Mach. Tools Manuf.* **50** 143.
[7] McNelley T R 2015 *Lett. Mater.* **5**(3) 246.
[8] Chai F, Zhang D, Li Y, Zhang W 2013 *Mater. Sci. Eng.* A **568** 40.
[9] Chen Y, Ding H., Li J, Cai Z, Zhao J, Yang W 2016 *Mater. Sci. Eng.* A **650** 281
[10] Chen X, Zhang Y, Cong M 2020 *Vacuum* **175** 1.
[11] Rudskoy A I, Naumov A A, Chernikov E V 2014 *Tsvetnye Metally* **4** 36 (in Russian).
[12] Mondolfo L F1979 *Structure and Properties of Aluminum Alloys* (Moscow: Metallurgy) p 640 (in Russian).

Structure and properties of Cu-based alloys diluted by Pd and Ag

E G Volkova, O S Novikova, A E Kostina, A V Glukhov and A Yu Volkov

Mikheev Institute of Metal Physics, Ural Branch, Russian Academy of Sciences, ul. S. Kovalevskoi 18, Ekaterinburg, 620137 Russia

E-mail: volkova@imp.uran.ru

Abstract. The microstructure, mechanical and electrical properties of Cu-Pd alloys (with Pd content not more than 8 at.%) and triple Cu-Pd-Ag alloys with 5 at.%Ag were studied. It was established that the strength and the recrystallization temperature of the Cu-Pd alloys increased with palladium content. A comparison of the electrical properties of these alloys when Pd content was increased from 0 to 8 at.%Pd was done. An anomaly was revealed in the Cu-5.9Pd and Cu-8Pd alloys: the plastic deformation does not lead to an increase in the electrical resistivity. It is due to the destruction of short range atomic order in the quenched Cu-Pd alloys under deformation. The alloying with silver leads to a further increase in strength with a constant electrical conductivity. It was found that annealing of the deformed Cu-Pd-Ag alloys at 200-300°C led to an increase in strength. It was shown that the effect was not related to the Pd-content and was due to silver only.

1. Introduction

Supersaturated solid solutions based on Cu-Ag alloys have high strength properties, acceptable ductility and low electrical resistivity. Such an optimal combination of functional properties allows using these alloys in practice. However, the search for ways to further improve the strength properties of copper-based alloys without losing their electrical conductivity is an important scientific and practical task.

Earlier, we showed [1] that the introduction of even a small amount of palladium (less than 10 at.%) into copper led to a significant increase in strength, which was caused by solid-solution hardening and grain refinement. Furthermore, the addition of palladium increases the corrosion properties of copper. Thus, it has been suggested that palladium in copper solid solutions can be used as a matrix for additional strengthening of copper with silver. The implementation of this idea can result in obtaining high-strength corrosion-resistant Cu-Pd-Ag alloys with a sufficiently high electrical conductivity that are in demand by the industry.

The purpose of this work was to study the structure and physical and mechanical properties of binary Cu-Pd alloys (with palladium content from 0 to 8 at.%) and ternary alloys based on them Cu-3Pd-5Ag and Cu-5Pd-5Ag (at.%).

2. Materials and methods

The alloys under investigation were prepared by melting Cu, Pd and Ag of 99.98, 99.99 and 99.99% purity, respectively. The metals were melted in vacuum of at least 10^{-2} Pa, and the alloys were poured into a graphite crucible. The chemical composition of samples was analyzed using a JEOL JCXA-733 microprobe. In total 5 binary Cu-Pd alloys were melted (with palladium contents 1.5, 3.0, 4.5, 5.9, and 8.0 at.% Pd) and two ternary alloys Cu-3Pd-5Ag and Cu-5Pd-5Ag (at.%). Further, for the sake of brevity, we omit "at.%" in the alloy names, for example: Cu-3Pd-5Ag. The pure Cu was also melted and used as a reference material.

The ingots 5 mm in diameter were homogenized at 800°C for 3 hours and then cut into two pieces. One piece of the ingot was to a 0.25 mm diameter thin wire (it was then cut to obtain samples for the resistometric study). The other piece of the ingot was rolled to obtain 0.3 mm thick plates (they were used as samples for the X-Ray analysis and microhardness measurements). We also carried out cryodeformation experiments. For this purpose, the sample was placed between two plates of stainless steel 18/10 and this "assembly" was then soaked for about 1 min in liquid nitrogen prior to each rolling pass. Although the rollers of the rolling mill had room temperature, we believe that the alloy samples inside this "assembly" were deformed at a temperature close to cryogenic one.

Some of these wires and plates were annealed at 700°C for 1 h and then were immediately quenched in cold salt water. As a result, we studied the samples that were in three structural states: hardened, deformed at room temperature and subjected to deformation at a low temperature.

The room-temperature electrical resistivity was measured by the four-point method at DC 20 mA. Vickers microhardness values were determined using the PMT-3 tester with 0.05 kg load and dwelling time of 30 s. Each microhardness value was the average of at least ten measurements.

3. Results and discussion

3.1. Binary Cu-Pd alloys

It is well known that plastic deformation leads to an increase in the electrical resistivity of metals and alloys. It is caused by the formation of different scattering centers such as dislocations, point defects, and stacking faults due to deformation [2]. As an example, curve 1 in figure 1 shows the dependence of the electrical resistivity of pure copper on the deformation strain (e). This experiment was carried out using wires and their deformation strain was defined by the common equation: $e = \ln (S_o/S_f)$, where S_o and S_f are the initial and final cross section areas, respectively [3].

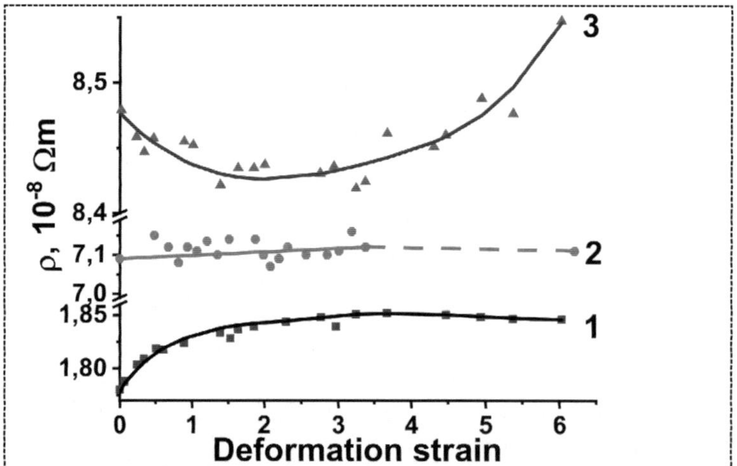

Figure 1. Dependences of the electrical resistivity on the deformation strain of pure copper (1), Cu-5.9Pd (2) and Cu-8Pd (3) alloys.

Because of severe plastic deformation (SPD), the electrical resistivity of copper increases approximately by 4%. It is clear that the deformation strain of copper higher than e≈4.0 does not lead to any change in its electrical resistivity. In this case, there is a dynamic equilibrium between the generation of defects through SPD and their disappearance due to the processes of dynamic recovery/recrystallization [4].

However, the dependences of the electrical resistivity obtained during the deformation of the initially quenched Cu-5.9Pd and Cu-8Pd alloys significantly differ from the behavior of copper. Indeed, the electrical resistivity of the Cu-8Pd alloy even slightly decreases at the initial stages of deformation (curve 3 in figure 1). The effect of SPD on the electrical resistivity of these alloys has never been investigated before. However, as it was established in [5], the electrical resistivity of the Cu-19.5Pd alloy abnormally decreases under plastic deformation.

The Cu-Pd alloys in vicinity of Cu_3Pd stoichiometry are known to form $L1_2$-superstructure and the Cu-19.5Pd alloy has the maximum rate of the disorder→order phase transformation among them [6]. Quenching of ordered alloys leads to the formation of a short range atomic order (SRO) in them. Clusters of SRO are effective centers of electron scattering, but plastic deformation destroys them. That is why the electrical resistivity of the quenched Cu-19.5Pd alloy is more than that of the deformed one [5].

The $L1_2$-type superstructure is also formed in Cu-8Pd alloy, but its ordering rate is significantly lower compared with the Cu-19.5Pd alloy. As shown in [7], the formation of the SRO in both Cu-8Pd and Cu-19.5Pd alloys increases their electrical resistivity as the result of quenching. The destruction of the SRO at the first step of plastic deformation leads to a decrease in the electrical resistivity of these alloys. When the deformation strain increases, the mean free path of the electron will decrease due to the rise in the number of different scattering factors such as grain boundaries, dislocations, vacancies, etc. As a result, the dependence of the electrical resistivity from the deformation strain of the initially quenched Cu-8Pd alloy has a parabolic shape.

It is shown in figure 1 (curve 2) that the electrical resistivity of the Cu-5.9Pd alloy remains almost constant during SPD (we had not sufficient quantity of this alloy, so this graph has no intermediate points). As follows from the phase diagram of the Cu-Pd system, only a solid solution of palladium in copper should form in the Cu-5.9Pd alloy. However, our results do not confirm that. The electrical resistivity of the solid solution should increase during the deformation, as it happens in pure copper [8]. However, the result is quite in agreement with our assumption about low degree of SRO, which forms in the Cu-5.9Pd alloy as the result of the quenching from high temperatures [7]. In this case, during the deformation of the quenched alloy, short-range order clusters are destroyed and new defects are generated. However, if the oppositely directed contributions from these processes to the electrical resistivity are comparable, it will remain almost unchanged.

The experiments carried out once again showed that the existing phase diagram of the Cu-Pd alloys needed further refinement. Indeed, our experimental data (see figure 1 and our works [1, 7]), as well as the results of the theoretical evaluation in [9], suggest that the region, where the $L1_2$-superstructure exists in the Cu-Pd system, is somewhat wider than it follows from the generally adopted phase diagram.

3.2. Triple Cu-Pd-Ag alloys

It was established before [10] that the preliminary cryogenic deformation suppressed the processes of dynamic recovery/recrystallization, increased the dislocation density and activated the mechanical twinning. It leads to the formation of an ultrafine-grained structure during annealing of the alloys that increases their mechanical properties. It was found in [11] that the application of moderate annealing after cryo-rolling allowed achieving an optimal combination of high strength and good ductility in the Cu-3Ag alloy. As these results may be applied for practical purposes, the study of cryo-deformed alloys is of doubtless scientific interest. However, until now, the effect of the preliminary cryo-deformation on the evolution of the structure and properties during annealing has been studied in detail mainly in pure metals (Cu, Al, Nb etc.). For example, changes in the structure, texture, and properties during long low-temperature annealing of cryo-deformed copper was studied in [12]. However, the efficiency of preliminary cryogenic deformation for increasing the strength of dilute Cu-Pd and Cu-Pd-Ag alloys is still not clear.

The study has shown that the introduction of silver into the copper-palladium matrix leads to a significant increase in microhardness (figure 2). Moreover, during the study, we found a temperature anomaly: the strength properties of pre-deformed Cu-Pd-Ag alloys grow after low-temperature annealing. This phenomenon is particularly pronounced after the preliminary cryodeformation. The study found that the detected anomaly did not depend on the amount of palladium in the alloy and was determined exclusively by the silver content. The nature of this anomaly is known: low-temperature annealing leads to the decomposition of the supersaturated solid solution with the precipitation of submicrocrystalline particles of the new phase on all defects (dislocations, twin boundaries, etc.) [13].

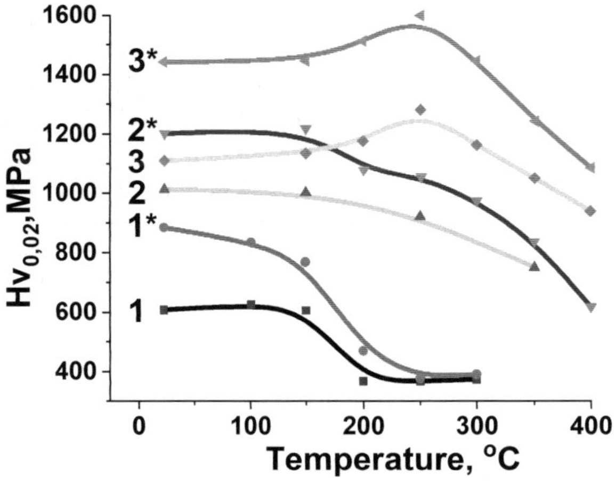

Figure 2. Dependences of microhardness on annealing temperatures of deformed copper samples (curves 1, 1*), Cu-5Pd (2, 2*) and Cu-5Pd-5Ag (3, 3*) alloys. All samples were deformed at room temperature, but the samples 1*, 2* and 3* were pre-soaked in liquid nitrogen. The duration of annealing was 1 hour.

We assumed hereabove that the $L1_2$-superstructure might be formed in dilute Cu-Pd alloys. However, our TEM investigation did not help to detect the ordered $L1_2$ phase due to its small content in the Cu-5.9Pd alloy. From this point of view, a detailed study of the microstructure of Cu-Pd-Ag ternary alloys is of particular interest. As is known, the simultaneous implementation of the atomic ordering and decomposition processes in ternary alloys leads to a significant acceleration of both reactions. We showed it earlier in Cu-Pd-Ag alloys heavily doped with palladium and silver [14]. A recent study of precipitate growth at reduced temperatures during ordering by $L1_0$-type in deformed Au-Cu-Ag alloys confirmed this conclusion [15]. Therefore, it is of interest to carry out a detailed study of the microstructure of dilute Cu-Pd-Ag alloys after long low-temperature annealing. Indeed, the antiphase domain boundaries will splash with the particles of the precipitated phase, facilitating the detection of the the ordered phase. Thus, it is necessary to continue the study of the structure and properties of dilute Cu-based alloys with Pd and Ag.

4. Conclusion

The results of this study give reason to doubt that the Cu-5.9Pd alloy is a solid solution. Since the short-range atomic order is observed in the quenched alloy, it is possible that the long-range atomic order can be formed in it. Thus, the results obtained in this work necessitate further studies of low-alloyed copper-palladium alloys in order to clarify the position of the thermal-concentration boundary of the existence of the ordered $L1_2$ phase on the phase diagram of the Cu-Pd system. The high physical and mechanical properties of Cu-Pd-Ag alloys make it possible to consider them as promising high-strength conductors of electric current. The detected anomalous increase in strength properties during low-temperature annealing of Cu-Pd-Ag alloys can be used in practice for their further strengthening. The preliminary cryodeformation causes an increase in the strength properties of all investigated alloys.

Acknowledgments
The work was carried out with the financial support of the Russian Foundation for Basic Research (project № 18-03-00532).

References

[1] Volkov A Yu, Kostina A E, Volkova E G, Novikova O S and Antonov B D 2017 *The Physics of Metals and Metallography* **118** 1236
[2] Mavlyutov A M, Bondarenko A S, Murashkin M Yu, Boltynjuk E V, Valiev R Z and Orlova T S 2017 *Journal of Alloys and Compounds* **698** 539
[3] Wang F, Wakoh K, Li Y, Ito S, Yamanaka K, Koizumi Y and Chiba A 2017 *Materials and Design* **126** 64
[4] Deng L, Han K, Hartwig K T, Siegrist T M, Dong L, Sun Z, Yang X and Liu Q 2014 *Journal of Alloys and Compounds* **602** 331
[5] Kim M J and Flanagan W F 1967 *Acta Metallurgica* **15** 735
[6] Subramanian P R and Laughlin D E 1991 *Journal of Phase Equilibria* **12** 231
[7] Volkova E G, Novikova O S and Volkov A Yu 2018 *IOP Conference Series: Materials Science and Engineering* **447** 012029
[8] Freudenberger J, Kauffmann A, Klaub H, Marr T, Nenkov K, Subramanya Sarma V and Schultz L 2010 *Acta Materialia* **58** 2324
[9] Li M, Guo C and Li C 2008 *Computer Coupling of Phase Diagrams and Thermochemistry* **32** 439
[10] Lapeire L, Sidor J, Verleysen P, Verbeken K, Graeve I D, Terryn H and Kestens L A I 2015 *Acta Materialia* **95** 224
[11] Gubicza J, Hegedus Z, Labar J L, Subramanya Sarma V, Kauffmann A and Freudenberger J 2014 *IOP Conference Series: Materials Science and Engineering* **63** 012091
[12] Konkova T, Mironov S, Korznikov A, and Semiatin S L 2011 *Materials Science and Engineering* A **528** 7432
[13] Zeng Z R, Zhu Y M, Bian M Z, Xu S W, Davies C H J, Birbilis N, Nie J F 2015 *Scripta Materialia* **107** 127
[14] Syutkina V I and Volkov A Yu 1992 *The Physics of Metals and Metallography* 73 205
[15] Garsia-Gonzalez M, Petegrem S V, Baluc N, Hocine S, Dupraz M, Lalire F and Swygenhoven H V 2019 *Scripta Materialia* **170** 129

IOP Conf. Series: Materials Science and Engineering 1008 (2020) 012027 doi:10.1088/1757-899X/1008/1/012027

Effect of accumulative high-pressure torsion on the structure of steel 10

I A Ramazanov[1], D V Gunderov[1,2], V V Astanin[1], V D Sitdikov[1], G T Zakiryanova[3], A V Sharafutdinov[1]

[1] Ufa State Aviation Technical University, K. Marx str. 12, Ufa, 450008, Russia
[2] Institute of Molecule and Crystal Physics, Ufa Federal Research Centre RAS, October Avenue 71, 450054, Ufa, Russia
[3] Bashkir State Medical University, Lenin str. 3, Ufa, 450008, Russia

E-mail: ilnar.ramazanov@ya.ru

Abstract. The effect of processing by accumulative high-pressure torsion on the structure of steel 10 is investigated. It is found that after processing by accumulative HPT, a stronger structure refinement takes place in steel 10 as compared to regular HPT processing for the same number of revolutions.

1. Introduction

In the past two decades, there has been a great interest in the new approach for increasing the properties of metallic materials by means of producing ultrafine-grained (UFG) structures in them using severe plastic deformation (SPD) processing [1]. Under UFG materials one means polycrystalline materials with grain sizes below 1 μm having predominantly high-angle grain boundaries [1]. The formation of such structures is possible only when very high strains ($\varepsilon \geq 4-6$) are attained at relatively low temperatures ($T_{SPD} \leq 0.4\, T_{melt}$) under high applied pressures. High-pressure torsion (HPT) enables producing the highest strains [1]. Previously there have been a number of studies on the effect of HPT processing on different steels, in particular [1–3]. Shear strain, γ, introduced during HPT can be calculated using the following equation: $\gamma = 2\pi nR/h$ [1]. However, it is known that in solid metallic materials the strain that can actually be produced in the process of HPT is much smaller than the expected one calculated using above equation, which is caused by the effect of "slippage" [4,5]. In [6] a new technique called "accumulative high-pressure torsion" (ACC HPT) that enables producing higher strains in hard metallic materials than the regular HPT technique was proposed. In this article, the accumulative high-pressure torsion method was first used for low carbon steel (C 0.1 wt.%, Russian steel 10).

2. Research methods and results

The aim of the present work is to study the structure of low-carbon steel 10 after processing by regular and accumulative HPT.

The chemical composition of the investigated steel 10 is as follows: Fe (basis), C 0.1 wt.%, other substances (Mn, Si, Ni) 0.5 wt.%.

HPT processing was conducted on anvils with a diameter of 10 mm, having a groove 0.3 mm deep, under a pressure of 6 GPa at room temperature. Two disks were subjected to regular HPT for 5 and 11 revolutions. One sample was subjected to accumulative HPT. At the first stage, the sample was subjected to HPT for n=2 anvil revolutions. Then the sample was broken into 4 parts, the parts were stacked onto the anvils, the stack was subjected to compression, and again to HPT for n=2. In this manner, 4 cycles were performed. As a result, the torsional strain was summed with the compression strain (about 75% at each stage). At the final stage, the sample was subjected to HPT for n=3. Thus, the total number of accumulative HPT revolutions was n=11 and an integral disk-shaped sample was obtained.

An analysis of the results of the transmission electron microscopy (TEM) studies for steel 10 after HPT processing for n=5, n=11 and accumulative HPT for n=11 reveals the following. In the electron diffraction pattern of steel 10 after regular HPT processing for n=5, n=11, there are several large blurred reflections on the ring, as well as individual smaller dots (figures 1 and 2). The general view of the electron diffraction pattern indicates the structure refinement. However, in the electron diffraction pattern for accumulative HPT processing for n=11 the reflections are smaller, more numerous and more uniformly distributed on the ring, which indicates a finer structure in the condition after accumulative HPT processing for n=11 (figure 3). The bright-field TEM image with smaller elements also indicates a finer structure in the condition after accumulative HPT processing for n=11.

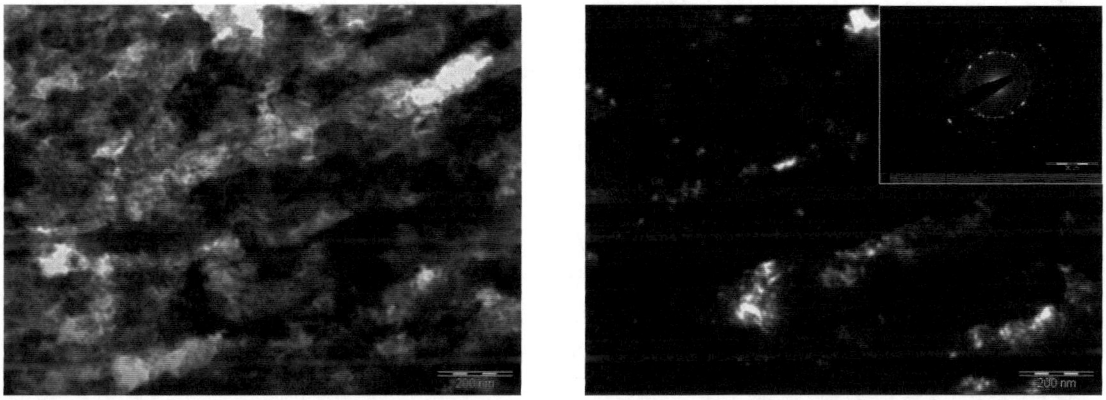

Figure 1. Structure of steel 10 after HPT for n=5. Bright- and dark-field images, electron diffraction pattern.

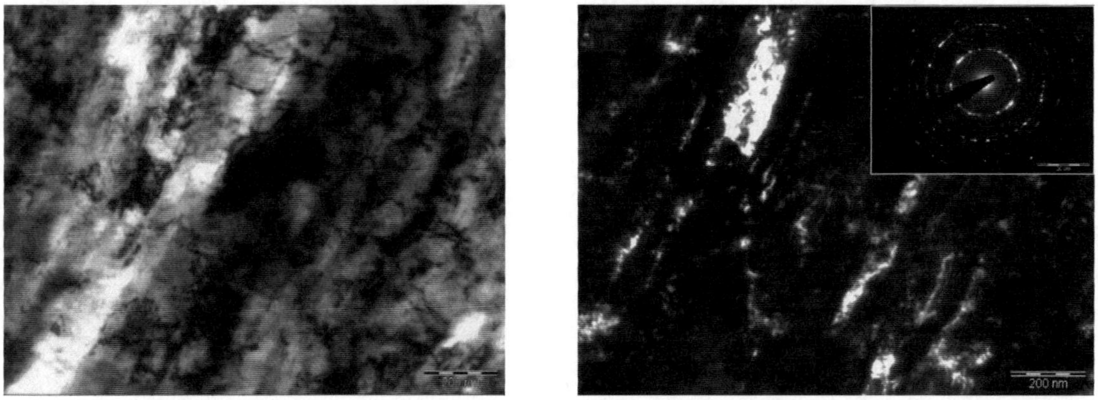

Figure 2. Structure of steel 10 after HPT for n=11. Bright- and dark-field images, electron diffraction pattern.

IOP Conf. Series: Materials Science and Engineering 1008 (2020) 012027 doi:10.1088/1757-899X/1008/1/012027

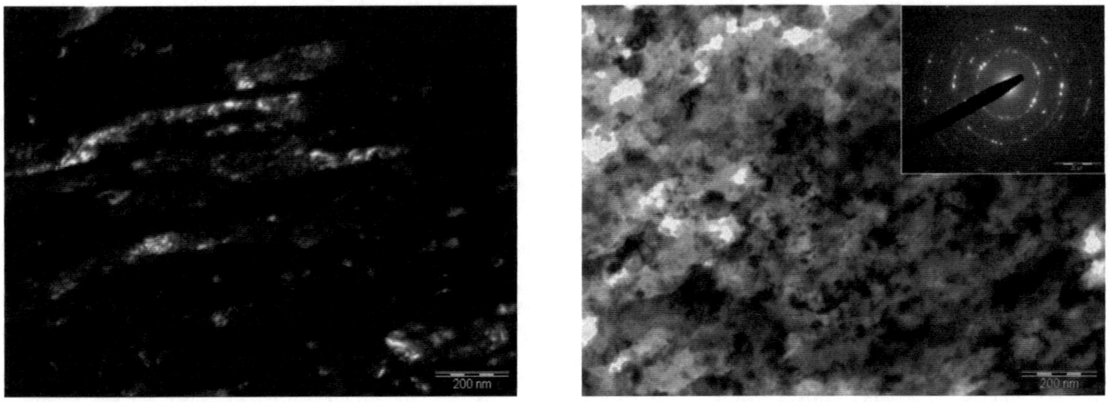

Figure 3. Structure of steel 10 after accumulative HPT for n=11. Bright- and dark-field images, electron diffraction pattern.

The dark-field TEM image of steel 10 processed by HPT for n=11 shows a refined structure with an individual fragment size of about 200 nm. It is difficult to analyze the structure since it is very refined, strongly cold-worked and inhomogeneous. However, in the dark-field TEM image for the condition after accumulative HPT processing for n=11 the grain size is about 50 nm.

The results of the measurements of Vickers microhardness are presented in table 1. The measurements were made on a DuraScan-50 automated microhardness tester under a load of HV 0.025 with the standard exposure time of 10 s.

Table 1. Microhardness test results.

Condition	Values of H_V microhardness		
	Sample center	Mid-radius	Sample periphery
Initial		227	
HPT n= 5 rev.	344	371	492
HPT n=11 rev.	425	662	735
ACC HPT n=11 rev.	729	736	807

An analysis of the test results shows that an increase in the number of revolutions leads to an increase in the microhardness values. The implementation of accumulative HPT instead of the classic procedure leads to an overall increase in Hv, and also enables increasing microhardness in the periphery and at mid-radius.

Figure 4 shows the X-ray diffraction patterns of steel 10 in the initial condition, after processing by accumulative and regular HPT.

Table 2. XRD results.

Condition	Lattice parameter, nm	CSR, nm	Lattice distortion, %	Dislocation density, 10^{15} m^{-2}
Initial	2.86918(2)	118	0.15	0.8
HPT 5 rev.	2.86991	25	0.35	2.8
HPT 11 rev.	2.87187(8)	24	0.62	5.6
ACC HPT 11 rev.	2.87279(7)	21	0.73	8.7

According to the XRD data, HPT processing for n=11 leads to structure refinement, coherent scattering region (CSR) reduction, and a considerable increase in dislocation density as compared to the initial condition and that after HPT processing for n=5. ACC HPT processing for n=11 leads to the formation of a CSR with a smaller size (21 nm) and a further increase in dislocation density ($8.7 \cdot 10^{15}$ m^{-2}). The XRD results are presented in table 2.

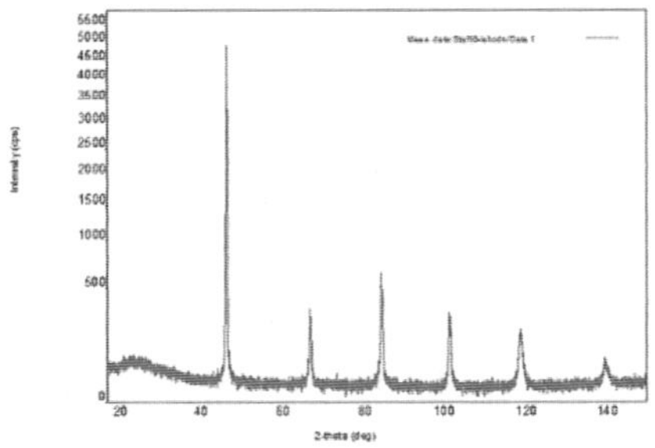

Initial

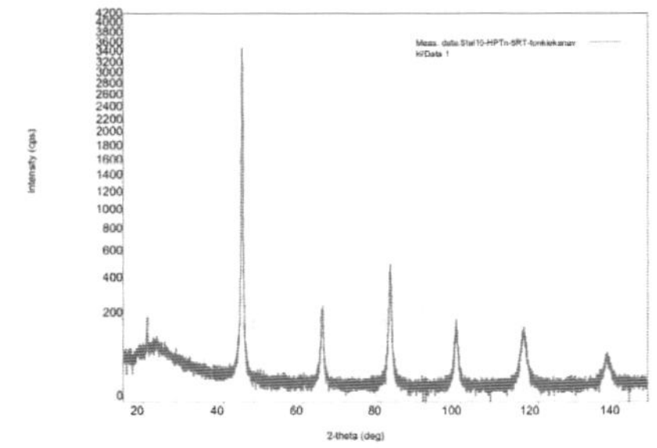

HPT n=5

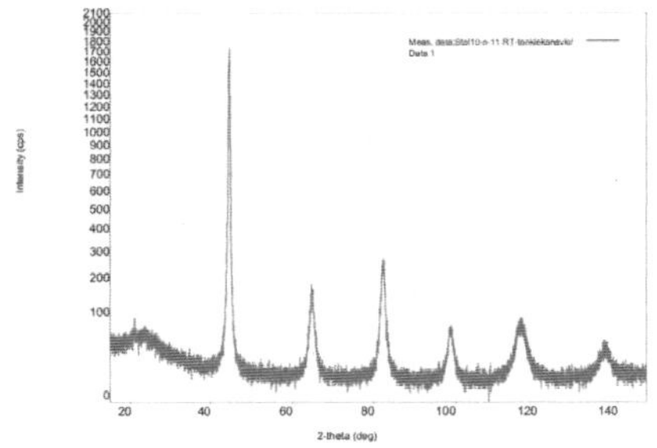

HPT n=11

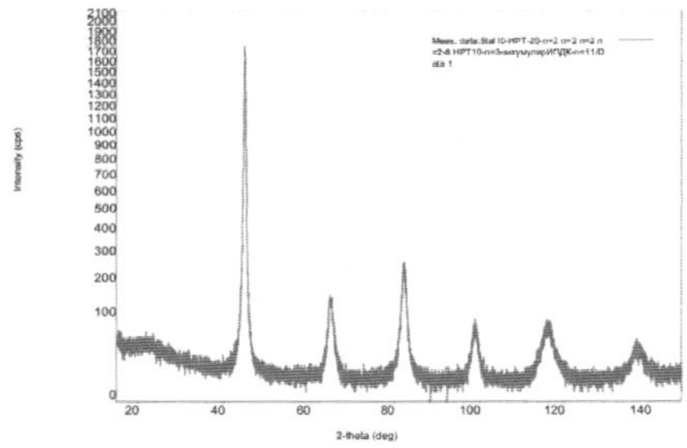

ACC HPT n=11

Figure 4. X-ray diffraction patterns of steel 10.

Hence, the data of X-ray diffraction analysis and transmission electron microscopy indicate that as a result of the accumulative HPT, a higher density of dislocations is accumulated, as well as a stronger grain refinement than in the case of using HPT according to the usual scheme.

3. Conclusions

During accumulative HPT processing a stronger structure refinement takes place in steel 10 as compared to regular HPT processing with the same number of revolutions. The grain (fragment) of steel 10 as a result of accumulative HPT is refined to about 50 nm (according to the TEM dark field), a higher dislocation density and a higher and more uniform microhardness across the disk are achieved than after conventional HPT.

Acknowledgments

This work was supported by Russian Foundation for Basic Research under grant No. 19-58-80018 (BRICS_t).

References

[1] Valiev R Z and Alexandrov I V 2000 *Nanostructured Materials Produced by Severe Plastic Deformation* (Moscow: Logos)
[2] Ivanisenko Y, Lojkowski W, Valiev R Z and Fecht H-J 2003 *Acta Mater.* **51** 5555–70
[3] Ganeev A V., Karavaeva M V., Sauvage X, Courtois-Manara E, Ivanisenko Y and Valiev R Z 2014 *IOP Conf. Ser. Mater. Sci. Eng.* **63** 012128
[4] Valiev R Z, Zhilyaev A P and Langdon T G 2013 *Bulk Nanostructured Materials* (Hoboken, NJ: John Wiley & Sons, Inc)
[5] Gunderov D V, Churakova A A, Boltynjuk E V, Ubyivovk E V, Astanin V V, Asfandiyarov R N, Valiev R Z, Xioang W and Wang J T 2019 *J. Alloys Compd.* **800** 58–63
[6] Gunderov D V, Churakova A A, Astanin V V, Asfandiyarov R N, Hahn H and Valiev R Z 2020 *Mater. Lett.* **261** 127000

Theoretical estimation, simulation and experimental evaluation of deformation-induced heating of magnesium during lateral extrusion

D A Komkova[1], J V Zamaraeva[1, 2], Yu N Loginov[1, 2], P S Koshevoy[1], A L Sokolov[1] and A Yu Volkov[1]

[1] M N Mikheev Institute of Metal Physics, Ural Branch, Russian Academy of Sciences, S. Kovalevskoi St. 18, Yekaterinburg, 620108 Russia
[2] Ural Federal University named after the first President of Russia B N Yeltsin, Mira St. 19, Yekaterinburg, 620002 Russia

E-mail: komkova_d@imp.uran.ru

Abstract. Theoretical estimation, computer simulation and experiment were performed to evaluate temperature of the deformation heating of magnesium during severe plastic deformation by lateral extrusion method. The simulation of the deformation process was carried out by DEFORM-3D. Experimental temperature measurement of magnesium deformation-induced heating at the deformation zone was made. According to the simulation, temperature of the magnesium billet has reached 114°C during deformation. Experimental results showed that the maximum temperature of deformation heating of the material is ~ 140°C.

1. Introduction
It is known that severe plastic deformation (SPD) methods, such as equal channel angular pressing (ECAP) or high pressure torsion (HPT), are used to obtain materials with ultrafine-grained structure and improved mechanical properties [1, 2]. SPD could be performed at different temperatures, either at higher or lower than the recrystallization temperature of material. However, to produce fine-grained structure material, it is recommended to deform metal at low temperature to prevent active recrystallization process. Nevertheless, even at low-temperature deformation, the heating could be possibly induced by large compressive stresses and by friction between the billet and the inner container wall [3]. In this case, the highest temperature would be in the deformation localization zone. On the one hand, the deformation heating has positive aspects, e.g., the plasticity of the deformable materials is increased and the load on the experimental tool is reduced. On the other hand, a rise of temperature during deformation can lead to the surface oxidation, the increase in the size of the new forming crystallites, and the development of the texture unfavorable for further deformation.

Earlier, in [4, 5], we revealed a new SPD-method called the lateral extrusion (LE). LE-method has allowed us to deform such low-plasticity metal as magnesium at room temperature without any pre-heating. As a result, we have obtained magnesium with fine-grained structure after one-pass deformation. Nonetheless, there is still an open question of the magnesium billet temperature in the deformation localization zone during the LE-process.

The aim of the study is to evaluate deformation-induced heating of magnesium at low-temperature SPD by theoretical estimation as well as computer simulation and experimental measurements.

2. Material and methods

2.1. Lateral extrusion method

To perform LE-deformation, a cylindrical billet of commercially pure magnesium MG90 was used as the initial material. The billet has a diameter of 42 mm and a height of 40 mm. The Mg-billet was placed in a steel container (figure 1a). At the base bottom of the container, there is a rectangular channel perpendicular to the container. The size of the channel is 42 mm in width and 1 mm in height. The billet is coated with wax and molybdenum disulfide lubricant to reduce friction [6]. The deformation process was performed at room temperature in one pass without pre-heating of the Mg-billet. Under load, the material has started to flow through the channel. Depending on the texture of the initial material, loading stress has reached 0.7 – 1.2 GPa [4]. Obviously, the deformation localization zone is in the intersection of the container and the channel (figure 1b). Magnesium flowed out in the form of a 1 mm thick plate. True strain after LE-deformation is $\varepsilon \sim 3.9$.

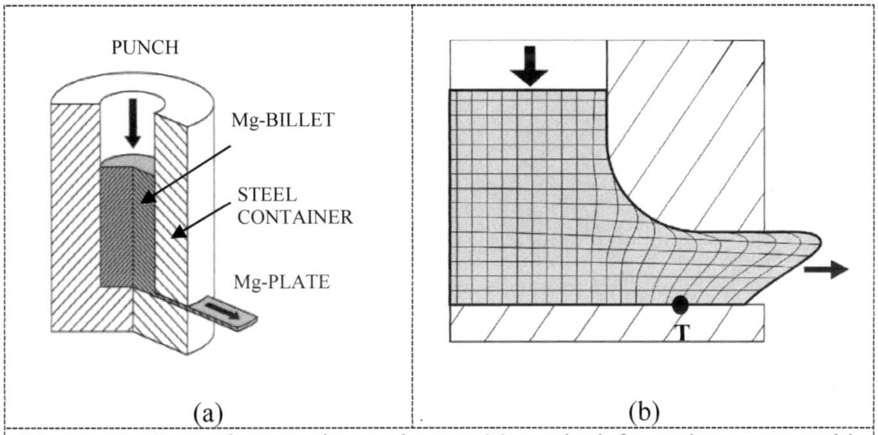

Figure 1. Lateral extrusion scheme (a) and deformation zone with indication of the location of the thermocouple T (b).

2.2. Theoretical estimation and 3D-simulation

To evaluate the distribution of temperature fields in magnesium at LE-process, computer simulation was performed by DEFORM-3D software. A finite element model of LE-equipment and the billet was created in DEFORM-3D pre-processor. The size of the model Mg-billet corresponds to the physical experiment. The boundary condition of the third kind was used for non-isothermal simulation. In this case, heat exchange between the billet and the steel container is characterized by a constant value of the thermal output. At the billet-container contact boundary, the Siebel friction law was used with the coefficient 0.2.

2.3. Experimental method

To measure magnesium temperature during LE, a hole was made in the bottom of the steel container, and a thermocouple was set in the intersection of the container and rectangular channel (figure 1b). The thermocouple was placed to be in direct contact with flowing magnesium plate. In our experiment, we used the type K thermocouple (the Chromel/Alumel thermocouple) with high linearity in the temperature range from 0 to 250°C. The analog signal from the thermocouple was digitized by a Metakon-613-T-TP-1 device and recorded by a computer with a frequency of one measurement per second.

3. Results and discussion

For theoretical estimation of deformation-induced heating of magnesium in the deformation-localization zone, the sizes of the zone, the friction forces and load stresses were taken into account.

Stress F of lateral extrusion is

$$F = S_0 \times q, \tag{1}$$

where S_0 is cross-sectional area of the billet, q is load on the punch.

Friction force F_{fr} at the moving billet is

$$F_{fr} = \pi \times D \times h \times f, \tag{2}$$

where D is diameter of the billet, h is the height of the billet, f is the friction coefficient of 0.5. In this case, the value of the force (F_{pd}) induced on plastic deformation in the steady-state process of LE is

$$F_{pd} = F - F_{fr}. \tag{3}$$

To estimate the work of plastic deformation, it is necessary to find the value of billet displacement (l) required to fill the deformation zone with metal in the steady-state LE-process:

$$l = \frac{V}{S_0}, \tag{4}$$

$$A = F_{pd} \times l, \tag{5}$$

where V is the volume of deformation zone (7.144×10^{-6} m^3).

The temperature of deformation-induced heating (ΔT) of magnesium in the deformation zone during LE is

$$\Delta T = \eta_h \times \frac{A}{V \times \rho \times c}, \qquad [7]\ (6)$$

where η_h is a heat output coefficient ($\eta_h = 0.9$), A is the work in the deformation localization zone, V is the volume in the deformation localization zone, ρ is Mg density, c is heat capacity of Mg. According to the results of mathematical estimation, the temperature (ΔT) of deformation-induced heating is about 52°C.

Therefore, magnesium temperature in the intersection zone of channels is ~ 75°C at LE (assuming that the temperature of Mg before deformation is the ambient temperature of 23°C).

3D-simulation has made it possible to evaluate the temperature field not only in the deformation zone but over the entire volume of the billet as well (figure 2). Generally, it is seen that the temperature distribution at SPD is non-uniform. It varies from 24°C to 114°C (figure 2a, b). Moreover, the edges of the plate have been found to be heated more than the central part (figure 2a). During LE, in the deformation zone (figure 1b), the microvolumes of the metal move along different paths. Thereby, accumulated shear deformations of different intensities arise in different areas. When magnesium flows in a rectangular channel, it is affected by the frictional forces of the channel surfaces which are directed against of the flow direction. This leads to the difference in rates of metal layers movement along the rectangular channel [6]. In the central part of the forming plate, only the upper and lower surfaces of the forming plate contact with the channel surface, due to what the middle layer of the forming plate flows faster. At the same time, the edges of the forming Mg-plate overcome the additional friction forces of the side surfaces of the channel. It probably causes an increase in both internal shear stresses in microvolumes of the metal at the edges of the forming plate and in internal energy. As a consequence, stronger heating of magnesium during deformation occurs.

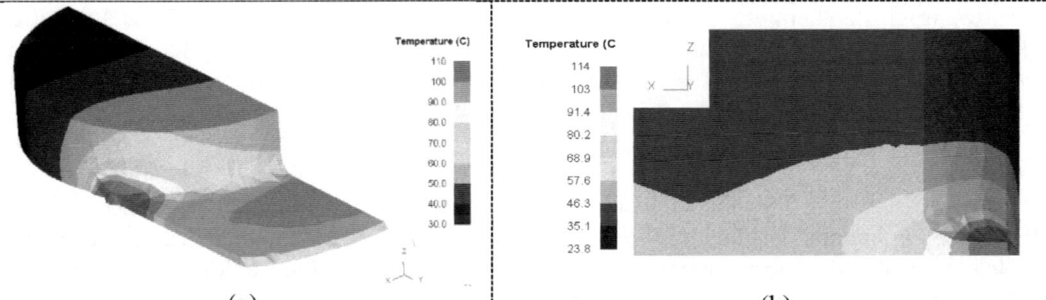

Figure 2. Temperature distribution of Mg in the deformation localization zone: XYZ-plane, at the begging of LE-process (a) and XY-plane, during LE-process (b).

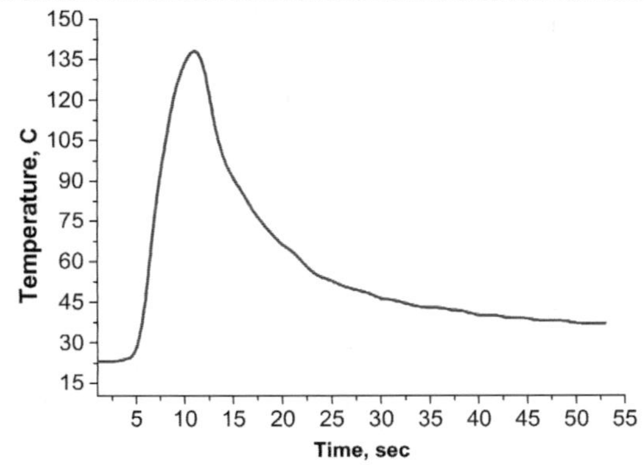

Figure 3. Plot of Mg-temperature (°C) vs. LE-process time (sec).

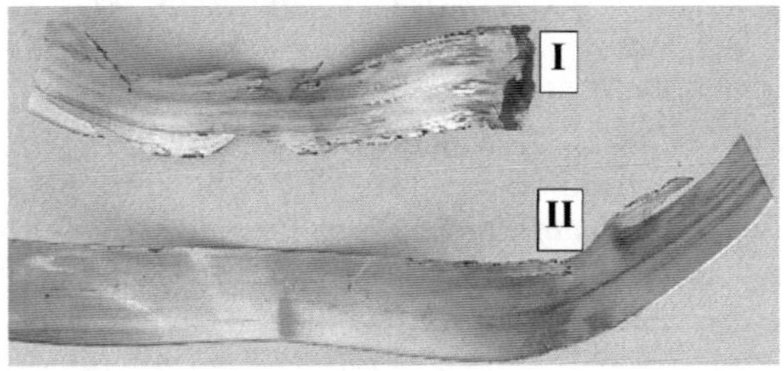

Figure 4. The obtained magnesium plate after lateral extrusion (LE): the beginning of the plate (I), the end of the plate (II).

Figure 3 shows the results of the experiment on measuring the temperature of magnesium during LE-deformation. The plot reveals that the temperature of magnesium at the beginning of the process was about 23°C (the ambient temperature). It is seen that metal has been gradually heated up to 140°C with deformation process.

The effect of the deformation-induced heating on the plasticity of magnesium can be judged by the shape and state of the obtained plate. Figure 4 demonstrates two parts of one Mg-plate (part I is the beginning of the plate, and was cut from the right side of part II). Part I has an irregular shape and severely cracked edges. At the same time, part II of the obtained plate seems to be smoother, without cracks at the edges. It can be concluded that at the beginning of deformation the metal has low plasticity due to low temperature of metal.

It is known that magnesium has hcp-lattice [8, 9]. This means that the metal has insufficient number of the slip systems for plastic deformation. Magnesium deformation occurs mainly along the basal planes <0001>. To activate the additional slip systems, deformation treatment is mostly performed at temperatures above 150°C. In our case, the temperature at which metal can flow easily through the rectangular channel is achieved during the deformation process. Gradual heating of the material takes place due to the large amount of energy that is transferred to the material during SPD. Temperature increase, especially in the zone of deformation localization, has a positive effect on the plasticity of the material and causes the migration of the boundaries of fragments, which especially occur in pure metals with a low melting point [11].

Conclusions

In this work, the deformation-induced heating of magnesium during LE-process was evaluated by both theoretical and experimental methods. For theoretical estimation, we have used mechanistic approach according to which, temperature of magnesium during LE is about 75°C. Computer simulation in DEFORM-3D software has shown that temperature of metal at SPD can reach up to 114°C. As for the experimental measurements, in this case, temperature of Mg in the deformation localization zone has grown during the LE-process up to ~ 140°C.

Acknowledgments
The work was carried out with the financial support of the Russian Foundation for Basic Research (RFBR). The experimental part and math estimation were performed within the framework of RFBR project No. 18-33-00474. Computer simulation was carried out within the framework of RFBR project No. 20-38-90051.

References

[1] Cao Y, Ni S, Liao X, Song M and Zhu Y 2018 *Mater. Sci. Eng.* R **113** 1
[2] Harsha R N, Kulkarni V M and Babu B S 2018 *Mat. Today: Proceed.* 2018 **5** 22340
[3] Utyashev F Z and Raab G I 2007 *Phys. Met. Metallorg.* **104** 605
[4] Antonova O V,. Volkov A Yu, Komkova D A and Antonov B D 2017 *Mater. Sci. Eng.* A **706** 319
[5] Komkova D A, Antonova O V and Volkov A Yu 2018 *Phys. Metals Metallorg.* **119** 1120
[6] Kamenetskiy B I, Volkov A Yu, Sokolov A L, Antonova O V and Klyukin I V 2014 *Method of foil Manufacturing from Magnesium* (RF patent No. 25630077) [in Russian]
[7] Konovalov Yu V, Ostapenko A L and Ponomarev V I 1986 *Estimation of Sheet Billet Parameters* (Moscow: Metallurgy) [in Russian]
[8] Raynor G R 1959 *The Physical Metallurgy of Magnesium and its Alloys* (London – NewYork: Pergamon Press)
[9] Mordike B L and Ebert T 2001 *Mater. Sci. Eng.* A **302** 37
[10] Botkin A V 2013 *Scientific and Methodological Foundations of the Design of Angular Extrusion Processes*: Doc. theses (Ufa, Ufa State Aviation Technical University) [in Russian]

High-pressure torsion of Zr-based bulk metallic glasses and amorphous melt-spun ribbons

D A Khasanova[1], D V Gunderov[1,2], V V Astanin[1], S D Gunderova[1], A A Churakova[1,2], A I Bazlov[3], D V Louzguine-Luzgin[4,5]

[1] Ufa State Aviation Technical University, K. Marx str. 12, Ufa, 450008, Russia
[2] Institute of Molecule and Crystal Physics, Ufa Federal Research Centre RAS, October Avenue 71, 450054, Ufa, Russia
[3] National University of Science and Technology «MISiS», 4 Leninskiy pr., Moscow, 119049, Russia
[4] WPI Advanced Institute for Materials Research, Tohoku University, Aoba-Ku, Sendai 980-8577, Japan
[5] Mathematics for Advanced Materials-OIL, National Institute of Advanced Industrial Science and Technology (AIST), Sendai 980-8577, Japan

E-mail: dimagun@mail.ru

Abstract. High-pressure torsion allows one to achieve high strains of bulk metal glasses without their destruction. The article presents the results of studies on the effect of high-pressure torsion on the amorphous alloys of the same composition but obtained in different ways: in a form of a bulk plate and a melt-spun ribbon. It is shown that the degree of nonequilibrium after high-pressure torsion, determined by the diffraction and calorimetry methods, increases for bulk metallic glass, approaching the values characteristic for melt-spun alloy.

1. Introduction

Amorphous metals and alloys are usually produced by melt spinning (MS) at high cooling rates (10^6 K/s) in the form of thin ribbons and at a cooling rate of about 10^2 K/s in the form of bulk metallic glasses (BMGs) [1,2]. As produced at different melt cooling rates, these materials differ in the level of nonequilibrium (excess energy) and properties [3].

The deformation of amorphous alloys at room temperature occurs through the formation of shear bands. A transformation of the amorphous structure takes place in shear bands and surrounding regions [4]. In many cases deformation leads to the formation of an excess volume in the amorphous phase (and softening of materials) [5,6]. High-pressure torsion (HPT) processing enables achieving high strains in brittle amorphous alloys, and consolidating amorphous ribbons [7,8]. HPT processing of amorphous alloys produces a high density of shear bands [9] and changes their structure and properties [10]. Partial nanocrystallization may take place in some alloys during the HPT processing [11,12]. Structural transformations as a result of HPT depend on the initial internal energy of the amorphous alloy. As a consequence, of great interest are the studies of the effect of HPT on the amorphous alloy with the same composition, but produced at different melt cooling rates and with different degrees of disequilibrium.

2. Material and methods

In this study, plates of the $Zr_{62}Cu_{22}Al_{10}Fe_5Dy_1$ at. % BMG (BMG Zr62) with sizes 60×10×2 mm^3 produced by casting at a cooling rate of 10^3 K/s and melt-spun $Zr_{62}Cu_{22}Al_{10}Fe_5Dy_1$ at. % (MS Zr62) ribbons with the thickness of 0.04 mm and a width of about 2 mm produced by melt -spinning at a cooling rate of 10^6 K/s were used.

The HPT was carried out on anvils with a diameter of 10 mm and groove depth of 0.3 mm under the pressure of 6 GPa at a rotation speed of 1 rpm at room temperature (RT). The structure was investigated by X-ray diffraction under Cu radiation on a Rigaku Ultima IV diffractometer. The parameters of broad diffraction peaks (amorphous halos) were evaluated by a PHILIPS ProFit software. The DSC tests were performed on a Netzsch DSC 204 F1 Phoenix calorimeter; the heating temperature was 520 °C (higher than the crystallization temperature in this BMG, MS) and the typical heating rate was 20 °C/min.

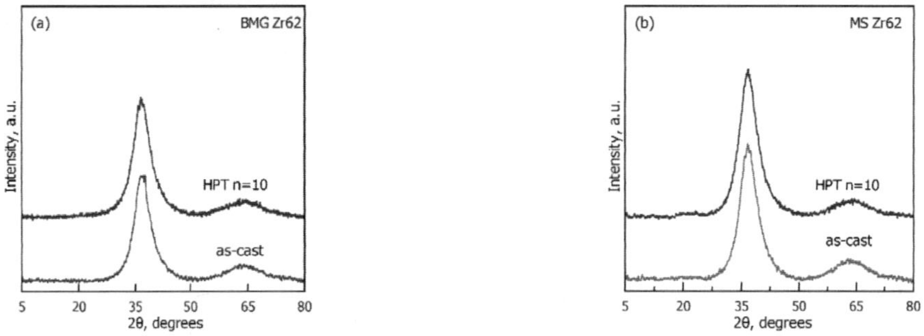

Figure 1. X-ray diffraction patterns of a) BMG $Zr_{62}Cu_{22}Al_{10}Fe_5Dy_1$ in the initial state and after HPT n=10; b) MS $Zr_{62}Cu_{22}Al_{10}Fe_5Dy_1$ alloy in the initial state and after HPT n=10.

3. Results and discussion

According to XRD results, initial BMG, MS, and the same materials after HPT processing are mostly amorphous (figure 1). The radius of the first coordination sphere, R_1, was determined based on the angular position of the amorphous halo maximum. R_1 of BMG increase as a result of HPT (table 1). The relative change in free volume (ΔV) can be calculated based on the change of R_1 [13,14]. According to the calculations, HPT n=10 of BMG leads to a 2% increase in ΔV. HPT also leads to an increase in the values of full width at half maximum (FWHM) of BMG, that also attributed to an increase in the disequilibrium of the amorphous alloy [13].

Table 1. XRD and DSC data for the BMG and MS $Zr_{62}Cu_{22}Al_{10}Fe_5Dy_1$ alloy in the initial state, and subjected to HPT n=10.

State	R_1, nm	FWHM, deg	ΔV, %	ΔFWHM, %	H_{rel}, J/g
BMG, initial	2.996	5.34		-	9
BMG, HPT n=10	3.017	5.70	2.1	6.8	18
MS alloy, initial	3.022	5.64		-	25
MS alloy, HPT n=10	3.015	5.76	-0.7	2.1	28

HPT n=10 also lead to increase in the values of FWHM of the MS alloy (table 1). However, the relative increment in the FWHM of the MS alloy resulting from HPT is noticeably smaller than increment in the FWHM for the BMG. The ΔV values of the MS alloy, calculated on the basis of the R_1 change, even slightly decrease as a result of the HPT. This may be due to a large error in the determining R_1 by XRD. On the whole, the FWHM parameters of the MS alloy and BMG become closer due to HPT processing.

Based on the DSC data, the H_{relax} of the initial MS Zr62 is higher than H_{relax} of the Zr62 BMG. This is attributed to the more disequilibrium of the MS alloy in comparison with the BMG. The HPT of the BMG leads to a notable increase in H_{relax} (from ≈ 9 to ≈ 18 J/g), which is the result of growth of the energy of amorphous alloy and ΔV [1].

The relative increase of H_{relax} in the MS resulting from the HPT, is not as large as that in the BMG resulting from the HPT. The gain is close to the measurement error of H_{relax} based on DSC. Thus, the following tendency can be observed: in the Zr62 BMG, as a result of HPT processing, the structure becomes more non-equlibrium, while in the MS Zr62 alloy, as a result of HPT processing, the state does not change that much, and on the whole, the characteristics (H_{relax}, FWHM) of the Zr62 BMG and MS Zr62 alloy become closer as a result of HPT processing. This is explained by the competition between the processes of deformation and relaxation during the HPT of amorphous alloys, and the intensification of relaxation with an increase in the disequilibrium of the amorphous alloy.

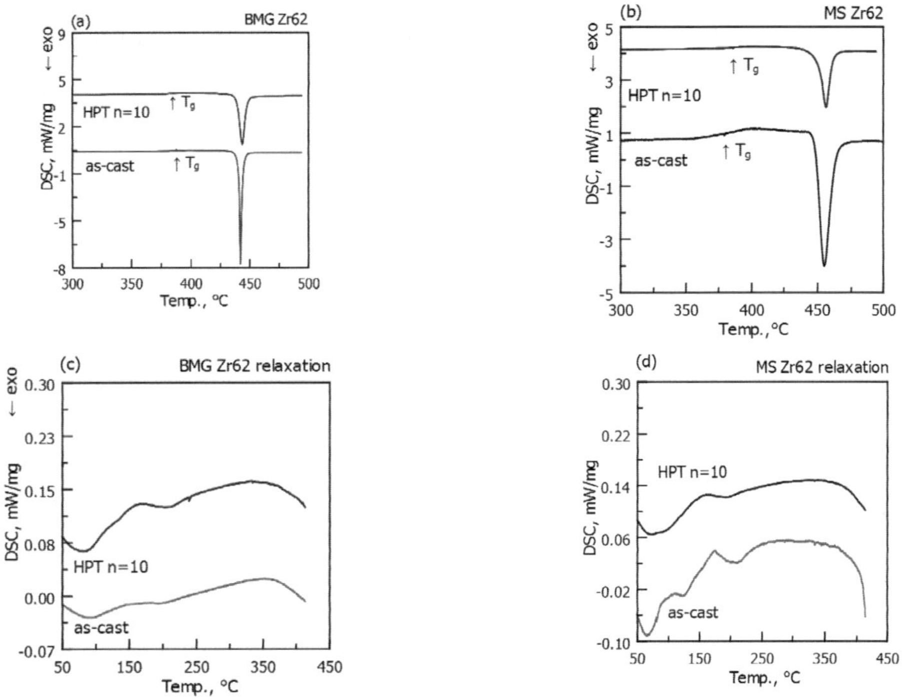

Figure 2. DSC of: a) BMG $Zr_{62}Cu_{22}Al_{10}Fe_5Dy_1$ in the initial state, subjected to HPT n=10, b) MS $Zr_{62}Cu_{22}Al_{10}Fe_5Dy_1$ alloy in the initial state and subjected to HPT n=1; c) relaxation of the BMG Zr62; d) relaxation of the MS Zr62.

4. Conclusions

As HPT of BMG $Zr_{62}Cu_{22}Al_{10}Fe_5Dy_1$ at. % results in a noticeable increase in the energy state, whereas in MS $Zr_{62}Cu_{22}Al_{10}Fe_5Dy_1$ at. % a change in the disequilibrium is not so manifested and, in the whole, the characteristics (H_{relax}, FWHM) of BMG and MS converge as a result of the HPT. This may be explained by the competition between the processes of deformation and relaxation during the HPT of amorphous alloys, and the intensification of relaxation with an increase in the disequilibrium of the amorphous alloy.

Acknowledgments
This work was supported by the Ministry of Education of the Russian Federation, project 0838-2020-0006, Russian Foundation for Basic Research under grant 20-08-00497 and by the Natural Science Foundation of China, grant No. 51520105001.

References
[1] Greer A L 1995 *Science* **267** 1947–53
[2] Louzguine-Luzgin D V. and Inoue A 2013 *Bulk Metallic Glasses Handbook of Magnetic Materials* Handbook of Magnetic Materials vol 21, ed K H J Buschow (Sendai, Japan: Elsevier B.V.) pp 131–71
[3] Hu X, Ng S C, Feng Y P and Li Y 2001 *Phys. Rev.* B **64** 172201
[4] Greer A L, Cheng Y Q and Ma E 2013 *Mater. Sci. Eng.* R **74** 71–132
[5] Pan J, Chen Q, Liu L and Li Y 2011 *Acta Mater.* **59** 5146–58
[6] Ketov S V V., Nguyen H K K, Trifonov A S S, Nakajima K and Louzguine-Luzgin D V V. 2016 *J. Alloys Compd.* **687** 221–6
[7] Valiev R Z, Zhilyaev A P and Langdon T G 2013 *Bulk Nanostructured Materials* (Hoboken, NJ: John Wiley & Sons, Inc)
[8] Zhilyaev A and Langdon T 2008 *Prog. Mater. Sci.* **53** 893–979
[9] Astanin V V, Gunderov D, Boltynjuk E, Ubyivovk E, Churakova A and Gunderova S 2019 *IOP Conf. Ser. Mater. Sci. Eng.* **672** 012019
[10] Meng F, Tsuchiya K, Seiichiro and Yokoyama Y 2012 *Appl. Phys. Lett.* **101** 121914
[11] Valiev R, Gunderov D, Zhilyaev A P, Popov A G and Pushin V 2004 *J. Metastable Nanocr. Mater.* **22** 21–6
[12] Gunderov D V. and Stolyarov V V. 2010 *J. Appl. Phys.* **108** 053901
[13] Cao Q P, Li J F, Zhou Y H, Horsewell A and Jiang J Z 2006 *Acta Mater.* **54** 4373–83
[14] Yavari A R, Moulec A Le, Inoue A, Nishiyama N, Lupu N, Matsubara E, Botta W J, Vaughan G, Di Michiel M and Kvick Å 2005 *Acta Mater.* **53** 1611–9

Surface polymorphism of silica nanoparticles

V V Syzrantsev[1(a)], E A Paukstis[2] and T V Larina[2]

[1] Institute of Materials Science of the Siberian Branch of the Russian Academy of Sciences, Ulan-Ude, Sakhyanovoy str. 6, 670070, Russia
[2] Institute of Catalysis of the Siberian Branch of the Russian Academy of Sciences, Lavrentieva av. 5, Novosibirsk, 630090, Russia

E-mail: vvveliga@mail.ru

Abstract. A comparative study was performed on the surface properties of silica nanoparticles synthesized by various methods using FTIR, UV-Vis DRS and X-ray diffraction. It is shown that different methods of synthesis affect the surface properties of nanoparticles while maintaining the phase composition of the material. It is also shown that the nanoparticles of each material have unexpected properties. In particular, a significant difference was observed in the concentration of Lewis centres for the samples. This variation in properties indicates the importance of accurately characterizing the surface properties of nanoparticles as they determine the interactions of nanoparticles with other materials.

1. Introduction

Nanoparticles are a special class of materials that are increasingly used in the development of new materials. The fundamental properties of bulk materials have been studied, but the uniqueness of their surface properties still needs further research. Although nanoparticles can have identical chemical and phase compositions, their interaction with different media can vary [1] and is determined by the conditions of the synthesis process [2, 3]. The method and conditions of synthesis determine the interactions of atoms and their associations during the formation of nanoparticles. Formation features can appear during nucleation, during growth of a molecular cluster, or upon the connection of nanoparticles [4].

It is known [2, 5] that a change in the silica synthesis method leads, for example, to a change from densely packed (aerosol) nanoparticles structure to a strip-like one (air gel). However, surface properties, including the content of various functional groups and their heterogeneity, are the most important factors determining the effectiveness of their application [6]. The contradictions in the data on the surface properties of silica and the interpretation of their donor-acceptor properties are also associated with different methods of synthesis. This inconsistency can be associated with the presence of other trace elements, such as aluminium or iron, which are usually concentrated on the surface. This can be determined by the distribution and intensity of the bridging and terminal -OH-groups, as well as other surface components.

Thus, to assess the possible interaction between nanoparticles and their environment, we performed a comparative analysis of the surface properties of nanoparticles synthesized by various methods.

2. Materials and Methods

For the studies, we used nanoparticles of silica produced by (1) liquid-phase synthesis (China), (2) the plasma-chemical method (Plasmotherm, Russia), (3) evaporation by an electron beam [7] and (4) high-temperature flame hydrolysis (Aerosil, Germany). The nanoparticles were of a spherical shape and had similar size distributions, and an average size of 25 nm. The size was determined by the optical method on a Shimadzu SALD 7500 device according to the standard method. For control and measurement of particle sizes, photographs taken by an electron microscope JEOL-2010 (Japan) were also used. Both measurements showed identical results.

The X-ray analysis of the structure of samples was carried out in [8, 9] and showed the presence of an identical short-range atomic order with a general X-ray amorphousness of nanoparticles.

The samples were compressed into tablets with a thickness (ratio of the tablet weight to its geometric area) of 15 to 49 mg / cm^2. The tablets were placed in a vacuum infrared cuvette and evacuated for 1 hour at a vacuum of 10^{-3} Pa. Subsequently, the samples, without contact with the atmosphere, were transferred to the measuring chamber, which made it possible to record spectra at temperatures down to 77 K. Then the basic spectra were recorded at room temperature and then at 77-110 K. The recording of the spectra was carried out on a Shimadzu IRTraser-100 spectrometer in the range of 600-6000 cm^{-1} with a resolution of 4 cm^{-1}. The spectra were accumulated every 100 scans, which ensured a signal-to-noise ratio of at least 75000. CO adsorption was carried out at doses, which provided a pressure in the cell of 0.1, 0.4, 0.9, 1.4 and 10 Torr at 77 K.

The optical properties of the materials were studied by diffuse reflectance electron spectroscopy (UV-Vis DRS) using a Shimadzu UV-2501 PC spectrophotometer with an ISR-240A diffuse reflectance attachment. The samples were placed in a quartz cell with a sample layer thickness of 2 mm. The spectra were recorded relative to the $BaSO_4$ reflection standard in the range of 190-900 nm (11000-54000 cm^{-1}), with a step of 2 nm. Reflection coefficients were converted into absorption coefficients using the Kubelka-Munk function $F(R) = (1-R)^2 / 2R$, where R is the diffuse reflection coefficient, depending on the wavenumber, expressed in cm^{-1}. The obtained data were presented in the coordinates of Kubelka - Munk function F (R) vs the wave number.

3. Results

3.1 IR spectroscopy

Figure 1 shows the spectra of the CO and OH groups of Sample 1 calcined at 550 ° C in vacuum. The spectrum of OH groups exhibits a band at 3745 cm^{-1} related to valent vibrations of SiOH groups. It also shows the absence of a weak broad band at about 3650 cm^{-1} produced by hydrogen-bonded OH groups. Upon CO adsorption, the 3745 cm^{-1} band shifts to the low frequency region down to 3666 cm^{-1}. The shift to low frequencies by 79 cm^{-1} is close to that observed for conventional aerosils and silica gels, but it means that the acidity of SiOH groups is lower than that of Sample 2.

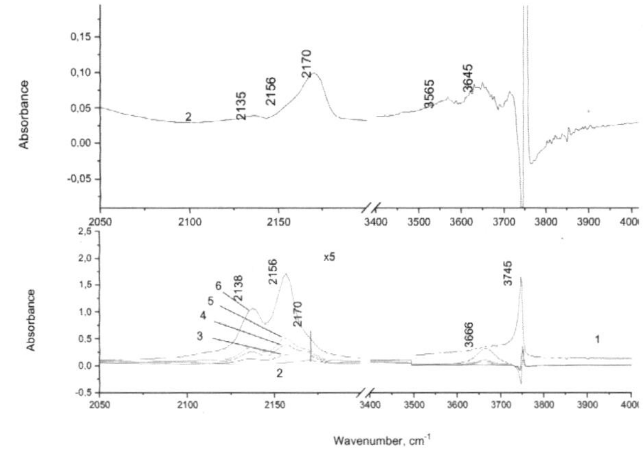

Figure 1. The IR spectra of CO adsorbed on Sample 1. 1- calcined sample (the thickness 16.3 mg / cm^2), 2-6 – the spectra taken after CO adsorption at pressures of 0.1; 0.4; 0.9; 1.4 and 10 torr.

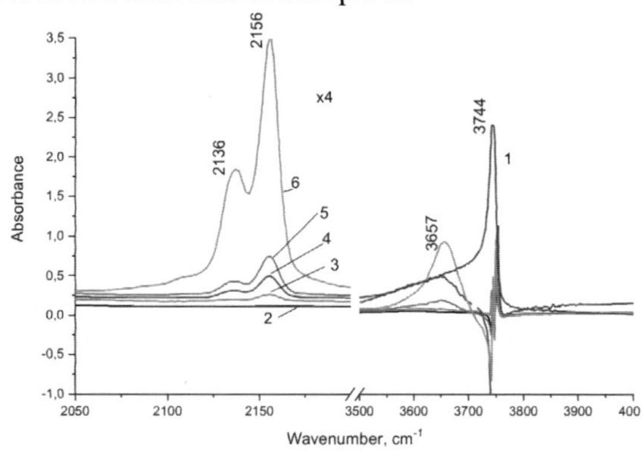

Figure 2. The IR spectra of CO adsorbed on Sample 1. 1- calcined sample (the thickness 22.8 mg / cm^2), 2-6 – the spectra taken after CO adsorption at pressures of 0.1; 0.4; 0.9; 1.4 and 10 torr.

The same complexes are characterized by the CO band at 2156 cm^{-1}. The 2138 cm^{-1} band has been produced by physically adsorbed CO. For Sample 1, upon adsorption of the first CO doses, bands of CO valent vibrations at 2170 cm^{-1} also appear. Bands in this range are specific to CO complexes with strong Bronsted acid sites. Curve 2 (upper part of the figure) clearly shows that CO adsorption shows two bands of hydrogen-bonded OH groups - 3565 and 3645 cm^{-1}, which have the shift of 180 and 100 cm^{-1}, respectively. Along with the appearance of the CO band at 2170 cm^{-1}, this can be interpreted as the introduction of small alumina clusters into the silica surface. The number of such clusters (based on the intensity of the 2170 cm^{-1} band) can be estimated as 45 μmol / g (0.1 wt%).

Figure 2 shows the spectra of CO and OH groups of Sample 2 calcined in vacuum at 550 °C. The spectrum of OH groups has a band at 3744 cm^{-1}, valent vibrations of SiOH groups and a weak broad band at about 3650 cm^{-1} of hydrogen-bonded OH groups. Upon CO adsorption, the 3744 band shifts to the low-frequency region by 87 cm^{-1} to 3657 cm^{-1}. The same complexes are characterized by a CO band at 2156 cm^{-1}, which is also typical for aerosils and silica gels. The 2138 cm^{-1} band has been related to the physically adsorbed CO.

Figure 3 shows a wide band at 927 cm^{-1} at room temperature, which belongs to strained Si-O-Si bridge bonds. As the temperature decreased, this band split up into two components 902 and 956 cm^{-1}. During the decrease in temperature from 110 to 77 K (not measured), the splitting of the band increased.

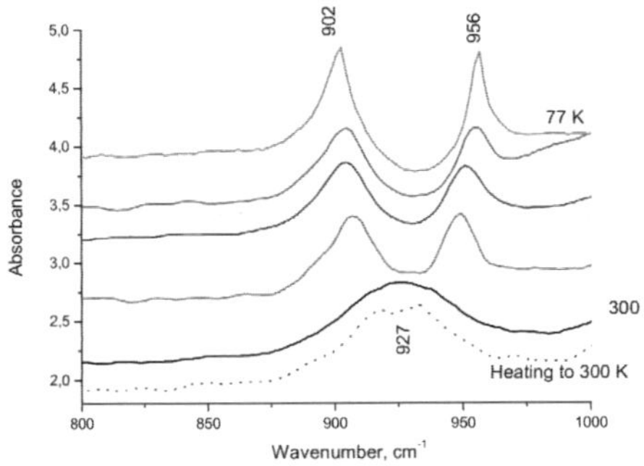

Figure 3. The IR spectra of Sample 1 (thickness 16.3 mg / cm^2) recorded in the region of absorption bands of strained bridges at different temperatures.

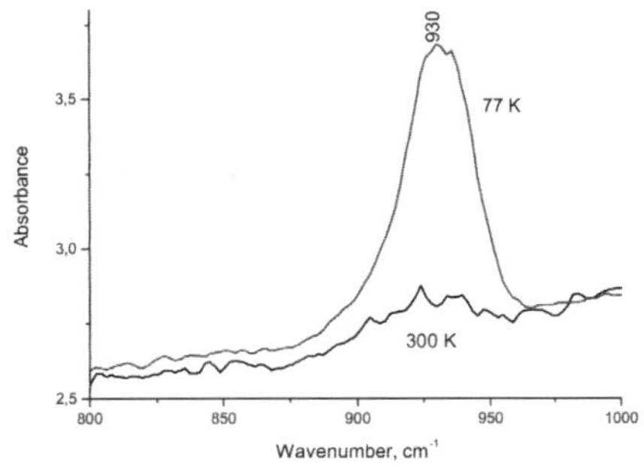

Figure 4. The IR spectra of Sample 2 (thickness 22.8 mg / cm^2) recorded in the region of absorption bands of strained bridges at different temperatures.

Subsequent heating leads to the restoration of the broad band at 927 cm^{-1}. Apparently, the presence of alumina clusters leads to the existence of other types of strained bridges. At room temperature, it can be Si-O-Si-O-Al formed due to the kinematic bond of fragments, or Si-O-Al with a common silicon atom; they make a wide band. At low temperatures, due to the general decrease in vibrational energy, the kinematic connection is weakened. Therefore, a high-frequency component at 956 cm^{-1} related to the Si-O-Si fragment and a low-frequency component of the weaker Si-O-Al bridge bonds are created. In figure 4, in the region of 800–1000 cm^{-1}, we observe a broad band at 930 cm^{-1}, which is caused by strained Si-O-Si bridge bonds. One should point out that an increase in the band's intensity with a decrease in temperature down to the liquid nitrogen one occurred. These effects have not been previously described.

Samples 3 and 4 show a different structure of bending vibrations. The bands in the region of 1090-1300 cm^{-1} belong to the antisymmetric valent vibrations of SiO_4 tetrahedra, and the band at 450 cm^{-1} belong to the symmetric vibration of the same fragment. The 810 cm^{-1} band is related to the bending vibrations of Si-O-Si. In addition, for Samples 3 and 4, bands at 886 and 946 cm^{-1} were found. They can correspond to bending vibrations of Si-OH or vibrations of strained bridge bonds of Si-O-Sifragments, the structure of which can be represented by the diagram 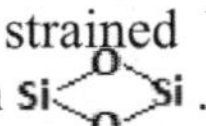.

In figure 5 the spectra show bands at 1626 and about 3400 cm^{-1}, ascribed to adsorbed water, band 1868 caused by vibrational overtones of SiO_4 and Si-O-Si. Bands in the 2147-2261 cm^{-1} region, close in position to the valent vibrations of silicon hydrides, were found only for Samples 3 and 4. A high water content is distinguished only for Sample 3. Sample 4 is characterized by the maximum number of SiOH groups, the band at 3746 cm^{-1}. Moreover, in contrast to Sample 3, the content of molecular water in this sample is insignificant; it turned out to be hydrophobic, despite the high concentration of silanol groups.

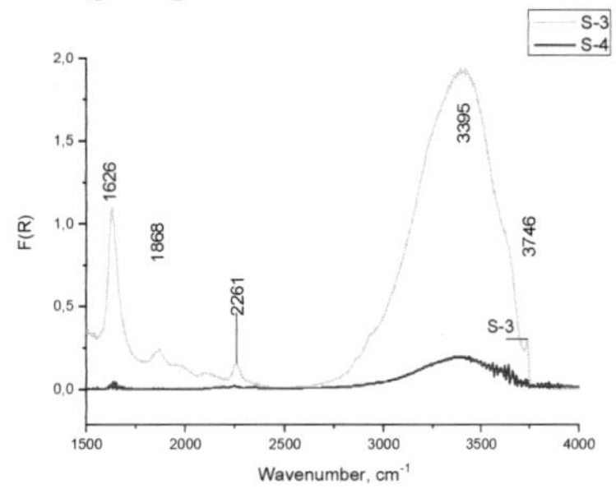

Figure 5. The IR spectra of diffuse reflectance of Samples 3 and 4 in the region above 1500 cm^{-1}.

The IR spectra of adsorbed CO on Samples 3 and 4 are shown in figures 6 to 9. The figures also show the concentrations of different types of OH groups and small amounts of concentrations of Lewis acid sites. It can be seen from the figures that both samples have a similar concentration of OH groups. Bands were found at 3750, 3690, and 3580 cm^{-1}. The first band belongs to isolated terminal SiOH groups, and the second and third bands to weak and strong hydrogen-bonded groups. During the adsorption of CO, interaction takes place only with the first band groups, which shifts to the low frequency region by 83 cm^{-1}. This is slightly less than the shift for the SiOH bands of the aerosil groups. From the CO spectra recorded at a pressure of 10 Torr, it is possible to conclude that the surfaces of Samples 3 and 4 are almost equivalent. The samples are distinguished only at lower pressures (figures 6 and 8). For Sample 4, there is a small content of Lewis centres characterized by a band at 2200 cm^{-1} (centres of medium strength). This may indicate a metal impurity, possibly aluminium; the total content of such impurities is estimated at 20-30 ppm. An analysis of the UV-Vis DRS spectra (figure 10) shows that the spectra of all samples are divided into two parts: intense absorption in the visible region of the spectrum up to 25,000 cm^{-1} and relatively strong absorption in the region above 30,000 cm^{-1}.

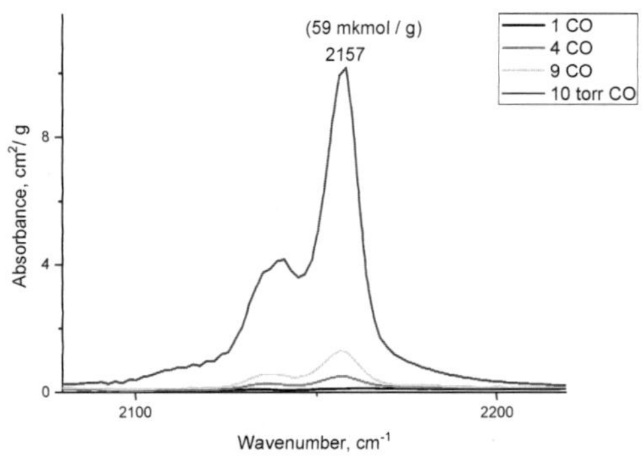

Figure 6. The IR spectra of adsorbed CO on Sample 3 at different portions of adsorption - 1 CO - 0.1 Torr, 4 CO - 0.4 Torr, 9 CO - 0.9 Torr.

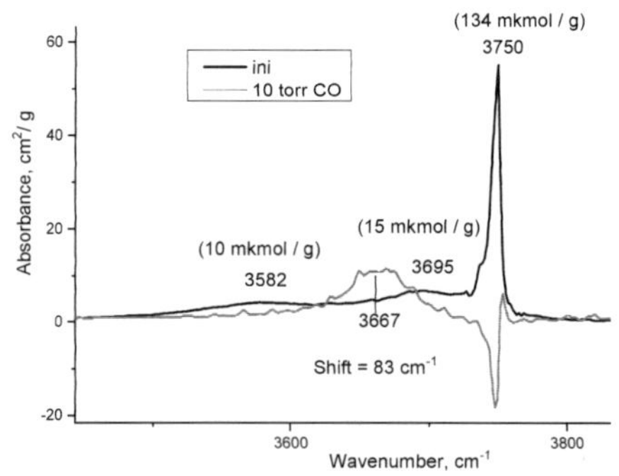

Figure 7. The IR spectra of OH groups of Sample 3 before (ini) and difference of spectra after adsorption at 10 Torr CO.

3.2 UV-Vis DR spectroscopy

Absorption in the visible region of the UV-Vis DRS spectrum is usually caused by the d-d-transitions of various impure d elements of 3-5 periods [10]. The intensity of absorption may depend on several factors. First, it depends on the amount of impure cations of d elements in any matrix, including the matrix containing oxygen. Second, various distortions of the octahedron or tetrahedron around the central cation of the d element increase the absorption intensity, since changes in the cation-anion distance in the first coordination sphere are observed when the influence of the ligands of the second and subsequent spheres on the central cation is considered. Third, the absorption intensity is inversely related to the size of the absorbing particles, i.e. the smaller the particle size, the more the light is scattered from them and appears in the absorption spectrum [11].

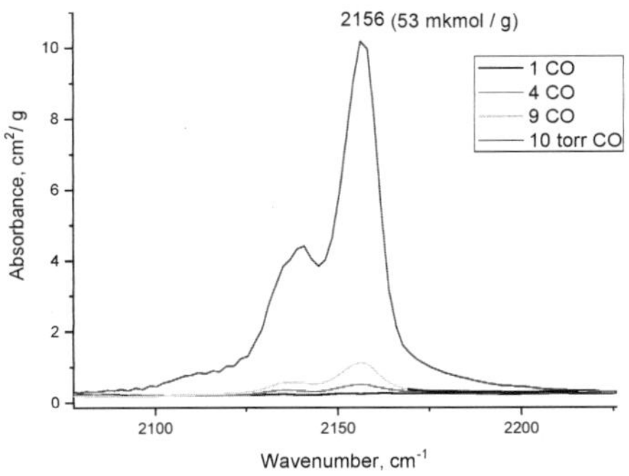

Figure 8. The IR spectra of adsorbed CO on Sample 4 at different portions of adsorption - 1 CO - 0.1 Torr, 4 CO - 0.4 Torr, 9 CO - 0.9 Torr.

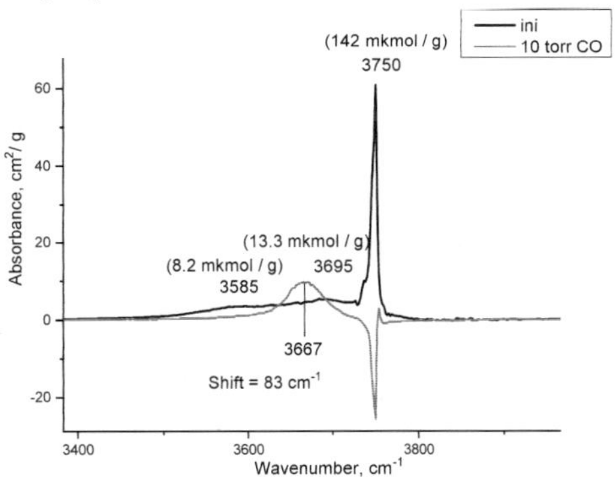

Figure 9. The IR spectra of OH groups of Sample 4 before (ini) and difference of spectra after adsorption at 10 Torr CO.

Based on these findings, we can conclude that all the samples contain at least two particle bands: small (about 5 nm) and large (up to 25 nm). In this case, the samples have the next particle size: 1> 3> 4> 2, consequently, an increase in intensity is observed: 1 <3 <4 <2. Thus, more absorption due to scattering is observed for Sample 2, which indicates that its particles are smaller than all the other samples. The smallest absorption occurs for Sample 1, which has the largest particles among the samples studied.

Usually, absorption is observed in the UV region of the UV-Vis DRS spectra, which is caused by the manifestation of ligand - metal charge-transfer bands (L – M PCR), which have a high intensity [10]. Also in the UV region of the UV-Vis DRS spectrum for massive, for example, oxygen-containing or other systems, as well as well-conducting systems, the fundamental absorption edge (FAE) is manifested, which corresponds to the bandgap (Eg) of semiconductors and dielectrics [12]. It is also known [13] that for many years the size of the oxide particles has been characterized by the FAE value. Moreover, it is known that the larger particles make the large red shift and decrease the Eg value [11, 14, 15].

Based on the analysis of absorption in the UV region of the UV-Vis DRS spectra, it can be concluded that the samples in the UV region are divided into two groups: Sample 1 and three other samples (Curves 2-4). The FAE value for Sample 1 is 4.4 eV, and for the other three, it is very close to 3.9 eV. A higher FAE value indicates a small size of large particles in Sample 1 compared to other samples. However, for the other three samples having close FAE values, there is a variation in the absorption intensity, which may indicate a variation in the texture of the particles themselves and the presence of pores in these particles (free volume and looseness of these particles). Moreover, for Sample 2, all these characteristics should be higher, i.e. its particles should be fluffier and should repel each other more strongly; therefore, the free volume between the particles becomes larger, which leads to an increase in the absorption intensity in the UV region of the UV-Vis DRS spectrum, owing to an increase in the distance between individual atoms.

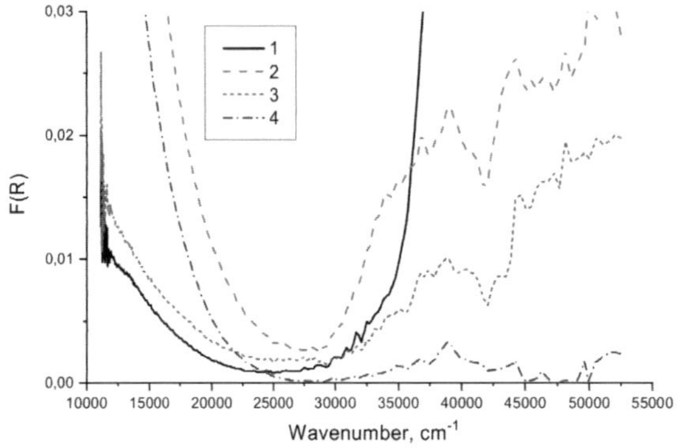

Figure 10. UV-Vis DRS spectra of the silica samples.

Conclusion

Thus, it has been shown that different conditions for the synthesis of silica nanoparticles form a surface cover with different types and concentrations of OH groups, Lewis acid sites, strained siloxane bridge bands and other characteristic elements. Specific differences in the UV-Vis DRS spectra due to the inclusion of $Fe2^{+}$ Oh cations were also noted. The possibility of varying the effect of nanoparticles on the material containing them, depending on the surface properties of the particles, has been demonstrated.

It is noted that the size distribution can have a different shape (for example, two peaks) under different synthesis conditions, which cannot be determined by the nitrogen absorption method.

Acknowledgments

The studies of UV-Vis DR, IR and PL spectroscopy of the Al_2O_3 samples are conducted using the equipment of the Center of Collective Use "National Center of Catalyst Research".

This work was supported by Ministry of Science and Higher Education of the Russian Federation (project No. AAAA-A17-117041110045-9 and project No. 0336-2019-0008).

References

[1] Syzrantsev V V, Zobov K V, Zavjalov A P and Bardakhanov S P 2015 *Dokl. Physics* **60** 46-8
[2] Sheka E F, Khavryuchenko V D and Markichev I V 1995 *Russ. Chem. Rev.* **64** 389
[3] Bardakhanov S P, Vasiljeva I V, Kuksanov N K and Mjakin S V 2010 *Adv. Mater. Sci. Eng.* **2010** 241695
[4] Gen M Y and Petrov Y I 1969 *Russ. Chem. Rev.* **38** 1007
[5] Khavryuchenko V D and Sheka E F 1994 *J. Str. Chemistry* **35** 291
[6] Fu S, Sun Z, Huang P, Li Y and Hu N 2019 *NanoMater. Sci.* **1** 12
[7] Bardakhanov S P, Korchagin A I, Kuksanov N K, Lavrukhin A V, Salimov R A, Fadeev S N and Cherepkov V V 2006 *Mate. Sci. Eng.: B* **132** 204
[8] Abzaev Y A, Syzrantsev V V and Bardakhanov S P 2017 *Phys. Sol. State* **59** 1874
[9] Syzrantsev V V and Abzaev Y A 2020 *Russ. Phys. J.* **63** 49
[10] Lever A B P *Inorg. Electr. Spectrosc.* 1987 (Amsterdam, Elsevier)
[11] Smirnov B M *Clusters and Small Particles: In Gases and Plasmas* 1999 (New York-Berlin, Springer Verlag)
[12] Mott N F and Davis E A *Electron Processes in Non-Crystalline Materials* 1979 (Oxford, Clarendon Press)
[13] Gnanam S and Rajendran V 2011 *J. Sol-Gel Sci. Techno.l* **58** 62–9
[14] Ross-Medgaarden E I and Wachs I E 2007 *J. Phys. Chem. C.* **111** 15089–99
[15] Weber R S 1995 *J. Catalysis* **151** 470–4

IOP Conf. Series: Materials Science and Engineering 1008 (2020) 012031 doi:10.1088/1757-899X/1008/1/012031

High density of shear bands in the Vitreloy bulk metallic glass subjected to high-pressure torsion

Vasily Astanin[1], Dmitry Gunderov[1,2], Zhi Qiang Ren[3], Ruslan Valiev[4] and Jing Tao Wang[3]

[1] Ufa State Aviation Technical University, K. Marx str. 12, Ufa, 450008, Russia
[2] Institute of Molecule and Crystal Physics, Ufa Federal Research Centre RAS, October Avenue 71, 450054, Ufa, Russia
[3] School of Materials Science and Engineering, Nanjing University of Science & Technology, Nanjing 210014, China
[4] Saint Petersburg State University, 28 Universitetskiy pr., Saint Petersburg, 198504, Russia

E-mail: v.astanin@gmail.com

Abstract. Plastic deformation by high-pressure torsion allows one to achieve high strains of bulk metallic glasses without their failure. It is shown that inclusions of the primary crystalline phase in the Vit105 alloy provoke the formation of cracks, and shear bands themselves have a complex composite structure and are prone to branching and merging during the development of deformation. The density of the bands reaches 1.76 μm^{-1}, with a distance between them 80-120 μm.

1. Introduction

Bulk metal glasses (BMGs) are apt to localized deformation at temperatures below the glass transition temperature T_g, since the main mechanism of their deformation is the formation of narrow shear bands [1]. Because of this feature, BMGs have low plasticity at room temperature and are, as a rule, brittle materials.

At the same time, under conditions of omnidirectional compression, large strains can be achieved in BMGs. For example, during severe plastic deformation by high-pressure torsion (HPT), a large number of shear bands are generated. To date, many articles have been published on the HPT of amorphous alloys [2–7] including several reviews [8,9]. When a shear band appears, a shear affected zone extends at some distance from it, in which shear bands no longer arise. Thus, the limiting band density achieved by the HPT method is of interest.

2. Material and methods

The Vit105 alloy ingots with a standard composition $Zr_{52.5}Cu_{17.9}Ni_{14.6}Al_{10}Ti_5$ were obtained by arc melting of pure components in an argon atmosphere. The material was remelted four times to achieve a chemical homogeneity. Then the melt was poured into a water-cooled copper mold forming a rod 10 mm in diameter. Disks with the thickness of 1 mm were then cut from the rod and deformed by HPT at room temperature under the pressure of 6 GPa. To form shear bands, the method described in [10] was used. Initial disks were cut into halves, the surface formed by the cut was polished, and the two halves were subjected to HPT together, as shown on figure 1, a, by ¼ turn in anvils with a rough groove 0.8 mm thick and 10 mm in diameter.

Figure 1. Scheme of joint HPT of two halves (a); appearance of the sample after HPT n = ¼ (b).

3. Results and discussion

Earlier it was shown that HPT treatment of such hard-to-deform materials as BMG leads to much less observed deformation [7,10] than the one predicted by the theory [11]. In this regard, rough-surfaced anvils were used. However, a large shift in the upper part of the sample relative to the lower one did not occur, and the assumed deformation was not achieved due to slipping. Nevertheless, a high density of shear bands is formed in the BMG during HPT, especially in the edge regions of the samples.

Shear bands form traces in the form of steps on the polished surface of the sample. There are two main systems of steps, radial and diagonal ones, the formation of both of them is associated with stress / strain distribution during the HPT deformation.

At the edges of the samples, in the region of the largest deformation, individual intense wide patterns in a bright contrast represent a combination of a very large number of smaller bands (figure 2, b). The amorphous matrix surrounding the set of bands is also covered with a system of shear bands in such a pattern that the intersecting shear bands divide the amorphous matrix into diamond-shaped regions, which in turn are divided into smaller diamond-shaped regions by bands of lower intensity. Apparently, a similar picture of the distribution of shear bands reflects the flow of the material at large deformations in confined conditions, at an increased hydrostatic pressure.

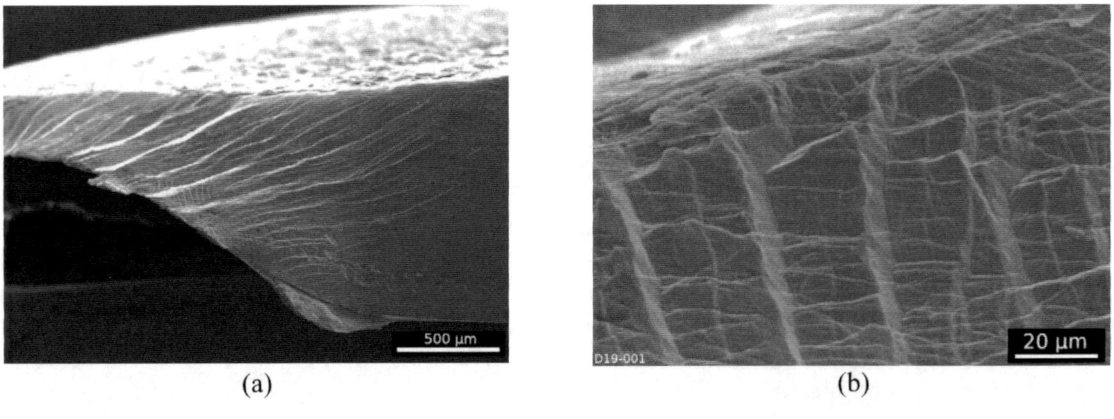

Figure 2. Edge section of the BMG specimen subjected to joint HPT of two halves for n = ¼ (a); the same sample, at higher magnification (b), SEM-BSE.

The steps have different scales in height, from tens of nanometers to fractions of a millimeter, while large steps are not monolithic, but are themselves divided by shear bands into many steps down to the nanometer scale. It should be noted that the density of shear bands estimated from their traces in the form of steps, only in a first approximation correlates with the actual density of shear bands in the material, since their spatial position is unknown. The distance between the largest steps is 80-120 microns, and, as their size decreases and the number increases, the interval decreases and reaches 100-150 nm. The density of the bands in this case is 1.76 μm^{-1}, which is significantly higher than the achieved density of 0.09-0.14 μm^{-1} during deformation by compression [12]. Nevertheless, in the literature there are data on an even denser arrangement of bands, with a step of the order of 30 nm [13], which significantly exceeds the band density observed in this study. On the other hand, in BMG samples subjected to rolling even to a large strains, a noticeably lower density of shear bands was observed [14].

A characteristic feature of high-strain shear bands achieved with HPT is their branching observed at both ends of shear bands. Apparently, deformations under the action of a stress field, as they develop, are localized in the form of shear bands, in which stress is relieved, and shear bands, in turn, can branch and merge (figure 3, b). Higher number of rotations do not lead to an increase of the shear band density, the overall picture of deformation is formed already in the early stages of processing. However, according to recent data, internal structure and properties undergo changes [4].

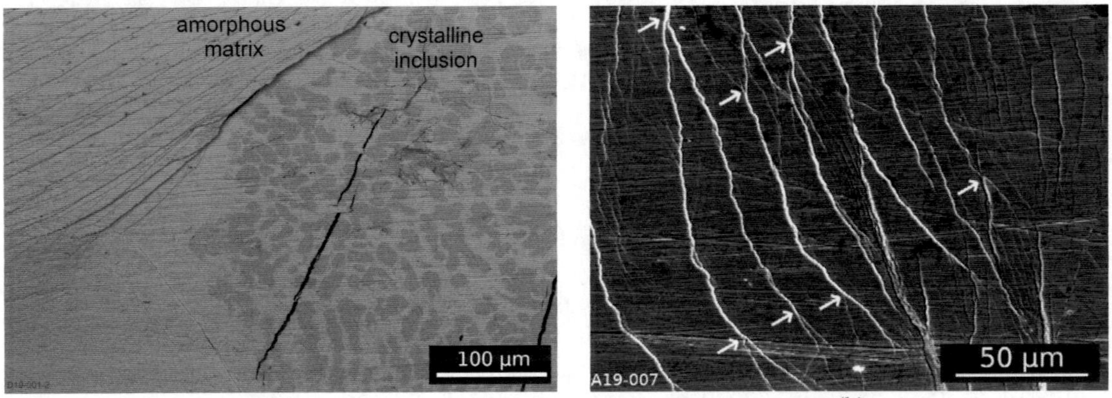

(a) (b)

Figure 3. Inclusions of the crystalline phase cause cracking upon deformation. Vit105, HPT n = 1/4, section 2/3 of the sample radius, SEM-BSE (a); shear band branching, HPT n = 5 + 1/12, section 2/3 of the sample radius, SEM-SE (b).

In not well homogenized samples, large clusters of a crystalline phase encapsulated in the amorphous phase were observed. The inclusions of this crystalline phase negatively affect the plasticity of the Vitreloy BMG. The origin of the crystalline phase can be associated with liquation in the liquid state. According to the results of energy dispersive spectroscopy, the inclusions are enriched in nickel and copper in comparison with the amorphous phase. Due to a deviation from the stoichiometric composition, the critical cooling rate was not reached, and a part of the material acquired a crystalline structure. The hard and brittle intermetallic crystalline phase, well distinguishable in the backscatter electron image mode, does not deform at the stresses achieved as a result of HPT (figure 3, a). The deformation is redistributed into the amorphous part of the material creating a characteristic pattern of "flowing" around the crystallite by shear bands, and the crystalline inclusion itself serves as a place for the nucleation of cracks, which, as a rule, do not propagate to the amorphous part of the material at the achieved strain, but can serve as stress concentrators upon further deformation.

It should be noted that the formation of shear bands in the amorphous phase when using the specified deformation scheme (figure 1, a) does not lead to an appearance of cracks, and, at the same time, inclusions of the crystalline phase provoke cracking of the material (figure 3, a). This is contrary to the observed toughening and ductilization effect of the crystalline phase in a uniform composite of bulk metallic glass with a ductile second phase [15], where the mechanically soft and ductile second phase acts as a stabilizing factor against the shear localization and critical crack propagation.

4. Conclusions

As a result of severe plastic deformation by high pressure torsion, a complex pattern of shear bands is formed in bulk metal glasses. Its feature is the existence of several preferential systems of orientations of shear bands, the composite nature of shear bands themselves, as well as their branching and merging during deformation. The plasticity of the amorphous part of the material allows one to deform it without cracking and destruction, while inclusions of the crystalline phase can provoke the formation of cracks. The density of the bands reaches 1.76 μm^{-1}, with a distance between them 80-120 µm.

Acknowledgments

This work was supported by the Ministry of Education of the Russian Federation, project No. 0838-2020-0006, the Russian Foundation for Basic Research under the grant No. 20-08-00497 and by the Natural Science Foundation of China, grant No. 51520105001.

References

[1] Suryanarayana C and Inoue A 2011 *Bulk Metallic Glasses* (Boca Raton: CRC Press)
[2] Gunderov D V., Boltynuk E V., Ubyivovk E V., Churakova A A, Lukyanov A V., Raab A G, Khasanova D A and Churyumov A Y 2016 *Lett. Mater.* **6** 322–6
[3] Boltynjuk E V., Gunderov D V., Ubyivovk E V., Lukianov A V., Kshumanev A M, Bednarz A and Valiev R Z 2016 *AIP Conf. Proc.* **1748** 030006
[4] Gunderov D V, Churakova A A, Astanin V V, Asfandiyarov R N, Hahn H and Valiev R Z 2020 *Mater. Lett.* **261** 127000
[5] Astanin V V, Gunderov D, Boltynjuk E, Ubyivovk E, Churakova A and Gunderova S 2019 *IOP Conf. Ser. Mater. Sci. Eng.* **672** 012019
[6] Joo S-H, Pi D-H, Setyawan A D H, Kato H, Janecek M, Kim Y C, Lee S and Kim H S 2015 *Sci. Rep.* **5** 9660
[7] Adachi N, Todaka Y, Yokoyama Y and Umemoto M 2015 *Mater. Sci. Eng.* A **627** 171–81
[8] Gunderov D and Astanin V 2020 *Metals (Basel).* **10** 415
[9] Révész Á and Kovács Z 2019 *Mater. Trans.* **60** 1283–93
[10] Gunderov D V, Churakova A A, Boltynjuk E V, Ubyivovk E V, Astanin V V, Asfandiyarov R N, Valiev R Z, Xioang W and Wang J T 2019 *J. Alloys Compd.* **800** 58–63
[11] Valiev R Z, Zhilyaev A P and Langdon T G 2013 *Bulk Nanostructured Materials* (Hoboken, NJ: John Wiley & Sons, Inc)
[12] Louzguine-Luzgin D V, Ketov S V, Wang Z, Miyama M J, Tsarkov A A and Churyumov A Y 2014 *Mater. Sci. Eng.* A **616** 288–96
[13] Zheng B, Zhou Y, Mathaudhu S N, Valiev R Z, Tsao C Y A, Schoenung J M and Lavernia E J 2014 *Scr. Mater.* **86** 24–7
[14] Liu J W, Cao Q P, Chen L Y, Wang X D and Jiang J Z 2010 *Acta Mater.* **58** 4827–40
[15] Szuecs F, Kim C P and Johnson W L 2001 *Acta Mater.* **49** 1507–13

IOP Conf. Series: Materials Science and Engineering 1008 (2020) 012032 doi:10.1088/1757-899X/1008/1/012032

Work function of chemical compounds of aluminum-magnesium system

R Kh Khisamov[1], A A Kistanov[2], K S Nazarov[1], R U Shayakhmetov[1], G F Korznikova[1], Yu M Yumaguzin[1,3], S V Dmitriev[1,4], R R Mulyukov[1,3,4]

[1] Institute for Metals Superplasticity Problems of Russian Academy of Sciences, 39 Stepan Khalturin st., Ufa, 450001, Russia

[2] Nano and Molecular Systems Research Unit, University of Oulu, Oulu, 90014, Finland

[3] Bashkir State University, 32 Zaki Validi st., Ufa, 450076, Russia

[4] Ufa State Petroleum Technological University, 1 Kosmonavtov st., Ufa, 450064, Russia

E-mail: andrey.kistanov@oulu.fi

Abstract. For pure metals aluminum and magnesium and for chemical compounds Al_3Mg, Al_2Mg, AlMg and $AlMg_3$ with crystallographic orientations (100), (110) and (111) the work function is calculated from the first principles in the framework of the density functional theory. It is shown that the dependence of the work function on the magnesium concentration in aluminium is nonlinear. The reasons for the work function differences of chemical compounds depending on the crystallographic orientation are analysed.

1. Introduction

Work function is the important emissivity characteristic for the cathodes that operate on the basis of thermionic and field emission, photo- and ion-induced electron emission phenomena. The lower work function the higher emission currents the cathode is able to produce. The emission ability of a cathode depends on the material from which it is made. For this reason, the cathode materials must have a low work function.

One of the ways to reduce the work function of the cathode is to introduce into the cathode material chemical elements with a low work function. In a rough approximation, the integral work function is the sum of the work function of chemical elements, taking into account the areas they occupy on the emitting surface of the cathode. Meanwhile, the introduced chemical elements can form chemical compounds with the cathode material, so that the work function of the compounds can be a non-additive sum relative to the work function of the cathode elements. In particular, such non-additive behavior of mechanical properties is observed in *in-situ* metal-matrix composites [1-6]. For a reasonable choice of the combination of chemical elements in the cathode material, it is necessary to know the work function of the cathode chemical compounds.

In this work, the work function of chemical compounds of the aluminum-magnesium system is obtained from ab initio calculations. The choice of aluminum and magnesium is associated with the fact that they are widely used as cathode materials, in particular, in glow gas discharge devices.

2. Simulation technique

The calculation of work function from the first-principles in the framework of density functional theory were performed using the plane-wave Vienna *ab-initio* simulation package (VASP) [7]. The first Brillouin zone was sampled with a 20 × 20 × 10 k-mesh grid and the vacuum space of 15 Å was set in the direction perpendicular to the surface plane. The kinetic energy cut-off for the plane wave expansion was set to 520 eV. All the considered structures were fully optimized until the atomic forces and total energy values were smaller than 10^{-3} eV/Å and 10^{-8} eV, respectively. The work function was calculated as the minimum energy required to remove an electron from a solid to a point remote from the solid. For the calculation of work function the atomic structure (the data is taken from the open sources) of pure aluminum and magnesium as well as atomic structures of various chemical compounds of aluminum-magnesium system with crystallographic orientations (100), (110) and (111) were used (see figure 1).

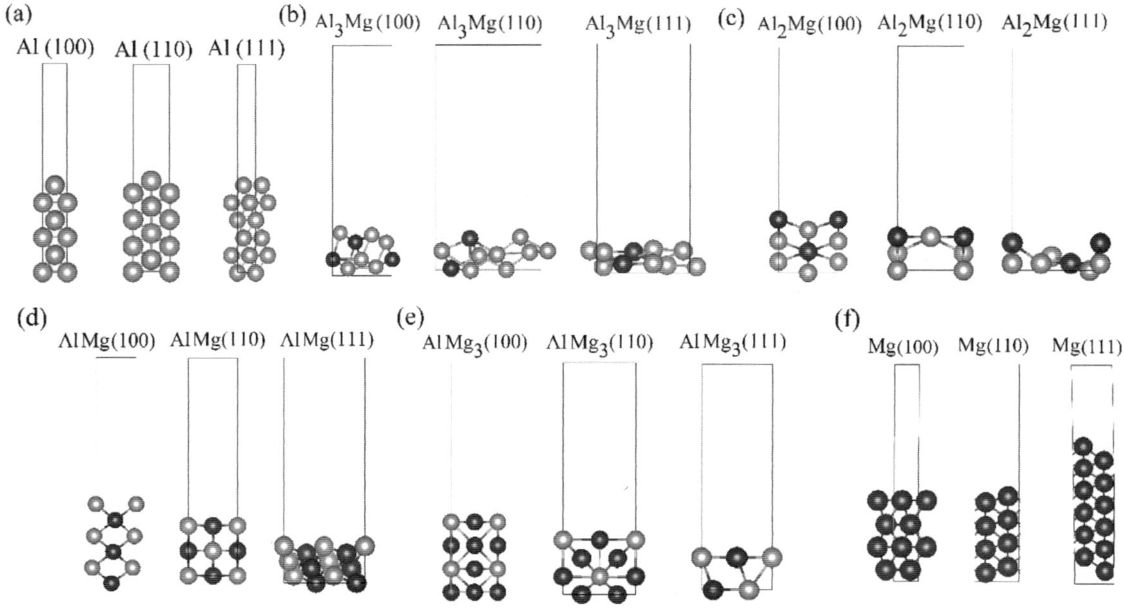

Figure 1. Schematic images of atomic structure for aluminum, magnesium as well as chemical compounds of aluminum-magnesium system with different crystallographic orientations: Al (a), Al_3Mg (b), Al_2Mg (c), AlMg (d), $AlMg_3$ (f), Mg (g).

3. Results and discussion

The calculated values of the work function for pure metals aluminum and magnesium with crystallographic orientations (100), (110) and (111) are presented in Table 1. The difference of the work function of pure metals with different crystallographic orientations is related to the different crystal face structure and this is explained by the electronic smoothing effect [8]. Electron gas on an atomically smooth surface creates a negative electric field, which makes it difficult for electrons to be emitted from the crystal. However, if the surface of the face is an atomically rough surface, i.e., the atoms protrude somewhat above the underlying plane of atoms, then the electron gas is redistributed in such a way that the surface becomes smooth. Electrical charge drains from the protruded atoms into cavities (underlying atomic plane). The positive charge forms on the protruded atoms while the negative charge forms on the cavities. As a result, a double electrical layer forms on the surface of a crystal face, a positively charge layer of which is directed to the vacuum. The electric field generated this way facilitates the emission of electrons from the crystal. For this reason, the work function φ of the crystal face with densely-packed atoms on the surface is higher than that of the face with the less dense atomic surface. For example, for fcc metals the following tendency is observed: φ(111)>φ(100)>φ(110) [9,10]. But for aluminum, as it was shown in experimental and theoretical works [11-14], such tendency is invalid. Our calculations showed that the (111) orientation had the lowest work function value. In the case of magnesium, the work function decreases from 3.67 eV for the (111) orientation to 3.39 eV for the (110) orientation.

The work function values for chemical compounds of the aluminum-magnesium system with different crystallographic orientations are presented in figure 2. A nonlinear decrease of the work function with an increase of magnesium concentration is observed. For the orientation (110) the work function decreases from 4.21 eV for pure aluminum to 3.74 eV for the Al_3Mg compound, then the work function increases up to 3.83 eV for the Al_2Mg compound. Further increase of the concentration of magnesium leads to a decrease of the work function to 3.39 eV that corresponds to pure magnesium. For the orientation (110) the value of the work function of the Al_3Mg compound is close that of pure aluminum which is equal to 4.20 eV. For the Al_2Mg compound the work function decreases to 3.66 eV. Further increase of the concentration of magnesium (the $AlMg_3$ compound) the work function slightly increases to 3.76 eV, then the work function decreases up to 3.39 eV that corresponds to pure magnesium. In the case of the orientation (111) the work function decreases from 4.08 eV (corresponds to pure aluminum) to 3.57 eV (Al_3Mg), then the work function increases up to 4.06 eV (Al_2Mg). After that the work function decreases to 3.29 eV that corresponds to AlMg compound. Then the work function increases to 3.8 eV for the $AlMg_3$ compound and decreases to 3.78 eV for pure magnesium.

The differences of the work function for chemical compounds with different crystallographic orientations may be due to, firstly, the surface crystal face contains the different relation of aluminum and magnesium atoms. Secondly, the differences of the work function similarly to pure metals may arise due to the different surface atomic roughness of chemical compounds and depending on the orientations. The atomically flat surface i.e., the surface crystal face with densely packed atoms has the higher work function than the atomically rough surface i.e., the surface crystal face with the low packing atoms.

Table 1. The work function of aluminum and magnesium for different crystallographic orientations.

Metal	Work function in our calculations, eV	Other theoretical results of the work function, eV	Experimental results of the work function, eV
Al (100)	4.20	4.41[11], 4.38[12]	4.41[13,14]
Al (110)	4.21	4.08[11], 4.30[12]	4.28[13]
Al (111)	4.08	4.36[11], 4.25[12]	4.24[13,14]
Mg (100)	3.6	–	–
Mg (110)	3.39	–	–
Mg (111)	3.67	3.86[15]	3.66[15]

4. Summary

The work function of pure aluminum and magnesium as well as chemical compounds of aluminum-magnesium system was calculated from the first principles in the framework of the density functional theory were performed. The effect of crystallographic orientation of the free surface was studied. The work function values for aluminum with orientations (100), (110) and (111) were found to be 4.2, 4.21 and 4.08 eV, respectively. The work function values for magnesium with the same orientations were found to be 3.6, 3.39 and 3.66 eV, respectively. The nonlinear decrease of the work function of chemical compounds was observed for all considered orientations. The highest deviation from the linearly decreasing work function was observed for the orientation (100) for the chemical compound Al_2Mg (~0.25 eV); for the orientation (110) for Al_3Mg (~0.25 eV) and for the orientation (111) for AlMg (~0.6 eV). The lowest work function, even lower than that of pure magnesium, has the AlMg chemical compound with the crystallographic orientation (111) (3.29 eV). Therefore, the formation of such a chemical compound in the cathode material can increase its emissivity.

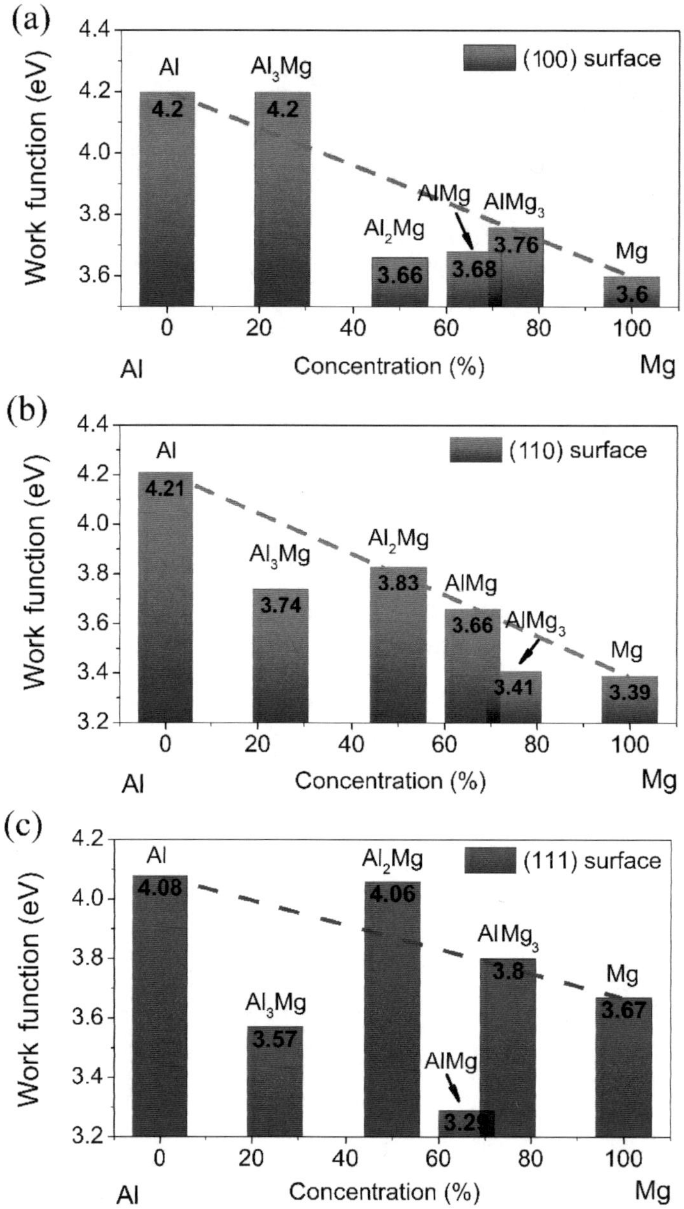

Figure 2. The work function calculated from the first principles in the framework of the density functional theory for chemical compounds of aluminum-magnesium system with crystallographic orientations (100) (a), (110) (b) и (111) (c).

Acknowledgments
The work was supported by the RSF grant № 18-12-00440 and partially supported by the state assignment of the IMSP RAS. A.A. Kistanov acknowledges the financial support provided by the Academy of Finland (grant № 311934).

References
[1] Svetlov I L, Karpov M I, Neuman A V and Stroganova T S 2018 *Russian Metall.* **4** 348

[2] Han J K, Herndon T, Jang J, Langdon T G and Kawasaki M 2020 *Adv. Eng. Mater.* **22** 1901289
[3] Castro M M, R. Pereira P H, Figueiredo R B and Langdon T G 2019 *Lett. Mater.* **9** 541
[4] Castro M M , Wolf W, Isaac A, Kawasaki M and Figueiredo R B 2019 *Lett. Mater.* **9** 546
[5] Korznikova G, Kabirov R, Nazarov K, Khisamov R, Shayakhmetov R, Korznikova E, Khalikova G and Mulyukov R 2020 *JOM* **72** 2898
[6] Danilenko V N, Korznikova G F, Zhilyaev A P, Sergeev S N, Khalikova G R, Khisamov R K, Nazarov K S, Kiekkuzhina L U and Mulyukov R R 2018 *IOP Conf. Ser.: Mater. Sci. Eng.* **447** 012021
[7] Kresse G and Furthmüller J 1996 *Physical Rev.* B **54** 11169
[8] Smoluchowski R 1941 *Phys. Rev.* **60** 661
[9] Trana R, Lia X G, Montoya J H, Winston D, Persson K A and Ong S P 2019 *Surf. Sci.* **687** 48
[10] Wang J and Wang S Q 2014 *Surf. Sci.* **630** 216
[11] Patra A, Bates J E, Sun J and Perdew J P 2017 *PNAS* **114** E9188
[12] Fall C J, Binggeli N and Baldereschi A 1998 *Phys. Rev.* B **58** R7544
[13] Grepstad J K, Gartland P O and Slagsvold B J 1976 *Surf. Sci.* **57** 348
[14] Michaelson H B 1977 *J. Appl. Phys.* **48** 4729
[15] Skriver H L and Rosengaard N M 1992 *Phys. Rev.* B **46** 7157

IOP Conf. Series: Materials Science and Engineering 1008 (2020) 012033 doi:10.1088/1757-899X/1008/1/012033

Modified hybrid CdSe/ZnS quantum dots and their size dependent unique characteristics

A F Ishankulov[1], R R Shamilov[2], Yu G Galyametdinov[2] and N K Mukhamadiev[1]

[1] Samarkand state university, 15 University boulevard, 140104 Samarkand, Uzbekistan
[2] Kazan National Research Technological University, 68 Karl Marx, 420015 Kazan, Russia

E-mail: ishankulov-alisher@mail.ru

Abstract. In this study, CdSe-based hybrid CdSe/ZnS quantum dots were synthesized. Their optical properties have been characterized using the absorption and luminescence spectroscopy methods. When the ZnS shell was grown on the surface shell of CdSe nanoparticles, the photoluminescence stability and the shelf life of the quantum dots increased significantly. The energy of the outer shell has increased. The absorption and luminescence spectra shifted to a shorter wavelength range. The size of the nucleus decreased from 3.7 nm to 2.8 nm. The surface was stabilized with oleic acid as an active ligand. After stabilization with a surfactant, the average total hydrodynamic size of the nanoparticles increased.

1. Introduction

Colloidal quantum dots are semiconductor nanocrystals stabilized by surfactant molecules. Typical dimensions of quantum dots are 1–10 nm, but these limits depend significantly on the nature of the material and the electronic properties of the particle structure. Quantum dots (QDs) are large associates consisting mainly of InAs, Si, InP, CdSe/ZnS and other simple semiconductor materials and containing tens and hundreds of atoms [1-3].

Cadmium chalcogenides, which have good luminescent properties over a wide range of the electromagnetic spectrum, are among the well studied nanoparticles [4]. The development of hybrid “core/shell” type quantum dots can lead to a further increase in the luminescence intensity. In particular, quantum dots consisting of elements of groups II and VI have a wide band gap and possess high photoluminescence efficiency at room temperature [5].

The synthesis of quantum dots with a core/shell structure is carried out by the growth of monolayers by adding precursors to the reaction medium. The thickness of the shell is one of the important parameters determining the properties of the crystal, which allows one to control the yield, stability, and other properties of quantum dots [6,7].

It is known that colloidal synthesis methods allow the production of high quantum efficient nanoparticles with a short size distribution. Depending on the synthesis conditions, stabilizers are selected from fatty acids, amines, thiols and various organic solvents that provide distribution in different environments. They are adsorbed on the phase boundary, creating a barrier that prevents the accumulation of nanoparticles and passivates the surface energy state of the nanocrystals, which in turn provides monodisperse-sized, stabilized quantum dots [8,9].

Changing the surface area of quantum dots is usually done by means of a ligand (stabilizer). For example, quantum dots with good luminescent properties obtained in an organic medium are transferred to an aqueous medium as a result of the replacement of surface stabilizers. Also, different molecules or compounds exhibit different beneficial biological properties on the surface due to changes in the surface condition [10,11].

Numerous studies on the changing of a hydrophobic ligand with a hydrophilic one have been reported in papers [12-15]. According to them, the transfer of the obtained quantum dots to an aqueous medium is easier in hydrophilic stabilizers.

In this paper, the task was set to develop a method for converting a hydrophobic ligand to a hydrophilic one in the process of the synthesis of hybrid quantum dots. Through this process, nanoparticles with multifunctional properties can also be obtained.

2. Material and methods

Cadmium oxide (99%, pure for analysis), selenium (Se), sulfur (S), zinc oxide (ZnO, 99%, pure for analysis), oleic acid (OA, 98%, pure), 1-octadecene (ODE), 90%, pure for analysis), dodecanethiol-1 (DT, 98%, Alfa Aesar, UK), dodecyl ether of dihydrolipoic acid (DDL, 96%, Alpha Aesar, UK), toluene (99%, chemically pure), ethanol (96%), acetone (99%, chemically pure), oleylamine (ODA, 96%) were used for the experiments.

CdSe nanoparticles were synthesized according to the methods described in the authors' work [16,17]. The process was as follows. The synthesis was provided with a three-neck glass flask, and the solution temperature was controlled by a thermometer immersed in the solution. A magnetic stirrer was used to achieve a uniformity of heating and distribution of the components in the dispersion medium. A condenser was used to maintain a constant volume of the solution. To prevent oxidation of the reagents, the synthesis was carried out in an inert gas argon medium. In high-temperature synthesis, 1-octadecene as an organic solvent and oleic acid as a stabilizer were used to obtain cadmium selenide nanoparticles. Initially, 5.04 ml of oleic acid and 20 ml of ODE mixture were dissolved in 0.256 g of cadmium oxide in a 3-neck glass flask to obtain cadmium precursors at 260°C for the required time. Se precursors were then prepared by dissolving 0.158 g of selenium and 10 ml of ODE at 200°C.

To prepare a solution of zinc oleate (0.1 M), 0.045 g of ZnO was prepared by heating a solution of 1.41 ml of oleic acid and 3.6 ml of ODE at a temperature of 260°C. Sulfur in the quantity of 0.016 g in 5 ml of ODE at 200°C was dissolved to obtain a sulfur precursor (0.1 M). A 20 ml solution was prepared by adding 5 ml of ODE to a solution of CdSe ($6.6 \cdot 10^{-6}$ mol / l) QDs in toluene. The toluene in the mixture was then heated to 200°C until it evaporated. Then 2 ml of oleylamine and 4.7 ml of zinc oleate solution were added and 4.7 ml of sulfur precursor solution was infused over a period of time using the SILAR method.

The photoluminescence and photoluminescence excitation spectra of QDs were detected by a Cary Eclipse (Varian) fluorescence spectrophotometer. Optical absorption spectra were measured with a Perkin Elmer Instrumental LAMBDA 35 two-beam spectrometer. QDs size were evaluated by dynamic light scattering with a Malvern Zetasizer Nano which was equipped with a narrow filter at 632.8 nm. The average size of the QDs was estimated to be 9.3 nm.

3. Results and Discussion

From the absorption spectra of the obtained hybrid CdSe/ZnS QDs (figure 1 a), it can be seen that as the ZnS shell grows, the maximum of the excitation peak in the spectrum shifts toward the short-wavelength region. The emission spectrum of CdSe QDs is symmetrical, with a peak near 587 nm (figure 1 a). We have obtained hybrid CdSe/ZnS QDs by a preliminary formation of CdSe core and growing a ZnS shell on them. The formation of the shell leads to a shift of the excitation peak of CdSe core to shorter wavelengths. According to calculations, during the formation of the ZnS layer, the size of the CdSe core decreases from 3.7 to 2.8 nm due to the replacement of cadmium atoms on the surface by zinc atoms. In this case, the QD luminescence peak is also shifted to the short-wavelength region (figure 1 b). The quantum yield, which was 4.9% for CdSe nanoparticles, increased to 19% after the growth of the ZnS shell. This indicates that the yield of the quantum dot obtained on the basis of hybrid CdSe/ZnS increases with decreasing particle size.

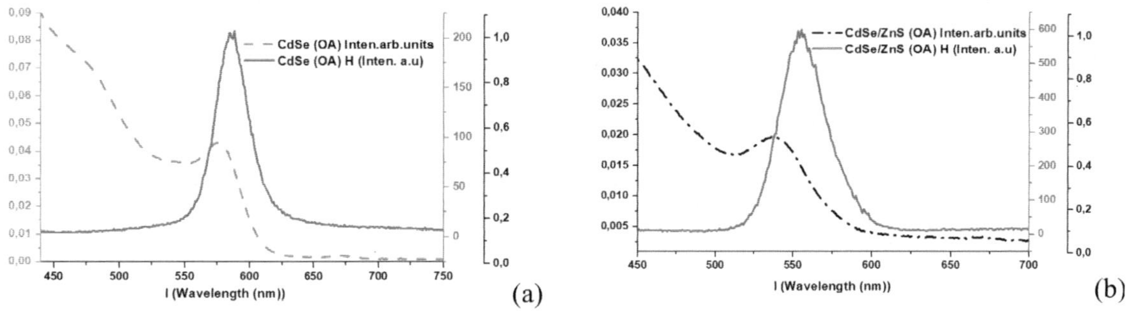

Figure 1. Absorption and emission spectra of CdSe (a) and CdSe/ZnS (b) QDs (λ_{ex} = 400 nm).

The absorption spectroscopy data also allow the calculation of the extinction coefficient and concentration of nanoparticles. The extinction coefficients for CdSe and hybrid CdSe/ZnS nanocrystals were determined by the following formula [18]:

$$\varepsilon = 5857\ (D)^{2{,}65},$$

where D is the diameter of the nanoparticle, ε the extinction coefficient ($M^{-1}\cdot cm^{-1}$). From this formula, the extinction coefficients of 187676.4 M^{-1} cm^{-1} for CdSe QDs and 89669.34 $M^{-1}cm^{-1}$ for hybrid CdSe/ZnS QDs were determined.

In turn, the concentration of nanoparticles in the solution was determined using the following formula based on the Lambert-Beer law [19]:

$$C_M = \frac{A_{max,abs}}{\varepsilon \cdot l},$$

where C_M is the quantum dot concentration (mol/l), $A_{max,\ abs}$ – the maximum absorbance, l – surface shell layer.

The average hydrodynamic size of the nanoparticles was determined basing on the method of dynamic scattering of light on a Malvern Zetasizer Nano instrument. The average hydrodynamic size of dodecanthiol-coated quantum dots is 8 nm, which is smaller than that of CdSe/ZnS quantum dots 9.3 nm. The use of DDL as a stabilizer has led to an increase in the particle size of quantum dots (up to 9.6 nm) (figure 2 b).

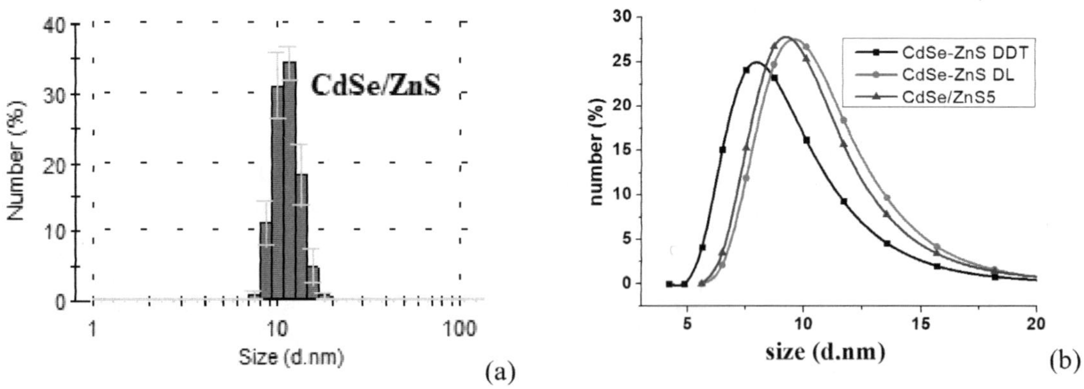

Figure 2. Mean hydrodynamic size distribution of hybrid CdSe/ZnS quantum dots (a) and size distribution diagram of CdSe/ZnS hybrid quantum dots coated with different stabilizers (b).

The mean hydrodynamic size of the synthesized hybrid CdSe/ZnS nanocrystals was 11.2 nm taking into account the stabilizer shell (figure 2 a). From this, it can be determined that the thickness of the ligand shell is 2.35 nm, given that the size of the core of the hybrid CdSe/ZnS QDs is 2.8 nm.

In the infrared spectra of CdSe/ZnS hybrid quantum dots stabilized with oleic acid, intense symmetrical and asymmetric stretching signals of the oleate anion COO^- group are present in the range from 1546 to 1456 cm^{-1} (figure 3). In turn, the presence of a strong signal with a value of 1734 cm^{-1} in the -COOR group in the DDL molecule and the bending mode intensity in C-O are in the range between 1249 and 1174 cm^{-1}. The stretching signals at 2922, 2854, and 1461 cm^{-1} belong to groups C–H. The ester bond signals in the DDL-modified CdSe/ZnS QDs correspond to 1734 cm^{-1} stretching mode signals of C-O and 1248 cm^{-1}, and 1171 cm^{-1} to the bending mode signals of C-O bond, respectively. In this case, the stretching signals of the oleate anion COO^- group are significantly reduced. In the IR spectrum of dodecanthiol-modified quantum dots, it was observed that the value of the vibration signals of the -COO^- group decreased in the same way. Thus, an exchange between the ligands occurred during the reaction, but only a small fraction of the oleic acid molecules remained unchanged. To cease the ligand exchange process, it is necessary to increase the concentration of the replaced ligand or repeat the reaction step several times.

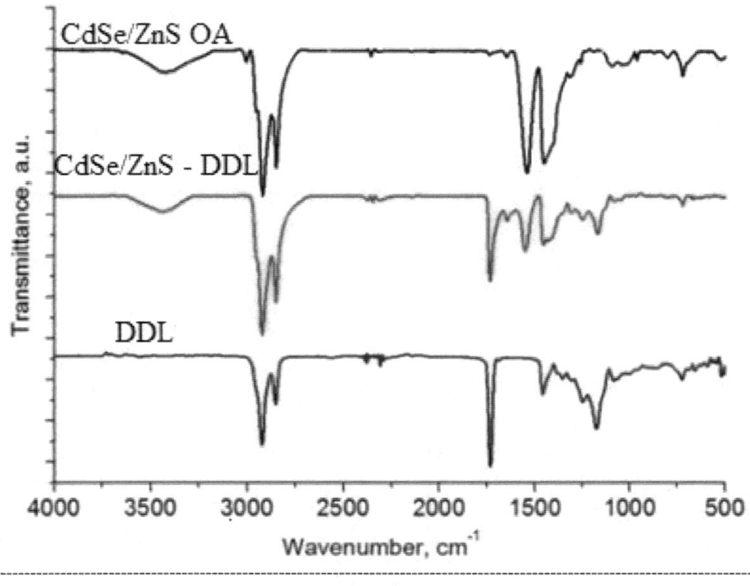

Figure 3. IR spectra of CdSe/ZnS QDs and DDL.

When the surface is modified with dodecanethiol and dodecyl dihydrolipoate, the QD surface remains hydrophobic. In this case, a high luminescence intensity is maintained and the peak position does not shift. According to IR spectra, a small part of oleic acid molecules on the QD surface remains unsubstituted. The hydrodynamic particle size is slightly reduced as a result of the modification, due to the smaller length of the stabilizer molecules in comparison with oleic acid.

4. Conclusions

Synthesis of oleic acid-stabilized CdSe and its hybrid CdSe/ZnS quantum dots was performed and their optical-dimensional properties were studied.

The absorption and luminescence spectra of the hybrid CdSe/ZnS quantum dots shifted to a shorter wavelength. The monodispersity of the obtained QDs and the radiation efficiency were found to depend on the molar ratio of cadmium to selenium. An excess of cadmium and selenium led to a high quantum yield. In this case, an excessive increase in the amount of selenium facilitated the synthesis of particles with high efficiency over a long period of time for the reaction. The luminescence quantum yield was found to increase from 4.9% to 19%.

Acknowledgments

This work was financially supported by the Russian Foundation for Basic Research and the Government of the Republic of Tatarstan in the framework of the scientific project No. 18-43-160009.

References

[1] Barve A V, Lee S J, Noh S K and Krishna S 2010 *Laser & Photonics Reviews* **4**(6) 738-50.
[2] Galyametdinov Y G, Sagdeev D O, Voronkova V K, Sukhanov A A and Shamilov R R 2017 *Materials Research Express* **5**(7) 075009
[3] Das D, Aiello A, Guo W and Bhattacharya P 2020 *IEEE Transactions on Nanotechnology* **19** 571-4
[4] Ishankulov A F, Shamilov R R, Halilov K F, Galyametdinov Yu G and Mukhamadiev N K 2020 *International Journal of Advanced Science and Technology* **29**(9) 2140-8
[5] Hernández-Gordillo A, Acevedo-Peña P, Bizarro M, Rodil S E and Gómez R 2018 *Journal of Materials Science*: *Materials in Electronics*, **29**(9) 7345-55
[6] Nayak A P et al 2015 *ACS Nano* **9**(9) 9117-23.
[7] Mikhailov I I, Tarasov S A, Solomonov A V, Matyushkin L B and Mazing D S 2014 *Functional Materials* **21**(2) 186-9
[8] Qin W, Shah R A and Guyot-Sionnest P 2012 *ACS Nano* **6**(1) 912-8
[9] Shamilov R R, Ishankulov A F and Galyametdinov Yu G 2020 *Technological University Bulletin* **23**(4) 19-22 [in Russian]
[10] Aderhold J, Davydov V Y, Fedler F, Klausing H, Mistele D, Rotter T and Graul J 2001 *Journal of Crystal Growth* **222**(4) 701-5
[11] Van Embden J, Jasieniak J, and Mulvaney P 2009 *Journal of the American Chemical Society* **131**(40) 14299-309
[12] Hofman E, Robinson R J, Li Z J, Dzikovski B and Zheng W *Chemistry of Materials* 2017 **139** 8878–85
[13] Guo W and Peng X 2003 *Chemistry of Materials* **15**(14) 2854-60
[14] Huang X, Parashar V K and Gijs M A 2018 *Langmuir* **34**(21) 6070-6
[15] Rogach A L, Kornowski A, Gao M, Eychmuller A and Weller H 1999 *The Journal of Physical Chemistry* **103**(16) 3065-9
[16] Farmonovich A I, Rashitovich R Sh, Faxritdinovich Q X, Gennadievich Yu G and Qurbonalievich N K 2020 *Jour of Advanced Research in Dynamical & Control Systems* **12**(7) 2201-5
[17] Ghosh Chaudhuri R and Paria S 2012 *Chemical Reviews* **112**(4) 2373-433
[18] Wang Y, Suna A, McHugh J, Hilinski E F, Lucas P A and Johnson R D 1990 *The Journal of Chemical Physics* **92**(11) 6927-39
[19] Kim D, Miyamoto M, Mishima T and Nakayama M 2005 *Journal of Applied Physics* **98**(8) 083514

IOP Conf. Series: Materials Science and Engineering 1008 (2020) 012034 doi:10.1088/1757-899X/1008/1/012034

Regimes of mass transfer in a 2D crystal in nonequilibrium states

E A Korznikova[1,2], I A Shepelev[3], E L Vardanyan[2], K N Ramazanov[2] and P V Zakharov[4,5]

[1]Institute for Metals Superplasticity Problems of RAS, 39 Khalturin St., 450001 Ufa, Russia
[2]Ufa State Aviation Technical University, 12 K. Marx St., 450008, Ufa, Russia
[3]Saratov Chernyshevsky State University, 83 Astrakhanskaya St., 410012 Saratov, Russia
[4]Peter the Great St. Petersburg Polytechnic University, 29 Politekhnicheskaya St., St. Petersburg, 195251, Russia
[5]Shukshin Altai State University for Humanities and Pedagogy, 53 Vladimir Korolenko St., Biysk, 659333, Russia

E-mail: elena.a.korznikova@gmail.com

Abstract. A variety of new technologies associated with the severe impact on materials have been developed recently. The promotion of these technologies is, in turn, associated with the stability of the structure and properties of materials under nonequilibrium conditions. It is well known that under such conditions new mechanisms of structure evolution can occur. The study of these mechanisms is an important task of modern materials science. In our work,a we present the results of an analysis of possible scenarios of impact induced mass transfer in a two-dimensional crystal with a Morse interaction potential including a topological compression soliton, a acrowdion complex and a moving vacancy – a voidion. The influence of the initial impact energy distribution is analysed. The obtained results contribute to the fundamental understanding of mass transfer processes under nonequilibrium conditions.

1. Introduction

Diffusion is a process basically responsible for all processes occuring in a material, including the evolution of a defect structure, phase transition, and others. Enhancement of diffusion-type processes is often observed in the case of a nonequilibrium state of materials associated with extreme external impacts. Generally, there are two main mechanisms of diffusion – the motion of vacancies and the motion of interstitial atoms; the first being more energy demanding [1]. However, both of these processes are discrete and can be related to the equilibrium state of the lattice. In case of nonequilibrium conditions associated with high energy impacts, new modes of collective atomic motion can be activated. Among them, one can recall the rapid diffusion of atomic clusters (islands) [2] and one-dimensional (1D) motion of atomic groups that can be realized by the motion of both compacted and rarefied lattice regions.

The case of a sparsed one-dimensional region, being a delocalized vacancy, is called a voidion, and this object has been less studied [3]. Recent studies have revealed that the mass transfer efficiency of the mentioned dynamic point defects depends essentially on their initial configuration and can increase considerably when not a single but several atoms are involved in one dimentional motion [4–7]. It was also demonstrated that when several crowdions propagate in neighboring parallel rows, they exchange energy with each other that inevitably influences the scenario of the evolution of the defect structure [8–

10]. This work is devoted to the analysis of the influence of the initial impact energy on the propagation scenario of crowdion ensemble excited in three parallel crystallographic rows.

2. Simulation details

The investigation was performed by the method of atomistic simulation due to the small spatial and temporal scale of the phenomena studied. Molecular dynamics has earlier been proved to be a powerful tool for the investigation of small scale phenomena associated with mass and energy transfer [4,7–9], nonlinear dynamics of the lattice [11–14], and the evolution of mechanical characteristics of the material upon external loading [15–20], analysis of phase transformations [21–24] and many other applications [25-26]. The calculations were done for the calculational cell forming a 2D lattice with a size of 100 x 300 interatomic distances. Boundary conditions are periodic in all directions. A Cartesian coordinate system is used, with the X-axis coinciding with one of the densely packed directions. The interaction between particles in a triangular lattice is determined using the Morse potential, described elsewhere [5,10].

The main attention is paid to the processes associated with the transfer of mass and energy in the crystal by means of the M-crowdion, which is initiated by applying the initial velocities simultaneously to three atoms in three adjacent closely packed crystallographic atomic rows, i.e. M = 3 (see figure 1 a). We have analyzed the development of the defect structure evolution scenario for several course groups of initiation velocity combinations resulting in different types of relaxation path. Thus, the major groups of principally different cases can be distinguished as follows (1) $v_1 = v_3 < v_2$; (2) $v_1 = v_3 > v_2$,(3) $v_1 = v_3 >> v_2$, (4) $v_2 > 0$, $v_1 = v_3 >> v_2$, (5) $v_2 < 0$, $v_1 = v_3 >> v_2$ marked hereinafter as A,B,C,D and E modes respectively. Each of these modes results in different combination of point defects emerged as the result of the initial impact.

3. Results and discussion

The main purpose of this work is the investigation and analysis of the effects arising upon the excitation of an M-crowdion cluster with different initial velocities in adjacent atomic rows. The initial kick of the atom results in its shift from the equilibrium position. The resulting scenario of both the excitation propagation dynamics and the resulting defect structure depends on two main parameters, namely the initial velocity of the kick and the ratio of initial velocities of the atoms involved in the M-crowdion (3 atoms in the considered case). The former defines the ability of the affected atoms to overcome the potential barrier of vacancy formation and initiation of a moving soliton type object with 1D locally increased density (crowdion). If the energy provided at the initial moment is below the potential barrier threshold, one can expect the formation of a topological soliton. Thus, due to the presence of inevitable cone type radiation, the crowdion motion is a subject of the permanent complicated self-regulated process of energy exchange defining the motion dynamics and corresponding stress-strain state in local areas affected by the impact. The major issue concerns the difference between the values of energy provided to atoms located in the middle and edge rows.

Table 1. Summarization of possible scenarios of crowdion propagation for different ratios and amounts of emerging energies.

A	Propagation of a compression soliton (without the appearance of atomic vacancies) or a crowdion (with the appearance of prismatic vacancies in 3 adjacent atomic rows)
B	Propagation of a soliton ⇒ the appearance of a vacancy at a certain distance from the starting point in the median row (voidion) and of a quasi-1D subsonic crowdion
C	Propagation of a supersonic crowdion with the emergence of two vacancies at the starting point in the edge rows ⇒ the appearance of a vacancy remote from the starting point in the median row (voidion) and of a M = 3-subsonic crowdion
D	Propagation of a supersonic crowdion with the emergence of two vacancies at the starting point in the edge rows ⇒ the appearance of a voidion and a M = 3-subsonic crowdion ⇒ their

	subsequent merger ⇒ forming a M = 2-subsonic crowdion propagating in another close-packed direction
E	Propagation of a supersonic M = 3-crowdion ⇒ the appearance of a voidion (V1), M = 3-subsonic crowdion (Cr1) and of the additional Frenkel pair (voidion (V2) + a quasi-1D subsonic crowdion (Cr2)) ⇒ subsequent merger of V 1 with Cr1 and of V 2 with Cr2 ⇒ forming a M = 2-subsonic crowdion propagating in the same or another close-packed direction
F	The same as for regime E but merging of V 2 and Cr2 is absent. V2 remains at the same place, while Cr2 moves to the starting place, loses their stability and transforms to an interstitial

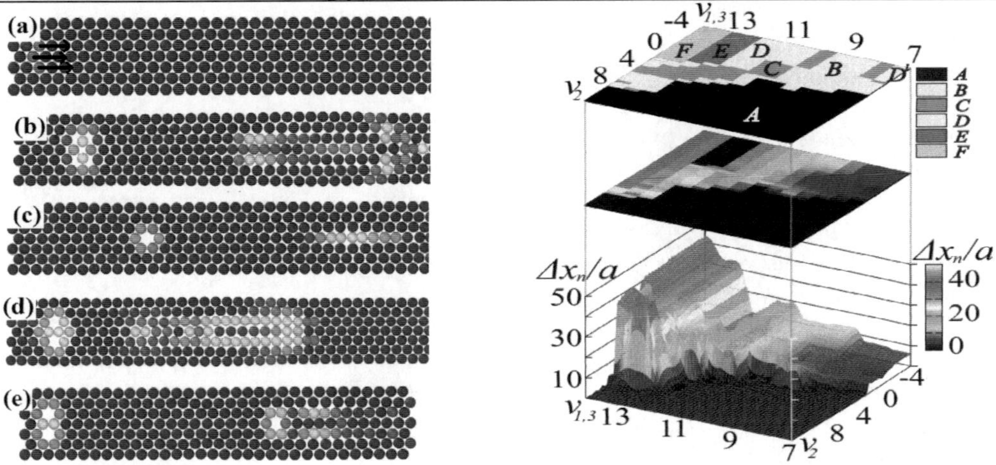

(f)

Figure 1. (a) Scheme of the initial crowdion complex excitation, propagation of the M-crowdion induced by the initial impact with velocities for the initial velocities (b) $v_{1,3} = 10$ and $v_2 = 8$ (regime A) (c) $v_{1,3} = 9$ and $v_2 = 0.5$ (regime B) (d) $v_{1,3} = 12$ and $v_2 = 4$ (regime C) (e) $v_{1,3} = 11$ and $v_2 = 3$ (regime D). (f) Summarizing map of modes and propagation distances of the voidion on the plane of initial velocities (v_2, v_1). The A mode on the mode map corresponds to the propagation of the M-soliton without the formation of a voidion, and the B - F modes correspond to the voidion propagation. The crowdion propagaton distance is indicated by the color scale bar on the bottom panel.

Table 1 presents the summarization of the possible scenarios of crowdion propagation for different ratios and amounts of emerging energies. The values of the applied velocities are presented in the regime map in fugure 1 f. Thus, one can see that the region of small velocities corresponds to the propagation of the compression soliton because the initial energy in this case is insufficient to overcome the potential barrier of bond breaking. An increase invelocity results in a switch to regimes B and C where the propagation of a supersonic crowdion with the emergence of two vacancies takes place (figure 1 c-d). One of the most interesting cases in this study is the situation of $v_1 = v_3 >> v_2, v_2 \leqslant 0$. In this case, due to energy exchange due to Cherenkov radiation 27, a vacancy is formed in the middle row, which, under the influence of the stress field of adjacent rows, moves forward, forming a voidion (figure 1 d-e). The reason for the emergence of a voidion with sufficiently large propagation distances is an open question and requires a detailed analysis.

4. Conclusions

We have performed a molecular dynamics study of the propagation regimes of the triple crowdion complex with variable initial velocities in all three rows. It was revealed that the scenario of the point defect evolution strongly depends on the velocities value and ratio. Three general scenarios, each having several features can be distinguished, namely (i) propagation of a atopological compression soliton, (ii) propagation of a 3-crowdion complex with a triple vacancy at the initiation site and (iii) propagation of two crowdions and a voidion in the middle row in case of zero or negative initial velocities. The obtained results contribute to a fundamental understanding of the defect structure evolution under nonequilibrium conditions.

Acknowledgments
I.A.Sh. acknowledges the support from the Russian Science Foundation, grant No. 19-72-00109 (obtaining of results). E.A.K. thanks the Council of the President of the Russian Federation for support of Russian scientists, grant No. MD-3639.2019.2 (writing the manuscript). Work partially supported by State Assignment of Russian Ministry of Science and Education.

References

[1] Fitzgerald S P 2018 *J. Micromech. Mol. Phys.* **03** 1840003
[2] Dudarev S L 2003 *Philos. Mag.* **83** 3577–97
[3] Mazilova T I, Sadanov E V, Voyevodin V N, Ksenofontov V A and Mikhailovskij I M 2018 *Surf. Sci.* **669** 10–5
[4] Babicheva R I, Evazzade I, Korznikova E A, Shepelev I A, Zhou K and Dmitriev S V 2019 *Comput. Mater. Sci.* **163** 248–55
[5] Chetverikov A P, Shepelev I A, Korznikova E A, Kistanov A A, Dmitriev S V and Velarde M G 2017 *Comput. Cond. Matt.* **13** 59–64
[6] Dmitriev S V, Korznikova E A and Chetverikov A P 2018 *J. Exp. Theor. Phys.* **126** 347–52
[7] Moradi Marjaneh A, Saadatmand D, Evazzade I, Babicheva R I, Soboleva E G, Srikanth N, Zhou K, Korznikova E A and Dmitriev S V 2018 *Phys. Rev.* E **98** 023003
[8] Shepelev I A, Korznikova E A, Bachurin D V, Semenov A S, Chetverikov A P and Dmitriev S V 2020 *Phys. Lett.* A **384**
[9] Shepelev I A, Dmitriev S V, Kudreyko A A, Velarde M G and Korznikova E A 2020 *Chaos, Solitons & Fractals* **140** 110217
[10] Korznikova E A, Shepelev I A, Chetverikov A P, Dmitriev S V, Fomin S Y and Zhou K 2018 *J. Exp. Theor. Phys.* **127** 1009–15
[11] Barani E, Korznikova E A, Chetverikov A P, Zhou K and Dmitriev S V 2017 *Phys. Lett.* A **381** 3553–7
[12] Zakharov P V, Korznikova E A, Dmitriev S V, Ekomasov E G, Zhou K 2019 *Surf. Sci.***679** 1–5
[13] Korznikova E, Bachurin D, Fomin S, Chetverikov A, Dmitriev S 2017 *Eur. Phys. J.* B **90** 1-8
[14] Korznikova E A, Shcherbinin S A, Ryabov D S, Chechin G M, Ekomasov E G, Barani E, Zhou K and Dmitriev S V 2019 *Phys. Stat. Sol.* B **256** 1800061
[15] Shepelev I A, Chetverikov A, Dmitriev S V, Korznikova E A 2020 *Comp. Mat. Sci.* **177** 109549
[16] Savin A V, Korznikova E A and Dmitriev S V 2019 *Phys. Rev.* B **99** 235411
[17] Rysaeva L K, Korznikova E A, Murzaev R T, Abdullina D U, Kudreyko A A, Baimova J A, Lisovenko D S, Dmitriev S V 2020 *Facta Universitatis, Ser.: Mechanical Engineering* **18** 1–12
[18] Babicheva R I, Dmitriev S V, Korznikova E A and Zhou K 2019 *J. Exp. Theor. Phys.* **129** 66–71
[19] Abdullina D U, Korznikova E A, Dubinko V I, Laptev D V, Kudreyko A A, Soboleva E G, *Computation* **8** 27
[20] Bhui A S, Singh G, Sidhu S S, Bains P S 2018 *Facta Universitatis, Ser.: Mech. Eng.* **16** 337
[21] Tsukanov A and Psakhie S 2016 *Facta Universitatis, Ser.: Mech. Eng.* **14** 269
[22] Li Y, Li Y, Han P, Wang S, Jiang Z and Wang X 2018 *Comput. Model. Eng. Sci.* **117** 109–23
[23] Chen H-Y and Tsou N-T 2019 *Comput. Model. Eng. Sci.* **120** 319–32
[24] Lee W-J, Lo Y-C, Yang A, Chen K, Chen N-Y 2019 *Comp. Mod. Eng. Sci.***120** 293–304
[25] Korznikova E, Sunagatova I, Bayazitov A, Semenov A, Dmitriev S 2019 *Lett. Mater.* **9** 386–90
[26] Baimova J 2016 *Lett. Mater.* **6** 31
[27] Koch V, Majumder A and Wang X-N 2006 *Phys. Rev. Lett.* **96** 172302

IOP Conf. Series: Materials Science and Engineering 1008 (2020) 012035 doi:10.1088/1757-899X/1008/1/012035

Synthesis of the catalysts based on oxides of some d-elements

Kh M Saidov and N K Mukhamadiev

Samarkand State University, 15 University Blvd., 140104 Samarkand, Uzbekistan

E-mail: xmsaidov@gmail.com

Abstract. In this work, nanocomposite materials were synthesized on the basis of a mixture of nitrates of Ni (II), Zr (IV), Mg (II) and $Al(OH)_3$ with glycine by the solution combustion method. The synthesized nanocatalysts with the composition $xNiO/yZrO_2/zAl_2O_3$ (TiO_2, MgO) are characterized by a high specific volume (0.80 $\pm$ 0.05 cm^3/g), an average size (1.85 $\pm$ 0.05 nm) of pores, a relative specific surface area (180 $\pm$8 m^2/g) and a big monolayer capacity (3.17 $\pm$ 0.24 mol/kg). A hysteresis loop is observed in isotherms during the absorption of benzene vapours, while in the case of hexane it is absent. The resulting nanocatalysts can be used as a selective catalyst for the conversion of methane in a flow reactor.

1. Introduction

There is a tendency of decrease of the reserves and production volumes of renewable energy sources, including light oils and natural gas in the world. That entails the need to deepen the processing of hydrocarbon raw materials. Therefore, it is necessary to create new or modernize existing processes of processing hydrocarbon raw materials. The most common process in oil and gas processing plants is the catalytic conversion of hydrocarbons raw materials. For a long time of its existence, the process has undergone changes both in the method of contact of the raw material and the catalyst, and in the catalysts used. These changes made it possible to obtain the maximum amount of products. The choice of the catalyst for the process plays an important role; the yield and quality of the target products depends on the chosen catalyst. The use of nanocatalysts leads to an increase in their selectivity and productivity due to the size effect [1]. In this regard, researchers are involved in the development of various nanocatalysts and their modification for processing natural gas [2].

The goal of this work is to synthesize nanocatalysts based on oxides of some d-elements and to study the phase composition, geometric and sorption characteristics.

2. Material and methods

For the preparation of the catalysts we used nickel (II) nitrate hexahydrate — $Ni(NO_3)_2{\cdot}6H_2O$ (extra pure, Reachem 98%), zirconium oxynitrate — $ZrO(NO_3)_2$ (extra pure, Reachem 98%), magnesium nitrate dihydrate — $Mg(NO_3)_2{\cdot}2H_2O$ (extra pure, Reachem 98%), aluminum oxide — Al_2O_3 (extra pure), titanium (IV) oxide — TiO_2 (extra pure, Reachem 98%).

Figure 1 shows schematically the process of synthesis of the nanocatalysts. In the process Al_2O_3 obtained from a local raw material, bentonite, was used [3]. To ensure the stability of the nickel catalyst, alkali [4], alkaline earth metals [5] and some oxides of d-elements are used as promoters. Rare metal oxides play an important role in catalytic processes. It has now been proven that the use of inexpensive two - and multicomponent metal oxides instead of rare metals used as additives to the support allows one to prepare high-quality and inexpensive catalysts. For this purpose, during the synthesis, magnesium (MgO) and titanium (TiO_2) oxides were used as additional promoters. According to the technology, metal nitrates were dissolved in water and glycine was added to the solution in a molar ratio of glycine / nitrate = 5. The obtained Al_2O_3 was calcined at 400 °C (programmed temperature up to 400 °C with a heating rate of 1-2 °C / min) for 4 hours. The obtained Al_2O_3 was added to the solution, the mixture was stirred for 60 min. Excess water was distilled off the mixture and the mixture was heated to 80 °C to obtain a gel. After that, the gel was calcined at a temperature of 400 °C. In this case, the resulting gel must boil with the formation of a foam in order to form a self-combusting powder. The resulting powder was kept in an air stream at 400 °C for 3 hours. As a result, a nanostructured catalysts of the composition $NiO/TiO_2/Al_2O_3$, $NiO/ZrO_2/TiO_2/Al_2O_3$ and $NiO/ZrO_2/MgO/Al_2O_3$ was formed [6].

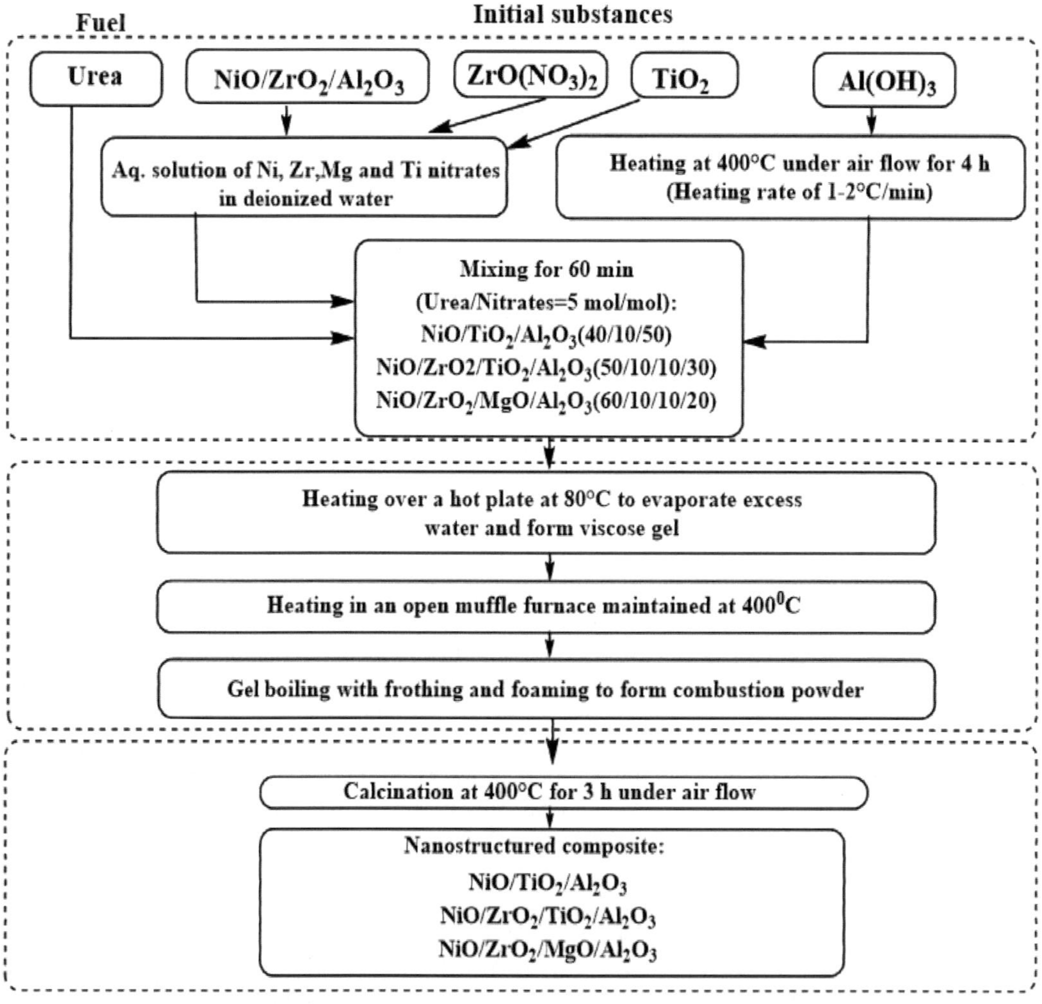

Figure 1. Technological scheme for the preparation of nanocatalysts

The structure of the catalysts was studied by powder X-ray diffraction on a Panalytical Empyrean (XRD) instrument, the geometric characteristics by SEM EVO MA 10 (Carl Zeiss), and the sorption isotherm — on a McBain - Bakr instrument. The results were processed by graphical and analytical methods of the statistical method.

3. Results and discussion

Phase analysis by X-ray diffraction (XRD) goes beyond elemental composition to characterize the crystalline forms present. An active metallic element might be present as pure metal, oxide, mixed oxide, carbonate, carbide, etc. Catalyst support materials can undergo changes under reaction conditions. If crystal structures are available, as they are for most inorganic phases, the Rietveld method can be used based on these structures.

The phase composition established by the diffractometry method is shown in figure 2. CuK_{α}-irradiation (β-filter, Cu, current mode 1.5406A°) was used to obtain the diffractogram. The voltage was 30 kV, the voltage of the current in the tube was 10-20 amps. The phase composition of the catalyst was examined by the International Center for Diffraction Data – ICDD.

Energy Dispersive X-ray analysis is an analytical technique used for the elemental analysis or chemical characterization of a sample. It relies on an interaction of some source of X-ray excitation and a sample. Its characterization capabilities are due in large part to the fundamental principle that each element has a unique atomic structure allowing a unique set of peaks on its X-ray emission spectrum. Figure 3 and the table 1 give the results of Energy Dispersive X-ray analysis for the nanocatalysts consisting of elements O, Ni, Zr, Al, (Mg, Ti) [7,8,9]. It can be seen that Al is the dominant element in the sample, although oxygenium is also found in the sample.

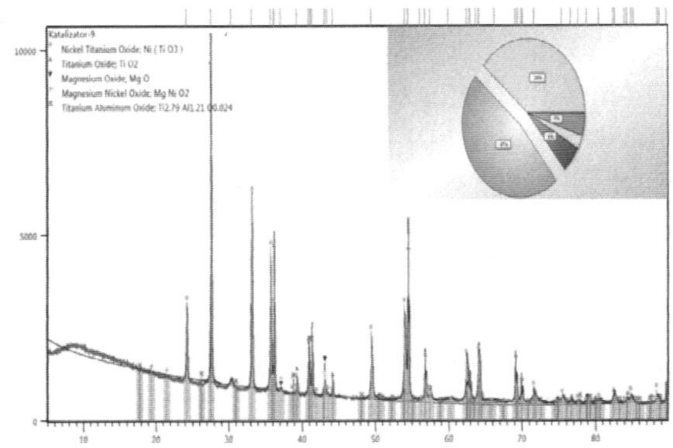

Figure 2. X-ray spectrum of the composite.

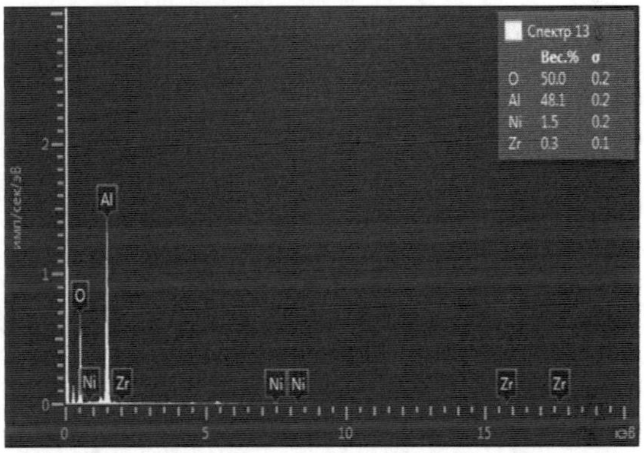

Figure 3. Elemental analysis of the catalyst.

Geometric characteristics of synthesized nanocatalysts are presented in the table.

Table 1. Geometric characteristics of synthesized nanocatalysts

Composite	Elemental composition	Pore volume V, cm^3/g	Average pore size, nm	Relative specific surface area (BET), S, m^2/g	Monolayer capacity, α_m, mol/kg
K1	Ni, Ti, Al, O	0.81	1.80	182	2.48 ± 0.18
K2	Ni, Zr, Ti, Al, O	0.77	1.87	190	3.34±0.36
K3	Ni, Zr, Mg, Al, O	0.85	1.90	197	3.7 ± 0.42

Determination of geometric characteristics and surface of the synthesized nanostructured catalysts show that the synthesized nanocatalysts are characterized by a large volume of pores (0.80 ± 0.05 cm^3/g), an average pore size (1.85 ± 0.05 nm), and a big monolayer capacity (3.17 ± 0.24 mol/kg) (Table 1).

It was found that a change in the composition of the catalyst has no noticeable effect on the porosity of the layer and the porosity of the particles. The specific total pore volume of the samples with added oxides of different metals is greater than that of the original one.

The sorption isotherm of hexane and benzene vapours in the nanocatalysts (see Figure 4,5) also confirms the above data.

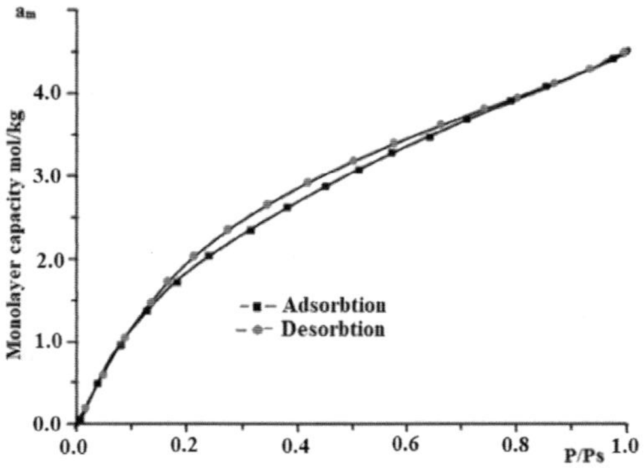

Figure 4. Sorption isotherm of hexane vapours in the nanocatalysts.

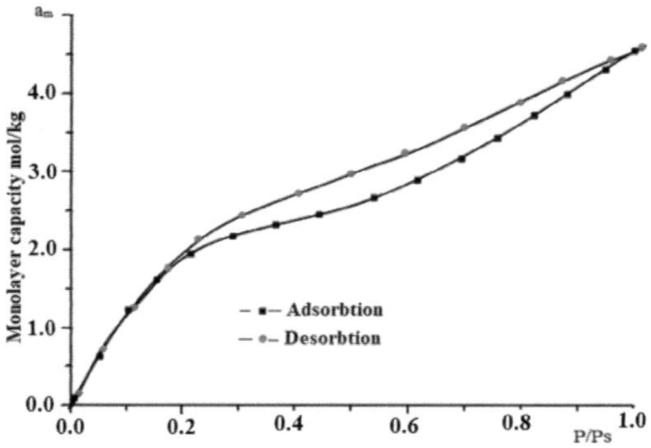

Figure 5. Isotherm of sorption of benzene vapours the nanocatalysts.

To study the texture and sorption properties of the composites, adsorption isotherms were obtained using benzene and hexane vapours. To obtain adsorption isotherms, a sensitive McBain-Bakr quartz spiral device was used. Benzene, used as an adsorbate, was purified in a vacuum by repeated freezing and defrosting by removing dissolved gases from the solvent. In each adsorption system, keeping in vacuum it continued to a residual pressure of $1.33 \cdot 10^{-3}$ Pa. Sorption isotherms of nanocatalysts were obtained by heating the samples at 473 K for 8 h by studying the adsorption of benzene and hexane vapors.

Based on the results obtained and Balandin's multiplet theory, the synthesized nanocatalysts can be used as a catalyst for methane conversion. The methane conversion reaction was carried out in a flow reactor. The progress of the reaction was monitored by gas chromatography.

4. Conclusions

The nanocomposites with the composition $xNiO/yZrO_2/zAl_2O_3(TiO_2, MgO)$ have been synthesized. The synthesized nanocomposites are characterized by a high volume (0.80 ± 0.05 cm^3/g) and an average pore size (1.85 ± 0.05 nm), a relative specific surface area (180 ± 8 m^2/g) and a big monolayer capacity (3.17 ± 0.24 mol/kg). A hysteresis loop is observed in isotherms during the absorption of benzene vapors, while in the case of hexane, it is absent. The resulting nanocatalysts can be used as a selective catalyst for the conversion of methane in a flow reactor.

References

[1] Zyryanova M M, Badmaev S D, Belyaev V D, Amosov Yu I, Snytnikov P V, Kirillov V A and Sobyanin V A 2014 *Catalysis in Industry* **3** pp 22-7

[2] Baláž P, Godočíková E, Kril'ová L, Lobotka and P Gock E 2004 *Materials Science and Engineering* A, **1-2** 442-6

[3] Mukhamadiev N K and Sayitkulov Sh M 2016 *Scientific Bulletin of Samarkand State University* **5** 48–51

[4] Osaki T and Mori T 2001 *Journal of Catalysis* **204(1)** 89–97

[5] Hou Z, Yokota O, Tanaka T and Yashima T 2003 *Applied Catalysis* A **253(2)** 381–7

[6] Baneshi J, Haghighi M, Jodeiri N, Abdollahifar M and Ajamein H 2014 *Ceramics International* **9** 14177-84

[7] Saidov Kh M and Mukhamadiev N K 2020 *Innovations in the Oil and Gas Industry, Modern Power Engineering and Actual Problems. Proc. Int. Conf. May 26, 2020* (Tashkent, Uzbekistan) p 534 [In Uzbek]
[8] Saidov Kh M, Sayitkulov Sh M and Mukhamadiev N K 2019 *Uzbek Chemical Journal* **4** 3-9
[9] Saidov Kh M and Muhamadiyev N K 2019 *Int. Conf. on Integrated Innovative Development of Zarafshan Region: Achievement, Challenges and Prospects, 27-28 Nov. Navoi, Uzbekistan* (Navoi, Uzbekistan) pp 274-6

IOP Conf. Series: Materials Science and Engineering 1008 (2020) 012036 doi:10.1088/1757-899X/1008/1/012036

Mass transfer in platinum by crowdion clusters initiated by the impact of a Pt_2 molecule

I A Shepelev[1], S V Dmitriev[2], A Yu Morkina[3], S Yu Fomin[3] and E A Korznikova[2,3]

[1]Saratov Chernyshevsky State University, 83 Astrakhanskaya St., 410012 Saratov, Russia
[2]Institute for Metals Superplasticity Problems of RAS, 39 Khalturin St., 450001 Ufa, Russia
[3]Ufa State Aviation Technical University, 12 K. Marx St., 450008, Ufa, Russia

E-mail: igor_sar@li.ru

Abstract. This work presents a molecular dynamics study of the propagation scenario of the crowdion complex initiated in the platinum crystal by means of atom or molecule bombardment with parallel and non-parallel orientations. The obtained results enable to conclude that the crowdion initiated by the Pt_2 molecule bombardment is significantly more effective for the mass transfer comparing to the crowdion excited by the single atom impact. It was detected that the atoms involved in the crowdion motion bear a vibrational mode with the frequency above the phonon spectrum. The obtained results contribute to the understanding of the mechanisms of mass transfer under equilibrium conditions.

1. Introduction
Crowdions - interstitial atoms located in closely-packed atomic rows, play an important role in relaxation processes in metals and alloys under severe external impacts, effectively transferring mass and energy. It is necessary to study the dynamics of crowdions in pure metals, as well as to analyze and compare the results obtained for various metals. The authors recently show that the concept of a supersonic crowdion can be extended to N-crowdion, where not one, but N atoms move at a high speed along a close-packed row. Experimental study of interstitial atoms moving along a crystal lattice at supersonic speeds encounters serious technical difficulties, that, in turn, promotes the study of these small scale phenomena by means of numerical methods including *ab-initio* simulations [1] and molecular dynamics. The latter has earlier been proved to be a powerful tool for investigation of small scale phenomena like phase transformations [2–5], mass and energy transfer [6–9], evolution of mechanical characteristics of the material subjected to various external impacts [10–15], analysis of nonlinear dynamics of the lattice [16–21], and many other structure evolution features. This work deals with the simulaiton of the one-dimensional (1D) propagation of 1- and 2-crowdions in fcc Pt lattice.

2. Simulation details
The simulation is performed for computational cells in the form of a parallelepiped having dimensions of 138×38×40 nm (Figs. 1). The cells consist of 14,400 atoms of Pt that form the fcc structure. Periodic boundary conditions are applied along the y- and z-axis of the computational cell, while in the x-direction, the free surface condition is adopted. The crystallographic directions of the Pt cell that correspond to the axes are shown in figure 1. We study the effect of a particle impact, that can be a single atom or a molecule consisting of two Pt atoms (Pt_2), on the free surfaces of the materials. The

surfaces of the cell are parallel to the <101> and <111> crystallographic planes of the fcc lattice and perpendicular to the direction of the flying particles. Such orientation of the cell with respect to the particles striking the surfaces is chosen along these crystallographic directions due to the distance between atoms of the lattices is the shortest, and hence, better mass transition can be expected in this case. Along the x-axis, the ideal fcc lattice has the ABAB type atom (spheres) packaging. Therefore, the third layer from the surface has the same arrangement of atoms as the first layer on the surfaces for this material. The situations, when the particles strike different sites of the surfaces are considered in the present study (indicated by numbers in figure1).

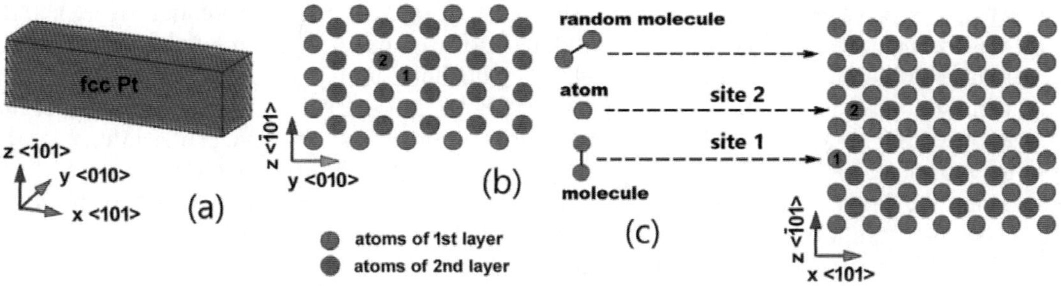

Figure 1. (a) Computational cell. Faces normal to x-axis are free surfaces. Along y and z-axes periodic boundary conditions are used. Panels (b) and (c) show crystal structure projected on (y, z) and (x, z) planes, respectively. Left side of the crystallite in (c) is the surface which is bombarded by single Pt atoms or by Pt_2 molecules. Random atom moves along x-axis and hits the surface at a randomly chosen location (not shown). Random molecule, having random orientation in 3D space, moves along x-axis without rotation and internal oscillation and hits the surface at a randomly chosen location.

The main results are obtained for bombardment by an atom or by a molecule oriented in parallel to the surface moving exactly along a close-packed atomic row. The atom can hit into an atom of the outer atomic layer (site 1) or into an atom of the second atomic layer (site 2), while the molecule hits atoms of the both layers simultaneously. The molecular dynamics (MD) simulations are conducted using the atomic/molecular massively parallel simulator (LAMMPS) program package [22]. The embedded-atom method [23] interatomic potentials obtained by Foiles et al. [24] are employed for the simulation of the particle impact on the Pt surface. Before the modeling, the computational cell is relaxed at the zero temperature to obtain the state of the minimum potential energy. The effect of the particle impact on the mass transfer in the system is studied at 10 K for different initial velocities of the striking molecule. The simulation is performed in the NVE ensemble (constant number of atoms, volume and energy). Further, the materials are equilibrated at the corresponding temperature for 5 ps. At 10 K, the atom perturbation is very negligible and almost does not affect the results.

3. Results and discussion

The main purpose of this work is the investigation and analysis of effects arising upon the excitation of a crowdion cluster under the influence of the bombarding Pt_2 molecule. The hit of this atom or molecule, shifts the surface atoms from their equilibrium positions. The resulting scenario of both excitation propagation dynamics and resulting defect structure depends on two main parameters, namely:

1)The initial velocity of the bombarding particle defines the ability of the affected atoms to overcome the potential barrier of vacancy formation and initiation of a moving soliton type object with 1D locally increased density (crowdion). If the energy provided at the initial moment is below the potential barrier threshold, one can expect the formation of a topological localized vibration mode.

2) The direction of the bombardment plays a crucial role in forming the crowdions. Our previous study [4] shows that the bombardment of a Pt_2 molecule along one of the close-packed atomic rows excites a N=2 crowdion when the molecule energy exceeds a certain threshold value of the initial energy. This crowdions move along only one close-packed atomic row. However, if the molecule orients in

parallel to the surface and hits atoms in the 1st and 2nd sites, then the appearance of so-called M=2 crowdion takes place. This mode is characterized by propagation along two adjacent rows simultaneously, namely, the mass and energy transport occurs in both these rows.

Now we explore the dynamics of this crowdion in two adjacent atomic rows, where the M-crowdion propagates. Figure 2 demonstrates normalized atomic displacements x_n/a, where x_n is atomic displacement along the x axis and a is the interatomic distance $a = 3.95$ Å at the initial velocity of the molecule equal to $v_0 = 39$ Å/ps during sufficiently long time (90 ps).

It is seen, that the first short-term stage of crowdion motion is supersonic. During this stage, crowdion propagates for a long distance and intensively emits its energy through phonons. Therefore, at a certain time moments, the crowdions energy becomes insufficient for overcoming the potential barrier and the crowdion transforms into a common soliton with the short lifetime. At the same time, the interstitial atoms with the small but non-zero kinetic energy are in both adjacent rows. Around these atoms a new localized vibration mode is formed with the frequency above the upper limit of the phonon spectrum. For evaluation the conditions of the appearance of an M-crowdion cluster under the influence of a bombarding molecule, we calculate distances of propagation of the excited crowdions (crowdion clusters), when the initial velocities (energies) of a bombarding molecule are varied. Besides, we compare these distances with the one of crowdions initiated by the bombarding of only one atom into site 1 and site 2. These dependencies are depicted in figure 3.

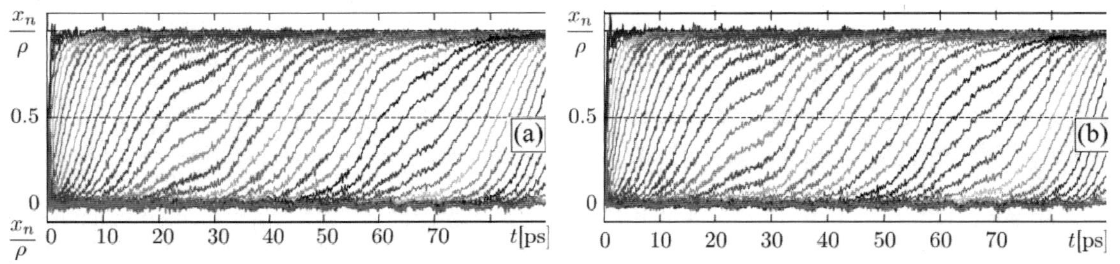

Figure 2. Normalized atomic displacements of the atoms in the close-packed atomic row initiated by a molecule impact simultaneously on site 1 and 2 as the functions of time for a row of the 1st site (a) and for a row in the 2nd site (b). Molecule initial velocities are $V_0 = 39$ Å/ps. Results for T = 10 K. Lines of different colors correspond to the displacements of different atoms.

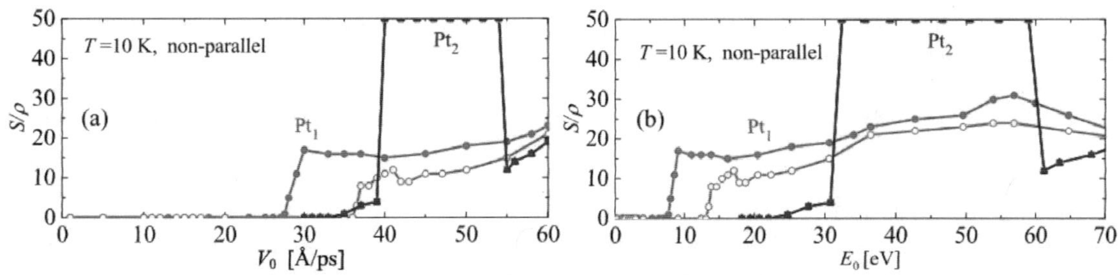

Figure 3. Distance traveled by crowdion initiated by bombardment of the atom (red line) hitting in the 1st site (red dots) and in the 2nd site (white dots) or the molecule (blue squares) bombarding simultaneously in the 1st and 2nd sites as the function of (a) initial velocity and (b) initial energy of atom or molecule. Results for T = 10 K.

The plots in figure 3 show that supersonic crowdions are initiated by an atom at a noticeably lower energy. At the same time, the distance of their propagation is sufficiently short. This causes forming only quasi-1D crowdions, which are not stable in time and transform into immovable interstitial atoms. Unlike the atom bombardment, the molecule hit can lead to the excitation of an M-crowdion, which propagates in two adjacent atomic rows. Its subsonic stage is characterized by the high-level stability.

Hence, the subsonic M-crowdion cluster can exist for a very long time. For this reason, it propagates along the whole lattice despite its slow velocity. However, exceeding a certain value of the initial energy of the molecule leads to the nonperformance of the self-focusing condition of the crowdion and the M-crowdion cluster does not emerge. Only the appearance of several 1D crowdions takes place.

4. Conclusions

The effect of atom/molecule induced 1D mass transfer in a fcc crystal of Pt has been investigated by means of atomistic simulations. It has been shown that the crowdion excitation under the bombardment of a Pt atom occurs for lower values of the initial energy than for the case of Pt_2 molecule hitting. At the same time, the latter can lead to forming the so-called M-crowdion cluster (M=2) in the cristall. The important features of this mode is the very long distance of propagations. Our study have shown that the subsonic M=2 craudion passes through the whole lattice under study. The reason of this phenomenon is that the subsonic M-crowdion cluster, which forms after the supersonic stage of the crowdion motion, is characterized by great stability in comparison with the 1D subsonic crowdion. Oscillations inside the M-crowdion have a frequency higher than the top limit of the phonon band, hence, the crowdion energy cannot be dissipated into the lattice via phonons and remains almost constant in time.

Acknowledgement

I.A.Sh. acknowledges the financial support from the Russian Science Foundation, grant No. 19-72-00109 (obtaining of results, analysis of simulation results). E.A.K. thanks for the financial support the Council of the President of the Russian Federation for state support of young Russian scientists, grant No. MD-3639.2019.2 (problem statement, discussion of model). Work partially supported by State Assignment of Russian Ministry of Science and Higher Education.

References

[1] Dudarev S L 2003 *Philos. Mag.* **83** 3577–97
[2] Tsukanov A and Psakhie S 2016 *Facta Universitatis, Ser.: Mech. Eng.* **14** 269
[3] Li Y, Li Y, Han P, Wang S, Jiang Z and Wang X 2018 *Comput. Model. Eng. Sci.* **117** 109–23
[4] Chen H-Y and Tsou N-T 2019 *Comput. Model. Eng. Sci.* **120** 319–32
[5] Lee W-J, Lo Y-C, Yang A, Chen K and Chen N-Y 2019 *Comput. Model. Eng. Sci.* **120** 293–304
[6] Babicheva R I, Evazzade I, Korznikova E A, Shepelev I A, Zhou K and Dmitriev S V 2019 *Comp. Mater. Sci.* **163** 248–55
[7] Moradi Marjaneh A, Saadatmand D, Evazzade I, Babicheva R I, Soboleva E G, Srikanth N, Zhou K, Korznikova E A and Dmitriev S V 2018 *Phys. Rev.* E **98** 023003
[8] Shepelev I A, Korznikova E A, Bachurin D V, Semenov A S, Chetverikov A P and Dmitriev S V 2020 *Phys. Lett.* A **384(1)** 126032
[9] Shepelev I A, Dmitriev S V, Kudreyko A A, Velarde M G and Korznikova E A 2020 *Chaos, Solitons & Fractals* **140** 110217
[10] Shepelev I A, Chetverikov A P, Dmitriev S V and Korznikova E A 2020 *Comp. Mater. Sci.* **177** 109549
[11] Savin A V, Korznikova E A and Dmitriev S V 2019 *Phys. Rev.* B **99** 235411
[12] Rysaeva L K, Korznikova E A, Murzaev R T, Abdullina D U, Kudreyko A A, Baimova J A, Lisovenko D S and Dmitriev S V 2020 *Facta Universitatis, Ser.: Mech. Eng.* **18** 1–12
[13] Babicheva R I, Dmitriev S V, Korznikova E A and Zhou K 2019 *J. Exp. Theor. Phys.* **129** 66–71
[14] Abdullina D U, Korznikova E A, Dubinko V I, Laptev D V, Kudreyko A A, Soboleva E G, Dmitriev S V and Zhou K 2020 *Computation* **8(2)** 27
[15] Bhui A S, Singh G, Sidhu S S and Bains P S 2018 *Facta Universitatis, Ser.: Mech. Eng.* **16** 337
[16] Barani E, Korznikova E A, Chetverikov A P, Zhou K and Dmitriev S V 2017 *Phys. Lett.* A **381** 3553–7
[17] Zakharov P V, Korznikova E A, Dmitriev S V, Ekomasov E G and Zhou K 2019 *Surf. Sci.* **679** 1–5

[18] Korznikova E A, Bachurin D V, Fomin S Y, Chetverikov A P and Dmitriev S V 2017 *Eur. Phys. J.* B **90** 1-8
[19] Korznikova E A, Shcherbinin S A, Ryabov D S, Chechin G M, Ekomasov E G, Barani E, Zhou K and Dmitriev S V 2019 *Phys. Stat. Sol.* B **256(1)** 1800061
[20] Korznikova E A, Kistanov A A, Sergeev K S, Shepelev I A, Davletshin A R, Bokii D I and Dmitriev S V 2016 *Lett. Mater.* **6(3)** 221-226
[21] Kistanov A, Korznikova E, Fomin S, Zhou K and Dmitriev S 2014 *Lett. Mater.* **4(4)** 315-318
[22] Plimpton S 1995 *J. Comp. Phys.* **117** 1–19
[23] Daw M S and Baskes M I 1984 *Phys. Rev.* B **29** 6443–53
[24] Mendelev M I, Han S, Srolovitz D J, Ackland G J, Sun D Y and Asta M 2003 *Philos. Mag.* **83** 3977–94

X-ray analysis of the Cu-Zn alloys system subjected to ECAP and subsequent rolling

L I Zaynullina[1] and I V Alexandrov[1]

[1] Ufa State Aviation Technical University, K. Marx str. 12, Ufa, 450008, Russia

E-mail: ZaynullinaLI@yandex.ru

Abstract. The results of X-ray diffraction analysis aimed to reveal the features of the evolution of microstructure parameters as a result of equal-channel-angular pressing and subsequent flat rolling of the Cu-Zn alloys system with different Zn (10 and 30 wt. %) contents and, as a result, different stacking fault energies are presented.

1. Introduction

The method of severe plastic deformation (SPD) makes it possible to form bulk ultrafine-grained (UFG) nanostructured states in various metallic materials, characterized by attractive structural and functional properties. Flat rolling is a method of deformation processing that allows one to continue improving the microstructure of these materials and to give blanks the geometric dimensions and shape required for the manufacture of various products [1]. Bulk UFG nanostructured alloys with a lower stacking fault energies (SFE) than that characteristic for the corresponding pure metals are characterized by even smaller grain sizes [2].

The results of microstructure investigations performed by transmission electron microscopy (TEM) as applied to the Cu-10 wt.% Zn and Cu-30 wt.% Zn alloys subjected to ECAP and subsequent flat rolling also indicate the formation of UFG states with nanosized grains and nanotwins [3]. SFE of these alloys differs by 2.5 times. X-ray structural analysis is a powerful tool for investigation of UFG nanostructured metallic materials, which makes it possible to estimate a number of important additional parameters in relation to those established by the TEM method [3].

The purpose of the investigations, the results of which are presented in this paper, is the application of the X-ray structural analysis method for the comparative characterization of the microstructure of Cu-10 wt.% Zn and Cu-30 wt.% Zn alloys subjected to ECAP and subsequent flat rolling.

2. Experimental

The Cu-Zn alloys with a content of 10 and 30 wt% Zn were selected as the material for research in this work. The SFE values of these alloys were ~ 35 $mJ \cdot m^{-2}$ and ~ 14 $mJ \cdot m^{-2}$, respectively. The prepared samples of Cu-10 wt.% Zn and Cu-30 wt.% Zn alloys were subjected to two passes of ECAP at T = 150, 300 °C, respectively, in a tool with an angle of intersection of the channels $\varphi = 90$ ° along the B_C route. Further, the obtained billets, heated to the temperatures selected for the corresponding alloys, were repeatedly passed between the rotating rolls of the rolling mill until a reduction degree of 95% was reached. The choice of different temperatures of deformation processing is associated with the requirement to preserve the integrity of the samples.

X-ray diffraction analysis was performed using a Rigaku Ultima IV diffractometer (the "Bragg-Brentano" scheme at an accelerating voltage of 40 kV and a current of 40 mA at room temperature). The X-ray tube anode was copper, and the corresponding characteristic radiation length was $K_{\alpha 1}$ = 0.15406 nm. X-ray measurements were carried out from a longitudinal horizontal section of the workpieces in the initial states and states after ECAP, and from the plane of the rolled samples.

Before measurements, the surface of the workpieces was polished with diamond suspensions. The MAUD software was used to calculate the crystal lattice parameter (a), the size of the coherent scattering domains (d_{XRD}), elastic microdistortions ($<\varepsilon^2>^{1/2}$) of the crystal lattice, and the dislocation density (ρ) [4]. The concentration of twins (the twins probability) β was estimated by determining the displacement of the diffraction peaks (111) and (200) on the X-ray diffraction patterns of the material under study relative to the initial state [5].

3. Results and analysis

Figure 1 a, b shows the X-ray diffraction patterns of the investigated states of the alloys. All X-ray patterns are characterized by a set of diffraction peaks characteristic of single-phase copper alloys with an FCC lattice. The relative intensity of the peaks changes as a result of broadening and displacement of the peaks due to changes in the microstructure depending on the nature of the deformation treatment.

Figure 1a shows the diffraction patterns of the Cu-10 wt% Zn alloy in the initial state, and states after two passes of ECAP and ECAP with subsequent flat rolling to a maximum reduction degree of 95%. Figure 1b shows the diffraction patterns of the Cu-30 wt% Zn alloy of similar states. As the degree of applied deformation increases, the intensity of the peaks changes. During SPD, the tendency of the change in the peaks (111), (200), and (220) is different for the alloys. In the Cu-10 wt% Zn alloy, the intensity of the (111) peak decreases in the ECAP state with flat rolling with reduction degree of 95%, while in the Cu-30 wt% Zn alloy it increases. The intensity of the (200) peak in both alloys during ECAP with flat rolling with a reduction degree of 95% decreases in comparison with the initial state. There is also a sharp increase in the intensity of the (220) peak in the alloys in comparison with the initial state. This is explained by possible changes in the crystallographic texture of the samples, apparently, the grains acquire preferential orientations.

It should be noted that, in both alloys, the peaks in the diffraction patterns with an increase in the deformation degree are shifted towards small values of the angles 2θ in comparison with the peaks for pure copper, which indicates an increase in the crystal lattice parameter (the lattice parameter of pure copper 0.36150 nm) due to the complete dissolution of Zn atoms in a copper matrix (table 1).

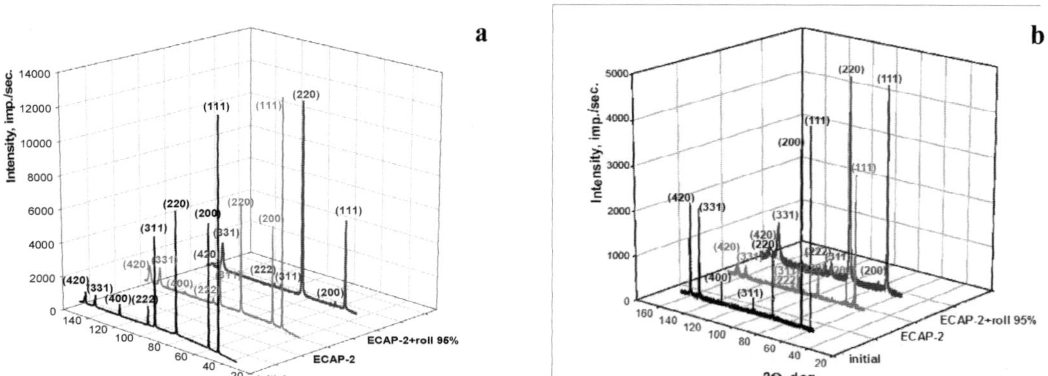

Figure 1. General view of diffraction patterns of Cu-10 wt% Zn alloy (a) and Cu-30 wt% Zn alloy (b) in different structural states.

The results of the following parameters such as a, d_{XRD}, $<\varepsilon^2>^{1/2}$, ρ and β for all realized deformation values are presented in tables 1 and 2.

In the initial coarse-grained state, the lattice parameter of the Cu-10 wt% Zn alloy was 0.36395±0.00003 nm (table 1). Carrying out ECAP and then subsequent flat rolling results in changes in the values of the lattice parameters almost within the measurement error. In this case, ECAP and subsequent flat rolling lead to a significant increase in the dislocation density. As the degree of deformation increases, the d_{XRD} values decrease, the obtained XRD results do not contradict the TEM results (d_{TEM}). There is also a decrease in the $<\varepsilon^2>^{1/2}$ values, which indicates the recovery processes. In [3], the TEM method revealed the formation of nanotwins in the structure as a result of SPD. X-ray diffraction data also indicate the formation of deformation twins in this alloy in various structural states (table 1).

Table 1. Lists of the XRD-measured lattice parameters (a), crystallite size (d_{XRD}), dislocation density (ρ), elastic microdistortion level ($<\varepsilon^2>^{1/2}$), twin probability (β) and average size of the structural elements (d_{TEM}) of the ECAP + flat rolling Cu–10 wt.% Zn sample.

	***a*, nm**	**d_{XRD}, nm**	**ρ, 10^{14} m^{-2}**	**$<\varepsilon^2>^{1/2}$, %**	**β, 10^{-4}**	**d_{TEM}, nm [3]**
Non rolled	0.36395±0.00003	-	0.142	0.031	46.5	$117\cdot10^3\pm5\cdot10^3$
ECAP – 2 p	0.36372±0.00007	180±5	1.0	0.13	41.5	800±66
ECAP – 2 p + roll 30%	0.36354±0.00012	142±5	1.7	0.18	37.8	-
ECAP – 2 p + roll 40%	0.36368±0.00010	79±2	2.9	0.17	40.7	-
ECAP – 2 p + roll 60%	0.36346±0.00011	73±1	3.6	0.19	36.2	-
ECAP – 2 p + roll 80%	0.36388±0.00012	54±1	3.9	0.15	45.0	375±49
ECAP – 2 p + roll 90%	0.36401±0.00009	50±1	3.4	0.13	47.5	292±21
ECAP – 2 p + roll 95%	0.36389±0.00007	45±1	3.6	0.12	45.1	220±30

The strength of the alloy in the initial state is 240 ± 30 MPa, after ECAP it is 413 ± 78 MPa, and after ECAP with subsequent flat rolling with reduction degree of 95%, the strength increases up to 670 ± 120 MPa [3]. Thus, the reason for the increase in the strength of the alloy can be grain refinement, increase of dislocation density, concentration of deformation twins, and the evolution of crystallographic texture with an increase in the strain during SPD.

In the Cu-30 wt.% Zn alloy the lattice parameter is higher than in the Cu-10 wt.% Zn alloy, which is apparently associated with an increased content of alloying atoms in the solid solution. With an increase in the reduction degree during rolling in general, the a value in the Cu-30 wt.% Zn alloy increases (table 2). Apparently, this may be due to the peculiarities of the forming microstructures, in which the d_{XRD} sizes have significantly decreased. The dislocation density increases with the applied strain. In the case of the Cu-30 wt% Zn alloy with a lower SFE after SPD, the twins probability is 2.8 times higher than that of the Cu-10 wt% Zn alloy.

The results of the performed investigations of alloys in UFG states, obtained by ECAP and subsequent flat rolling, indicate an increased probability of twins in an alloy with a high Zn content. This observation correlates with the known results obtained for coarse-grained Cu-Zn alloys with different SFE. At the same time, it is known that twinning processes play an important role in the deformation of coarse-grained Cu-Zn alloys with a lower SFE [6]. In our case, it was found that rolling of alloys in the UFG state does not lead to an increase in the probability of twins (tables 1 and 2). This phenomenon can be explained by the activation of detwinning processes. A similar process, which consists in the transformation of atomically flat coherent twin boundaries into incoherent ones, and then, as the applied strain increases, the transformation of incoherent boundaries into high-angle grain boundaries was previously observed as a result of investigation a Cu-Zn alloy subjected to SPD by high pressure torsion [7]. More solid conclusions on this issue will be made based on the results of planned studies of the evolution of the crystallographic texture in these alloys in the UFG states during flat rolling.

The $<\varepsilon^2>^{1/2}$ values for a given alloy increase with an increase in the deformation degree and remain higher than for a Cu-10 wt% Zn alloy.

The sizes of structural elements (d_{TEM}) estimated by TEM decrease during SPD in comparison with the initial state. However, the difference in the data obtained by TEM and XRD is very significant. This is due to the fact that the X-ray diffraction analysis method determines the d_{XRD} values associated with the inner regions of grains with a slightly distorted crystal lattice, while the TEM method measures the total grain size, which includes highly distorted near-boundary regions [8]. Also, the anisotropy of the grain shape can significantly affect the results obtained by TEM and XRD. The TEM method is used to measure the grain size in a plane parallel to the surface of the sample, and by XRD, the grain size is measured in the direction perpendicular to this plane.

Table 2. Lists of the XRD-measured lattice parameters (a), crystallite size (d_{XRD}), dislocation density (ρ), elastic microdistortion level ($<\varepsilon^2>^{1/2}$), twin probability (β) and average size of the structural elements (d_{TEM}) of the ECAP + flat rolling Cu–30 wt.% Zn sample.

	a, nm	d_{XRD}, nm	ρ, 10^{14} m^{-2}	$<\varepsilon^2>^{1/2}$, %	β, 10^{-4}	d_{TEM}, nm [3]
Non rolled	0.36770±0.00008	-	0.052	0.003	121.1	$93\cdot10^3\pm3\cdot10^3$
ECAP – 2 p	0.36762±0.00019	111±6	2.4	0.19	122.7	290±55
ECAP – 2 p + roll 30%	0.36810±0.00015	79±2	4.3	0.25	132.8	-
ECAP – 2 p + roll 40%	0.36775±0.00014	85±1	2.7	0.17	125.4	-
ECAP – 2 p + roll 60%	0.36782±0.00018	66±2	5.9	0.29	126.8	-
ECAP – 2 p + roll 90%	0.36783±0.00017	38±1	10.3	0.29	127.0	161±7
ECAP – 2 p + roll 95%	0.36793±0.00014	33±1	10.3	0.25	129.2	130±10

The revealed structural changes lead to an increase in strength properties of the Cu-30 wt.%Zn alloy from 320 ± 40 MPa in the initial state to 590 ± 89 MPa after ECAP and 812 ± 130 MPa after ECAP and flat rolling with reduction degree of 95% [3].

4. Conclusions

As a result of the application of X-ray diffraction analysis, the features of the microstructure of the Cu-10 wt% Zn and Cu-30 wt% Zn alloys subjected to ECAP and subsequent flat rolling and having different SFE are revealed.

With an increase in the applied deformation degree, the lattice parameter a of the Cu-10 wt% Zn alloy does not change, while the lattice parameter a of the Cu-30 wt% Zn alloy increases insignificantly. At the same time, the d_{XRD} values in both alloys decrease, while in the Cu-30 wt% Zn alloy the values remain lower. The dislocation density ρ values during ECAP and ECAP with subsequent flat rolling increase for both alloys. The dislocation density in all structured states of Cu-30 wt% Zn alloy is higher. Values $<\varepsilon^2>^{1/2}$ during ECAP increase for Cu-10 wt% Zn and Cu-30 wt% Zn alloys. Subsequent flat rolling changes the data only slightly. However, the values are higher for Cu-30 wt% Zn alloy. These differences can be associated with the temperature conditions during deformation and the SFE values. The values of twin probabilities β of the alloys differ by about 3 times. The data correspond to the previously obtained results of TEM investigations and practically do not depend on the deformation degree. It can be concluded that in the process of SPD and subsequent flat rolling, the processes of twinning and detwinning take place in parallel.

Acknowledgments

Authors are grateful to be supported by the RFBR, project number 19-33-90109 and the Ministry of Science and Higher Education of the Russian Federation within the framework of the project 0838-2020-0006 "Fundamental research of new principles of creating promising electromechanical energy converters with characteristics higher than the world level, with increased efficiency and minimum specific indicators, using new high-performance electrical materials".

References

[1] Valiev R, Islamgaliev R and Alexandrov I 2000 *Progr. Mater. Sci.* **45** 103
[2] Zhao Y, Zhu Y, Liao X, Horita Z and Langdon T 2007 *Mater. Sci. Eng.* A **463** 22
[3] Zaynullina L, Alexandrov I and Wei W 2019 *IOP Conf. Ser.: Mater. Sci. Eng.* **672** 012067
[4] Lutterotti L, Matthies S and Wenk H 1999 *Proceeding of the 12th International Conference on Textures of Materials* **1** p 1599
[5] Cohen B and Wagner C 1962 *J. Appl. Phys.* **33** 2073
[6] Zhao Y, Horita Z, Langdon T and Zhu Y 2008 *Mater. Sci. Eng.* A **474** 342
[7] Wang Y, Liao X, Zhao Y, Lavernia E, Ringer S, Horita Z, Langdon T and Zhu Y 2010 *Mater. Sci. and Eng.* A **527** 4959
[8] Valiev R, Zhilyaev A and Langdon T 2013 *Bulk Nanostructured Materials: Fundamentals and Applications* (John Wiley & Sons)

Calculation of hardening contributions of the TiNi alloy undergoing martensitic transformations in a free state

A A Churakova[1,2], D V Gunderov[1,2], N E Tolstov[2], D K Magomedova[3]

[1]Institute of Molecule and Crystal Physics, Ufa Federal Research Center RAS, Ufa, Russia
[2]Ufa State Aviation Technical university, Ufa, Russia
[3]Saint-Petersburg State University, Saint-Petersburg, Russia

E-mail: churakovaa_a@mail.ru

Abstract. The paper reports on the studies of the effect of thermal cycling on the microstructure of the Ti - 50.8 at. % Ni alloy in coarse-grained and ultrafine-grained states, mechanical properties, and calculation of contributions to hardening in the studied states. The comparison of the calculated and experimental values of the yield stress in the TiNi alloy was carried out.

1. Introduction

The TiNi shape memory alloys are widely used due to their excellent functional properties [1,2]. TiNi alloy products and elements usually undergo a cyclic phase (martensitic) transformation between B2 (A) austenite and B19' (M) martensite during use. Consequently, it can be said that the stability of the phase transformation during multiple transformations is of decisive practical importance. The defects, mainly dislocations, which tend to suppress the martensitic transformation (MT), are introduced during the cyclic thermally induced transformation A↔M [3–5]. A shift of the MT temperatures towards lower ones is often observed during thermal cycling [6, 7]. After a certain number of thermal cycles, the transformation behavior can be stabilized by saturating the defects caused by thermal cycling [5].

The design of products with SME imposes certain requirements on the physical, mechanical, and functional properties and their stability. It is possible to additionally improve the properties of shape memory alloys by forming an ultrafine-grained (UFG) state in them by the methods of severe plastic deformation (SPD), in particular, by equal-channel angular pressing (ECAP), by the method of high pressure torsion (HPT), or by a combination of various methods [8-13]. Both deformation and cyclic effects have a significant effect on the microstructure and, accordingly, on the mechanical behavior of the material and on the yield stress σ_{YS}. To analyze the mechanical properties with different microstructures in alloys, the calculation of the hardening contributions according to the known equations are performed. In particular, the Hall - Petch relationship is used, which is well studied for single-phase and two-phase alloys [14-15], but was little analyzed for materials with different grain sizes that undergo stress-induced martensitic transformation during tensile tests before the onset of plastic deformation. Several studies on alloys undergoing stress induced transformation report on the stress required to trigger the transformation (trigger stress) [16–18]. But it is worth noting that the stress required for transformation under load is significantly less than the yield stress, and probably should not affect the Hall-Petch relationship.

In this work, the contributions to hardening are considered for the example of the Ti - 50.8 at. % Ni alloy with a high nickel content undergoing cyclic martensitic transformation.

2. Experimental

The material for the study was a binary alloy of the TiNi - Ti - 50.8 at. % Ni. To obtain a solid solution, quenching was carried out from the homogeneity region (heating at a temperature of 800 ° C in a Nabertherm muffle furnace for 1 hour) into water. The average grain size of the hardened alloy was about 20 ± 2 microns. An ultrafine-grained (UFG) structure with an average grain size of 350 ± 10 nm was obtained by the ECAP method (USATU equipment) according to the mode - 8 passes, route B, T = 450 ° C, the angle of intersection of the channels (φ) - 120 ° [19].

Cyclic martensitic transformations were carried out according to the regime described in [20-22]. X-ray diffraction studies of the samples were carried out on a Rigaku Ultima IV diffractometer (U = 40 kV and I = 35 mA) at room temperature in the angle range 2θ = 30 - 120 °. The dislocation density was calculated by processing the X-ray structural study data using the MatLab software. The study of the microstructure was carried out on a JEOL JEM-2100 transmission microscope, foils for the study were obtained using a Tenupol-5 setup in a solution of 10% perchloric acid and 90% butanol. Mechanical tensile tests were carried out at a strain rate of 10^{-3} s^{-1} at room temperature. In the dislocation theory, the main hardening mechanisms providing an increase in the plastic flow stress are divided into [23]: hardening by dissolved interstitial or substitutional atoms, hardening by dislocations, hardening by grain and subgrain boundaries, hardening by dispersed particles. E. Orowan [24] found that the superposition of each of the hardening mechanisms with the lattice friction stress is linearly additive; the yield point of a hardened material is the sum:

$$\sigma_{0,2} = \sigma_0 + \Delta\sigma_{s.s} + \Delta\sigma_{\text{d.p.}} + \Delta\sigma_{\text{d}} + \Delta\sigma_g, \quad (1)$$

where σ_0 is the frictional stress of the crystal lattice; $\Delta\sigma_{s.s.}$– increase in yield strength due to solid solution hardening; $\Delta\sigma_d$ - increase in the yield strength due to dislocation (strain) hardening; $\Delta\sigma_g$ is the increase in the yield point due to grain boundary hardening; $\Delta\sigma_{d.p.}$ - an increase in the yield stress due to precipitation hardening [23].

The frictional stress in the crystal lattice (Peierls-Nabarro stress) is calculated by the formula:

$$\sigma_0 = [2G/(1-\nu)]\exp[-2\pi/(1-\nu)], \quad (2)$$

where ν is the Poisson's ratio, G is the shear modulus of the matrix [23].

Solid solution hardening is described by the Mott-Nabarro mechanism according to the formula:

$$\Delta\sigma_{\text{s.s.}} = 2{,}5G\delta_L^{3/4}C_L, \quad (3)$$

where δ_L is the parameter of the dimensional mismatch between the atoms of the dissolved element and the matrix, C_L is the atomic concentration of the alloying element [23].

Strengthening in the case of incoherent precipitates is described by the Orowan mechanism and is calculated by the formula:

$$\Delta\sigma_{\text{d.p.}} = 0{,}85\frac{Gb}{l-D}, \quad (4)$$

where b is the Burgers vector, l is the average distance between the centers of particles, D is the average size of particles [23].

The qualitative dependence of the plastic flow stress on the dislocation density is described by the following formula:

$$\Delta\sigma_{\text{d}} = \alpha mGb\sqrt{\rho_{\text{d}}}, \quad (5)$$

where α is the parameter of interdislocation interaction, m is the orientation coefficient, G is the shear modulus of the matrix, ρ_d is the dislocation density [23, 25].

A decrease in the grain size contributes to an increase in the density of grain boundaries in polycrystalline materials, that prevent the development of plastic deformation, since grain boundaries are effective barriers to the movement of dislocations. Grain-boundary hardening is described by the Hall-Petch relationship:

$$\Delta\sigma_g = k_3 d^{-m}, \quad (6)$$

where k_g is the coefficient of grain-boundary hardening, d is the average grain size, m is the coefficient of the angle of misorientation.

3. Results and discussion

According to the obtained TEM data in the CG state without thermal cycling grain boundaries and triple junctions of grains free from dislocations are observed in the microstructure of the alloy, the average grain size is about 20 ± 5 μm (Figure 1, a). Thermal cycling with the maximum number of heat changes retains the dislocation structure in the form of clusters and disordered walls and dislocation tangles (Figure 1, b). After applying the ECAP method, the transformation of the initial coarse-grained structure into an inhomogeneous grain-subgrain ultrafine-grained structure with an increased density of dislocations is observed (Figure 1, c). As the number of martensitic transformation cycles increases, grains with predominantly nonequilibrium boundaries are observed in the structure, which may indicate a high imperfection of the structure, in the form of an accumulation of dislocations; after the maximum number of thermal changes, it decreases to 260 ± 20 nm (Figure 1, d).

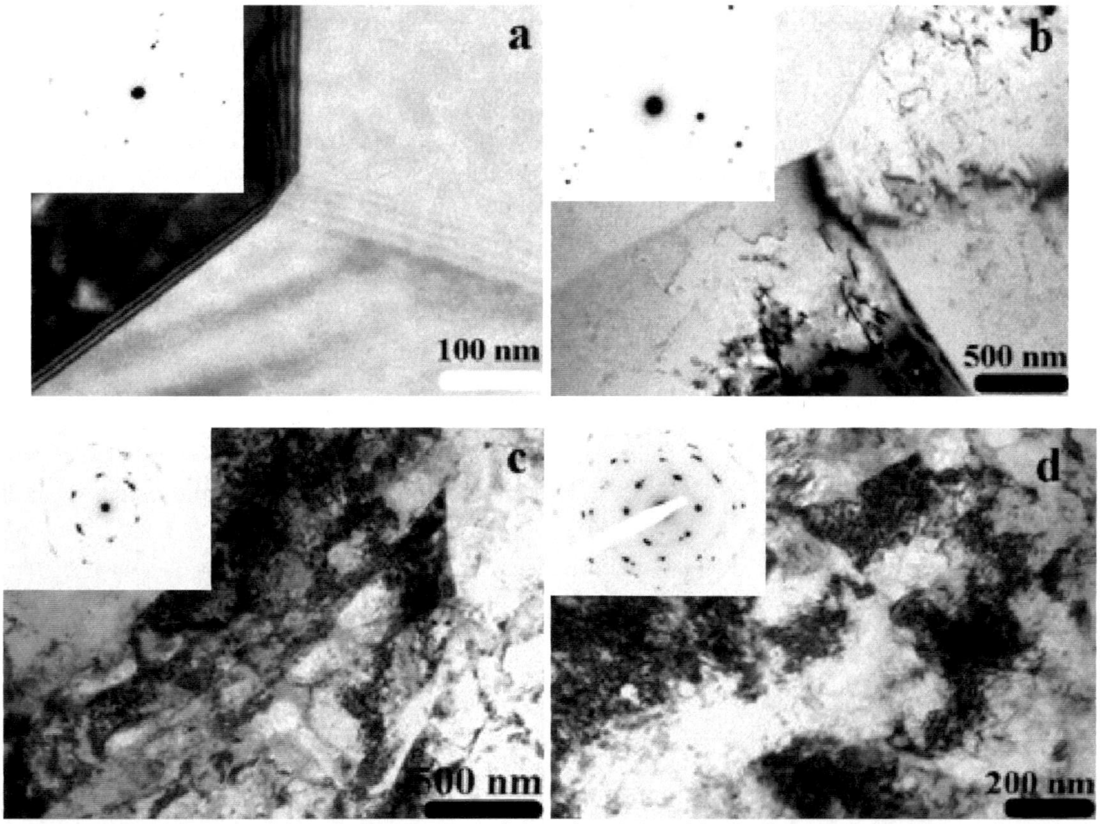

Figure 1. TEM images of the microstructure of the Ti - 50.8 at.% Ni alloy in the coarse-grained (a, b) and ultrafine-grained states (c, d): a, c - n = 0 cycles, b, d - n = 250 cycles.

Analysis of the structural parameters showed that in both states there is a decrease in coherent scattering regions (CSR) values, an increase in internal microdistortions, and an increase in the dislocation density associated with them. The increase of the dislocation density in the ultrafine-grained state is higher than in the coarse-grained state, which indicates that a high density of grain boundaries and a smaller grain size contribute to the intensity of defect accumulation.

Table 1.- Parameters of the structure of the Ti - 50.8 at.% Ni alloy in various states

State	Parameters of lattice a, Å	CSR, nm	$<\varepsilon^2>^{1/2}\times10^{-4}$	$\rho\times10^{15}$, m^{-2}
CG	3,013±0,001	550	0,8	0,5
CG+TC	2.895±0,001 (monoclinic)	320	2,2	1,6
UFG	3,011±0,003	170	2,7	5,3
UFG+TC	3,013±0,001	120	3,4	7,1

Table 2 shows the data of mechanical tensile tests and calculations of contributions to hardening by formulas (1-6). Since the concentration of alloying elements is exceedingly small, the value of solid solution hardening is equal to 0 MPa, similarly, the value of the contribution from incoherent particles is equal to zero. Therefore, the main contributions to hardening are the Peierls-Nabarro stress, dislocation hardening, and grain-boundary hardening described by the Hall-Petch relationship.

Table 2. Results of mechanical tensile tests of the Ti - 50.8 at.% Ni alloy in various states *

State	Ultimate strength σ_{UTS}, MPa	Calculated yield strength, σ_{YS}, MPa			Yield strength, σ_{YS}, MPa	Phase yield stress, σ_M, MPa	Estimated reactive stress σ_r, MPa	Elongation δ,%	Reversible** deformation ε_{rev},%
		σ_0	σ_d	σ_g					
CG	940	5.146	92.809	292.609	495	320	175	50	4,0
		390.5							
CG+TC n=250	1075	5.146	166.023	383.614	685	505	180	55	5,1
		554.7							
UFG	1155	5.146	302.166	526.314	690	295	340	30	4,2
		833.6							
UFG+TC n=250	1075	5.146	349.733	626.439	765	460	305	40	5,7
		981.3							
* Errors in determining σ_{YS} - ± 5%, σ_{UTS} - ± 5%, σ_m - ± 5%, δ - ± 5%, ε_{rev} - ± 10%									
** Estimated - along the length of the plateau at the stage of phase yield stress									

The most sensitive characteristic to thermal cycling is the yield stress. In both states of the alloy, it increases with an increase in the number of cycles. The UFG state before TC is characterized by higher values of strength and yield stress due to the contribution of grain boundary hardening. The ultimate strength in the coarse-grained state increases to 150 cycles, then the values decrease. This is probably since after a certain number of cycles (in this state n = 150), saturation occurs in the material and no further increase in characteristics is observed. In the UFG state, with an increase in the number of cycles of multiple transformations, the values of the ultimate strength increase to 100 thermal cycles, then a decrease in the parameter is observed, which can be explained by the inhomogeneity of the formed structure.

The greater difference between the calculated value of the yield stress and the results of mechanical tests in states after thermal cycling may be associated with a more complex structure formed during thermal cycling. The differences for the ultrafine-grained state can also be explained by the fact that a complex defect structure is formed in the samples, which complicates the flow of dislocations during tensile tests.

4. Conclusions

As a result of thermal cycling in the coarse-grained and ultrafine-grained states, an increase in the density of defects and a decrease in the size of structural elements took place. These factors have affected the mechanical properties of the TiNi alloy. The calculation of the contributions to strengthening and comparison with the test results showed that for the coarse-grained state the difference in theoretical and predicted values is not so significant, while for the ultrafine-grained state the difference between the calculated values and the experimental ones is more than 20%.

Acknowledgments The study is supported by the Russian Science Foundation, project No. 20-72-00075.

References

[1] Van Humbeeck J 1999 *Mater. Sci. Eng.* A **273-275** 134–48
[2] Otsuka K and Kakeshita T 2002 *MRS Bull.* **27** 91–8
[3] Zhang J, Somsen C, Simon T, Ding X, Hou S, Ren S, Ren X, Eggeler G, Otsuka K and Sun J 2012 *Acta Mater.* **60** 1999–2006
[4] Simon T, Kroger A, Somsen C, Dlouhy A and Eggeler G 2010 *Acta Mater.* **58** 1850–60
[5] Miyazaki S, Igo Y and Otsuka K 1986 *Acta Metall.* **34** 2045–2051
[6] Wang X, Verlinden B and Van Humbeeck J 2015 *Intermetallics* **62** 43–9
[7] Wagner M.F.-X, Dey S R, Gugel H, Frenzel J, Somsen C and Eggeler G 2010 *Intermetallics* **18** 1172-79
[8] Valiev R Z and Alexandrov I V 2007 (Moscow: Akademkniga) [in Russian]
[9] Valiev R Z, Islamgaliev R K and Alexandrov I V 2000 *Progr. Mater. Sci.* **45** 103–89
[10] Tatyanin E V, Kurdyumov V G and Fedorov V B 1986 *Fizika Metallov i Metallovedenie* **62(1)** 133-37 [in Russian]
[11] Valiev R Z, Gunderov D V and Pushin V G 2005 *J. Metast. Nanocr. Mater.* **24–25** 7–12
[12] Prokoshkin S D, Khmelevskaya I Yu, Dobatkin S V, Trubitsyna I B, Tatyanin E V, Stolyarov V V and Prokofiev E A 2004 *Fizika Metallov i Metallovedenie* **97(6)** 84–90 [in Russian]
[13] Valiev R Z, Gunderov D V, Lukyanov A V and Pushin V G 2012 *J. Mater. Sci.* **47(22)** 7848–53
[14] Hall E O 1951 *Proc. Phys. Soc.* **64** 747
[15] Petch N J 1953 *J. Iron Steel Inst.* **174** 25
[16] Sure G N and Brown L C 1984 *Metall. Trans.* A **15** 1613
[17] Leherute P, Eberhardt A and Phlippe M J 2005 *Mater. Sci. Eng.* **396** 223
[18] Montecinos S, Cuniberti A and Sepulveda A 2008 *Mater. Charact.* **59** 117
[19] Gunderov D, Churakova A, Lukyanov A. Prokofiev E, Pushin V, Kreitcberg A and Prokoshkin S 2017 *Mate. Today: Proc.* **4(3)** 4825-29
[20] Churakova A A, Gunderov D V and Dmitriev S V 2018 *Materialwissenschaft und Werkstofftechnik* **49 (6)** 769-83
[21] Churakova A and Gunderov D 2017 *Mater. Today: Proc.* **4(3)** 4732-4736.
[22] Churakova A and Gunderov D 2020 *Metals* **10(2)** 227
[23] Prikhodko V M 2003 *Metallophysical Bases of Development of Hardening Technologies* (Moscow: Mechanical Engineering) [in Russian]
[24] Orowan E 1954 *Dislocations in Metals* ed. J S Koehler, F Seitz, W T Read, W. Shokley and E Orowan (New York: AIME) p 69
[25] Goldstein M I, Litvinov V S and Bronfin B M 1986 *Metal Physics of High-Strength Alloys* (Moscow: Metallurgy) [in Russian]

IOP Conf. Series: Materials Science and Engineering 1008 (2020) 012039 doi:10.1088/1757-899X/1008/1/012039

Mechanical properties and structure of welded joints of VT6 alloy through nanocrystalline interlayer of VT22

M Mukhametrakhimov

Institute for Metals Superplasticity Problems, Russian Academy of Sciences, Ufa, Russia

E-mail: msia@mail.ru

Abstract

Experimental studies of the solid-phase weldability of the VT6 alloy through an interlayer of VT22 alloy in the temperature range of the low-temperature superplasticity (T=820 °C) of one of the elements to be joined have been carried out. Welding of dissimilar titanium alloys allows us to significantly reduce the temperature of pressure welding due to the use of alloys alloyed with different elements. The VT22 with nanocrystalline structure interlayer allows localizing deformation at the junction zone due to a significant difference between the values of the interlayer flow stress and welded billets. To improve the mechanical properties, the welds were heat-treated at T=900 °C. The optimal properties of welded joints are increased after the heat treatment of researched titanium alloys. This leads to the activation of diffusion processes and the disappearance of micropores. Thus, it can be concluded, that getting dissimilar titanium alloys high-quality solid-phase joint depends not only on the size of the interlayer grains but also on the material in the joint zone chemical and phase composition.

1. Introduction

An increase in demand for new technology and the development of modern mechanical engineering are associated with the search for new cost-effective technologies for forming parts of complex configuration from traditionally difficult materials. Pressure welding combined with superplastic forming (SPF / SD) is one of the perspective technological areas of mechanical engineering [1, 2].
The use of the phenomenon of low- temperature superplasticity [3] opens up a way for lowering the temperature of the thermal deformation treatment of metals. The problem of reducing the treatment temperature is especially urgent in the production of articles from titanium alloys.

The solution to this problem is possible by using a process of solid-phase deformation welding of bulk billets in a nanostructured state [4] based on the effect of low-temperature superplasticity. Nanostructuring of titanium alloy VT6 allows to significantly reduce the temperature of superplastic deformation [5, 6] and observe the phenomenon of "low-temperature superplasticity". The lower temperature boundary of manifestation of superplasticity depends on the grain size of the treated alloy and decreases upon its decease. The creation of nanocrystalline structure in the titanium alloys can be accompanied by the manifestation of low-temperature superplasticity [7, 8]. Isothermal rolling of two-phase titanium alloy VT22 in low-temperature superplasticity ensured nanostructuring directly during sheet production without preliminary deformation treatment [9, 10]. In this connection, the creation of high mechanical properties in a joint obtained by pressure welding in the state of superplasticity (SP), which depends both on the temperature and on the structural state of the treated alloy, is an interesting metal science problem due to the experimental results.

The aim of the present work is considered in evaluating the mechanical properties of specimens obtained by joining a typical two-phase titanium alloy VT6 through the alloy interlayer VT22 in the temperature range of the low-temperature superplasticity.

2. Materials and experimental procedure

The rod, 30 mm in diameter out of two-phase titanium micro-grain alloy VT6, with an average grain size of α-phase 3-5 μm, standard chemical composition was used in the experiment. Sheets of nanostructured VT22 alloy with a thickness of 2 mm, containing structural elements with dimensions of not more than 0.3-0.5 μm [11] were obtained by isothermal rolling. The chemical composition of alloys is given in table 1.

Table 1. Chemical composition of the studied alloys.

Brend	Ti	Al	Mo	V	Zr	Cr	Fe	O	Si	C	N	H
VT6	basic	5.3÷6.5	-	3.5÷5.1	0.3	-	0.25÷0.3	0.15÷0,2	-	0.1	0.05	0.015
VT22	79.4-86.3	4.4-5.7	4-5.5	4-5.5	0.3	0.5-1.5	0.5-1.5	0.18	0.15	0.1	0.05	0.015

The mechanical properties of allows in initial conditions at room temperature are presented in table 2.

Table 2. Mechanical properties of alloy VT6 and VT22 at T= 20°C.

State of the alloy	σ_u, MPa	$\sigma_{0.2}$, MPa	δ, %	ψ, %
VT6	1080	1010	12	18
VT22	1420	1380	7.5	14

The microstructure of the analyzed titanium alloys in initial states is shown in figure 1.

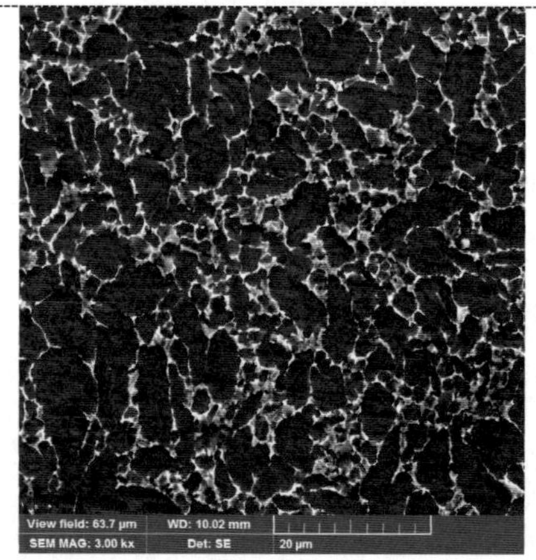

a

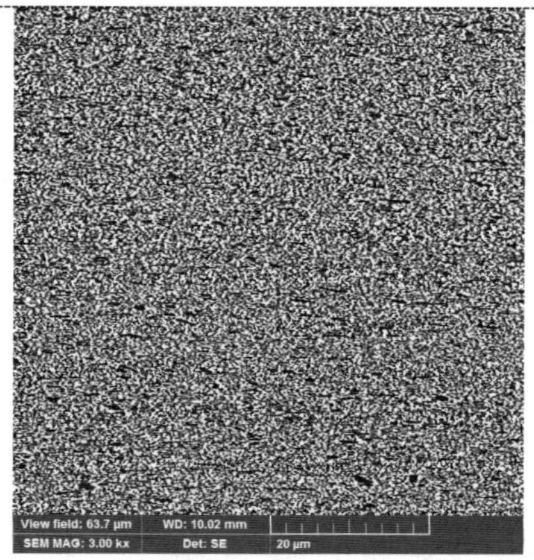

b

Figure 1. VT6 alloy microstructure in the initial state (a) and VT22 alloy microstructure after isothermal rolling (b).

Pressure welding of specimens with a total length of 58 mm from alloy VT6 through interlayer VT22 was carried out in a vacuum furnace under conditions of low-temperature SP (figure 2*a*). Welding was carried out at T=820 °C for 120 minutes with constant pressure (P = 3 MPa) and followed by heat treatment at T=900 °C. The vacuum depth during the experiment was not less than P = 2.0×10^{-3} Pa. The method of welding of the studied materials is described in [12].

The quality of the joint was estimated metallographically with the help of a TESCAN MIRA3 LMU scanning electron microscope in terms of the relative volume fraction of pores in the cross-section of the weld zone and using the results of tensile mechanical tests.

The surface of the joint was oriented perpendicularly to the tension axis in the central part of the specimen. The strength characteristics of the welded specimens were evaluated not only after pressure welding but also after heat treatment. Specimens with a diameter of 3 mm and a length of 10 mm were used to study mechanical properties (figure 2*b*). The mechanical properties of the alloy were determined by tensile tests at least three specimens in the "Instron 5982" testing machine at room temperature and deformation rate 1 mm/min.

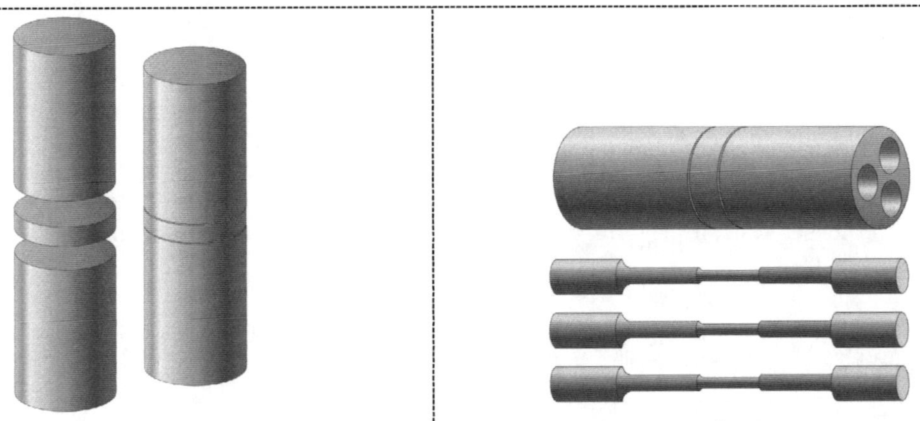

a b

Figure 2. Sample pressure welding diagram through interlayer (a) and cut after welding (b).

3. Results and discussions

As a result of pressure welding, a solid-phase joint was formed, two interfaces can be distinguished, - between the titanium alloys VT6 and VT22 (zone VT6 + VT22) and between the titanium alloys VT22 and VT6 (zone VT22 + VT6) (figure 3). The interface can be identified due to the different size of the grains in the rod and the interlayer.

Metallographically studies in the solid-state zones of the specimen after pressure welding revealed pores, at the degree of deformation ε = 1-2%. At the α- grain boundaries, separate pores and a chain of pores are visible (figure 3*a*). The relative long-range areas of the pores are 0.32. As a result of heat treatment at T=900 °C of these specimens, the relatively long-range areas of the pores decreased to 0.26 (figure 3*b*). The decrease in porosity in the bonding zone is probably due to the activation of diffusion, in particular caused by the structural gradient at the interface of the welded joints. The intensity of diffusion processes is evidenced by the significant growth of grains in the interlayer.

a b

Figure 3. Microstructure of solid-state zones of the VT6 specimens obtained by pressure welding at T=820 °C (a) through VT22 interlayer and the further heat treatment during 2 hours at T=900 °C (b).

IOP Conf. Series: Materials Science and Engineering 1008 (2020) 012039 doi:10.1088/1757-899X/1008/1/012039

The results of mechanical tensile welded specimens through a nanolayer are given in table 3.

Table 3. Tensile strength indices welded specimens through a nanolayer after mechanical tensile tests at room temperature

$T_{joining}$, °C	σ_u (MPa)	δ,%	ψ,%	Note
T=820°C	975	4,52	2, 1	VT6 +VT22+VT6
T=820°C + heat treatment at T= 900°C	1108	4,61	4,2	

As the results of mechanical tensile tests after pressure welding at temperature T = 820 °C indicates, the strength of the solid-phase joint was 975%, and after heat treatment increasing to 1108%. After tensile tests, neck formation is practically not observed, as confirmed by the small value of relative necking.

The fractograms of the central part of the failure are presented in figure 4.

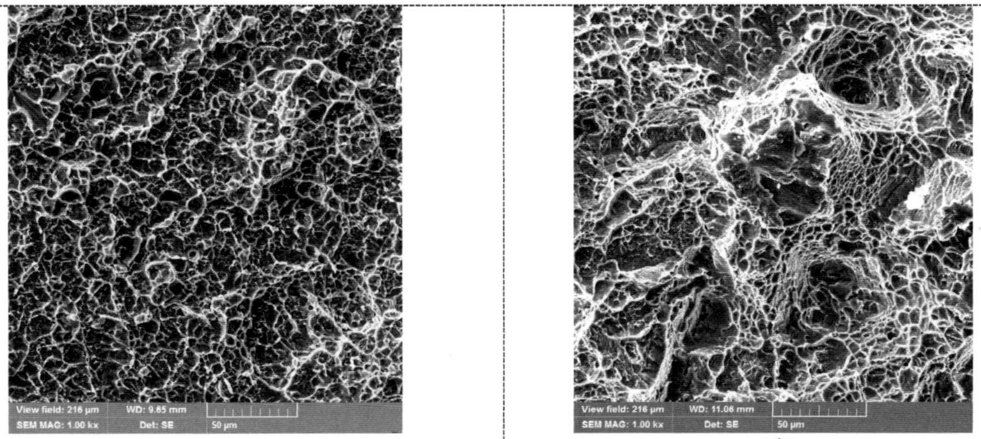

a b

Figure 4. Fraktogramma of the specimens obtained by pressure welding at T=820 °C (a) and the further heat treatment during 2 hours at T=900 °C (b) after mechanical tensile tests.

A fractographic study of specimen fractures indicates, after pressure welding at a temperature of 820°C, the crack initiation zone is an intergranular fracture with small pits on the fracture facets. The size of the pits in the fractures correlates with the sizes of the structural components in the samples. After heat treatment, the nature of the fracture changes sharply (intergranular - intragranular): the general view of the fracture has a mixed structure, including areas with flat fragments - quasi-cleavage facets and viscous fracture areas. And this leads to an increase in the strength of the welded joint.

Thus, the possibility of obtaining an equal strength solid-state of specimens from a bulk industrial titanium alloy VT6 through a layer of alloy VT22 at a relatively low temperature (T = 820 °C) and further heat treatment at 900 °C was shown.

4. Conclusions

The possibility of obtaining of equal hardness solid-state joints of the VT6 specimens obtained by pressure welding through VT22 interlayer in temperature – strain rate regime of low-temperature superplasticity is experimentally shown. To improve the complex of mechanical properties of solid-state joints, specimens were heat-treated at T=900 °C. This leads to the activation of diffusion processes and the disappearance of micropores.

The work was carried out within the framework of the state assignments of the IMSP RAS No. AAAA-A17-117041310221-5.

References

[1] Mulyukov R R, Imayev R M, Nazarov A A, Imayev M F and Imayev V M 2014 *Sverkhplastichnost' Ul'tramelkozernistykh Splavov: Eksperiment, Teoriya, Tekhnologii* (Moscow: Nauka) p 284 [in Russian]
[2] Mulyukov R, Safiullin R, Kruglov A, Lutfullin R, Inozemtsev A, Koksharov N and Kharin S 2009 *Nanotechnology and Nanomaterials in Perm Region* (Perm: CNTI) p 61 [in Russian]
[3] Kaibyshev O, Salishchev G, Galeyev R, Lutfullin R and Valiakhmetov O 1998 Patent PCT/US97/18642, WO 9817836, 30.04
[4] Lutfullin R, Kaibyshev O, Valiakhmetov O, Mukhametrakhimov M, Safiullin R and Mulyukov R 2003 *J Adv. Mater.* **4** 326
[5] Salischev G, Galeyev R and Valiakhmetov O 2001 *J. Mater. Process. Technol.* **116** 165
[6] Lutfullin R, Kruglov A,. Safiullin R, Mukhametrahimov M and Rudenko O 2009 *Mater. Sci. Eng.* A **503** 52
[7] Klassman E, Klassman P and AstaninV 2011 *Perspektivnye Materialy* **12** 218 [in Russian]
[8] Klassman E, Astanin V and Klassman P 2009 *Perspektivnye Materialy* **7** 14 [in Russian]
[9] Lutfullin R, Kruglov A, Mukhametrakhimov M, Klassman P and Astanin V 2013 *Perspektivnye Materialy* **15** 228 [in Russian]
[10] Lutfullin R and Mukhametrakhimov M 2009 *J. Adv. Mater.* **7** 189
[11] Klassman E and Astanin V 2020 *Lett. Mater.* **10** (1) 10
[12] Ganeeva A, Lutfullin R and Kruglov A 2009 *Perspektivnye Materialy* **7** 79 [in Russian]

IOP Conf. Series: Materials Science and Engineering 1008 (2020) 012040 doi:10.1088/1757-899X/1008/1/012040

Silver nanoparticles formation over Ti-Zr-Nb alloy *via* polyol synthesis

T O Teplyakova, D V Popova, S D Prokoshkin and A S Konopatsky

National University of Science and Technology "MISIS", Leninskiy prosp. 4, Moscow 119049, Russia

E-mail: teplyakova.to@mail.ru

Abstract. Ag nanoparticles (AgNPs) were formed over the surface of novel superelastic Ti-Zr-Nb alloy via polyol synthesis at room temperature. Two synthesis media were used for AgNPs formation: ethylene glycol (EG) and polyethylene glycol (PEG-400). AgNPs formation rate in different polyols was evaluated by UV-VIS spectroscopy. Based on the obtained results higher formation rate was ascribed to PEG-400. Morphology of the material's surface, as well as chemical composition, were analyzed by scanning electron microscopy (SEM) and energy dispersive spectroscopy (EDS). It was demonstrated that the usage of PEG-400 as synthesis media facilitates the formation of more monodisperse AgNPs evenly distributed over the alloy's surface.

1. Introduction

Shape memory alloys (SMA) draw significant attention as the materials for bone tissue replacement [1,2]. These alloys demonstrate superelastic behavior at room temperature (RT) as well as high corrosion resistance which makes them perfect candidates for bone tissue replacement material [3,4]. Recently a novel SMA Ti-18Zr-15Nb with enhanced functional properties was developed [5].

The success of implantation is determined by several factors, the most important of which is the absence of inflammatory processes caused by a bacteria activity at the interface between the bone tissue and the metallic material [6]. For this reason, the development of ways to provide the surface of the implant material with antibacterial properties is an important scientific task.

Silver nanoparticles are well-known for their antibacterial properties. There are many methods of silver coatings and NPs formation over various substrates such as magnetron sputtering [7], electrochemical methods [8], plasma spraying [9], acid- or alkali-heat treatment [10], and so on. In the case of superelastic titanium alloys implemented as a substrate, it is essential to develop specific AgNPs formation methods eliminating the needs of elevated temperatures and even strong reducing agents. Such a feature is explained by the fact that the superelastic properties of Ti alloys depend strongly on their structure and phase composition. For many prospective systems such as Ti-Zr-Nb secondary phases precipitation such as ω-phase takes place at a quite low temperature (300 °C) [11, 12]. Uncontrolled ω-phase formation leads to embrittlement of the material and superelastic properties decrease.

Polyol synthesis is a method of AgNPs formation where the reduction process can be controlled solely by organic compounds – polyols. This method is extensively used for the synthesis of various transition metals nanoparticles and bimetallic systems [13, 14]. The development of this method enabled many synthesis possibilities to control metallic NP's structure, morphology, concentration on the surface of the substrate, and their chemical composition [15]. This makes polyol synthesis an attractive technique for the task of AgNPs formation over the surface of superelastic Ti alloys.

Synthesis media and synthesis time have a significant influence on the structure and the properties of the final product. For that reason, in our current work, we investigate the influence of different synthesis media such as ethylene glycol (EG) and low-molecular polyethylene glycol (PEG-400) and synthesis time on the structure of the AgNPs deposited over the surface of superelastic Ti alloy. One of the most prospective alloys of the Ti-Zr-Nb system Ti-18Zr-15Nb was selected as the substrate for the current study.

2. Materials and methods

Small plates (10x5x2 mm) of the superelastic Ti-18Zr-15Nb alloy were used as a substrate for the formation of silver nanoparticles. Plates were subjected to thermomechanical treatment (cold rolling with 25% thickness reduction rate and post-deformational annealing at 550°C for 30 min) enhancing superelastic properties of the material [16]. Surface preparation was carried out by grinding with abrasive paper (up to P2000) and polishing on a polishing fabric using a diamond suspension. Obtained in such manner samples were subjected to sonication in an ultrasonic bath filled with isopropanol alcohol before further manipulations.

For typical synthesis, the glass with a certain amount of the polyol was placed on a magnetic stirrer at 1000 rpm. The substrate was immersed in the media and a freshly prepared $AgNO_3$ solution was quickly added. The synthesis continued for a required period, after which the substrate was taken out and placed in a glass with distilled water. The glass was placed in an ultrasonic bath for 30 minutes to remove residual polymer and secondary synthesis products from the surface of the alloy. Obtained in such manner samples were dried overnight at ambient air. Table 1 collects the sample description.

Table 1. Description of the obtained samples.

Sample	Synthesis media	Time, min
TiAg_EG_1	EG	1
TiAg_EG_10	EG	10
TiAg_PEG_1	PEG-400	1
TiAg_PEG_10	PEG-400	10

XRD analysis was performed on the Dron-3 diffractometer. $Cu_{k\lambda}$ radiation with a wavelength of 1.5406 Å was used. The influence of synthesis time on AgNPs formation rate was evaluated by optic spectroscopy using OceanOptics Flame equipment. The surface structure of the obtained samples was analyzed using a scanning electron microscope (SEM) JEOL 5600. The chemical composition of the material was studied by energy dispersive spectroscopy (EDS).

3. Results and discussion

Figure 1 shows the typical XRD pattern of Ti-18Zr-15Nb alloy after AgNPs deposition. It can be seen that the alloy contains only a high-temperature β-phase, which can be determined by four lines (110), (200), (211), (220).

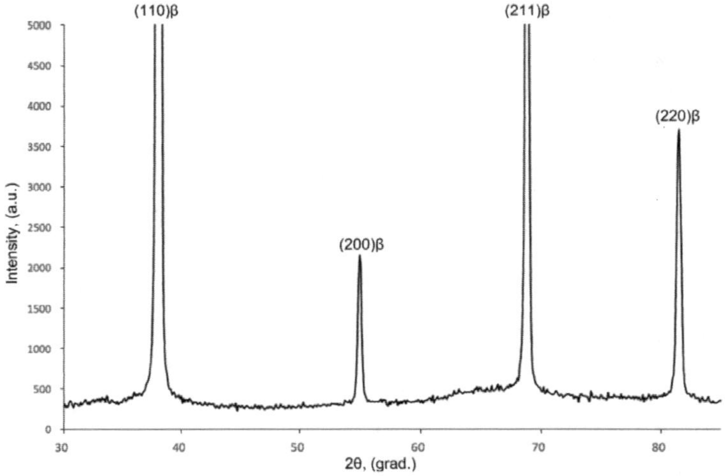

Figure 1. XRD pattern of Ti-18Zr-15Nb alloy after AgNPs deposition.

The absence of signals characteristic for ω-phase confirms that the selected synthesis scheme does not lead to changes in the phase composition of the alloy. It should be noted that the signal from AgNPs can not be seen either. It points to the fact that the synthesis time is not high enough for thick Ag layer formation.

UV-VIS spectra of the synthesis media corresponding to different samples are given in figure 2.

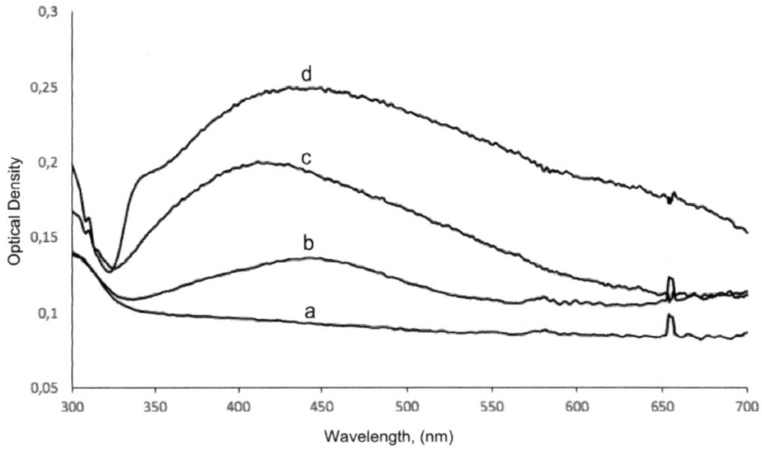

Figure 2. Absorption spectra of different synthesis media corresponding to the samples: (a) TiAg_EG_1; (b) TiAg_EG_10; (c) TiAg_PEG-400_1; (d) TiAg_PEG-400_10.

It can be seen from figure 2a that in the case of EG and 1 min of the synthesis time no plasmon peak appears in the studied range of wavelength. It indicates that at the given time EG is not efficient enough as a reducing agent hence the formation of a noticeable amount of AgNPs does not take place. On the other hand, at 10 min of synthesis, a pronounced peak is observed at 450 nm (figure 2b). It should be noted that the plasmon peak position shifts to larger values of the wavelength for NPs with larger sizes. It can be concluded that while the AgNPs formation process occurs, the stabilizing effect of the EG is rather weak and NPs growth rate is relatively high. In the case of PEG-400 significant signal can be observed starting from the 1 min of synthesis (figure 2c). What is more interesting is that the peak is located at a lower wavelength value (415 nm) which corresponds to smaller sizes of AgNPs. Such a feature should be ascribed to a more pronounced stabilizing effect of PEG-400. At higher synthesis time the peak position shifts to the 435 nm reflecting the AgNPs growth process. Nonetheless, the maximum is still at the lower wavelength as compared to the EG_10 sample. It should be noted that in the case of synthesis media of the PEG-400_10 sample a new signal occurs at approximately 640 nm. It can be suggested that AgNPs of a new morphology are formed as synthesis time rises [17].

Electron images of the surfaces of all obtained samples are shown in figure 3. Silver nanoparticles were formed on the surfaces of all samples as a result of synthesis. It is important to note that there is no contradiction between the results obtained for the TiAg_EG_1 sample by optical spectroscopy and SEM since the former describes synthesis media exclusively. NPs formation process is facilitated significantly on the surface of the substrate. So while there is not a sufficient amount of AgNPs in the whole volume of the synthesis media they can be found on the surface of the alloy immersed into the media.

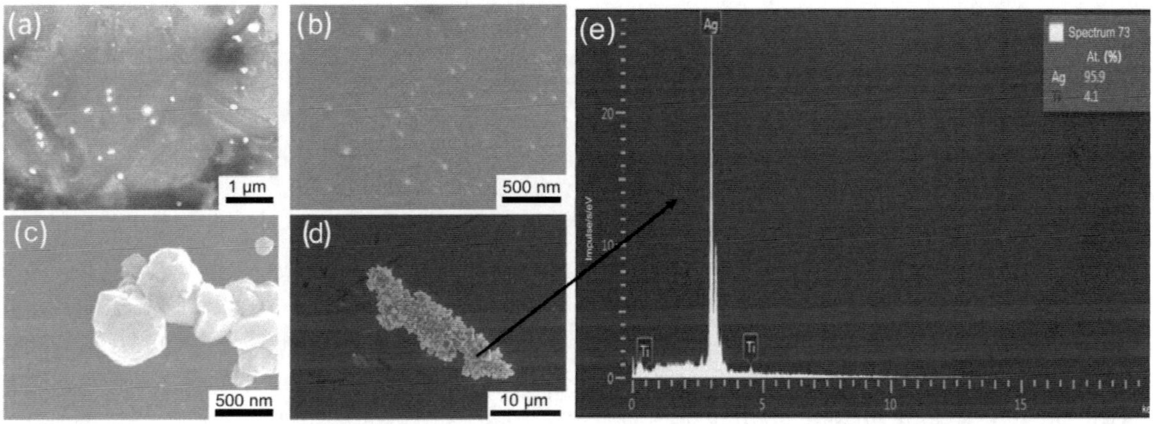

Figure 3. Electron images of the surface: (a) TiAg_PEG-400_1; (b) TiAg_PEG-400_10; (c) TiAg_EG_1; (d) TiAg_EG_10; (e) EDS analysis TiAg_EG_10.

It can be seen from figure 3a, b that AgNPs are distributed evenly on the surface of TiAg_PEG-400_1 and TiAg_PEG-400_10 samples and their size rarely exceeds 100 nm. In general similar picture is observed for TiAg_EG samples but occasionally agglomerates of larger AgNPs can be found on the surface of the alloy along with the smaller nanoparticles. It can be seen from figure 3c, d that the size of the agglomerate increases as synthesis times rises. The fact that the observed agglomerates consist of AgNPs was further supported by the results of the EDS analysis (figure 3e). It can be suggested that the observed difference in the surface structure of the obtained samples is related to the different stabilizing properties of the media. According to the obtained results, PEG-400 facilitates the formation of more monodisperse AgNPs uniformly distributed over the Ti-18Zr-15Nb alloy as compared to EG.

Figure 4 depicts the results of the EDS analysis conducted for the TiAg_PEG-400_1 sample.

Based on SEM studies it was suggested that AgNPs on the surface of the TiAg_PEG-400_1 sample have a core-shell structure. To confirm this suggestion EDS profile was obtained for one of the larger AgNPs (figure 4a). According to the obtained results signal from alloy's components decreases significantly as the Ag signal increases which confirms that the observed NPs are silver. What is more important is that the intensity of the C signal increases noticeably as well. Comparison of electron image and concentration profiles for Ag and C allows us to suggest that there is a shell around the Ag NP which consists of an intermediate compound containing both polymer and silver. It can be suggested that the observed feature of AgNPs structure in the TiAg_PEG-400_1 sample is due to the specific interaction of polyol macromolecules and $AgNO_3$ at the early stages of the synthesis.

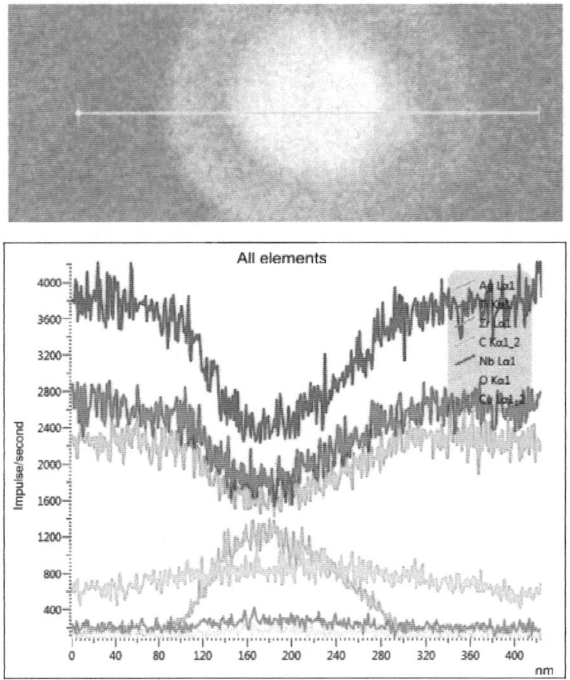

Figure 4. Results of EDS analysis TiAg_PEG-400_1.

4. Conclusion

AgNPs were successfully deposited on the surface of Ti-18Zr-15Nb alloy *via* polyol synthesis. Two synthesis media were used for the deposition process: EG and PEG-400. It was demonstrated that PEG-400 allows us to obtain more monodisperse and evenly distributed AgNPs on the surface of the alloy which makes it more preferable synthesis media as compared to EG. Results of SEM and EDS studies elucidated that at early synthesis stages AgNPs deposited on the surface of Ti-18Zr-15Nb alloy in PEG-400 have a core-shell structure. It was suggested that the shell consists of an intermediate polymeric-Ag compound. Obtained results demonstrate that polyol synthesis can be potentially used for providing the surface of novel superelastic Ti alloys with antibacterial properties.

5. Acknowledgment

The present work has been carried out with the financial support of the Russian Science Foundation (Project №. 19-79-10270).

References

[1] Brunette D M, Tengvall P, Textor M and Thomsen P 2012 *Titanium in Medicine: Material Science, Surface Science, Engineering, Biological Responses and Medical Applications* (Springer Science & Business Media).
[2] Ramires P A, Romito A, Cosentino F and Milella E 2001 *Biomaterials* **22(12)** 1467-74.
[3] Miyazaki S, Kim H Y and Hosoda H 2006 *Materials Science and Engineering* A **438** 18-24.
[4] Zhukova Y S, Pustov Y A, Konopatsky A S and Filonov M R 2014 *Journal of Alloys and Compounds* **586** S535-8.
[5] Kim H Y, Fu J, Tobe H, Kim J I and Miyazaki S 2015 *Shape Memory and Superelasticity* ***1*(2)** 107-16.
[6] Besinis A, Hadi S D, Le H R, Tredwin C and Handy R D 2017 *Nanotoxicology* **11(3)** 327-38.
[7] Chen W, Liu Y, Courtney H S, Bettenga M, Agrawal C M, Bumgardner J D and Ong J L 2006

Biomaterials **27(32)** 5512-17.
[8] Seuss S, Heinloth M and Boccaccini A R 2016 *Surface and Coatings Technology* **301** 100-5.
[9] Li B, Jia J, Han M, Gao Y, Wang W and Li C 2017 *Surface and Coatings Technology* **324** 552-9.
[10] Todea M, Vulpoi A, Popa C, Berce P and Simon S 2019 *Journal of Materials Science & Technology* **35(3)** 418-26.
[11] Pang E L, Pickering E J, Baik S I, Seidman, D N and Jones N G 2018 *Acta Materialia* **153** 62-70.
[12] Konopatsky A S, Dubinskiy S M, Zhukova Y S, Sheremetyev V, Brailovski V, Prokoshkin S D and Filonov M R 2017 *Materials Science and Engineering* A **702** 301-11.
[13] Fiévet F, Ammar-Merah S, Brayner R, Chau F, Giraud M, Mammeri F and Viau G 2018 *Chemical Society Reviews* **47(14)** 5187-233.
[14] Luo C, Zhang Y, Zeng X, Zeng Y and Wang Y 2005 *Journal of Colloid and Interface Science* **288(2)** 444-448.
[15] Fereshteh Z, Rojaee R and Sharifnabi A 2016 *Superlattices and Microstructures* **98** 267-75.
[16] Konopatskii A S, Dubinskii S M, Zhukova Y S, Inaekyan K, Brailovskii V., Prokoshkin S D and Filonov M R 2019 *Metal Science and Heat Treatment* **61(5-6)** 333-9.
[17] Lee G P, Bignell L J, Romeo T C, Razal J M, Shepherd R L, Chen J and Wallace G G 2010 *Chemical Communications* **46(41)** 7807-09.

Features of structure and physical and mechanical properties of metal-ceramic composite material depending on the content of boron nitride

B O Bolshakov[1,2] and A M Smyslov[2]

[1]Scientific and Production Enterprise "Uralaviaspetstekhnologiya", 5 Tramvainaya st., building 1, Ufa, 450027, Russia
[2] Ufa State Aviation Technical University, 12 K. Marx st., Ufa, 450008, Russia

E-mail: bobolshakov@gmail.com

Abstract. In this paper, the problem of structure formation in powder compacts of PCh13M2-BN obtained by solid-phase sintering is considered. A phenomenological model of the formation of regions with increased concentration of ceramic BN particles in the volume of compacts is developed. The influence of the concentration of boron nitride in the initial charge on physical and mechanical characteristics of compacts is determined.

1. Introduction

Powder metallurgy technologies are actively used in development of self-lubricating composite materials, which are based on interaction of filler elements belonging to the category of solid lubricants, such as graphite, MoS_2, BaF_2, CaF_2, Sn, Ag and h-BN with metal matrix used for manufacturing of various machine parts (bearings, gears, seals and connectors) [1-3].

It is known [4-9] that introduction of solid lubricant components into the volume of materials can reduce a friction coefficient, but their efficiency depends significantly on operating conditions of a finished product. The presence of moisture has a positive effect on reduction of friction in materials containing graphite, and negatively affects materials with MoS_2 additives [10]. The negative effect of high humidity on performance was also established for materials with additives of BaF_2, CaF_2, which are subject to hydrolysis. In addition, use of one or another component as a filler is limited by temperature of its destructuration. Of the whole variety of non-metallic compounds related to solid lubricants, h-BN possesses good lubricating ability in conditions of high humidity and applicability in the temperature range up to 700 ° C [11, 12].

There are works [15-19], in which authors evaluated the effect of concentration of hexagonal boron nitride on mechanical and tribological properties of composite materials on different bases. It was found that introduction of boron nitride into materials leads to decrease in friction at high loads and speeds [13, 15], h-BN has a low adhesive relationship with the matrix [13], its lubricity is lower than that of graphite due to higher van-der-Waals force between the layers of the crystal lattice of h-BN [15, 17]. A number of works [19] provide information on the achievement of a unique combination of mechanical and service properties of composite materials with boron nitride additives, which make them to be considered as seal materials for a flow path of steam turbines.

However, the formation of structure of composite materials with BN additives and its relationship with performance remain not thoroughly studied and are the subject of discussion among scientists.

2. Materials and procedures

For research was taken a powder of corrosion-resistant steel Fe-13% Cr-2% Mo with grain size distribution from 10 to 80 microns was chosen as base material, and was processed in accordance with GOST 13084-84. A decrease in the friction coefficient and better abrasion of the composite material were achieved by introducing a finely dispersed boron nitride powder BN according to TU.U 26.8-00222226-007-2003. The boron nitride particle diameter was from 1 to 10 μm, the particle thickness was from 0.7 to 2 μm. The effect of the concentration of BN powder in the charge on structure formation in compacts, their physical and mechanical properties, was studied on the samples of the following compositions:

1) Composition "1" - pressed specimens from pure powder PH13M2;
2) Composition "2" - powder PH13M2 + 0.5% BN;
3) Composition "3" - powder PH13M2 + 1% BN;
4) Composition "4" - powder PH13M2 + 2% BN;
5) Composition "5" - powder PH13M2 + 4% BN;
6) Composition "6" - powder PH13M2 + 4.5% BN;
7) Composition "7" - powder PH13M2 + 5% BN;
8) Composition "8" - powder PH13M2 + 6% BN.

The original powder components were thoroughly mixed. The obtained powder mixture was cold pressed into rectangular-shaped specimens of 50x10x10 mm. The compacts were sintered in electric furnaces in air atmosphere at 1200 ° C. The sintering time was 2 h.

The microstructure of samples was studied using a JSM-6490LV scanning electron microscope. The density of samples was determined by hydrostatic weighing. The hardness of samples was determined by the Brinell method in accordance with GOST 9012. The static stretching was carried out on an Instron 5982 universal testing machine at an assumed operating temperature of seals (600 °C) in accordance with GOST 9651. The crosshead movement speed was 1 mm / min. The impact strength was determined on an Instron Ceast 9350 drop tower. The stored impact energy was 50 J.

3. Results and discussion

Figure 1 shows the microstructure of composite compacts with different contents of boron nitride. Pores of various shapes and sizes are observed on the polished surface of all studied samples. In the structure of a compact without boron nitride (figure 1, a), the pores have a globular shape and are predominantly located along grain boundaries, which indicates that the applied force during pressing is insufficient to form a denser structure due to high friction force between metal particles and, as a consequence, limited flow of diffusion processes during subsequent sintering.

The introduction of boron nitride into the initial charge leads to formation of extended grain-boundary pores in the microstructure (figure 1, b-d). The higher is the boron nitride content, the larger are the volume fraction and pore sizes. The mechanism of formation of the observed extended grain-boundary pores differs from the classical concept described in [20-23]. Due to different bulk densities of metal powder PKh13M2 and boron nitride, the volume of the mixed components in the charge is comparable. Taking this into account together with the chemical inertness and low adhesion of boron nitride, it can be assumed that the extended grain-boundary pores observed in the microstructure are areas with increased concentration of solid lubricant, which is removed along with the abrasive in preparation for metallographic analysis. An increase in the volume fraction of pores with an increase in the boron nitride content in the initial charge from 0 to 6% by weight is from 3.5% to 17%, respectively (figure 3).

The investigation of microstructure of compacts with boron nitride at higher magnification revealed the presence of dispersed nanosized 20 ... 500 nm thick boron nitride interlayers both along grain boundaries and inside grains formed during sintering (figure 2). The discovered structural features of compacts with addition of BN indicate the development of processes of diffusion mass transfer between metal particles during sintering by various mechanisms. In addition, BN particles are observed along the edges of extended grain-boundary pores, which confirms the proposed assumption about the nature of their formation.

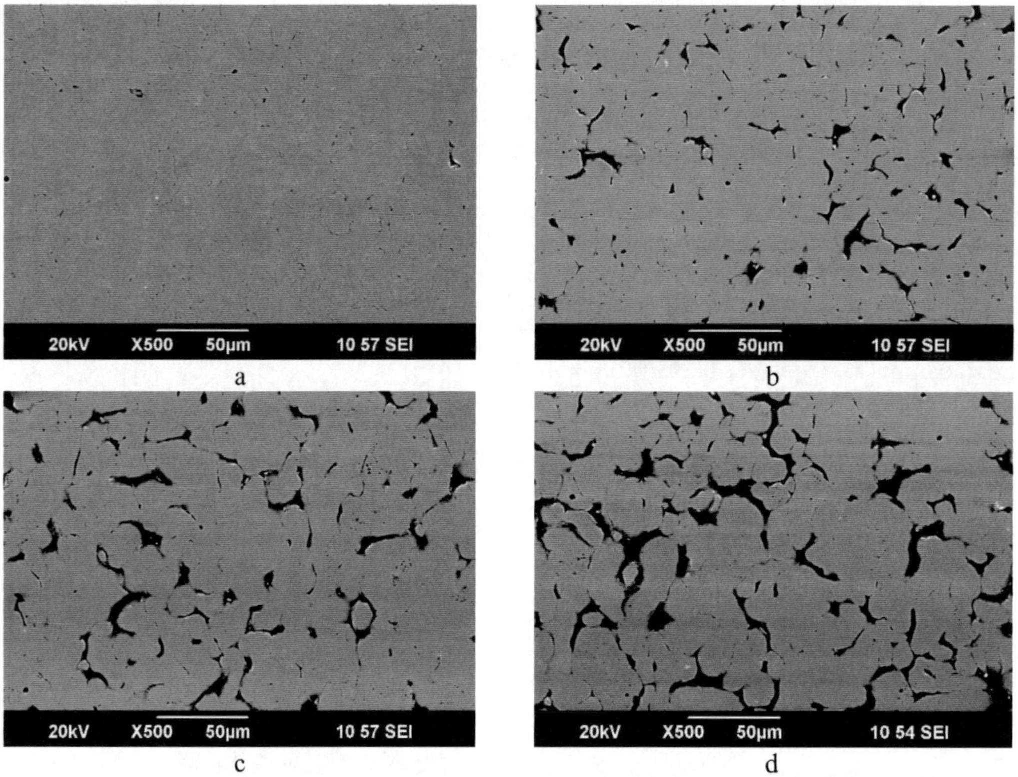

Figure 1. Microstructure of the studied compacts with different content of boron nitride: *a* – 0% BN; *b* – 2% BN; *c* – 4% BN; *d* – 6% BN.

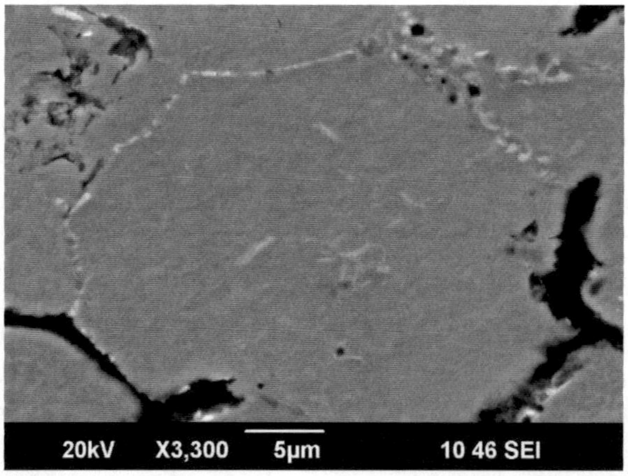

Figure 2. View of dispersed boron nitride layers

The evaluation of hardness and density of the compacts (figure 3) showed their almost monotonic dependence on the boron nitride content. Thus, with an increase in the BN content, the density and hardness change from 6.7 g / cm^3 and 92 HB typical of a compact without boron nitride additives, to 6.0 g / cm^3 and 61 HB typical of samples with 6% BN.

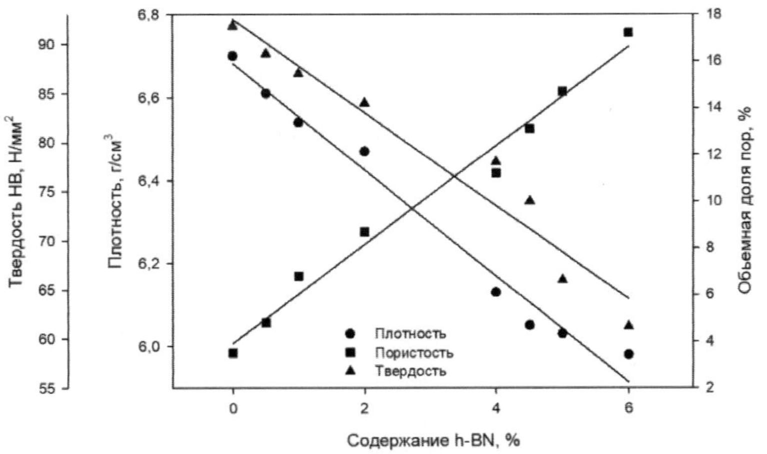

Figure 3. Basic physical and mechanical properties of compacts depending on the BN content

Table 1 shows the results of determining the physical and mechanical properties of the studied powder compacts. Despite the monotonous gradual decrease in the hardness of compacts with an increase in the boron nitride content, the strength and impact strength are of extreme nature with the maximum values observed at 2% BN content. The maximum strength is 161 MPa, and the impact strength reaches 114 kJ / m^2.

Table 1. Mechanical properties of studied powder compacts.

Test material	Tensile strength $\sigma_в^{600}$ (MPa)	Impact strength KC (kJ/m^2)	Hardness HB
Composition 1	134	99	91.8
Composition 2	139	102	89.2
Composition 3	148	107	87.1
Composition 4	161	114	84.3
Composition 5	100	42	78.5
Composition 6	87	34	74.3
Composition 7	72	25	65.8
Composition 8	64	18	61.2

The obtained data on the change in the microstructure of compacts with the boron nitride content growing allow considering the issue of its formation by first dividing the task into 2 parts: 1 - the effect of deformation during pressing, 2 - the effect of high temperature during sintering.

The boron nitride particles, due to their natural inertness, have a low adhesion ability [24] and cannot react with metal powder PH13M2. During pressing, part of BN particles uniformly distributed in the charge volume has a "lubricating effect", contributing to tighter adhesion of metal powder particles to each other. At the same time, the boron nitride particles under conditions of gradual compaction of material are partially subject to either direct rapid displacement into more free zones of reduced density, where the development of deformation processes did not begin due to insufficient applied force, or gradual slow displacement due to destruction of the relationship of layers bordering on metal particles (delamination of boron nitride particles (figure 5)). Ultimately, the directed motion of boron nitride particles during pressing leads to formation of closed grain-boundary porosity filled with BN particles (regions with increased BN concentration) in the volume of compacts. In this case, the higher is the boron nitride content in the initial charge, the greater is the volume fraction of grain boundary pores.

During sintering, the structure is formed according to the model described in [24]. At the interfaces, where there are point contacts of metal particles with each other, or the thickness of a nitride layer is so small that due to thermal expansion of PH13M2 powder, they may come into contact with each other, the processes of diffusion mass transfer develop, sintering bridges are formed, and metal particles merge with formation of coarser grains. In this case, partial absorption of BN particles by growing metal grains is possible (Figs. 2 and 6, b). At the interfaces with a high content of ceramic particles and in areas with an increased concentration of boron nitride, the development of mass transfer and recrystallization processes is blocked. As a result, weakened boundaries are formed in the volume of compacts, which determine high abradability properties [13].

Thus, the introduction of finely dispersed boron nitride powder into the initial charge makes it possible to control the strength properties of grain boundaries formed during sintering by means of directed blocking of the development of diffusion processes during sintering, and as a consequence, to achieve the combination of strength and abradability of the composite material PH13M2-BN as required by operating conditions.

4. Conclusions

The introduction of finely dispersed BN with a graphite-like structure into the composition of powder compacts PH13M2 leads to formation of an extended closed grain-boundary porosity filled with boron nitride particles during deformation.

Deformation treatment of the initial charge of powder compacts PH13M2-BN leads to formation of dispersed nanosized 20-500 nm thick boron nitride interlayers along grain boundaries due to partial delamination of boron nitride particles.

During sintering, due to the development of diffusion mass transfer processes, the growing metal grains absorb dispersed BN particles.

The change in the boron nitride content in powder compacts ПХ13M2-BN allows varying their physical and mechanical properties in a wide range.

The presented phenomenological models of influence of deformation and heat treatment describe the process of structure formation of composite materials on a metal base with additives of components belonging to the category of solid lubricants.

Acknowledgments

The research part of the work was performed using the equipment of the Nanotech Center for Collective Use of Ufa State Aviation Technical University.

References:

[1] Ding C H, Liu C H, Yang Z M, Wang Y P, Sun Z B and Yu L 2010 *Compos. Sci. Technol.* **70** 1000
[2] Kato H, Takama M, Iwai Y, Washida K and Sasaki Y 2003 *Wear* **255** 573
[3] Dhanasekaran S and Gnanamoorthy R 2007 *Mater. Des.* **28** 1135
[4] Sutor P 1991 *MRS Bull.* **16**(5) 24
[5] Clauss F J 1972 *Self-Lubricating Solids* (Amsterdam: Elsevier)
[6] Lancaster J 1988 *Solid Lubricants* in: *CRC Handb. Lubr.* (Boca Raton: CRC Press)
[7] McMurtrey E L 1985 *Lubrication Handbook for the Space Industry - Part A, Solid Lubricants* (Marshall Space Flight Center)
[8] Sliney H E 1982 *Tribol. Int.* **15** 303
[9] Lansdown A R 1999 *Molybdenum Disulphide Lubrication Tribology* (Amsterdam: Elsevier)
[10] Lansdown A R 1994 *High Temperature Lubrication* (Mechanical Engineering Publications)
[11] Saito T, Honda F 2000 *Wear* **237** 253
[12] Du L, Huang C, Zhang W, Li T and Liu W 2011 *Surf. Coat. Technol.* **205** 3722
[13] Bolshakov B O, Galiakbarov R F, Smyslov A M, Kuleshov A S and Usachev K M 2020 *Elektricheskie Stancii* **5** 25 (in Russian)
[14] Chen B, Bi Q, Yang J, Xia Y and Hao J 2008 *Tribol. Int.* **41** 1145
[15] Tyagi R, Xiong D, Li J and Dai J 2010 *Tribol. Lett.* **40** 181

[16] Mahathanabodee S, Palathai T, Raadnui S, Tongsri R and Sombatsompop N 2014 *Wear* **316** 37
[17] Hammes G, Mucelin K J, Gonçalves P C, Binder C, Binder R, Janssen R, Klein A N and De Mello J D B 2017 *Wear* **376-377** 1084
[18] Bolshakov B O, Galiakbarov R F, Smyslov A M, Lisjanskij A S and Usachev K M 2019 *Vestnik-UGATU* **23**(4) 11 (in Russian)
[19] Exner H and Petzow G 1975 *Syntering and Catalysis, Materials Science Research* vol 10, ed. G C Kuczynski (New York: Plenum Press) p 279
[20] Shinohara N, Okumiya M and Hotta T 2000 *J. Amer. Ceram. Soc.* **87**(7) 1633
[21] Balshin M Ju 1948 *Poroshkovoe Metallovedenie* (Metallurgizdat) (in Russian)
[22] Liniger E and Raj R 1987 *J. Amer. Ceram. Soc.* **70**(11) 843
[23] Kostoglou N, Polychronopoulou K and Rebholz C 2015 *Vacuum* **112** 42
[24] Ting Li, Danqing Yi, Jun Hu, Jiao Xu, Junlei Liu and Bin Wang 2017 *J. All. Compd.* **723** 345

Study of the deformation behavior and structure formation of Ti-Zr-Nb shape memory alloy

V A Sheremetyev[1], O Akhmadkulov[1], V S Komarov[1], A V Korotitsky[1], S P Galkin[1], V A Andreev[2,3] and S D Prokoshkin[1]

[1]National University of Science and Technology "MISiS", Moscow, Russia
[2]Baikov Institute of Metallurgy and Materials Science, RAS, Moscow, Russia
[3]MATEK-SMA Ltd., Moscow, Russia

E-mail: sheremetyev@misis.ru

Abstract. The deformation behavior and structure formation of the Ti-19Zr-14Nb (in at. %) shape memory alloy have been studied. The as-cast ingot was subjected to multi-axis forging at a billet heating temperature of 1050 °C and then to rotary forging at 600 °C. Plastometric compression tests of the forged alloy were carried out at a strain rate 0.1 s^{-1} at temperatures of 600-1000 °C, the strain degree is not more than 0.7. The obtained flow curves and the results of studying the grain structure and hardness were used to determine the features of the thermomechanical behavior of the material. The study of the microstructure showed that under these processing conditions, the processes of dynamic polygonization and dynamic recrystallization occur simultaneously. According to the obtained hardness measurement results, it can be assumed that in order to achieve a low Young's modulus and perfect superelastic behavior, deformation must be carried out at temperatures of 700-800 °C.

1. Introduction

Currently, one of the most intensively developing areas of scientific and technical activity is the development of technologies for the manufacturing of metallic materials for medical application. The need for biomaterials with high strength, low elastic modulus of elasticity and high biocompatibility has led to the development of new titanium-based alloys. Shape memory alloys (SMA) based on Ti-Zr-Nb system are promising materials for the bone implants fabrication due to their high biomechanical compatibility, which is due to the low Young's modulus and superelastic behavior similar to the mechanical behavior of bone tissue [1-4].

Thermomechanical treatment (TMT) of these alloys is a key tool for control the structural-phase state and thus improves the functional properties of these alloys [3, 5-7]. It was shown that the low-temperature TMT of the Ti-18Zr-14Nb alloy results in the formation of polygonized substructure of β-phase with an average subgrain size of 100-300 nm. The alloy in this structural state shows an improved combination of static functional properties, including a low Young's modulus (37 GPa) and a large difference between dislocation and transformation yield stresses [8]. On the other hand, this alloy with a mixed dynamically polygonized and dynamically recrystallized structure induced by hot deformation, demonstrates a 1.5-times or more increased functional fatigue resistance compared to the structure formed as a result of static processes during low-temperature TMT [9].

The TMT technology of Ti-Zr-Nb alloys which makes it possible to obtain semi-products suitable for the manufacturing of bone implants, needs to be developed and optimized. This can be most effectively implemented by using a combination of physical and mathematical modeling of the TMT processes [10-12]. Physical modeling makes it possible to obtain rheological models of material deformation behavior under various temperature and strain rate conditions, while mathematical modeling is very effective for studying the features of plastic deformation using a specific method of [11-13]. The study of the processes of structure formation in the course of physical modeling provides additional information that allows predicting the structural-phase state of the material.

The purpose of this work was to study the deformation behavior and structure formation of the Ti-Zr-Nb alloy under various temperatures simulating metal forming processes to identify favorable TMT conditions.

2. Experimental

A 30 kg ingot of the Ti-19Zr-14Nb (in at. %) alloy produced by vacuum arc remelting was used. The melting of the alloy provided low amount of impurities ($O < 0.05$, $C < 0.01$, $N < 0.01$, $H < 0.01$ in wt. %). To eliminate the initial cast structure, the ingot was subjected to multi-axis forging at a billet heating temperature of 1050 °C. Then a 6 mm bar stock was fabricated by multi-pass rotary forging at 600 °C with a total cumulative strain $e \approx 2.5$ using an "*RKM-2*" forging machine. Samples with a diameter of 5 mm and a height of 10 mm were cut from the bar stock.

Plastometric tests with a true strain of $e=0.7$ were performed in the temperature range of 600-1000 °C at strain rate of 0.1 s^{-1} using the complex of physical modeling of plastic deformation. To reduce the influence of surface friction forces, a lubricant based on graphite and boron nitride was used. To ensure deformation under isothermal conditions, samples were heated and deformed in steel cups with tungsten carbide strikers and a hole for a thermocouple. Following the deformation, samples were immediately cooled in water. Processing of obtained data, building of flow curves and their adjustment for friction, temperature and strain rate was carried out using the AUK specialized software.

The deformed samples were cut along the axis of the load direction, mechanically polished, and then etched in a Kroll etching solution with 10% HF, 30% HNO_3, and 60% H_2O_2 and were prepared for metallographic analysis. The microstructure was studied using a *Versamet-2 Union* light microscope in different deformation zones. The obtained images were used to calculate the average grain size by the linear intercept method. The hardness of the samples was measured using a *Metkon Metallography* hardness tester for at least 5 measurements per each deformation zone under a load of 1 kg and holding time 10 s.

3. Results and discussion

3.1. Plastometric tests

The flow curves of the Ti-Nb-Zr alloy at a strain rate of 0.1 s^{-1} and in the temperature range under study are shown in figure 1a. At these temperatures, the alloy is located in a single β-phase region before compression deformation. The shape of the flow curve can be used to define the process of dynamic softening. At temperature of 600 °C, the curves show an unclear smooth maximum. An increase in temperature in the 700 to 900 °C range leads to an early achievement of the critical deformation e_{max} which corresponds to the maximum flow stress (figure 1b). It can be assumed, that the recovery and/or polygonization processes predominantly occur at 600 °C, and the recrystallization process at 700-900 °C. In this context, the recovery is regarded as a combination of point defect redistribution and annihilation, and redistribution of dislocations without formation of new subboundaries; the polygonization is regarded as a formation of new perfect subboundaries, subgrains, and subgrain growth. During deformation at 1000 °C the e_{max} increase. It should be noted, however, that in all cases except 1000 °C a barrel-shaped specimen is formed as a result of deformation (figure 1b). As a result of deformation at 1000 °C, a specimen is formed with broadenings closer to the ends and a thick white-gray oxide layer. These circumstances prevent direct comparison of the flow curve with the curves obtained at other temperatures. Note that according to the flow curves, in all cases, no distinct steady-state stage was revealed. This thermomechanical behavior of the alloy may indicate its tendency to localize deformation during TMT.

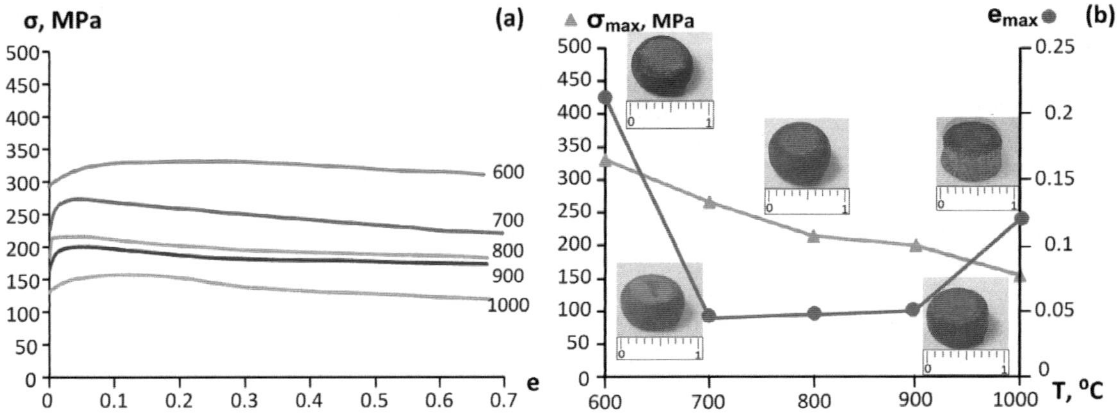

Figure 1. Flow curves at different temperatures at the strain rate of 0.1 s^{-1} (a); the dependence of the maximum deformation stress σ_{max} and the critical strain degree e_{max} on the deformation temperature (b).

3.2. Microstructure examination

The microstructure of the initial (as-forged) and deformed Ti-Zr-Nb alloy according to three different zones of the sample is presented on figure 2. In initial state, a mixed dynamically polygonized and dynamically recrystallized structure with the average grain size about 15 μm is observed (Figure. 2a), which correlates well with the data of [9]. The plastometric tests lead to a monotonic increase in the average grain size from 15-20 to 175-260 μm and from 10-15 to 70-85 μm perpendicular to the compression direction and along the compression direction, respectively, with increasing temperature from 600 to 1000 °C (figures 2 and 3). The elongated and equiaxed grains are observed in the structure of central part of the localized deformation zone after tests at all temperatures. It can be assumed that the process of dynamic polygonization in the alloy during deformation proceeds up to high temperatures. It also should be noted that at these temperatures, static recrystallization and grain growth can occur during heating.

3.3. Hardness measurements

Figure 4 represents the results of hardness measurements of samples in the initial forged state and after plastometric tests. In the initial as-forged state, the alloy hardness is around 210 HV. After compression tests at 600 °C, the hardness does not change. An increase in the deformation temperature has led to reduction of hardness down to about 195 HV in the 700-900 °C temperature range. It is associated with dynamic softening processes (recovery, polygonization and recrystallization) with an increase in the deformation temperature. Another reason for the HV reduction can be related with the $\alpha''\leftrightarrow\beta$ transformation phenomena and the corresponding softening of the crystal lattice which contributes to a decrease in Young's modulus and improvement of superelastic behavior [5, 9]. The transformation behavior requires additional study of the alloy after treatments at these temperatures. After deformation at 1000 °C, the hardness tends to increase in two zones of the samples located close to the surface (figure 4a,c). This tendency is explained by saturation with gas-forming impurities in near-surface layer, which can lead to embrittlement.

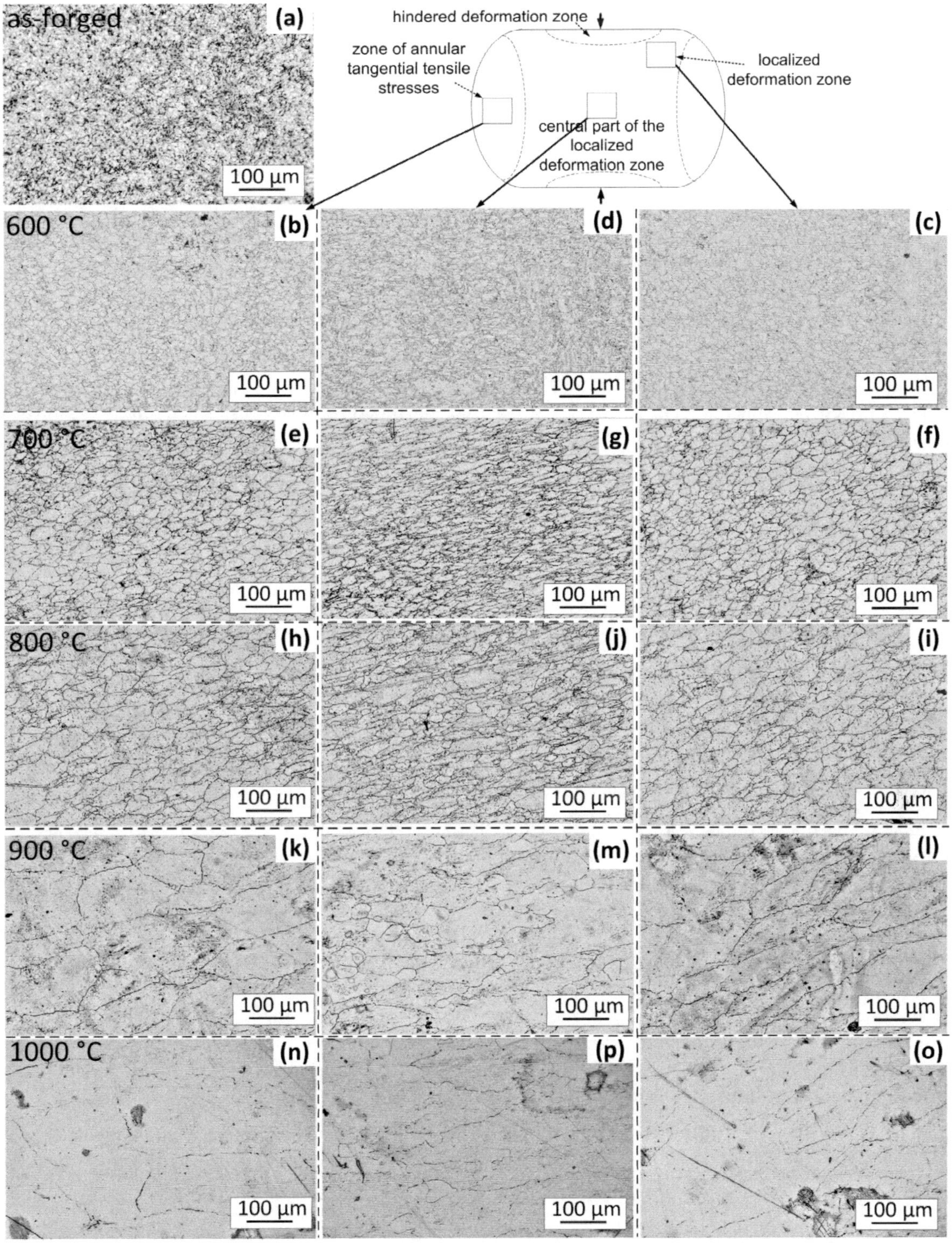

Figure 2. Microstructure of samples initial state (a) and after deformation in 3 zones: zone of annular tangential tensile stresses (b, e, h, k, n), central part of the localized deformation zone (d, g, j, m, p), localized deformation zone (c, f, i, l, o).

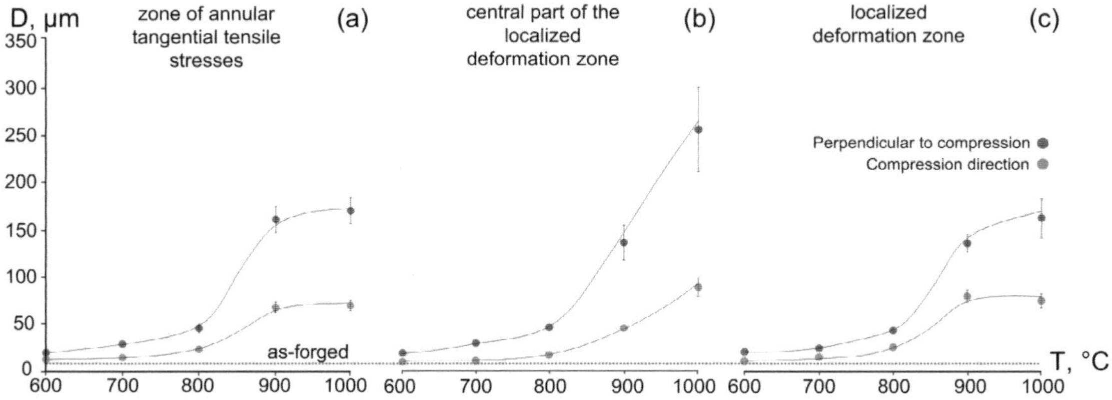

Figure 3. The distribution of the average grain sizes *vs.* deformation temperature: zone of annular tangential tensile stresses (a), central part of the localized deformation zone (b), localized deformation zone (c).

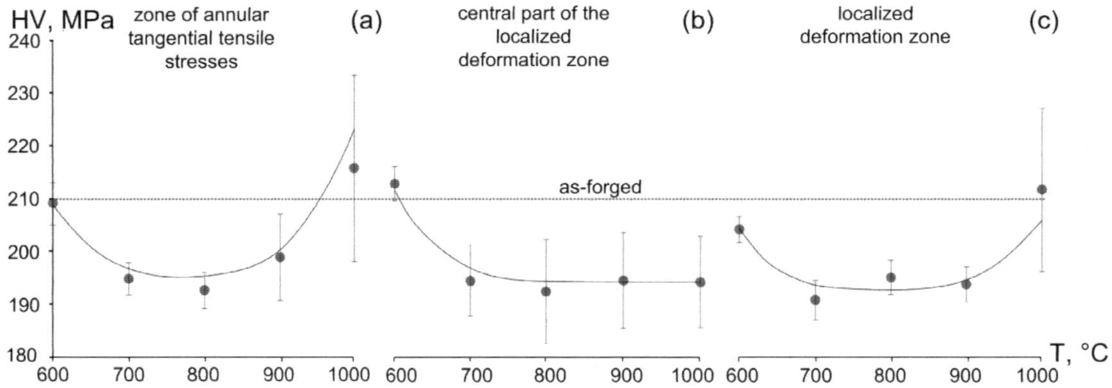

Figure 4. Hardness of samples *vs* deformation temperature: zone of annular tangential tensile stresses (a), central part of the localized deformation zone (b), localized deformation zone (c).

4. Conclusion

The deformation behavior and structure formation of the Ti-19Zr-14Nb (at.%) alloy have been studied in the compression temperature range of 600 to 1000 °C at strain rate of 0.1 s^{-1}. It is shown that the process of dynamic polygonization occurs in the entire deformation temperature range. In the temperature range of 800-1000 °C, dynamic recrystallization is a main mechanism of microstructural evolution, accompanied by dynamic polygonization. Based on the hardness measurements, it can be assumed that to get a low Young's modulus and perfect superelastic behavior, deformation should be carried out at temperatures of 700-800 °C.

Acknowledgements

The present work has been carried out under financial support the Russian Science Foundation (project №20-63-47063).

References

[1] Geetha M, Singh A K, Asokamani R and Gogia A K 2009 *Progress in Materials Science* **54(3)** 397
[2] Kim H Y, Fu J, Tobe H, Kim J I and Miyazaki S 2015 *Shape Memory and Superelasticity* **1** 107

[3] Sheremetyev V A, Kudryashova A A, Dinh S T, Galkin S P, Prokoshkin S D and Brailovsky V 2019 *Metallurg* **1** 45 (in Russian)
[4] Laheurte P, Prima F, Eberhardt A, Gloriant T, Wary M and Patoor E 2010 *Journal of the Mechanical Behavior of Biomedical Materials* **3(8)** 565
[5] Prokoshkin S, Brailovski V, Dubinskiy S, Zhukova Y, Sheremetyev V, Konopatsky A and Inaekyan K 2016 *Shape Memory and Superelasticity* **2(2)** 130
[6] Nunes A R V, Borborema S, Araújo L S, Malet L, Dille J, and de Almeida L H 2020 *Journal of Alloys and Compounds* **820** 153078
[7] Hao Y L, Niinomi M, Kuroda D, Fukunaga K, Zhou Y L, Yang R and Suzuki A 2003 *Metallurgical Materials Transactions* A **34(4)** 1007
[8] Sheremetyev V, Kudryashova A, Dubinskiy S, Galkin S, Prokoshkin S and Brailovski V 2018 *Journal of Alloys and Compounds* **737** 678
[9] Kudryashova A, Sheremetyev V, Lukashevich K, Cheverikin V, Inaekyan K, Galkin S, Prokoshkin S and Brailovski V 2020 *Journal of Alloys and Compounds* **843** 156066
[10] Akopyan T K, Gamin Y V, Galkin S P, Prosviryakov A S, Aleshchenko A S, Noshin M A, Koshmin A N and Fomin A V 2020 *Materials Science and Engineering* A **786** 139424
[11] Hadasik E, Kuziak R, Kawalla R, Adamczyk M, Pietrzyk M 2006 *Steel Research International* **77** 927
[12] Gamin Y, Akopyan T, Koshmin A, Dolbachev A, Aleshchenko A, Galkin S P and Romantsev B A 2020 *International Journal of Advanced Manufacturing Technology* **108(3)** 695
[13] Xuan T D, Sheremetyev V A, Kudryashova A A, Galkin S P, Andreev V A, Prokoshkin S D and Brailovski V 2020 *Izvestiya vuzov. Non-ferrous metallurgy* **2** 22 (in Russian)

Nanowhiskers for lithium battery anode

E Podlesnov and M V Dorogov

International Research Center of Functional Materials and Devices of Optoelectronics, ITMO University, Kronverksky Pr. 49, St. Petersburg, 197101, Russia

E-mail: epodlesnov@itmo.ru

Abstract. The paper presents the prospect of using copper oxide nanowhiskers as an anode material in a lithium-ion battery. Technology for the synthesis of copper oxide nanowhiskers by electrochemical deposition and thermal oxidation was developed. The stress state of CuO nanowhiskers under lithiation/delithiation was considered. A test was carried out of lithium batteries with anodes containing copper oxide nanowhiskers with various binders.

1. Introduction

The key trend of our time is the transition to alternative energy sources. Some countries use wind, sun, and water energy as a resource. An important problem is the storage of electricity; this requires high-capacity batteries that are stable at high and low operating temperatures, as well as capable of withstanding a large number of charge-discharge cycles. The lithium-ion battery is the most popular and efficient energy storage device, but it has a limited lifespan and does not withstand critical temperatures. In addition to the usual usage of lithium-ion batteries in portable devices, they are also used in electric vehicles and as energy storage devices in the smart home, which will solve environmental and economic problems. Therefore, research is being carried out on new materials for electrodes. The use of nanowhiskers as part of the anode in a lithium battery is promising since this can lead to an increase in the capacitive characteristics of the battery due to the large surface of the active mass of the substance, as well as to an increase in recharge cycles due to the mechanical properties of the material; also, nanowhiskers provide a short path for the diffusion of lithium ions. During the passage of the charge-discharge cycle, such materials do not experience a change in volume and the electrolyte does not recover on their surface with the formation of passive films. For research, we have chosen copper oxide due to its low cost, ease of production, strength, electrical properties, and high theoretical capacity [1].

A nanowhisker is a one-dimensional nanomaterial, the length of which significantly exceeds the other dimensions, which do not exceed several tens of nanometer. Nanowhiskers are highly durable, flexible, and resistant to aggressive environments. In this work, the use of copper oxide nanowhiskers as a material for lithium-ion batteries was considered. Batteries with high-performance characteristics will solve environmental problems: reduce the use of natural resources for energy generation, reduce environmental pollution [2].

The main components of a lithium-ion battery are the cathode, anode, electrolyte, separator, and current collector. Typically, carbonaceous materials such as graphite are used as the base of the anode materials. However, graphite anodes are prone to various types of degradation, such as structural damage to the surface during cycling [3], a decrease in capacity [4, 5], and delamination due to joint intercalation of solvents [6]. To solve these problems, research is being carried out on composite materials (for example, nanowhiskers and carbon) to improve the mechanical and electrical properties of graphite anodes.

There are several methods for the synthesis of CuO nanowhiskers, they include "wet" chemical [7], electrochemical and hydrothermal methods [8], as well as methods of thermal and plasma oxidation [9-12]. Thermal oxidation is one of the most widely used methods for producing CuO nanowhiskers due to its simplicity, mass production, and relatively low cost.

2. Experimental

In this work, for the synthesis of copper oxide nanowhiskers, the method of electrochemical deposition from an aqueous solution followed by thermal oxidation was chosen. The ratio of the compounds of electrolyte, the deposition time, and the annealing temperature noticeably affect the size and structure of the synthesized CuO nanoparticles.

To prepare the electrolyte, distilled water and 250 g/l of $CuSO_4{\cdot}5H_2O$ was used with constant stirring and heating. To increase the electrical conductivity of the electrolyte, as well as dissolve the salt, 90 g/l of H_2SO_4 was added to the electrolyte. All reagents were used with an analytical grade at least.

The steel mesh (type AISI 304) was used as the cathode. The mesh is preliminarily cleaned in distilled water, then in an ultrasonic bath, after which it is dried and placed in an electrolyte solution opposite the anode - a copper plate, between them there is a copper reference electrode. The electrochemical cell was connected to the potentiostat (Ellins P45) and then an overvoltage of 100 mV was applied, the deposition time was 40 minutes. After the electrodeposition process, the mesh with copper coating was washed in distilled water and dried. The samples in the form of mesh with copper coating are placed in a muffle furnace for 4 hours at a temperature of 400°C. Thermal oxidation of copper leads to the growth of CuO nanowhiskers.

3. Results and discussion

Scanning electron microscope investigation shows that the deposited copper particles have pentagonal symmetry (figure 1), which means they contain disclination-type defects, the average particle size is 10 μm.

Thermal oxidation of copper leads to the growth of CuO nanowhiskers. Figure 1b shows that the particles have lost their faceting and have a spherical shape; nanowhiskers with a length of up to 5 μm and up to 100 nm in diameter have formed on their surface.

Presumably, the formation of CuO nanowhiskers occurs as a result of a reaction at the CuO / Cu_2O interface, which creates compressive stress in the CuO layer and leads to forced diffusion of copper ions outward along the CuO grain boundaries [13].

In [14] was described and tested batteries with anodes based on copper oxide nanowhiskers with various binders as sodium carboxymethylcellulose (CMC) and polyvinylidene fluoride (PVDF). After 100 charging-discharging cycles, the electrode containing CMC maintains a stable capacity of at least 200 mAh/g, and in the electrode with PVDF, the capacity drops to 60 mAh/g. Presumably, the CMC binder has better adhesion to the surface of Cu_2O/CuO nanowhiskers, adjusting to volumetric expansion during cycling.

Let us consider in more detail the chemical processes taking place on the electrodes. As lithium ions flow into the CuO nanowhiskers, the concentration distribution is continuous, so the transformation strain also varies continuously with position. The central core of the nanowhisker has tension during lithiation.

Lithium ions diffuse in or out of the electrode during charging or discharging. For dilute solutes within the nanowhiskers, the lithium-ion concentration is governed by the diffusion equation:

$$\frac{\partial C}{\partial t} = D\nabla^2 C = D\left(\frac{\partial^2 C}{\partial r^2} + \frac{1}{r}\right) \qquad (1)$$

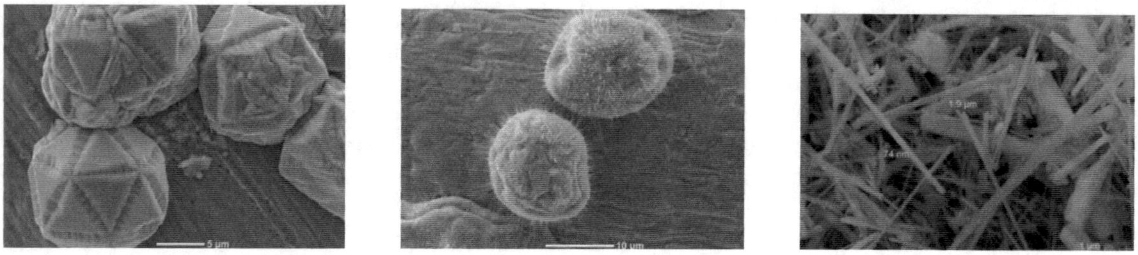

a) b) c)

Figure 1. Micrographs of copper particles deposited on a steel mesh (a) and nanowhiskers grew on particles after annealing (b), (c).

where C is the molar concentration, D is the diffusion coefficient of the lithium-ion.

It should be noted there are two possible ways of operating a battery: potentiostatic or galvanostatic regimes: In the first case the electrode is surrounded by a constant lithium-ion concentration. At galvanostatic regime, the current, thus the ionic flux at the surface of the electrode, is a constant. In our studies [14] the battery is charged/discharged with a constant current density J. Hence, the ionic flux at the surface of the electrode is constant. We assume the initial lithium-ion concentration inside the electrode is C_0 and the concentration at the center is finite. The appropriate boundary and initial conditions to solve the diffusion equation can be written as

$$\begin{cases} C(r,0) = C_0 \\ D\left.\dfrac{dC(r,t)}{dr}\right|_{r=R} = \dfrac{J}{F} \end{cases} \tag{2}$$

where F is Faraday's constant.

The diffusion equation (1) can be non-dimensionalized. Concentration-distance curves in non-dimensionalized forms, plotted from the numerical solution of equation (1) are shown in figure 2.

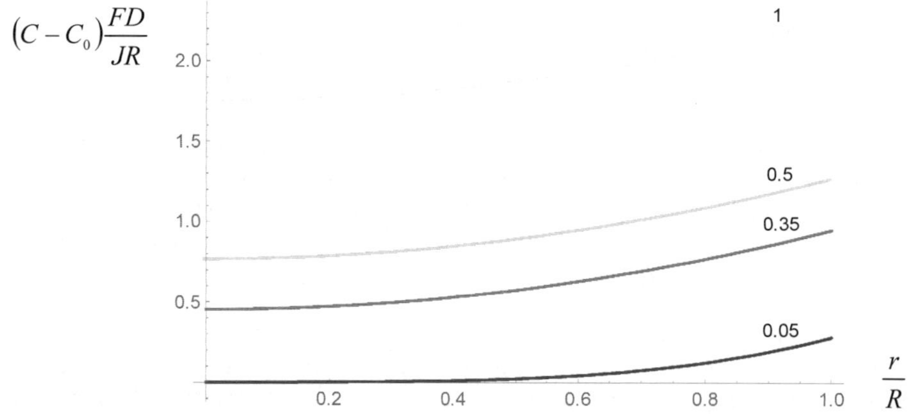

Figure 2. Concentration distribution of lithium ions in the CuO nanowhiskers for constant flux $\frac{J}{F}$ at the surface. The numbers on curves are the value of $\frac{Dt}{R^2}$.

Diffusion-induced stresses can cause the nucleation and growth of cracks, leading to mechanical degradation of the active electrode materials.

The stress state of nanowhiskers during lithiation/delithiation can be written using the analogy between thermal stresses [15]:

$$\begin{cases} \varepsilon_{rr} - \dfrac{\Omega}{3}C = \dfrac{1}{E}\left(\sigma_r - v(\sigma_\theta + \sigma_z)\right) \\ \varepsilon_{\theta\theta} - \dfrac{\Omega}{3}C = \dfrac{1}{E}\left(\sigma_\theta - v(\sigma_r + \sigma_z)\right) \end{cases} \tag{3}$$

where E is Young's modulus, C is the molar concentration, and Ω is the partial molar volume of the lithium ion, σ_r, σ_θ, σ_z are radial, tangential, axial stresses, ε_{rr} and $\varepsilon_{\theta\theta}$ are radial and tangential strains.

In our case, nanowhiskers electrodes were long whiskers grown on a substrate and thus one of the ends of the whiskers is always attached to the substrate. Such long nanowhiskers with a fixed end are under the generalized plane strain condition.

Atomic diffusion in solids is a much slower process than elastic deformation; mechanical equilibrium is established much faster than that of diffusion. So, mechanical equilibrium one can be treated as a static equilibrium problem:

$$\frac{d\sigma_r}{dr} + \frac{\sigma_r + \sigma_\theta}{r} = 0 \tag{4}$$

For infinitesimally small deformation, the radial and tangential strain of cylindrical particles can be related to radial displacement. There is no displacement at the center, furthermore, the radial stress has to satisfy mechanical equilibrium at the surface. Using these conditions, the stress components in dimensionless form were plotted and shown in figure 3.

The radial stress at the center initially increases, reaches a maximum 0.35, and then remains constant until the electrode reaches the saturation concentration. The radial stress is always tensile while tangential stress is tensile at the center and compressive at the surface. The magnitude of tangential stress is the same at the center of the whisker. The tangential stress maximum occurs at the surface at time zero. It should be noted the equation system is symmetric concerning lithiation and delithiation, and mirror opposite conclusions (i.e., replacing tensile stress by compressive stress).

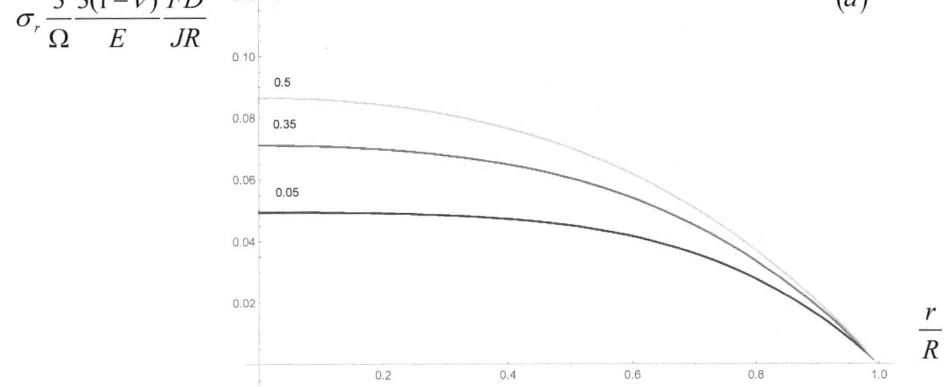

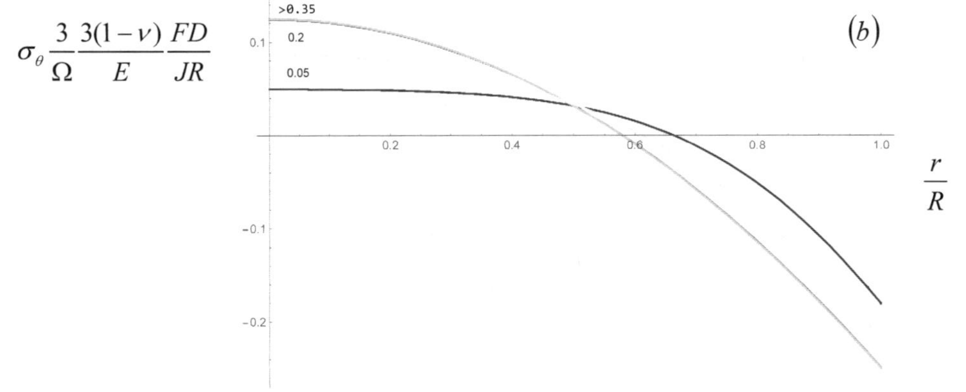

Figure 3. Radial (a) and tangential (b) stresses inside whisker at different radial locations and times of lithation/delithiation.

4. Conclusions

Copper (II) oxide nanowhiskers were synthesized by the method of electrochemical deposition of copper with subsequent thermal oxidation. These materials are used as active components of anodes of lithium-ion batteries.

The analysis of stress, strain, and strain energy in nanowhiskers electrodes, can be used to optimize the nanowhiskers size depending upon the properties to maximize the battery life.

Such CuO nanowhiskers can accommodate large strain without degradation in size. The previous experiment [14] also confirms the better behavior of nanowhiskers electrodes than bulk electrodes.

Acknowledgments
The authors are thankful to the Russian Science Foundation (Grant No. 19-72-10112) for support of this work.

References

[1] Odynets L L, Pergament A L, Stefanovich G B and Chudnosky F A 1995 *Fizika Tverdogo Tela* **37**(7) 2215 (in Russian)
[2] Guoxiu W, Xiaoping S and Jane Y 2009 *J. Power Sources* **189(**1) 543
[3] Hardwick L, Marcinek M, Beer L, Kerr J B and Kostecki R *J. Electrochem.* Soc. **155** A442
[4] Zhang Q and White R 2008 *J. Power Sources* **179** 793
[5] Aurbach D, Markovsky B, Rodkin A, Cojocaru M, Levi E and Kim H J 2002 *Electrochim. Acta* **47** 1899
[6] Zane D, Antonini A and Pasquali M 2001 *J. Power Sources* **97** 146
[7] Zhu J, Bi H, Wang Y, Wang X, Yang X and Lu L 2008 *Mater. Chem. Phys.* **109** 34
[8] Tiwari J, Tiwari R and Kim K 2012 *Progr. Mater. Sci.* **57**(4) 724
[9] Cheng G, Wang S J, Cheng K, Jiang X H, Wang L X, Li L S, Du Z L and Zou G T 2008 *Appl. Phys. Lett.* **92** 223116
[10] Zhang K, Rossi C, Tenailleau C, Alphonse P and Chane-Ching J Y 2007 *Nanotechnology* **18** 275607
[11] Sahoo S, Husale S, Colwill B, Lu TM, Nayak S, Ajayan P M 2009 *ACS Nano* **3**(12) 3935
[12] Farbod M, Narges MG and Iraj K 2014 *Ceram. Int.* **40** 517
[13] Yuan L, Wang Y, Mema R and Zhou G 2011 *Acta Mater.* **59** 2491
[14] Hu P, Wang B, Xiao D and Aifantis K E 2019 *ChemElectroChem* **6** 3139
[15] Carslaw H S, Jaeger J C 1959 *Conduction of Heat in Solids*, second ed., (Clarendon Press, Oxford) p 348

Copper oxide nanowhiskers: structure, growth mechanisms, and associated stresses

E Podlesnov[1], N V Churkunova[1,2] and M V Dorogov[1]

[1] International Research Center of Functional Materials and Devices of Optoelectronics, ITMO University, Kronverksky Pr. 49, St. Petersburg, 197101, Russia
[2] Department of General Physics, Togliatti State University, Belorusskaya St. 14, Togliatti, 445670, Russia

E-mail: maxim@tltsu.ru

Abstract. A detailed study of the structure of nanowhiskers and the processes accompanying their growth during thermal oxidation are presented. The structure of nanowhiskers is studied by scanning and transmission electron microscopy and X-ray diffraction. The role of internal stresses in the substrate on the growth of nanowhiskers is shown, and the stress state for various configurations of the copper/copper oxide system is calculated. The sorption, catalytic, electrochemical, and mechanical properties of oxide nanowhiskers are described. The areas of the practical application of copper oxide nanowhiskers such as water purification systems and lithium batteries are shown.

1. Introduction

First reports on the observation of the growth of copper oxide nanowhiskers during the thermal oxidation of copper can be found in the works of the 1950s [1 - 2]. This phenomenon turned out to be a simple and effective way of producing copper oxide nanowhiskers, which are already widely used in various fields, such as photocatalysis [3], solar panels [4], lithium batteries [5], etc. However, the issue of a selective method for producing such nanowhiskers has not yet been resolved. To develop a selective method (for example, in terms of size) for producing nanowhiskers by thermal oxidation, a clear understanding of the nature of the processes accompanying the formation and growth of nanowhiskers is required. Many phenomenological models are describing the growth of whiskers: dislocation, diffusion, stress-stimulated, etc. [6 - 9]. Summarizing the known facts and factors indicated in the aforementioned models, it should be noted that thermal oxidation is a complex process, and to understand its nature and modeling, it is necessary to take into account the interrelation of individual processes (diffusion, thermal expansion, phase transformations, and stress state). The chemomechanical approach to this problem makes it possible to take into account the effect of stress state on chemical processes, in particular, phase transformations: copper - copper oxide.

In our previous works [3], [6], we estimated the stress state during annealing for continuous electrolytic coatings deposited on a mesh substrate as well as individual particles. Note that the growth of whiskers is observed experimentally not only on cylindrical or spherical objects of the core-shell type. Therefore, in this work, we have estimated the stresses in flat copper layers subjected to thermal oxidation.

2. Material and methods

A rolled plate of the stainless steel mesh (type AISI 321) with thickness 700 μm was used as a substrate. The metal electrodeposition method was employed to coat the plate with a bulk copper layer as described in [10]. The copper deposition was conducted from the sulphuric electrolyte in a potentiostat regime. The duration of the electrodeposition was 30 minutes under the overvoltage of 200 mV. The regime resulted in a continuous coating with a thickness of 50-60 μm that was subsequently mechanically separated from the substrate. Thermal oxidation of the samples was conducted in a muffle furnace with a programmable process controller under a heating rate of 15 °C/min.

Morphology of copper and oxide layers was investigated by scanning electron microscopy (SEM, TESCAN Mira 3, and JEOL JCM 6000). The experimental analysis of the phase transformations accompanying the growth of nanowhiskers was carried out by X-ray diffraction (X-ray diffractometer Shimadzu XRD 7000). To study all samples diffraction patterns were taken in the range 20 ° –120 ° with a step of 0.02 ° and a scanning speed of 2 °/min. An X-ray tube with a copper anode with a power of 1.5 kW was used while the tube voltage and the current were set to 30 kV and 30 mA respectively.

3. Results and discussion

During the oxidation process, the oxide film and substrate can be modeled as a thin film/substrate system with a moving boundary, as shown in figure 1. To simplicity and due to the symmetry of the problem, we considered half systems.

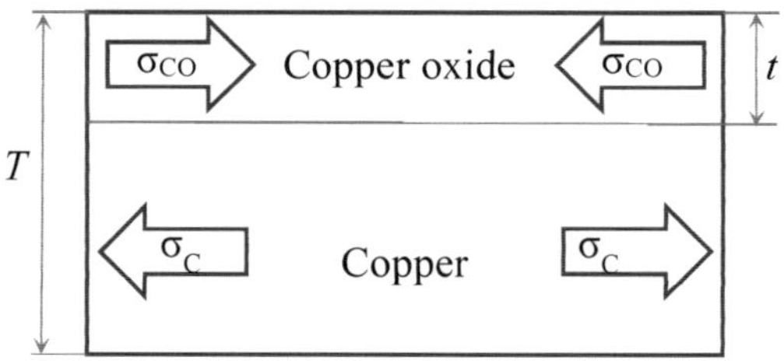

Figure 1. Scheme of the stress-strain state in the plane copper oxide/copper system, t is the thickness of the oxide film as a function of time, σ_{CO} and σ_C is the average biaxial stress in the oxide layer and copper, T is the initial thickness of the samples relative to the symmetric axis.

It is assumed that the oxide film is well bonded to the substrate. Taking into account the geometric symmetry of the system, the deformation in the oxide and the substrate satisfies the compatibility condition.

After analyzing the three stages of oxidation, the chemical mechanic's equation is proposed in a dimensionless form:

$$\begin{cases} \dot{\tilde{t}} = \dfrac{B\beta\Delta C}{\tilde{t}} e^{\tilde{\alpha}\tilde{\sigma}} \\ \left(M(1-\tilde{t})+\tilde{t}\right)\dot{\tilde{\sigma}} + \left(A(1-\tilde{t}) + \dfrac{\tilde{\sigma}}{1-\tilde{t}}\right)\dot{\tilde{t}} = 0 \end{cases}$$

where $M = \dfrac{M_C}{M_{CO}}$, $A = \dfrac{M_C D_{CO} T}{\sigma_0}$, $B = \dfrac{R_S^2}{T^2}$, $\tilde{\alpha} = \dfrac{\alpha\Omega\sigma_0}{R\theta}$, $\tilde{\sigma} = \dfrac{\sigma_{CO}}{\sigma_0}$, $\tilde{t} = \dfrac{t}{T}$, M_C and M_{CO} are the biaxial moduli of the copper base and the copper oxide layer, respectively, D_{CO} is the coefficient depending on the microstructure of the oxide, β is the coefficient taking into account the substance and the reaction (copper oxidation), ΔC is the difference in oxygen concentration at the boundary of gas/oxide and oxide/copper, α is a positive dimensionless constant, Ω is the molar volume of oxygen, R is the gas constant, θ is a temperature.

The initial conditions of the problem are written in the form:

$$\begin{cases} \tilde{l} = 0 \\ \tilde{\sigma} = 0 \end{cases}$$

For numerical calculations, the following parameters were selected: $M = 0.8$, $A = 500$, $B = 500$, $\tilde{\alpha} = 2.5$, $\beta\Delta C = 1.6 \cdot 10^{-7}$. The calculation results are shown in figure 2. It should be noted, one can take into account the defect structure by writing down the initial conditions, for example, indicate the stresses of the defects present in the sample.

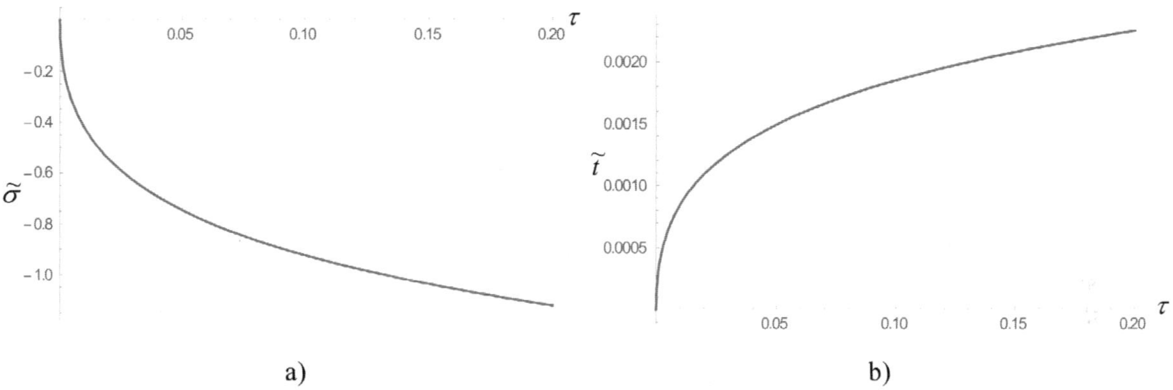

Figure 2. Kinetics of oxidation of a flat copper sample, *a* and *b* show the dependencies of the dimensionless stress in the oxide layer and the dimensionless thickness of the oxide layer on the dimensionless time respectively.

According to the theory of reactive diffusion [11], when copper is oxidized in air, from the beginning of the process, a two-phase diffusion zone is formed, consisting of layers of Cu_2O and CuO. During the formation of a two-phase diffusion zone, the thickness of the CuO phase layer is much less than the thickness of the Cu_2O layer and the CuO layer grows at a rate almost an order of magnitude lower than the growth rate of the Cu_2O layer. This is probably the reason that at the initial stages of thermal oxidation when the thickness of the diffusion zone is small, it is not possible to detect the CuO phase by X-ray phase analyzes as shown in figure 3. However, the presence of the CuO phase on the surface, starting from the beginning of oxidation, can be judged by the black color of the sample characteristic of CuO. This color of the surface is retained during long-term oxidation when the presence of the CuO phase is also detected by X-ray phase analysis.

4. Conclusions

Thus, it can be argued the oxidation of copper at atmospheric air pressure begins with the formation of the thinnest layer of the CuO phase on the sample surface. This is followed by the formation of the Cu_2O phase between the CuO layer and the copper substrate, which in all cases is the main oxidation product and grows at a much higher rate than the CuO phase.

It was shown that a noticeable yield of bivalent copper oxide CuO at a heating rate of 15 ° / min begins to form at a temperature of 400 ° C. At this time, in addition to the growth of whiskers, active growth of the CuO layer occurs. However, the temperature of the beginning of the formation of whisker structures is 250 ° C.

In general, the temperature and duration of the heat treatment process have a significant effect on the oxidation of copper. It has been revealed that, upon oxidation of copper in air, from the beginning of the process, a two-phase diffusion zone is formed, consisting of Cu_2O and CuO layers. Of these, the primary is the CuO phase. In this case, the oxidation of copper at the initial stage is accompanied by

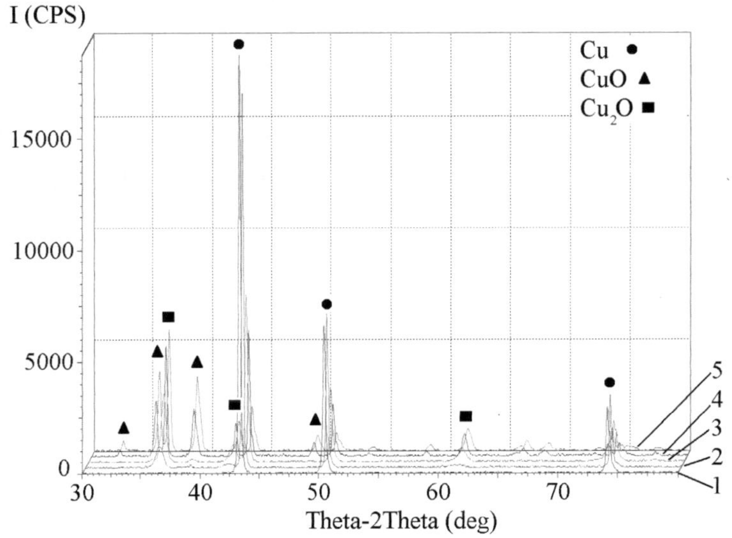

Figure 3. X-ray diffraction pattern of the thermal oxidation of a copper plate heated up to 400 °C and holdup time 4 hours, 1 is original copper, 2, 3, and 4 are heating in air to 250, 300, and 400 °C respectively, 5 is annealing at 400 °C for 6 hours.

an intensive growth of only the Cu_2O phase. Compressive stresses formed in the oxide layer are the driving force behind the release of copper cations to the surface, where these are oxidized with oxygen and ensure the growth of nanowhiskers.

The proposed equations of chemomechanics make it possible to estimate the stress state of the substrate during the growth of nanowhiskers in the process of thermal oxidation.

Acknowledgments
The authors are thankful to the Russian Science Foundation (project No. 19-72-10112) for support of this work.

References

[1] Cowley J M 1954 *J. Electrochem. Soc.* **101** 277
[2] Gulbransen E A, Copan T P and Andrew K F 1961 *J. Electrochem. Soc.* **108** 119
[3] Dorogov M, Kalmykov A, Sorokin L, Kozlov A, Myasoedov A, Kirilenko D, Chirkunova N, Priezzheva A, Romanov A and Aifantis E C 2018 *Mater. Sci. Technol.* **34** 2126
[4] Zhang Q, Zhang K, Xu D, Yang G, Huang H, Nie F, Liu C and Yang S 2014 *Prog. Mater. Sci.* **60** 208
[5] Hu P, Dorogov M, Xin Y and Aifantis K E 2019 *ChemElectroChem* **6** 3139
[6] Dorogov M V et al. 2015 *Appl. Surf. Sci.* **346** 423
[7] Mema R, Yuan L, Du Q, Wang Y and Zhou G 2011 *Chem. Phys. Lett.* **512** 87
[8] Lindborg U 1976 *Acta Metall.* **24** 181
[9] Eshelby J D 1953 *Phys. Rev.* **91** 755
[10] Vikarchuk A, Gryznova N, Dovzhenko O, Dorogov M and Romanov A 2014 *Adv. Mater. Res.* **1013** 205
[11] Park J H and Natesan K 1993 *Oxid. Metals* **39** 411

IOP Conf. Series: Materials Science and Engineering 1008 (2020) 012045 doi:10.1088/1757-899X/1008/1/012045

Influence of free torsion deformation modes on the microstructure and mechanical properties of commercially pure copper

I S Kodirov, G N Aleshin and A G Raab

Institute of Physics of Advanced Materials, Ufa State Aviation Technical University, 12 Karl Marx street Ufa, 450008, Russia

E-mail: ilyas-kodirov@mail.ru

Abstract. The paper investigates the effect of the free torsion rate of cylindrical specimes of commercially pure copper (grade M1) on their deformed state, microstructure, and mechanical properties. The study of the microstructure was carried out by means of transmission electron microscopy and X-ray diffraction analysis in the central region and at the periphery of the deformed rod. The mechanical properties were evaluated by measuring the microhardness. It was found that an increase of the torsion rate, other things being equal, promotes more intensive strengthening and refinement of the microstructure.

1. Introduction

Recently, it has become more and more obvious that traditional methods of thermomechanical treatment have been developed so completely that it is impossible to expect a significant improvement in the properties of the resulting bars [1-3]. The use of new methods of plastic deformation is one of the most promising ways in the creation of materials with an improved set of mechanical properties. With various combined processing schemes, processes are implemented that make it possible to improve the operational properties of materials by purposefully changing their microstructure [1-4].

Studies [4-14] have shown that the use of a simple shear in various deformation schemes allows realizing a special state of intensified motion of defects, which, in turn, leads to the formation of equiaxed fine grains with a small number of defects in its body. One of the promising methods of deformation for obtaining such a state in metal workpieces is the free torsion method. This method provides a significant non-monotonicity and, at the same time, a high value of the accumulated degree of shear strain in the metal, which leads to the evolution of the dislocation structure, activation of new slip systems, misorientation of microvolumes inside the grain and rearrangement of the formed fragments into ultrafine grains with nonequilibrium high-angle boundaries [1,6, 15,16].

The aim of this work was to study the effect of the rate of deformation processing by the method of free torsion on the microstructure and mechanical properties of commercially pure copper (grade M1).

2. Experimental

The studies were carried out on rods with a length of 250 mm and a diameter of 10 mm. The deformation was carried out at rates of rotation of the gripper - 12.5, 20 and 40 rpm until the failure of the rods (16,

IOP Conf. Series: Materials Science and Engineering 1008 (2020) 012045 doi:10.1088/1757-899X/1008/1/012045

12 and 9 revolutions, respectively). The accumulated shear strain before failure was determined by the expression [15]:

$$e = \frac{\pi d N}{h},$$

where d is the sample diameter, N is the number of revolutions before failure, h is the sample length.

After deformation processing, the specimens were cut at an angle of 15 ° to the axis of the rod. The mechanical properties were evaluated by the microhardness measurement method (Buehler Micromet 5101) at a load of 100 g during 10 s. Microstructural studies were carried out in the center and periphery of the beveled samples by transmission electron microscopy (JEM JEOL 2100; accelerating voltage 200 kV) and X-ray diffraction analysis (Rigaku Ultima IV) using CuKα radiation. The X-ray diffraction patterns were calculated using the full-profile Rietveld analysis in the MAUD software [17]. The dislocation density was calculated using the expression given in [18, 19]:

$$\rho = \frac{2\sqrt{3}\sqrt{\langle \varepsilon^2 \rangle}}{d \cdot b},$$

where b is the value of the Burgers vector for FCC copper alloys, d is the size of the coherent scattering domains (CSD) and ε is the value of the microstrain.

3. Results and discussion

The calculation of the accumulated strain at rotation rates of 12.5, 20, and 40 rpm showed that the workpieces accumulate shear strain $e = 2.5$, 1.9, and 1.4, respectively, before failure. Fig. 1 demonstrates the effect of deformation processing at different rates on the microhardness value. It can be concluded that, in general, an increase in the rotation rate of the lathe gripper leads to an increase in the microhardness value. In the initial state, the average microhardness was (102 ± 7) HV. The deformation leads to an increase in this parameter up to (109 ± 9 HV), (120 ± 6 HV) and (123 ± 9 HV) at gripper rotation speeds of 12.5, 20 and 40 rpm, respectively.

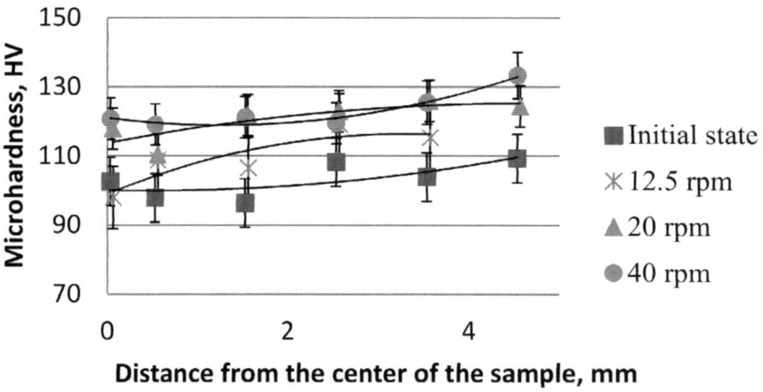

Figure 1. Microhardness distribution on the cross section of the samples in the initial state and after processing at different strain rates.

TEM studies have shown that in the central zone of the sample after processing at a speed of 12.5 rpm, the dislocation substructure is a weakly oriented cellular structure with a fragment thickness of about 300 nm. An increase in the deformation rate to 40 rpm (Fig. 4) led to a thinning of the fragments

IOP Conf. Series: Materials Science and Engineering 1008 (2020) 012045 doi:10.1088/1757-899X/1008/1/012045

to 200 nm and to a further set of misorientations between them, which is noticeable by the blurring of reflections in the electron diffraction patterns (Fig. 2 and Fig. 4). In the peripheral region, after free torsion at 12.5 rpm (Fig. 3), the dislocation substructure has the form of a disoriented cellular structure with partial fragmentation with a cell size of 300-600 nm and a fragment thickness of 200 nm. An increase in the deformation rate to 40 rpm (Fig. 5) led to a further refinement of the structural elements — cells to 250 nm and the thickness of fragments to 150 nm.

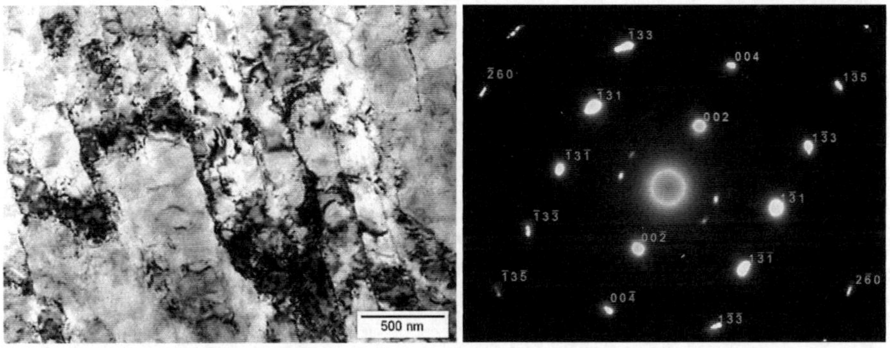

a b

Figure 2. Microstructure after processing: a - bright field,
b - electron diffraction pattern from the corresponding area (zone axis 310).
12.5 rpm, center

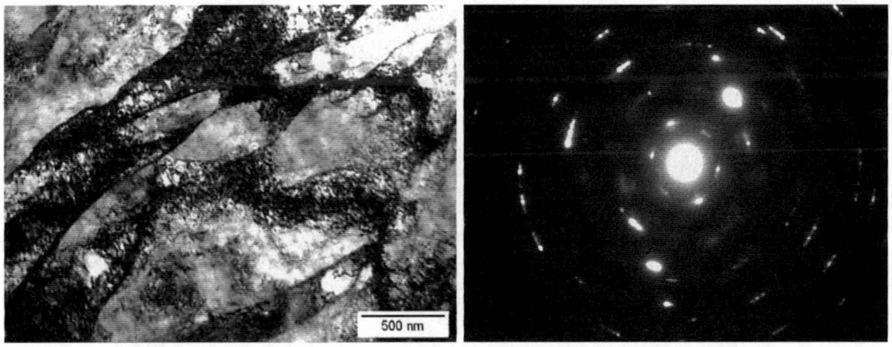

a b

Figure 3. Microstructure after processing: a - bright field,
b - electron diffraction pattern from the corresponding area.
12.5 rpm, periphery

The XRD study (table 1) showed that the intensity of the peaks in the peripheral region of the samples decreases relative to the center and their broadening occurs, which indicates an increase in the defectiveness of the structure. The calculation of diffraction patterns showed that the dislocation density at the periphery of the studied samples was higher than at the center, the CSD was lower, and the microstrain value was non-monotonic, which is associated with the rearrangement of the dislocation structure and relaxation of internal stresses.

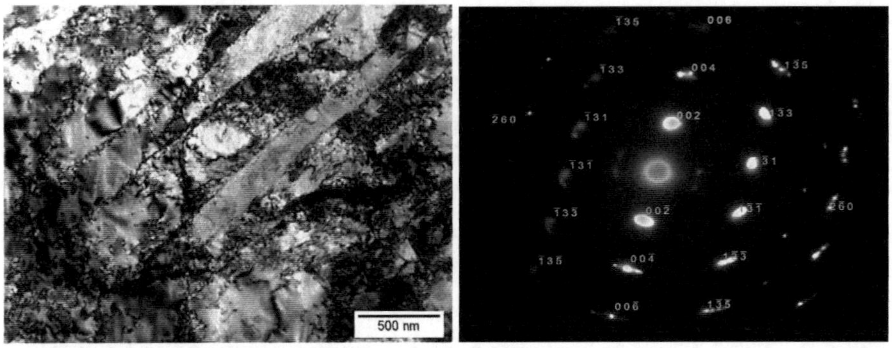

a b

Figure 4. Microstructure after processing: a - bright field, b - electron diffraction pattern from the corresponding area (zone axis 310). 40 rpm, center

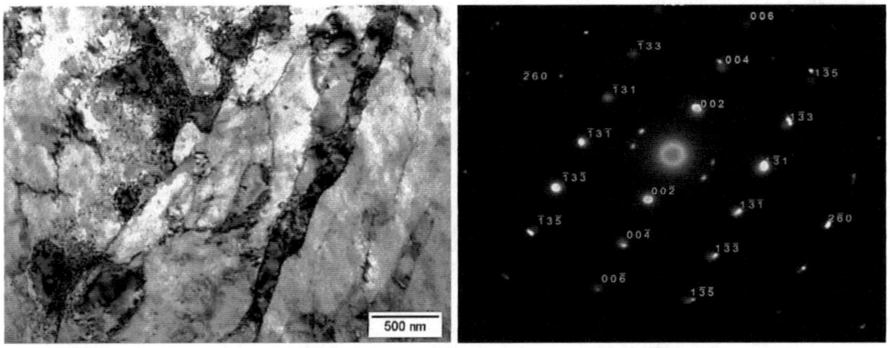

a b

Figure 5. Microstructure after processing: a - bright field, b - electron diffraction pattern from the corresponding area (zone axis 310). 40 rpm, periphery

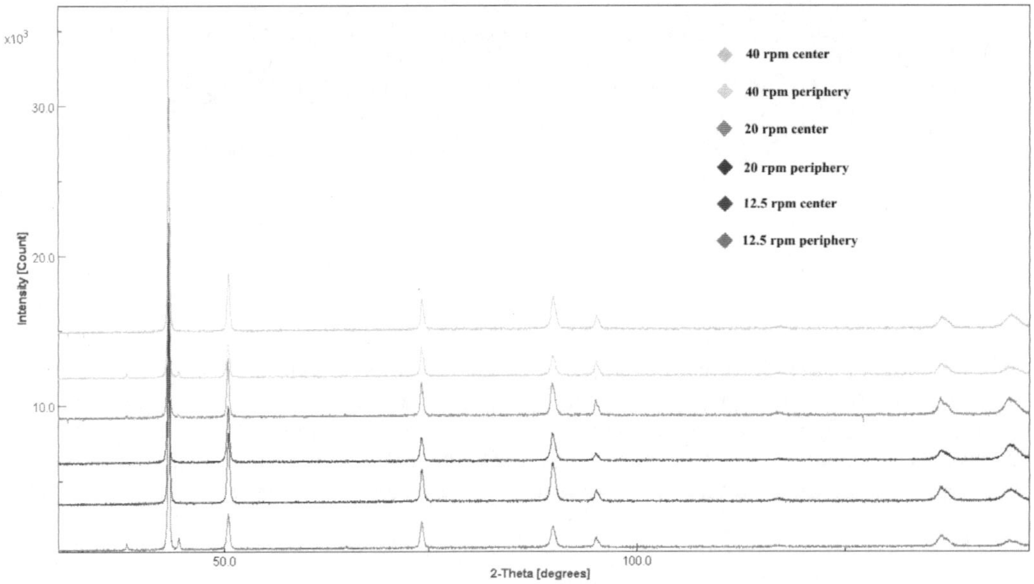

Figure 6. Diffraction patterns of deformed samples.

Table 1. XRD analysis results

State	Lattice parameter, Å	CSD size, nm	Microstrain, %	Dislocation Density, m^{-2}
13 rpm center	3,6168	83	0,13	2,1E+14
13 rpm periphery	3,6167	62	0,11	2,3E+14
20 rpm center	3,6172	96	0,10	1,5E+14
20 rpm periphery	3,6160	68	0,12	2,4E+14
40 rpm center	3,6160	87	0,12	1,9E+14
40 rpm periphery	3,6169	71	0,12	2,3E+14

4. Summary

It was found that the deformation by the method of free torsion leads to the hardening of the workpiece practically without changing the initial geometric parameters, and more intensively in its peripheral layers. Thus, an increase in the rotation rate provides a more intense hardening, but leads to a decrease in the plasticity resource at the moment of failure, therefore, at a speed of ω = 12.5 rpm, the accumulated deformation before fracture was 2.5, and at ω = 40 rpm - 1.4. Despite the decrease in the accumulated strain to fracture with an increase in the rate, the average value of the microhardness continues to increase and reaches 123 HV at 40 rpm, which is 20% higher than the initial value.

An increase of the gripper rotation rate leads to the refinement of structural elements at the periphery (from 600 nm to 250 nm) and in the central region (from 300 to 200 nm) of the metal. There is also a rearrangement of the dislocation substructure from a weakly misoriented cellular to a misoriented cellular one.

The CSD size and dislocation density reflect the distribution pattern of the accumulated strain during free torsion, since the most deformed peripheral layers have a higher dislocation density and, accordingly, a smaller CSD parameter.

References

[1] Haritonov B and Usanov M 2015 *Obrabotka Sploshnykh i Sloistykh Materialov* **1** 50 (in Russian)
[2] Utyashev F and Raab G 2013 *Deformatsionnyye Metody Polucheniya i Obrabotki Ul'tramelkozernistykh i Nanostrukturnykh Materialov* (Ufa: Gilem) (in Russian)
[3] Mulyukov R, Imaev R and Nazarov A 2013 *Nauchno-Tekhnicheskiye Vedomosti SPbGPU. Fiziko-Matematicheskiye Nauki* **4(1)** 190
[4] Pashinskaya E 2009 *Fiziko-Mekhanicheskiye Osnovy Izmel'cheniya Struktury pri Kombinirovannoy Plasticheskoy Deformatsii* (Donetsk: Veber) (in Russian)
[5] Segal V, Reznikov V and Kopylov V 1994 *Protsessy Plasticheskogo i Strukturoobrazovaniya Metallov* (Minsk: Nauka i Tekhnika) (in Russian)
[6] Berdin V, Karavaeva M and Syutina L 2003 *Metal Science and Heat Treatment* **45** 11
[7] Berdin V and Karavaeva M 2001 *Metally* **3** 35
[8] Berdin V, Karavaeva M, Syutina L and Nurieva S 2004 *Metal Science and Heat Treatment* **46** 1
[9] Kaibyshev O, Berdin V, Karavaeva M, Kashaev R and Syutina L 2001 *Mater. Sci. Forum* **357–359** 225
[10] Zhilyaev A, Raab A, Raab G and Kodirov I 2019 *Letters on Materials* **9** (4s) 571
[11] Berdin V, Karavaeva M and Nurieva S 2001 *Metallovedenie i Termicheskaya Obrabotka Metallov* **7** 16 (in Russian)
[12] Berdin V and Karavaeva M 2001 *Metally* **3** 35 (in Russian)
[13] Berdin V, Karavaeva M and Nurieva S 2001 *Metal Science and Heat Treatment* **43** 7
[14] Korchunov A, Chukin M, Polyakova M and Emaleeva D 2011 *Vestnik Magnitogorskogo Gosudarstvennogo Tekhnicheskogo Universiteta im. G.I. Nosova* **1** 43 (in Russian)
[15] Valiev R, Islamgaliev R and Alexandrov I 2000 *Prog. Mater. Sci.* **45** 103
[16] Valiev R, Zhilyaev A and Langdon T 2014 *Bulk Nanostructured Materials: Fundamentals and Applications* (New Jersey: John Wiley & Sons, Inc)
[17] Lutterotti L, Bortolotti M, Ischia G, Lonardelli I and Wenk H 2007 *Z. Kristallogr. Suppl.* **26** 125
[18] Smallman R and Westmacott K 1957 *Philos Mag* **2** 669
[19] Williamson G and Smallman R 1956 *Philos Mag* **1** 34

Superplastic behaviour and solid state weldability of a polycrystalline heavily alloyed nickel-base superalloy

Sh Kh Mukhtarov[1], V M Imayev[1], A A Ganeev[1] , R V Shakhov[1] and A V Logunov[2]

[1] Institute for Metals Superplasticity Problems of Russian Academy of Sciences, 39 Stepan Khalturin Street, Ufa, 450001, Russia
[2] PJSC "UEC-Saturn", 163 Lenin Avenue, Rybinsk, 152903, Russia

E-mail: shamilm@imsp.ru

Abstract. Effective thermomechanical treatment has been developed for the heavily alloyed Ni-base superalloy produced via ingot metallurgy with a high content of γ'-forming elements. It allowed us to eliminate dendritic segregation and to obtain a uniform fine-grained structure with a γ grain size of d=5-15 μm. In the fine-grained condition the superalloy exhibited superplastic properties at 1125-1175°C (δ>300%, σ=10-27 MPa, m>0.3). Pressure welding experiments were performed at 1150°C for samples with the fine-grained structure. The solid-phase joint was formed only after compression to the strain of 50%. It was shown that insertion of a nanocrystalline interlayer of the same superalloy between the welded samples facilitated formation of the solid-phase joint.

1. Introduction

The development of aircraft gas turbine engines requires the development of new advanced heat-resistant materials, such as nickel-base superalloys, capable of operating at higher loadings and temperatures [1,2]. Computer methods are used to design new superalloys. One of the designing ideas in the development of new nickel based superalloys is associated with an increase of the γ'-forming elements (Al, Ti, Nb, Ta) content and alloying with substitution elements like rhenium, which could effectively strengthen the γ matrix [3]. Rhenium in an amount of typically 2-6 wt. % is mainly used as an alloying element in single crystalline superalloys of 2nd-6th generations [4-6]. In recent years, several polycrystalline powder metallurgy superalloys containing rhenium have been developed [7,8]. Data on mechanical behavior and technological properties of the rhenium alloyed nickel base polycrystalline superalloys are not readily available in the literature.

The present work was aimed to study the microstructure and technological properties of a heavily alloyed rhenium-containing nickel based superalloy produced via ingot metallurgy and subjected to thermomechanical treatment. The superplastic properties and solid state weldability of the superalloy in the fine-grained condition obtained by thermomechanical treatment were investigated.

2. Material and methods

The heavily alloyed nickel based superalloy with the nominal chemical composition Ni-30(Cr,Co,Re)-12(Al,Ti,Ta,Nb)-7.5(W,Mo) (in wt.%) manufactured by ingot metallurgy was taken as a starting material. The superalloy composition measured by energy dispersive X-Ray analysis was found to be very close to its nominal composition. The γ′ solvus temperature for the superalloys was defined as T_s=1220°C. Thermomechanical processing for this superalloy was developed earlier in our previous work [9].

Microstructure examination was performed using scanning (SEM, BSE and SE) and transmission electron microscopy (TEM). The fine-grained forgings obtained by thermomechanical treatment were used to prepare flat samples with a gauge section 10×3×2 mm^3 for tensile tests. The tensile tests were carried out in air without any protection against oxidation in the temperature range of T=1100-1200°C with an initial strain rate of $\dot{\varepsilon}=5\times10^{-4}$ s^{-1}. Elongation to rupture, *δ*, the true stress value σ_{30} corresponding to 30% of elongation and true stress-elongation curves *σ-e* were determined. The true stress was calculated taking into account the graduate narrowing of the sample cross section during tensile testing. The strain rate sensitivity coefficient *m* was defined at 1125-1175°C by varying the strain rate employed.

Solid state pressure welding experiments were carried out using the fine-grained cylindrical specimens Ø6 mm × 8 mm, which were cut out of the forged workpieces. Prior to welding experiments the specimen surfaces were mechanically polished so that the roughness of the welded surfaces was the same. Pressure welding experiments were conducted under isothermal conditions in the argon atmosphere at T=1150°C using compression with the initial strain rate of $\dot{\varepsilon}=5\times10^{-4}$ s^{-1} to an engineering strain of *ε*=20 and 50%. The pressure welding experiment under the same conditions was also performed using a foil (with a thickness of 0.2 mm) of the superalloy as an interlayer, which was obtained by high pressure torsion (HPT) at room temperature by 5 turns of a thin fine-grained specimen with an initial size of Ø10 mm × 1 mm [10].

3. Results and discussion

3.1. Microstructure characterization

The study of the microstructure evolution during thermomechanical treatment of this superalloy showed that the formation of a fine-grained structure occurred as a result of continuous dynamic recrystallization of the γ phase [9]. After thermomechanical treatment, the microstructure was mostly recrystallized with a γ grain size of 5-15 μm. There were also individual large grains with a size of 30-100 μm. In the course of the thermomechanical treatment, precipitation and coarsening of incoherent primary γ′ phase occurred along γ grain boundaries as well. The size of the primary γ′ phase was $d_{\gamma'}$=1-5 μm, their volume fraction was defined as about 15% (figure 1a). Dispersed coherent particles of the secondary γ′ phase with a mean size of about 0.2 μm were precipitated within γ grains during cooling after thermomechanical treatment (figure 1). The volume fraction of the γ′ phase was defined as 68% [9]. A high volume fraction of the γ′ phase resulted from a high content of the γ′-forming elements. The carbide particles were sometimes observed, their volume fraction was about 1.5%. HPT of the fine-grained state led to the formation of a nanocrystalline (partially fragmented) microstructure [10]. The nanocrystalline foil obtained via HPT was used as the interlayer to facilitate the diffusion bonding during pressure welding.

3.2. Superplastic behaviour

Figure 2 demonstrates the results of tensile tests. Superplastic elongations (δ>300%), low flow stresses (10-27 MPa) and a higher strain rate sensitivity coefficient (m>0.3) typical of superplasticity were obtained at T=1125-1175°C and $\varepsilon'=5\times10^{-4}$ s^{-1}. The highest elongation (δ≈400%) was reached at 1150 and 1175°C. Microstructure examination of the sample tensile strained at 1175°C detected γ grain growth up to d_γ=15-20 μm (figure 3), which occurred as a result of grain boundary sliding and is typical of the superplastic behavior of fine-grained alloys. After tensile testing at 1125-1150°C the γ grain size was not appreciably changed. This suggests that dynamic recrystallization (first of all in relatively large γ grains) occurred during superplastic flow at 1125 and 1150°C along with the γ grain growth resulted from grain boundary sliding. Therefore, the flow stresses at 1125 and 1150°C slightly decreased as the elongation increased from 25 to 150%. Note that the flow stress changes could also be associated with minor dissolution and precipitation of the γ′ phase during tensile testing. This might occur because the initial strain rate decreased gradually with increasing the sample length during tensile testing. The tensile test at 1200°C led to a fast γ grain growth already prior to testing because of dissolution of the γ′ phase. That resulted in rapid fracture of the sample without any narrowing (figures 2).

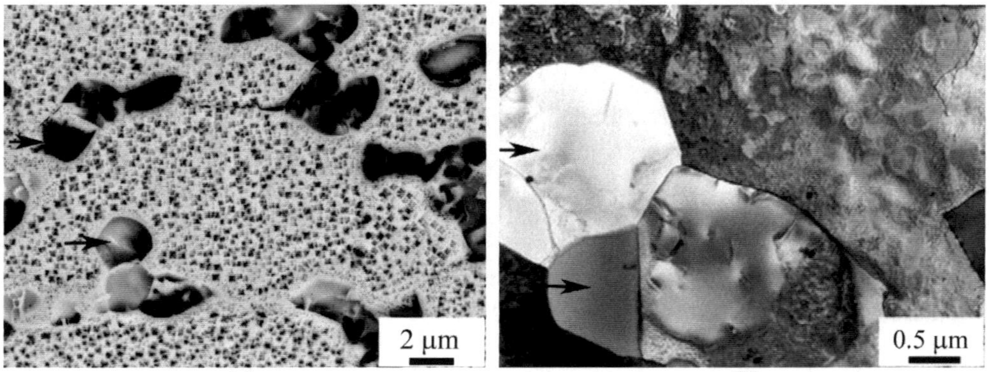

(a) (b)

Figure 1. The microstructure images of the Ni-30(Cr,Co,Re)-12(Al,Ti,Ta,Nb)-7.5(W,Mo) superalloy in the fine-grained condition obtained by thermomechanical treatment: (a) SEM, BSE, (b) TEM, bright field. The arrows show the primary γ′ phase.

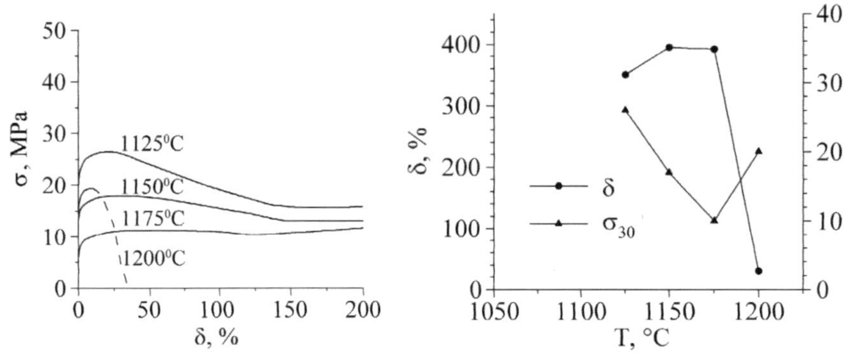

(a) (b)

Figure 2. The temperature dependences of (a) the true stress vs. elongation and (b) the total elongation and the flow stress obtained as a result of tensile tests at T=1125-1200°C ($\dot{\varepsilon}=5\times10^{-4}$ s^{-1}).

3.3. Pressure welding experiments

Pressure welding experiments showed that the solid-phase joint was formed at a temperature of T=1150°C only after compression to the strain of 50% (figure 4a), whereas the strain of 20% did not result in bonding formation. Even after compression to the strain of 50% separate voids were distinguished within the bonding zone (figure 4a). At the same time, the use of the nanocrystalline foil as the interlayer between the welded samples allowed decreasing the strain value required for bonding formation. After compression to the strain of 20% the diffusion bonding was formed, although some voids were observed along the welding joint (figure 4b). During pressure welding of samples with the nanocrystalline interlayer, a noticeable growth of γ grains and coarsening of γ′ particles occurred in the vicinity of the bonding zone that can be attributed to extensive operation of grain boundary sliding in the interlayer during compression. The growth of γ grains and coarsening of γ′ phase in the vicinity of the bonding zone occurred faster than in the base fine-grained material because of the strain localization within the nanocrystalline interlayer during compression. Apparently, this was a result of a high thermodynamic non-equilibrium of the nanocrystalline interlayer obtained via HPT at room temperature.

Figure 3. The BSE image obtained from the gauge area near the fracture zone of the superalloy sample after tensile testing at T=1175°C ($\dot{\varepsilon}=5\times10^{-4}$ s^{-1}).

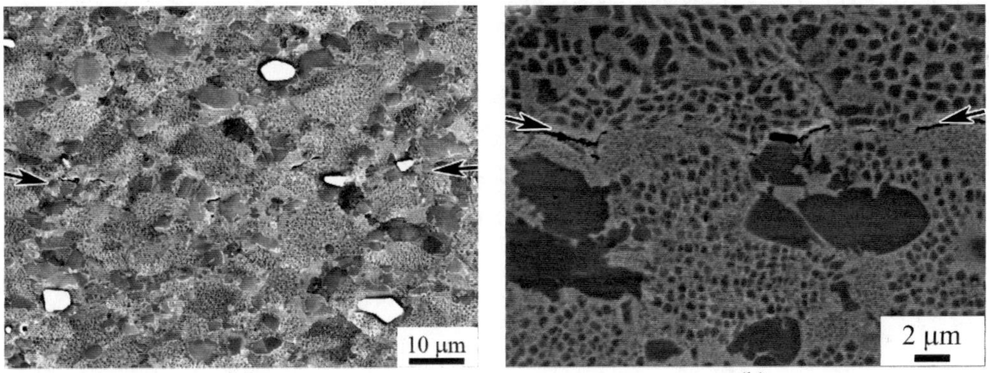

(a) (b)

Figure 4. The BSE images of the solid-phase joints obtained by pressure welding of the superalloy samples at T=1150°C ($\dot{\varepsilon}=5\times10^{-4}$ s^{-1}): (a) ε=50%; (b) ε=20% with using the nanocrystalline interlayer. The solid-phase joints are arrowed.

4. Conclusions

In the present work superplastic behavior and solid state weldability were investigated for the heavily alloyed superalloy Ni-30(Cr,Co,Re)-12(Al,Ti,Ta,Nb)-7.5(W,Mo) (wt.%) in the fine-grained condition obtained via thermomechanical treatment. The Ni-30(Cr,Co,Re)-12(Al,Ti,Ta,Nb)-7.5(W,Mo) superalloy in the fine-grained condition demonstrated superplastic properties in the temperature range of 1125-1175°C (by 45-95°C below the γ′ phase solvus temperature). High elongations (δ≈350-400%), low flow stresses (10-27 MPa) and increased strain rate sensitivity coefficient (m>0.3) typical of superplastic behavior of fine-grained alloys were observed in this temperature range.

The pressure welding experiments performed in the argon atmosphere for the fine-grained superalloy samples revealed extremely difficult formation of a sound solid-phase joint. More or less acceptable joint was attained only after compression at T=1150°C ($\dot{\varepsilon}=5\times10^{-4}$ s^{-1}) to the strain of ε=50%. It was revealed that the use of the nanocrystalline interlayer (produced by HPT at room temperature) between the welded fine-grained samples improved the weldability and reduced the strain value required for the formation of the solid-phase joint down to ε=20%.

Acknowledgments
The present work was funded by the Russian Foundation for Basic Research (Research Project No 18-08-00997-a). The work was performed using the facilities of the shared services center «Structural and Physical-Mechanical Studies of Materials» at the Institute for Metals Superplasticity Problems of the Russian Academy of Sciences.

References

[1] Reed R C 2006 *The superalloys: Fundamentals and Applications* (Cambridge University Press).
[2] Shakhov R V, Ganeev A A, Mukhtarov S K and Logunov A V 2018 *Letters on Materials* **8**(4) 494
[3] Zavodov A V, Petrushin N V and Zaitsev D V 2017 *Letters on materials* **7**(2) 111
[4] Li J R, Zhong Z G, Tang D Z, Liu S Z, Wei P, Wei P Y, Wu Z T, Huang D, Han M 2000 *Proc. of the 9th Int. Symp. Superalloys 2000* (Warrendale, Pa.: TMS) p 777
[5] Huang M and Zhu J 2016 *Rare Met.* **35**(2) 127
[6] Pollock T M and Tin S 2006 *Journal of propulsion and power* **22**(2) 361
[7] Mourer D P, Huron E S, Bain K R, Montero E E, Reynolds P L and Schirra J J Patent US 6,521,175 B1 Assignee: General Electric Co, Raytheon Technologies Corp Publ. 18.02.2003
[8] Mourer D P, Huron E S, Blackman D G, Bain K R, Reynolds P L, Schirra J J and Gabb T P Patent EP 1 195 446 A1 Assignee: General Electric Co Publ. 04.10.2000
[9] Mukhtarov S K, Imayev V M, Ganeev A A, Shakhov R V and Logunov A V 2019 *Journal of Physics: Conf. Series.* 1270 012052
[10] Mukhtarov S K, Shakhov R V and Ganeev A A 2019 *IOP Conf. Series: Mater. Sci. Eng.* 672 012048

IOP Conf. Series: Materials Science and Engineering 1008 (2020) 012047 doi:10.1088/1757-899X/1008/1/012047

Research and features of preliminary graphitizing processing of melt of iron with silicon carbide on the structure and properties of cast iron castings

D.A. Boldyrev[1], S.I. Platov[2], N.V. Urtsev[2], D.V. Terentyev[2] and O.R. Latypov[2,3]

[1]Togliatti State University, 14 Belorusskaya Street, Togliatti City, Samara Region, Russian Federation, 445020

[2]Nosov Magnitogorsk State Technical University, 38 Lenin Street, Magnitogorsk City, Chelyabinsk Region, Russian Federation, 455000

E-mail: latolegraf@list.ru

Abstract. The materials used for preliminary graphitizing treatment of cast iron melt (pre-modification), of which silicon carbide has the greatest technical and economic advantage, are considered. The features of the graphitizing effect of silicon carbide during the preparation of a gray cast iron melt on its structure and properties in standard cylindrical samples and clutch plate castings are analyzed. It is shown that its graphitizing effect on the cast iron melt begins to partially manifest itself already at the stage of introduction into the furnace and enhances the action of the main graphitizing modifier during subsequent ladle treatment as a result of a decrease in the content of interdendritic graphite in the microstructure.

1. Introduction

The problem of "necrosis" of the cast iron melt during the manufacture of castings often arises at industrial iron foundries. By "necrosis" is meant a significant deterioration in the susceptibility of the cast iron melt to graphitizing modification with a decrease in the initial number of nucleation centers of graphite inclusions (TCGI). "Necrosis" occurs due to its overexposure in the holding-furnace for more than 2 hours because of many organizational and technical problems [1-6].

In practice (when iron melting under the conditions of foundry of JSC AVTOVAZ), the consumption of the graphitizing modifier during ladle treatment of the melt from an induction crucible furnace (ICA) is 0.3%. The consumption is doubled during ladle treatment of the melt from an electric arc furnace (EAF). This difference is associated with the peculiarities of smelting processes in EAF (loops, gases in the metal, inhomogeneous overheating of the charge, reducing the amount of TCGI, etc.), which reduce the tendency of cast iron to modify and, as a consequence, cause the appearance of interdendritic graphite and carbides.

"Necrosis" is noticeably manifested in the one-stage graphitizing inoculation in the gray cast iron melt, therefore, the solution of this problem is extremely important and urgent, since it directly affects the quality of the resulting cast iron: casting defects (shrinkage and gas porosity, gas pockets), strength and structural indicators. An overexposed cast iron melt, modified according to any of the existing technologies, needs a "revitalization" operation. There are in-stream inoculation in the process of casting molds and in-stream inoculation during pouring from the holding-furnace into the casting ladle

[7]. The latter technology is more traditional and most widely used in most iron foundries. The main disadvantage of this technology is the withering of the effect of graphitizing modification in the process of casting molten cast iron into foundry molds. The problem is solved by treating the cast iron melt with pre-modifiers. Usually, a material based on FS65 Preseed® (Elkem, Norway) with additions of zirconium, aluminum and calcium is used. In the CIS countries, a material is used on the basis of FS45 Refloy® (NPP Tekhnologiya LLC, Russia), where magnesium, rare earth metals and fluorspar are introduced instead of zirconium [8,9].

However, the most economically and technically feasible is the use of metallurgical silicon carbide (SiC) to "revive" overexposed cast iron melt, in particular, the waste from the main production in the form of crushed and scrap carborundum stones, which is a secondary raw material with valuable metallurgical properties at a relatively low price [10].

The paper presents a study of the use of silicon carbide at the stage of finishing the cast iron melt in the duplex process during its overflow from the EAF to the holding-furnace (ICA).

2. Material and Methods

The study of the graphitizing ability of silicon carbide in the form of breakage and scrap of carborundum stones with a pure SiC content of 95% was carried out at the stage of finishing the cast iron melt in the duplex process during its overflow from the EAF to the holding-furnace (ICA). The execution time of each melting operation and the composition of the charge are given in table 1.

When overflowing, 50 kg of 95% silicon carbide was introduced into the gutter. There is a visual end of the process of its assimilation after overflow without an increase in slag formation. After that, the bath level of the cast iron melt was cleaned from slag and pieces of electrode breakage weighing 8-12 kg were introduced to reduce carbon waste and exclude the possibility of saturation of the surface layers of the metal with oxygen by creating a carbon film. The furnace and ladle chemical composition of cast iron is given in table 2.

To assess the microstructure of the smelted cast iron, a standard cylindrical sample ∅ 25-30 mm was poured after finishing the cast iron melt on the trough during overflow from EAF to ICA with silicon carbide in an amount of 50 kg at 1500°C.

In order to create stable TCGI with the elimination of the time factor of the smelted metal being in the holding-furnace, the use of silicon carbide makes it possible to obtain a uniform distribution of lamellar graphite in the microstructure of cast iron with its deterioration with a decrease in the sample cross section.

After lowering the temperature of the cast iron melt in an induction furnace to 1470°C and finishing it with FS45 ferrosilicon to a silicon content of 1.85%, it was cast into ladles with microalloying with copper, tin and graphitizing treatment with 0.32% of the Barinoc®75 modifier (fr. 2- 6 mm). The temperature of the cast iron melt in the ladle is 1410°C, the chill of the wedge sample is 1.8 mm. The chemical composition of cast iron in a ladle sample is given in table 2.

Table 1. The execution time of each melting operation and the composition of the charge

Operation, time	Charge bucket, kg	
Raw materials preparation, 30 min	Bucket I: return of own production 7500, low-alloy steel 9500, alloy steel 3000, pig iron (conversion) 2000, crushed graphite 500	Bucket II: return of own production 7500, low-alloy steel 9500, alloy steel 3000, pig iron (conversion) 2000, crushed graphite 500, ferrosilicon FS45 500
Bucket charging (I), 5 min		
Metal melting, 1 h		
Bucket charging (II), 5 min		
Metal melting, 1 h		
Removing slag from EAF, 5 min		
Finishing of the chemical composition, 25 min		
Pouring into an induction 20-ton furnace, 10 min		
Total: iron smelting 5 h 40 min	Total: 45,500 excluding waste and error batch weighing	

Table 2. The chemical composition of cast iron in a ladle sample

Sampling location	Chemical composition, %								
	C	Si	Mn	P	S	Cr	Ni	Cu	Sn
EAF	3,28	1,57	0,31	0,026	0,025	0,065	0,027	0,15	0,016
ICA	3,32	1,72	0,31	0,027	0,025	0,065	0,027	0,15	0,016
Ladle	3,33	1,99	0,32	0,026	0,026	0,067	0,027	0,55	0,085

The measurement of mechanical properties (tensile strength of cast iron) was carried out in accordance with GOST 1497 on the sample with a diameter of 14 mm with an estimated length of 70 mm.

3. Results

The microstructure of cast iron (×100) along the edges (in the area of the ∅25 mm feeder, from the opposite edge ∅ 30 mm) and in the middle (∅ 30 mm) of a standard cylindrical sample is shown in figure 1.

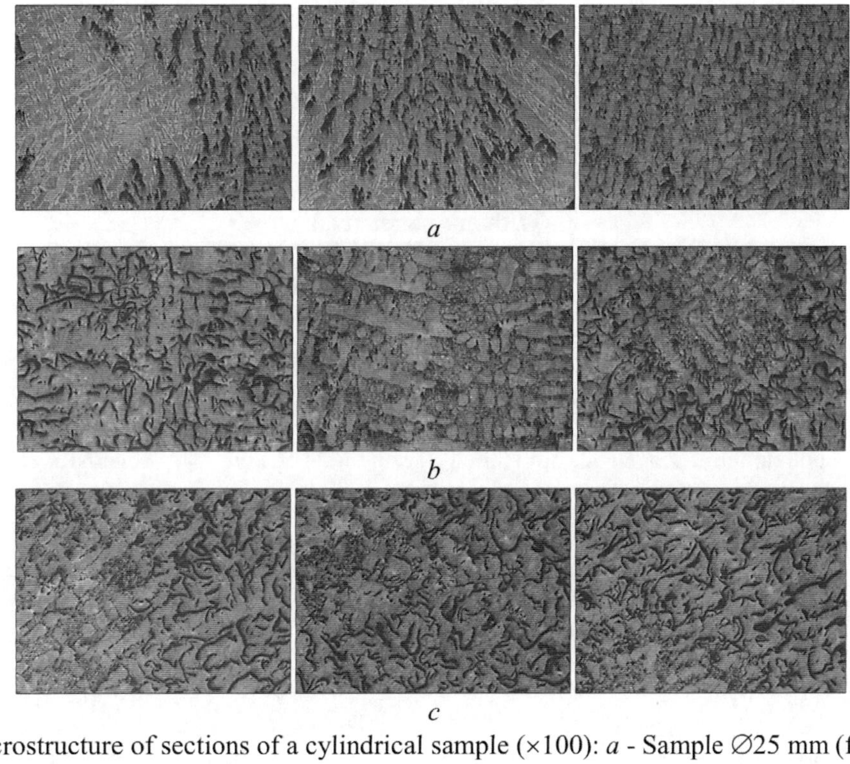

Figure 1. Microstructure of sections of a cylindrical sample (×100): *a* - Sample ∅25 mm (feeder zone) is characterized by a pearlite-cementite microstructure with interdendritic distribution of graphite types D, E; *b* - Sample ∅30 mm (middle) is characterized by a pearlite microstructure with a distribution of type A and D graphite. Ferrite up to 4% is present only between plates of type D interdendritic graphite; *c* - Sample ∅30 mm (edge) is characterized by a pearlite microstructure with a distribution of graphite of types A, D. Ferrite up to 4% is present both between plates of interdendritic graphite of type D and in zones with uniformly distributed graphite of type A.

Thus, finishing of the cast iron melt over silicon (ladle modification and microalloying with copper and tin, carried out after processing the cast iron melt with silicon carbide on the trough during overflow) allows providing a pearlitic metal base of cast iron and a uniform distribution of type A graphite in all sections of the cylindrical sample.

To assess the microstructure of cast iron in castings L 01202-0711-00 "Clutch disc", samples were cut along the "eye" from the side of the feeder and opposite it. Metallographic examination revealed the same content and distribution of ferrite in the area of the feeder and opposite it.

In the friction zone and the main section of the casting, graphitization is improved: the distribution of graphite is almost completely represented by type A, the content of types D, E is minimized (figure 2, 3). The rupture strength of cast iron in castings ranged from 260 to 290 MPa with a standard of at least 250 MPa.

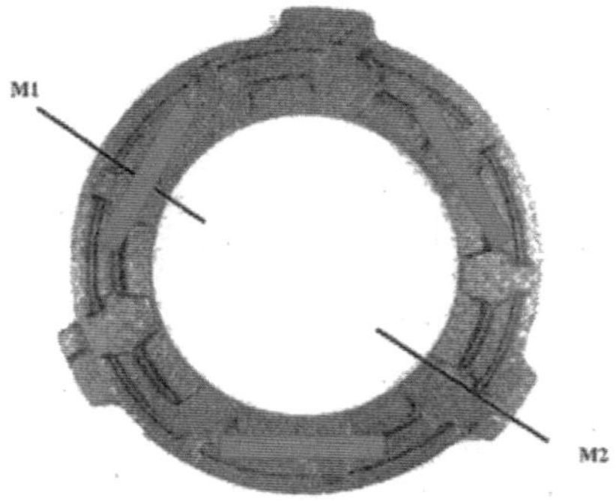

Figure 2. Scheme of cutting out samples from casting L 01202-0711-00 "Clutch disc" to assess the microstructure and determine the ultimate tensile strength

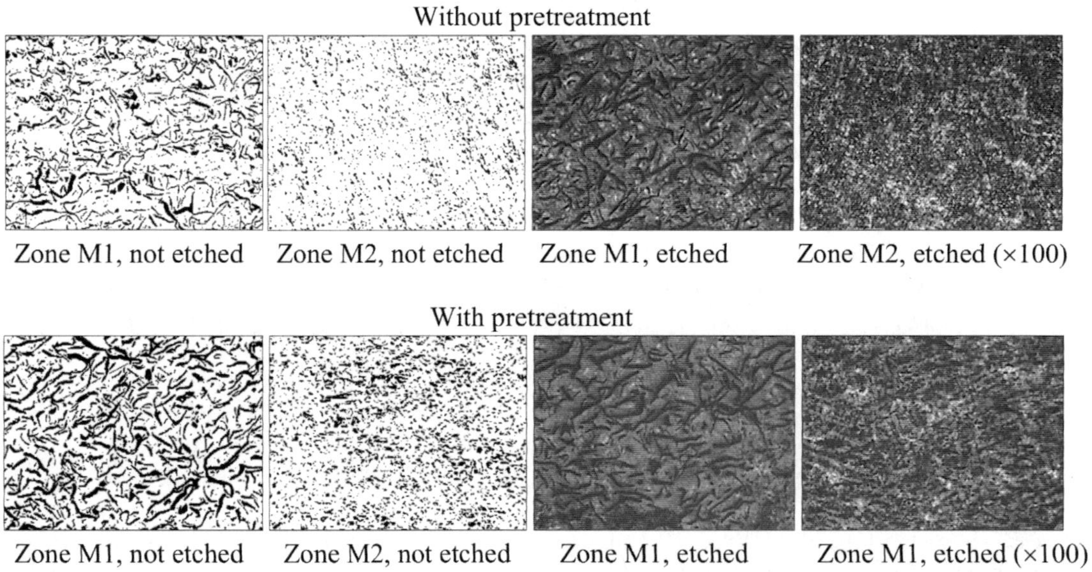

Figure 3. Morphology of graphite and microstructure of gray iron in castings L 01202-0711-00 "Clutch disc" without pretreatment / with pretreatment of the melt with silicon carbide

Conclusions

Compared with the results obtained by the serial technology, additional processing of the cast iron melt with silicon carbide on the trough during its overflow from the arc furnace into the induction furnace made it possible to improve graphitization (the number of graphite inclusions): in thin sections of the ear, the maximum content of interdendritic distributions of graphite types D, E decreased from 100% to 60%.

Acknowledgements
This work is carried out within a framework of the government order (No. FZRU-2020-0011) of the Ministry of Science and Higher Education of the Russian Federation.

References

[1] Joseph R D 1996 *ASM Specialty Handbook: Cast Irons* (ASM International)
[2] Girshovich N G 1978 *Handbook of Iron Casting* (Moscow: Mashinostroenie) [in Russian]
[3] Skaland T 1999 *Litejnoe Proizvodstvo* **6** 11 [in Russian]
[4] Litovka V I, Dubrovin A S, Venger V V 1987 *Litejnoe Proizvodstvo* **6** 11 [in Russian]
[5] Josan A, Bretotean C P and Raţiu S 2019 *IOP Conf. Ser.: Mater. Sci. Eng.* **477** 012033
[6] Bucur F, Socalici A, Josan A and Putan V 2020 *J. Phys.: Conf. Ser.* **1426** 012049
[7] Turkdogan E T 1985 *Physical Chemistry of High-Temperature Processes* (Moscow: Metallurgiya) [in Russian]
[8] Boldyrev D A., Kulakov B A, Tokarev A A, Prasolov S G, Popova L I 2020 *Litejshchik Rossii* **4** 18 [in Russian]
[9] Chajkin V A, Boldyrev D A and Chajkina N V 2007 *Vestnik of NMSTU* **4** 53 [in Russian]
[10] Boldyrev D A, Dema R R and Kalugina O B 2019 *Materials Today: Proc.* **11** 510

IOP Conf. Series: Materials Science and Engineering 1008 (2020) 012048 doi:10.1088/1757-899X/1008/1/012048

Structure of copper-containing nanocomposites obtained by electrochemical synthesis

I Yu Isaeva[1], G Yu Ostaeva[1], I V Odinokova[1], E A Eliseeva[1] and V I Busko[2]

[1] Moscow Automobile and Road Construction State Technical University (MADI), Leningradskii pr. 64, Moscow 125319, Russia

[2] D. Mendeleev University of Chemical Technology of Russia, Miusskaya Square 9, Moscow 125047, Russia

E-mail: irina-razumova-xim@yandex.ru

Abstract. The properties of copper nanocomposites obtained by electrochemical reduction in an aqueous-alcohol electrolyte solution ($CuSO_4$) with the addition of poly(N-vinylpyrrolidone) were investigated. Electrolysis was carried out at direct current using copper electrodes: the anode is made as a hollow cylinder, and the cathode in the form of a bundle of insulated wires with open cross sections, placed inside the cylinder with a working solution. The stabilizing properties of the polymer and the influence of the cathode on the electrochemical deposition processes were shown. The composition, morphology and size characteristics of synthesized copper-containing nanocomposites were determined. The resulting nanoparticles have a crystal cut, and also a high copper content.

1. Introduction

There are a large number of methods that allow obtaining metal ultrafine powders by electrochemical methods [1-6]. The proposed methods have a number of disadvantages: high cost, labor intensity technology, low efficiency and heterogeneity of the final product composition, etc. [7,8]. In this matter, it is particularly important to find fairly simple, effective, economical and environmentally friendly ways to synthesize ultrafine materials. One of the most environmentally friendly and economically justified ways to obtain fine-dispersed metal-containing materials is electrochemical reduction of metal ions. Also, with electrochemical reduction, it is possible to control the process by varying the conditions. For example, the nature and concentration of the electrolyte and solvent, the value of the cathode potential, the current density, the time of the experiment, etc.

Most often, nanoscale copper and its compounds are used to create new generation materials. However, the properties of nanostructured systems strongly depend on their dimensional characteristics and the uniformity of the final product composition.

Earlier, we obtained and studied copper-containing nanoparticles. The synthesis of copper nanocomposites was carried out by reducing Cu^{2+} ions in aqueous media using poly(N-vinylpyrrolidone (PVP) [9,10]. The reduction was performed by *tert*-butylamine borane at room temperature under air. The particles obtained in this way had a spherical shape and a narrow size distribution. The nanoparticles protected from aggregation by macromolecular screens polymeric protector of poly(N-vinylpyrrolidone). The phase composition of the synthesized copper nanoparticles was determined as a mixture of copper and its oxide (Cu^0 ~75 mass%), (Cu_2O~25 mass%).

IOP Conf. Series: Materials Science and Engineering 1008 (2020) 012048 doi:10.1088/1757-899X/1008/1/012048

This work focuses on the study of the mechanism of electrochemical processes occurring in a system containing Cu^{2+} ions in an aqueous alcohol solution of an electrolyte with the addition of PVP when using electrodes with an increased surface area, the structure of the particles obtained in this way.

2. Experimental

Copper sulfate was used to prepare a working solution, as a stabilizing component of PVP, ethyl alcohol (95%). The water was distilled twice. The concentrations of copper salt and PVP varied from 0.01 to 0.02 mol/L. Ethyl alcohol was added at a rate of 15 mf based on the concentration of the aqueous solution containing the copper salt.

The working solution is electrolyzed by direct current in an electrolysis cell with a copper cathode and an anode with the deposition of copper particles.

A copper anode in the form of a copper foil with an area of 1 dm^2 was immersed in a glass with a capacity of 1 L. The cathode, a section of insulated copper wire connected to the cathode of the power source, was placed in the center of the glass. Open sections of wire with a diameter of 0.5 mm, distributed over the volume, served as a working surface The current was brought to 1 A. The electrolysis time was 1 hour. The copper accumulated in the form of black globular growths at the ends of the wires, broke off under the influence of gravity and sank to the bottom of the glass.

The morphology of the samples was examined in a dual beam scanning electron microscope/ focused ion beam (SEM/FIB) Versa 3D (FEI, USA), equipped with a concentric backscatter detector (CBS). The samples were prepared by decanting the resulting sediment, followed by centrifugation and drying at room temperature.

3. Results and discussion

The copper nanoparticles obtained in this way have a crystalline faceting with a particle size distribution of up to 100 nm. The number of agglomerates is insignificant. Figure 1 depicts SEM and EDX images of the final product.

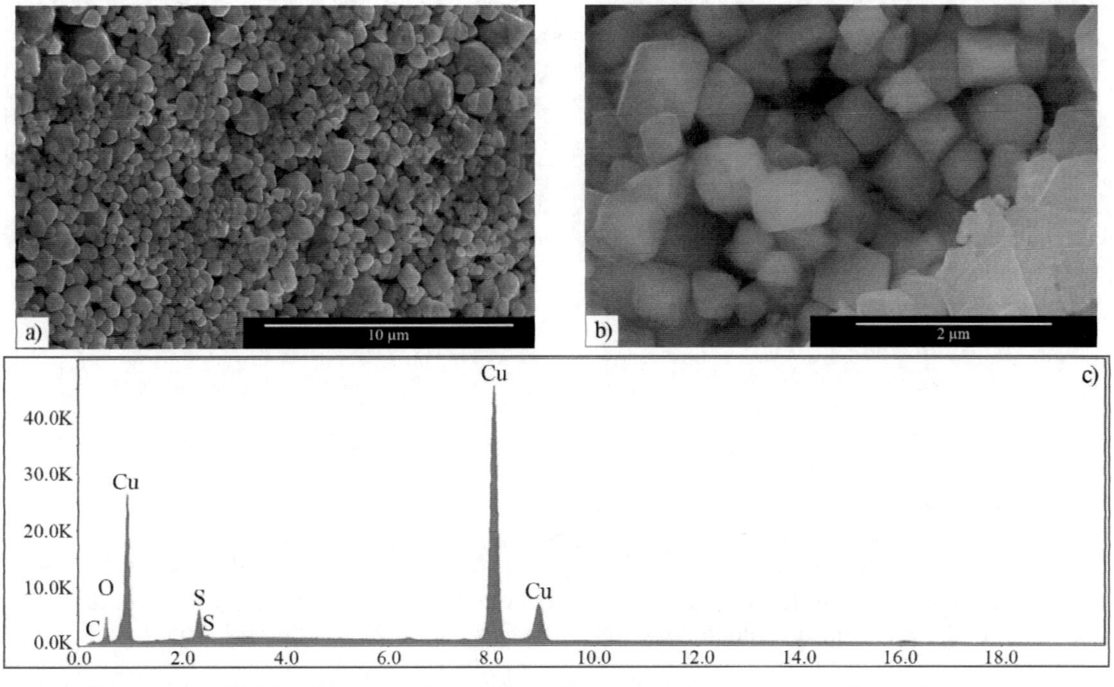

Figure 1. (a, b) SEM image and (c) EDX of copper-polymer nanoparticles obtained by electrochemical reduction of copper ions in an aqueous alcohol solution with PVP.

The composition of the samples is identified as Cu metal (~87 mass%). This is explained by the features of the electrochemical synthesis used in this work, which is that the anode was made in the form of a hollow cylinder, and the cathode was made in the form of a bundle of several copper insulated wires evenly distributed in the volume of the working solution inside the hollow cylinder (anode). This version of the cathode allows obtaining a total current density of up to 100 A / cm^2, which contributes to the production of nanoscale copper particles. Due to the proposed form of the cathode migration and diffusion of cations to such a cathode is carried out from all sides, that is, it is three-dimensional. It should be noted that the anode, made in the form of a hollow cylinder, reduces the migration time of the cation from the anode to the cathode.

In addition, a small area of the cathode facilitates the interaction of copper particles with the polymer, which in turn is not only a "regulator" of the size of nanoparticles in the process of electrochemical synthesis, but also a stabilizer from aggregation. In other words, the nanoparticle screened with a polymer chain. This is due to the fact that the polymer macromolecules interact with the growing nanoparticle during the reduction of metal ions and control the size of the metal particles, forming a polymer screen. [11].

The presence of ethyl alcohol in the system provides the most efficient and stable performance of the electrochemical recovery process. The main advantage is to slow down the competing hydrogen ion release reaction, which increases the current output for copper. Optimal conditions for diffusion transfer of copper ions to the cathode are made. This allows getting a fine-dispersed, uniform sediment.

Thus, the optimal composition of the electrolyte was obtained, which contributes to the production of highly dispersed copper precipitation.

A sample of copper nanoparticles obtained without adding ethyl alcohol, but with the same process characteristics is shown in figure 2.

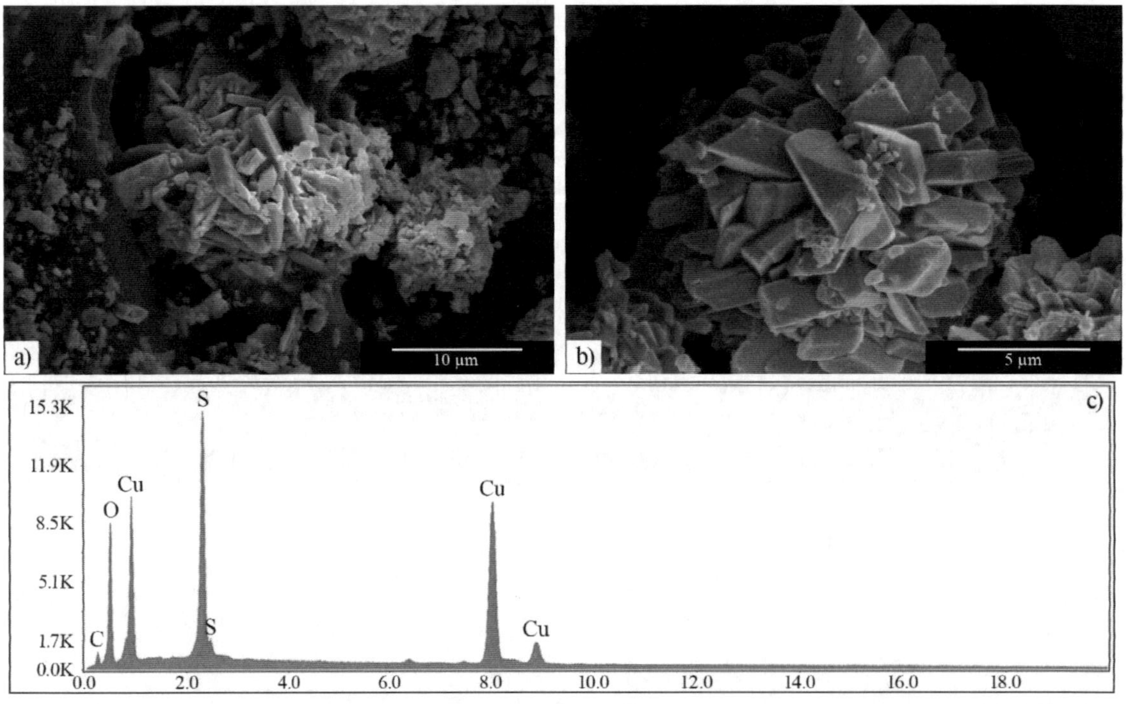

Figure 2. (a, b) SEM and (c) EDX image of the product of copper nanoparticle synthesis obtained by electrochemical reduction of copper ions in an aqueous polymer solution.

The samples are mostly irregularly shaped particles with crystal facets. In addition to the Cu^0 metal (~60% by mass), sulfur and oxygen are also present. This result is due to the absence of ethyl alcohol in the working solution. Alcohol has a weakly alkaline environment, which slows down the competing reaction of hydrogen release. In this system, there is no alcohol, which eventually leads to a low Cu^0 content in the final product.

4. Conclusions

The new technology including the reduction of copper ions by electrochemical reduction in an aqueous -alcohol electrolyte solution ($CuSO_4$) with the addition of poly(N-vinylpyrrolidone) at room temperature in air is applied for the synthesis of copper nanoparticles. It is shown that the addition of ethyl alcohol led to significant changes in the composition of the final product. The addition of ethyl alcohol leads to an increase in the concentration of Cu metal (~87 mass%). Thus, the optimal concentration of reagents of the working mixture and the selected conditions for electrochemical reduction of the most highly dispersed and homogeneous particles are formed. The Cu-containing particles up to 100 nm in size have a uniform shape with crystalline faceting. The features of the electrodes used, as well as the presence of ethyl alcohol in the electrochemical synthesis of copper-containing nanocomposites, allowed obtaining a product that can be used in various fields of industry, such as pharmacology, mechanical engineering, energy, etc.

Acknowledgment
This work was supported by the Ministry of Science and Higher Education of the Russian Federation (project no. FSFM-2020-0010) within the framework of a basic part of the state assignment in the field of scientific research.

References

[1] Chulovskaya S A and Parfenyuk V I 2008 *Electronic Processing of Materials* **1** 58
[2] Chulovskaya S A and Parfenyuk V I 2007 *Chemistry and Chemical Technology* **50** 49
[3] Chulovskaya S A, Kuzmin S M and Parfenyuk V I 2009 *Electronic Processing of Materials* **5** 24
[4] Noskov A V, Chulovskaya S A, Balmasov A V and Parfenyuk V I 2010 *Chemistry and Chemical Technology* **53** 43
[5] Anand V, Harshavardhan and Srivastava V C 2015 *J. of Nano RESEARCH* **31** 81
[6] Tesakova M V, Parfenyuk V I and Godlevsky V A 2008 *Electronic Processing of Materials* **6** 56
[7] Kreicberg G N, Golikov I V, Zavoistii I V, Gracheva I E and Kreicberg O G 2011 Pat. of the Russian Federation No 2410471 appl. 01.09.2009, publ. 27.01.2011
[8] Kreicberg G N, Golikov I V, Zavoistii I V, Gracheva I E and Kreicberg O G 2011 Pat. of the Russian Federation No 2410472 appl. 01.09.2009, publ. 27.01.2011
[9] Ostaeva G Yu, Isaeva I Yu, Morenko I V, Eliseeva E A and Litmanovich A A 2019 *Polymer Science, Series B* **61** 254
[10] Ostaeva G Yu, Isaeva I Yu, Grushina V V, Stuzhuk A N and Odinokova I V 2018 *Polymer Science, Series B* **60** 455
[11] Litmanovich O E 2008 *Polymer Science, Series C* **50** 63

Experimental study of the influence of different carbon content on the shrinkage kinetics and structure evolution of ultralow-cobalt hard alloys during spark plasma sintering

E A Lantsev, A V Nokhrin, M S Boldin, N V Malekhonova, P V Andreev, K E Smetanina and A A Popov

Lobachevsky State University of Nizhny Novgorod, 23 Prospect Gagarina (Gagarin Avenue), Nizhny Novgorod, 603950, Russia

E-mail: elancev@nifti.unn.ru

Abstract. The features of the compaction of plasma-chemical WC - (0.3; 0.6; 1) %Co nanopowders with the addition of 0.3 and 0.5% graphite during high-speed spark plasma sintering have been investigated. It was shown that an increased concentration of oxygen adsorbed on the surface of plasma-chemical nanoparticles during mixing with graphite, as well as the effect of graphite itself, which leads to a decrease in the activation energy of sintering due to a decrease in the intensity of the formation of particles of the η-phase in the "oxidized" WC-Co nanopowders, are the main structure features of ultralow-cobalt hard alloys. Samples of fine-grained ultralow-cobalt hard alloys with increased hardness and fracture toughness were obtained (for a WC-0.6% Co-0.3% C hard alloy with an average grain size of ~ 1-1.5 μm, the hardness is H_v = 20.2-20.5 GPa with a minimum crack resistance coefficient K_{IC} = 9.2-10.4 MPa·m$^{1/2}$).

1. Introduction

Currently, the effect of carbon on the behavior of hard alloys with additions of 8-12 wt.% cobalt during high-temperature sintering has been studied in sufficient detail. It has been found that at a temperature of T > 900°C diffusion of carbon and tungsten into cubic cobalt occurs [1], while the diffusion coefficient of tungsten in cobalt (γ-phase based on Co) is several orders of magnitude lower than that of carbon in cobalt [2]. In this regard, an increase in the carbon concentration in the composition of the initial powders leads to a slowdown in the compaction processes of WC-Co powders, as well as a shift in the temperature of the onset of shrinkage to the region of higher values [3]. It is interesting to note the works [4, 5], which show that the compaction of powders and grain growth slows down in hard alloys with a reduced content of free carbon due to the increased concentration of tungsten in the cobalt γ-phase, the creep rate of which controls the process of solid-phase sintering of hard alloys , and when a liquid phase appears, it determines grain growth according to the Wagner-Lifshitz-Slezov theory or the Ostwald coalescence rate due to changes in its diffusion permeability [1].

The aim of this work is to study the effect of carbon on the kinetics of high-speed sintering of fine-grained hard alloys with small additions of cobalt, as well as to study the effect of carbon additions on the structure and properties of ultralow-cobalt hard alloys obtained by the spark plasma sintering (SPS) method.

2. Methods and materials

The research objects of this work were α-WC nanopowders obtained by the plasma-chemical method, in which various concentrations of cobalt (0.3, 0.6, and 1 wt% Co) were deposited by the deposition method. Free carbon in the form of colloidal graphite was mixed into WC-Co nanopowders using a Fritsch-Pulverisette 6 planetary mill in a grinding jar with a hard alloy lining and tungsten carbide balls 1 mm in diameter. The powders were mixed in isopropyl alcohol for 2 h at a speed of 150 rpm.

The compaction of samples with a diameter of 12 mm and a height of h = 4 mm was carried out by the SPS method using a Dr. Sinter model SPS-625. The samples were sintered in a continuous heating mode, in vacuum (2-5 Pa) and heated at a given rate of 50 °C/min to the sintering temperature T_s, followed by cooling in a free mode. There was no holding at a temperature T_s. The value of the uniaxial pressure applied to the powders during heating was 70 MPa. The oxygen and carbon content was determined with an accuracy of ± 0.01 wt% using a LECO TC-600 analyzer.

The structure of the samples was studied using a JEOL JSM-6490 scanning electron microscope with an Oxford Instruments INCA 350 energy-dispersive microanalyzer. X-ray phase analysis (XRD) of the samples was carried out using a Shimadzu XRD-7000 diffractometer (CuK_α, scanning step 0.04, which corresponds to the exposure time 2 sec). Qualitative phase analysis was carried out using the Diffrac.EVA program. Quantitative analysis followed the Rietveld method. The accuracy in determining the volume fraction of α-WC and η-phase particles was ± 0.5%.

Sample density (ρ) was measured by hydrostatic weighing using a Sartorius CPA 225D scales. Hardness H_v and fracture toughness K_{IC} were measured using a Qness A60+ hardness tester with a load of 2 kg.

3. Results

From the obtained micrographs of the structure of the initial powders, it can be seen that the WC-Co nanopowders after the cobalt reduction operation contain single large agglomerates, which are absent in nanopowders with the addition of graphite. The size of the agglomerates depends on the cobalt content. In WC-0.3Co nanopowders the size of the agglomerates ranges up to 1–2 μm, while in nanopowders with 0.6 and 1% Co, it varies from 5 to 20 μm. The absence of agglomerates in WC-Co-C nanopowders is associated with the use of a mixing operation, during which the agglomerated regions are "broken" into smaller ones, which leads to a more uniform particle size distribution. Separately located cobalt particles were not detected in the structure of powder compositions, which allows us to conclude that ultrathin cobalt layers are uniformly deposited on the surface of α-WC nanoparticles.

The process of mixing nanopowders in a planetary mill leads to an increase in the oxygen concentration from 0.514 to 0.748 wt%, from 0.475 to 0.605 wt%, and from 0.469 to 0.623 wt% for nanopowders with the addition of 0.3, 0.6, and 1% Co respectively. A decrease in the oxygen concentration in WC-0.6% Co and WC-1% Co nanopowders as compared to WC-0.3% Co nanopowders is due to the larger volume fraction of cobalt, which has a lower oxygen concentration as compared to the α-WC tungsten monocarbide nanopowder.

XRD results show that the nanopowders contain hexagonal tungsten carbide α-WC, as well as cubic and hexagonal phases of cobalt. No peaks corresponding to the W_2C phase, WO_x and α-W oxides were found. The peaks corresponding to tungsten monocarbide α-WC are broadened and partially overlap with each other in the region of large diffraction angles. After mixing, the X-ray peaks of cobalt are practically invisible, which, in our opinion, indicates its uniform distribution over the surface of tungsten monocarbide nanoparticles. Note also that the mixing process led to a decrease in intensity and additional broadening of the peaks of tungsten monocarbide, which may indicate deagglomeration of WC-Co nanopowders during grinding, as well as the formation of additional internal stresses in α-WC nanoparticles [1].

Figure 1 shows the dependence of shrinkage L and shrinkage rate S on the heating temperature (T) of plasma-chemical WC-Co nanopowders with different carbon content. The character of the L (T) dependence for WC-0.3%Co nanopowders has a three-stage character: insignificant compaction (Stage

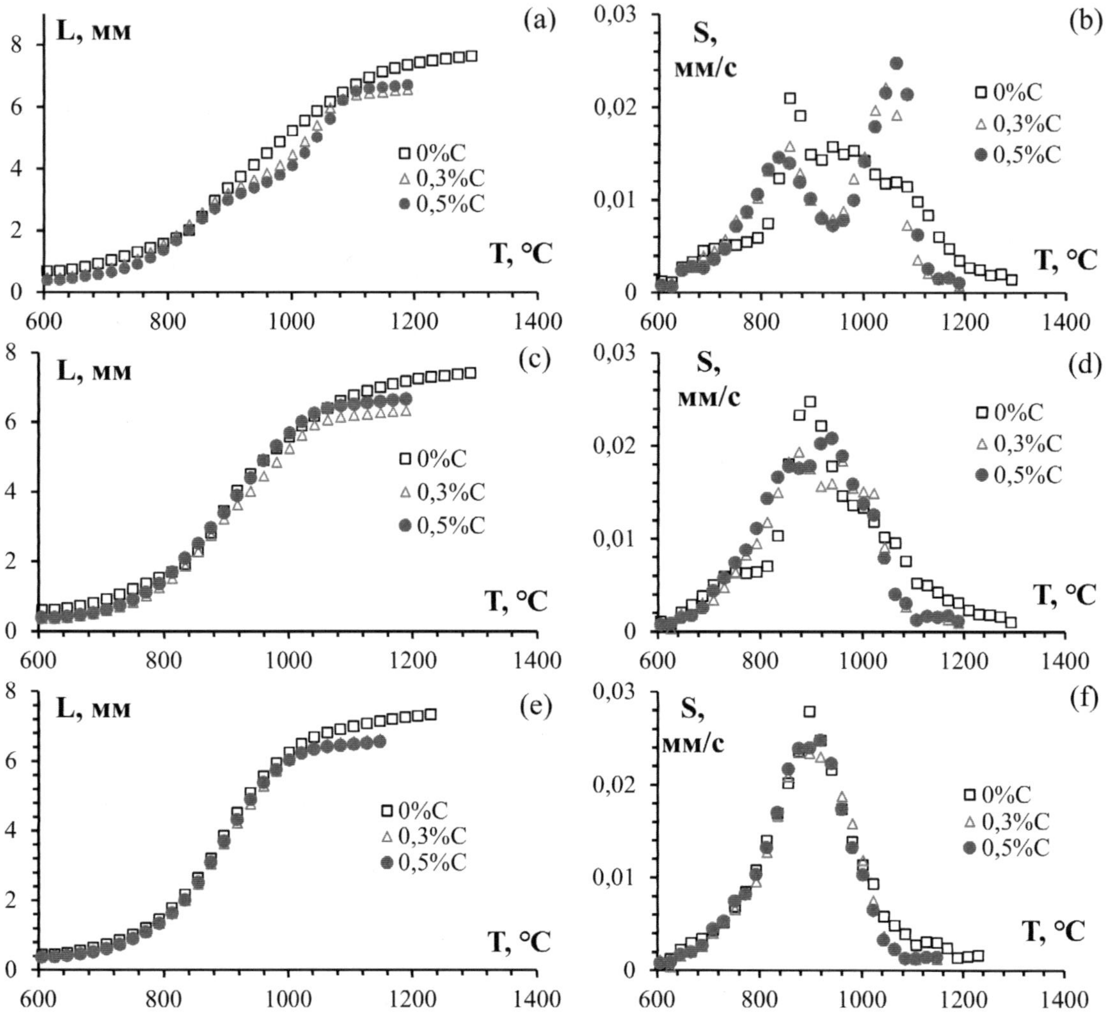

Figure 1. The dependence of shrinkage L (a, c, e) and shrink rate S (b, d, e) on temperature T of nanopowders WC-0.3% Co (a, b), WC-0.6% Co (c , d) and WC-1% Co (e, f)

I) at low temperatures (up to 700-800°C), Stage II, at which intense shrinkage is observed in the region of medium temperatures (from 700- 800°C to 1100°C) and, finally, the Stage III region of high temperatures (over 1100°C) in which the shrinkage rate decreases. It is noticeable that the addition of graphite to WC-0.3%Co nanopowders leads to a decrease in the temperature of the onset of stage III towards lower temperatures. The maximum shrinkage values for WC-0.3%Co nanopowders with the addition of graphite are lower than for the initial nanopowder without the addition of graphite, as can be seen from fig. 1a, the maximum shrinkage for WC-0.3%Co nanopowder after heating to a temperature of 1300°C is L_{max} = 7.67 mm, while for WC-0.3Co-0.3C and WC-0.3Co-0.5C nanopowders it is ~ 6.6-6.7 mm after heating to a temperature of 1200°C. Note also that the shrinkage curves L (T) for WC-0.3%Co nanopowders with the addition of graphite at the stage of intensive compaction are non-monotonic, which suggests a two-stage nature of the stage of intensive compaction of WC-0.3Co-C nanopowders. As can be seen from figure 1, the inflection in the L (T) curves is observed at a temperature of ~ 1000°C.

4. Discussion

4.1. Influence of C on the structure and properties of ultralow-cobalt hard alloys

Analysis of the data shows that with an increase in the content of graphite in the composition of plasma-chemical WC-Co nanopowders, a decrease in the absolute and relative density of hard alloys obtained by the SPS method is observed. With an increase in the graphite content from 0 to 0.5%, the relative density of ultralow-cobalt hard alloys decreases by ~ 2-3% at all cobalt contents. The result obtained is in good agreement with the results of studies of the effect of graphite on the sintering kinetics of WC-Co nanopowders - the introduction of graphite leads to a decrease in the maximum shrinkage of nanopowders L_{max} and a simultaneous "shift" of the L (T) dependences towards lower heating temperatures. This allows us to conclude that the introduction of graphite leads to a simultaneous slowdown of shrinkage and a shift in the characteristic temperatures of the end of the stage of intense shrinkage (Stage II) to a lower temperature region.

The results of electron microscopic studies of the microstructure of hard alloys showed that with an increase in the content of C in the composition of plasma-chemical WC-Co nanopowders, the intensity of the formation and growth of abnormally large grains of tungsten carbide decreases. For all three cobalt concentrations (0.3, 0.6, and 1 wt%), it was found that an increase in the graphite content from 0 to 0.5 wt% leads to a decrease in the average grain size from ~ 8 μm to 1–2 μm. This allows us to conclude that the introduction of free carbon stabilizes the grain structure of ultralow-cobalt hard alloys, thereby making the grain size distribution more homogeneous.

Analyzing the data, it can be seen that an increase in the graphite addition from 0.3 to 0.5% leads to a decrease in the microhardness and an increase in the minimum crack resistance of sintered specimens by an average of 2-3 GPa and 1-2 MPa·$m^{1/2}$, respectively. We assume that this is due, first of all, to a decrease in the volume fraction of the η-phase particles in the structure of sintered hard alloys, which, as is known [1], negatively affects their physical and mechanical properties.

4.2. Analysis of sintering mechanisms

To analyze the kinetics of compaction of nanopowders at the stage of intense shrinkage, we use the Young – Cutler model [6], which describes the nonisothermal process of sintering of spherical particles under conditions of simultaneous grain-boundary and/or bulk diffusion and/or plastic flow. In accordance with [6, 7], the slope of the dependence of the relative shrinkage (ε) on temperature in the coordinates ln ($T\partial\varepsilon$ / ∂T) - Tm / T corresponds to the effective activation energy of the SPS (mQ_{s2}). Figure 2a shows the dependences ln ($T\partial\varepsilon$ / ∂T) - Tm / T for plasma-chemical nanopowders WC-Co with different contents of cobalt and graphite. The ln ($T\partial\varepsilon$ / ∂T) –Tm / T dependences for WC – Co nanopowders with a content of 0.6% and 1% cobalt have the usual two-stage character with a maximum, which, according to [8], indicates a change in the sintering mechanism. The temperature corresponding to the maximum of the ln ($T\partial\varepsilon$ / ∂T) - Tm / T dependence decreases from 1050°C to 950°C with an increase in the cobalt concentration from 0.3 to 1 wt%. The value of the coefficient *m* can be taken equal to $m = 1$, which corresponds to the case of plastic flow of cobalt [8, 9]. The performed estimates show that the activation energy of SPS of WC-0.3%Co nanopowders at the stage of intense shrinkage is Q_{s2} ~ 11 kT_m. The activation energy of SPS of WC-0.3%Co nanopowders with an addition of 0.3 and 0.5% graphite, determined from the slope of the ln ($T\partial\varepsilon$ / ∂T) –Tm / T dependence near the first maximum, is ~ 7 and ~ 8 kT_m, respectively. Similar values of the activation energy of SPS for Stage II were obtained for hard alloys with 0.6% and 1%Co.

The result obtained allows us to conclude that the shrinkage rate at Stage II is determined by the rate of plastic flow of cobalt. The influence of graphite manifests itself by changing the phase composition of the γ-phase – an increase in the concentration of carbon in cobalt and a decrease in the volume fraction of the η-phase particles, which prevent the plastic flow of the γ-phase based on cobalt.

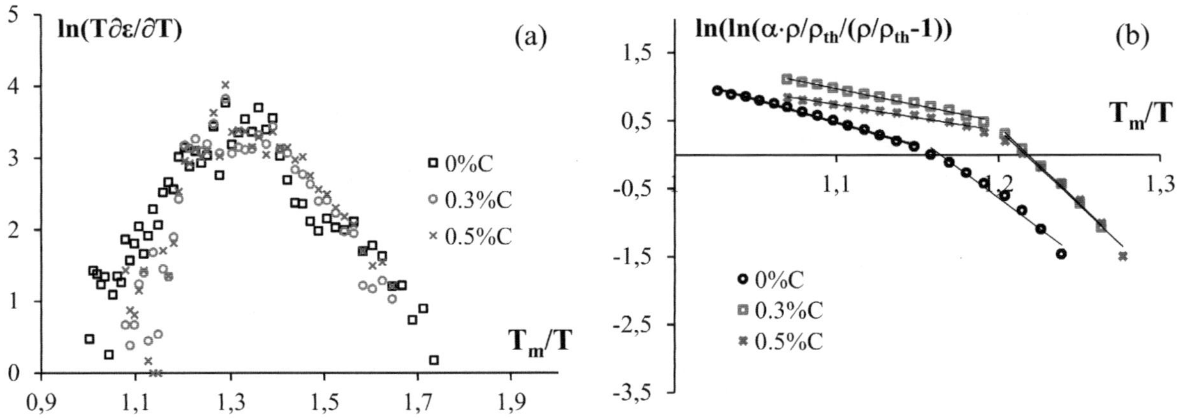

Figure 2. The dependences ln(T∂ε/∂T) – Tm/T (a) and ln(ln(α·ρ/ρth/(ρ/ρth-1)) – Tm/T (b) for plasma-chemical nanopowders WC-0.6% Co with various C content

The beginning of the third stage of ultralow-cobalt hard alloys sintering begins when the density reaches ~ 75-80%, at which, according to [10], the porosity becomes closed, and the pores themselves are located mainly at grain boundaries or at triple joints of tungsten carbide grain boundaries. This makes it possible to calculate the activation energy of this sintering stage using a model describing the diffusion dissolution of pores located near grain boundaries [11]. The activation energy within this model can be found from the slope of the dependence $\ln(\ln(\alpha\cdot\rho/\rho_{th}/(\rho/\rho_{th}-1))$, where $\alpha = 0.5$ is the compacting coefficient of the compact, on the inverse homological temperature Tm / T , where the value of ρ can be calculated using the formula $\rho(T) = \rho_{max}L_{max} / (L_0\text{-}L(T))$ (here L_0 is the initial pressing height, L_{max} is the maximum shrinkage of the sample, ρ_{max} is the experimentally measured density of the sample).

Figure 2b shows the dependences in coordinates $\ln(\ln(\alpha\cdot\rho/\rho_{th}/(\rho/\rho_{th}-1))$ for WC-0.6%Co nanopowder compositions with different carbon and graphite contents. Dependences $\ln(\ln(\alpha\cdot\rho/\rho_{th}/(\rho/\rho_{th}-1))$ - Tm / T for ultralow-cobalt hard alloys have a two-stage character, and the slopes of this dependences in the regions of "low" temperatures (Stage III-1) and in the region of "high" temperatures (Stage III-2) differ markedly. Analyzing the calculated values of the activation energy, it can be concluded that the addition of carbon increases the activation energy sintering at Stage III-1 by approximately ~ 5-6 kT_m for hard alloys with 0.6 and 1% Co, and by ~ 20 kT_m for hard alloy WC-0.3%Co. Energy of activation of SPS at the final stage of compaction (Stage III-2) turns out to be noticeably lower and with an increase in the graphite concentration it decreases by ~ 2-3 kT_m for all ultralow-cobalt hard alloys.

5. Conclusions

It is shown that the introduction of graphite leads to the formation of a fine-grained structure (suppression of anomalous grain growth) and a decrease in the intensity of the formation of η-phase particles. It is noted that the formation of particles of the η-phase of the composition Co_3W_3C is observed in the surface layer of the sintered samples, while in the central layers of the samples by XRD, particles of the "low-temperature" η-phase of the composition Co_6W_6C were found. It has been suggested that the formation of particles of the η-phase of composition Co_6W_6C in hard alloys with the addition of graphite is due to a decrease in the characteristic sintering temperatures, as well as a decrease in the total duration of the sintering process as compared to the traditional vacuum annealing of plasma-chemical nanopowders.

Features of the evolution of the structure of ultralow-cobalt hard alloys with the addition of graphite are due to the simultaneous effect of (i) an increased concentration of oxygen adsorbed on the surface of plasma-chemical WC-Co nanoparticles during mixing, and (ii) the effect of graphite, which leads to a decrease in the activation energy of sintering due to a decrease in the intensity of η-phases particle formation in "oxidized" WC-Co nanopowders, as well as the formation of a fairly uniform fine-grained structure.

It is shown that due to the combined use of the technology of plasma-chemical synthesis of tungsten carbide nanoparticles, the chemical-metallurgical method of obtaining nanocomposite WC-Co particles and the technology of spark plasma sintering, samples of ultralow-cobalt hard alloys with the addition of graphite were obtained, which have a homogeneous fine-grained structure (average grain size ~ 1-2 μm), increased density (not less than 98%) and enhanced mechanical properties. For a specimen of hard alloy WC-0.6%Co-0.3%C, the hardness value is H_v = 20.2-20.5 GPa with the value of the minimum crack resistance coefficient K_{IC} = 9.2-10.4 MPa·m$^{1/2}$.

Acknowledgments
The authors wish to thank the Russian Science Foundation (grant No. 18-73-10177).

References

[1] Kurlov A S and Gusev A I 2013 *Tungsten Carbides (Springer Series in Materials Science vol 184)* (Cham: Springer International Publishing) p 242
[2] Lavergne O, Robaut F, Hodaj F and Allibert C H 2002 *Acta Materialia* **50(7)** 1683
[3] Bounhoure V, Lay S, Charlot F, Antoni-Zdziobek A, Pauty E and Missiaen J M 2014 *Int. Journal of Refractory Metals and Hard Materials* **44** 27
[4] Petersson A 2004 *Int. Journal of Refractory Metals and Hard Materials* **22** 211
[5] Konyashin I, Hlawatschek S, Ries B, Lachmann F, Dorn F, Sologubenko A and Weirich T 2009 *Int. Journal of Refractory Metals & Hard Materials* **27 (2)** 234
[6] Young W S and Cutler I B 1970 *Journal of American Ceramic Society* **53 (12)** 659
[7] Johnson D L and Cutler I B 1963 *Journal of the American Ceramic Society* **46 (11)** 541
[8] Kumar A N, Watabe M and Kurokawa K 2011 *Ceramics International* **37 (7)** 2643
[9] Lantsev E A, Chuvil'deev V N, Nokhrin A V et al 2020 *Inorganic Materials: Applied Research* **11** 586
[10] Rahaman M H 2003 *Ceramic Processing and Sintering. Second Edition* CRC Press (New York: Marcel Dekker, Inc) p 875
[11] Chuvil'deev V N, Boldin M S, Dyatlova Ya G, Rumyantsev V I and Ordanyan S S 2015 *Inorganic Materials* **51 (10)** 1047

IOP Conf. Series: Materials Science and Engineering 1008 (2020) 012050 doi:10.1088/1757-899X/1008/1/012050

Superplastic forming of hemispheres out of AZ61 magnesium alloy

A A Kruglov[1], O P Tulupova[2], N V Olimpieva[2], V R Ganieva[2] and F U Enikeev[2]

[1] Institute for Metals Superplasticity Problems of Russian Academy of Sciences, 39 Khalturin St., 450001 Ufa, Russia
[2] Ufa State Petroleum Technological University, 1 Kosmonavtov St., Ufa, 450064, Russia

E-mail: alexform1960@mail.ru, box_mail_2011@mail.ru

Abstract. The paper proposes a method for calculating the rheological parameters included in the standard power ratio of superplasticity based on the results of test forming of hemispheres using a cylindrical matrix. The method takes into account the value of the entry radius of the matrix. The analytical model of the process is developed based on the main assumptions of the theory of thin shells, the finite element model is executed using the ANSYS10 ED software package. As compared to the traditional approach, the proposed method allows improving the modeling accuracy with an error less than 1%, not only for the forming time, but also for the thickness at the pole of the dome.

1. Introduction

The increased interest in the superplastic forming (SPF) process in recent years is associated with the possibility of manufacturing hollow structures from difficult-to-machine alloys based on titanium, aluminum and magnesium for the aerospace industry. [1].

Application of computer models makes it possible to replace mechanical experiments with computational ones that leads to savings in time and material costs. In addition, conducting a computational experiment based on a computer model allows visualizing the process of forming. However, to create such a model, it is necessary to know the values of the rheological parameters (material constants) K and m, which are included in the constitutive relations. The standard power-law relation for superplasticity is as follows:

$$\sigma = K\xi^{m} \tag{1}$$

where σ is the flow stress; ξ is the deformation rate, K and m are material constants.

Rheological parameters are usually determined from the results of uniaxial tensile tests, however, large differences between the values of numerical calculations and experimental data are often observed. For this reason, the results of biaxial tension tests were used for identification of the material constants that provided acceptable modeling accuracy for engineering calculations [2]. In this paper, not only the rheological parameters were obtained using the test forming of hemispheres, but the entry radius of the cylindrical matrix was also taken into account in the calculations.

The common approach to define the rheological parameters K and m does not consider the dependence of the thickness at the pole of the hemisphere's dome on the forming time, only the height of the dome is taken into account. Therefore, the problem of developing a method for identification the values of parameters K and m, which will increase the accuracy of computer modeling of SPF of hemispheres (dependencies of both the height and the thickness at the pole of the dome on the forming time), is still relevant.

2. Fitting the material constants

The method for calculating the rheological parameters K and m included in the standard power relation (1) consists of three steps and is described in detail in [3].

In this paper, the method was tested using the data obtained from the SPF of hemispheres of the magnesium alloy AZ61 (Mg–6% Al–1% Zn in wt.%) [4]. A sheet perform of AZ61 alloy with a thickness of s_0=0.7 mm and a diameter of d=50 mm was placed in a cylindrical matrix with an entry radius of r_0=1 mm. The SPF temperature was 400°C. The first set of experimental values was received by forming a hemisphere with a height of H_1=14.94 mm and a thickness at the pole of the dome s_1=0.2 mm during the time t_1=840 s under a constant gas pressure p_1=0.8 MPa. The second set of experimental values was received by forming a hemisphere with a height of H_2=15 mm and a thickness at the pole of the dome s_2=0.21 mm during the time t_2=250 s under a constant gas pressure p_2=1.2 MPa.

Using these data, the rheological parameters were found to be K_0=166.3 and m_0=0.34. They were used to create a finite element model of the SPF hemisphere. The results of calculations in the ANSYS 10ED software package are presented in table 1.

Table 1. Experimental thickness of the hemisphere at the pole of the dome ($s_{\_Exp}$) and forming time ($t_{\_Exp}$) vs calculated ones ($s_{\_Ans}$, $t_{\_Ans}$)

Constant pressure, [MPa]	Height of the dome, [mm]	Thickness of the dome, [mm]			Forming time, s		
		$s_{\_Exp}$	$s_{\_Ans}$	Δ, %	$t_{\_Exp}$	$t_{\_Ans}$	Δ, %
0.8	14.94	0.2	0.2	~0.1	840	934.03	~11.2
1.2	15	0.21	0.21	~0.1	250	250.12	~0.4

In accordance with Ref. [5], the accuracy of the method can be improved by adjusting the rheological parameters K and m in such a way as to obtain convergence in values of the forming time and the thickness at the pole of the dome, at least for one of two sets of experimental values. For further calculations, an experimental set of values (i=1) was selected, which was obtained as a result of the SPF of hemisphere at a constant gas pressure p_1=0.8 MPa. After adjusting the m value, both the calculated thickness at the pole of the dome and the experimental thickness were approximately the same, $s_{\_Ans} \approx s_{\_Exp}$=0.2 mm. Table 2 present the results of the ANSYS simulation of SPF of the hemisphere up to a height H_1=14.94 mm with a constant pressure p_1=0.8 MPa.

Table 2. Calculated thickness of the hemisphere at the pole of the dome ($s_{\_ANS}$) for different values of the rheological parameter m

m	0.32	0.335	0.34	0.35	0.36	0.4
$s_{\_ANS}$, mm	0.210	0.213	0.216	0.219	0.222	0.246

Linear interpolation of the m values from table 2 revealed the value of the adjusted rheological parameter m^*=0.32. The calculated values of thickness at the pole of the dome and the forming time were compared with the experimental ones in table 3.

Table 3. Calculated thickness of the hemisphere at the pole of dome ($s_{\_ANS}$) and the forming time ($t_{\_Ans}$) vs experimental ones ($s_{\_Exp}$, $t_{\_Exp}$) for the adjusted value of the rheological parameter m*

m^*	$s_{\_Ans}$, mm	$s_{\_Exp}$, mm	$t_{\_Ans}$, s	$t_{\_Exp}$, s
0.32	0.21	0.21	875.28	840

At the second step of calculations, the calculated thickness at the pole of dome of the hemisphere was equal to the corresponding experimental value, but there was a difference in the forming times.

In order to increase the accuracy in simulation of the SPF time, a third calculation step was suggested. At this step, the value of the rheological parameter K was adjusted to match the experimental forming time. The adjusted value of the parameter K was found to be K^*=148.256 MPa·s^m.

Table 4. Comparison of the results of finite element simulations with experimental values from [4]

Constant pressure, [MPa]	Thickness of hemisphere at the pole of dome, mm			Forming time, s		
	$s_{\_Exp}$	$s_{\_Ans}$	Δ, %	$t_{\_Exp}$	$t_{\_Ans}$	Δ, %
0.8	0.2	0.2	≈0.1	840	839.61	≈0.4
1.2	0.21	0.184	≈12	250	246.87	≈1.3

As can be seen from table 4, the difference between the results of calculations and the first set of experimental values (i=1) is less than 0.5%, while for the second set of values it exceeds 1%. This difference can be explained by the fact that the standard power ratio SPF (1) for the same values of rheological parameters K and m and different values of constant pressure does not allow obtaining correctly similar modeling for such parameters as the forming time and the thickness of hemisphere at the pole of the dome.

3. Conclusions

The use of rheological parameters K and m, which are included in the standard power relation and are determined according to the proposed method taking into account the entry radius of the matrix, makes it possible to create an adequate model of the SPF of hemisphere. The difference between the calculated and experimental values of the forming time and the thickness of hemisphere at the pole of the dome does not exceed 1%.

4. Acknowledgments

The present work was supported by the state assignment of the Institute for Metals Superplasticity Problems of RAS.

References

[1] Huang J, Chuang T 1999 Progress on superplasticity and superplastic forming in Taiwan during 1987-1997 *Mater. Chem. Phys.* **57** 195

[2] Aksenov S, Chumachenko E, Kolesnikov A, Osipov S 2015 Determination of optimal gas forming conditions from free bulging tests at constant pressure *J. Mater. Proc. Technol.* **217** 158

[3] Tulupova O, Ganieva V, Kruglov A and Enikeev F 2017 A new method of identification of constitutive equations according to the results of technological experiments *Lett. Mater.* **7** 68

[4] Chung S, Higashi K, Kim W 2004 Superplastic gas pressure forming of fine-grained AZ61 magnesium alloy sheet *Mater. Sci. Eng.* A **372** 15

[5] Cornfield G, Johnson R 1970 The forming of superplastic sheet materials *Int. J. Mech. Sci.* **12** 479

IOP Conf. Series: Materials Science and Engineering 1008 (2020) 012051 doi:10.1088/1757-899X/1008/1/012051

Effect of the structural elements size of crumpled graphene on hydrogen sorption: atomistic simulation

N G Apkadirova[1] **and K A Krylova**[1,2]

[1]Bashkir State University, 32 Z. Validy St., Ufa 450076, Russia
[2]Institute for Metals Superplasticity Problems of the Russian Academy of Sciences, 39 Khalturin st., Ufa 450001, Russia

E-mail: apkadirovanerkes@gmail.com

Abstract. The effect of the size of carbon flakes of crumpled graphene on the physical hydrogen sorption is considered using molecular dynamics simulation. Hydrogenated crumpled graphene was subjected to hydrostatic compression up to $\varepsilon = 0.77$, hydrogen dynamics in the dense structure is studied at 77 and 300 K. It has been found that the hydrogen sorption of crumpled graphene depends on its specific surface area. The highest sorption capacity was observed for crumpled graphene with carbon nanoflakes 20 nm in diameter. The difference in volumetric sorption capacity of flakes with the diameters of 9.5 and 20 nm was about 10% for a compression strain of $\varepsilon = 0.77$ at 20 ps at 77 K.

1. Introduction

Carbon materials are widely used in many industries due to their unique properties such as high strength, electrical and thermal conductivities, etc [1, 2]. In this connection, today a lot of theoretical and experimental studies are devoted to the investigation of such materials. 3D graphene-based materials, including crumpled graphene (CG), have a high specific surface area, porosity, and low weight [3, 4]. These properties open up the possibility of using such structures in hydrogen power engineering as a medium for hydrogen storing and transportation [5].

Many theoretical and experimental studies have shown that hydrogen sorption of CG can be increased due to a hydrostatic compression and a decrease in temperature [6, 7, 8]. In the experimental work [7], it was found that the hydrogen sorption of graphene is 1.7 wt.% under 1 atm. at 77 K, and it is less than 1 wt.% at 300 K but without compression [9]. A similar increase in the sorption capacity of CG was observed in theoretical studies using molecular dynamics simulation [6, 8, 10, 11, 12]. However, in addition to the above factors, the size of the structural elements of the CG can also affect the adsorption capacity, since the relative porosity of the structure changes by varying the diameter of the carbon nanoflakes constituent.

Thus, the possibility of increasing the hydrogen sorption of CG by changing the diameter of carbon nanoflakes formed from carbon nanotubes is investigated using molecular dynamics simulation.

2. Simulation details

The initial structure of CG with flakes of different diameters is shown in figure 1 (a) and (b). Carbon flakes were created based on 2.5 nm long carbon nanotubes with chirality (7,7) and (15,15), which corresponds to nanotubes 9.5 and 20 nm in diameters. According to the diameter of structural elements, we denote the studied structures of crumpled graphene - $D_{9.5}$ and D_{20}. The simulation cell contained

3×3×3 graphene flakes (GFs) with a total number of atoms 13 770 (7 722 carbon atoms and 6 048 hydrogen atoms) and 26 649 (16 335 carbon atoms and 10 314 hydrogen atoms), in figure 1 (a) and (b), respectively. Atomic hydrogen was placed inside the GFs since in this case, the energy of the C-H bond will be lower [13] and, therefore, this bond is formed more easily. Note that increasing the size of the simulation cell does not affect the final simulation results.

To obtain a solid CG structure, hydrostatic compression at 0 K is used. Figure 1 (c) shows a change in the relative density ρ/ρ_0 (where ρ_0 is the initial density) of the CG structures with an increase of compression strain ε. The compressive strain of the CG was calculated from the change in the structure volume: $\varepsilon = (V_0 - V) / V_0$ (where V_0 is the initial volume). The maximum deformation of the $D_{9.5}$ and D_{20} achieved using the AIREBO potential was estimated to be 77% and 83% respectively. In figure 1 (a) and (b), it can be seen that at ε = 0.77 a large number of pores is present that allows the structure to be compressed. In $D_{9.5}$, there are almost no pores, which complicates its further deformation. Note that further investigation of the hydration process of CG at 77 and 300 K was carried out at ε = 0.0 and 0.77.

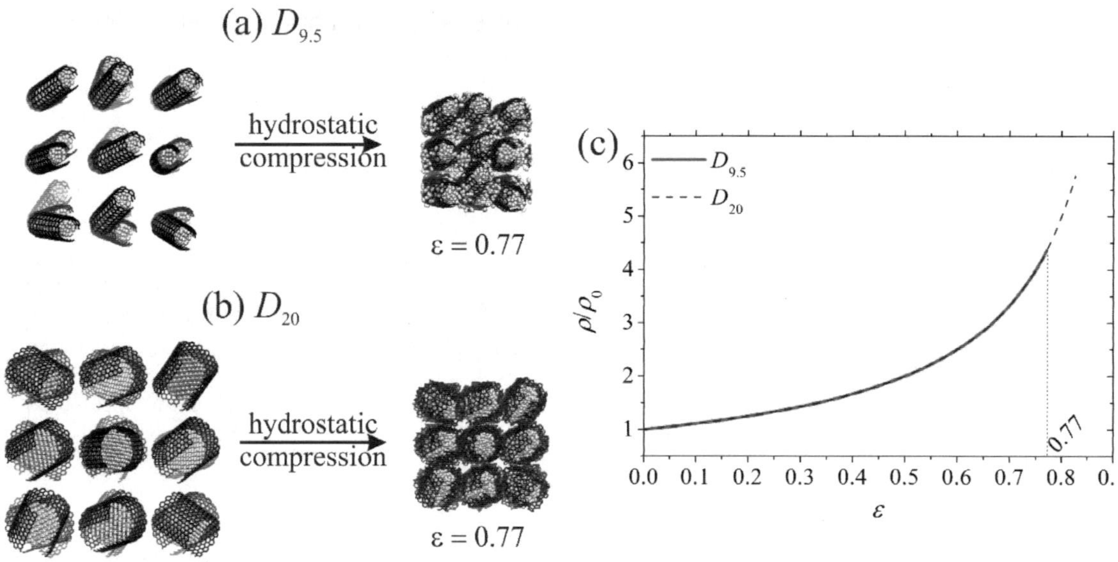

Figure 1. Initial structure of crumpled graphene with structural elements 9.5 nm ($D_{9.5}$) (a) and 20 nm (D_{20}) in diameters (b) before and after hydrostatic compression ($\varepsilon = 0.77$). Graphene nanoflakes are black, hydrogen atoms are white. (c) Density-strain curves for hydrogenated $D_{9.5}$ (solid line) and D_{20} (dashed line).

Molecular dynamics simulation is carried out using the publicly available and widely used LAMMPS package. To describe the interatomic interaction in CG, the adaptive intermolecular reactive empirical bond order (AIREBO) potential [14] was used. It was successfully applied in the study of deformation processes in 3D carbon structures [15, 16], as well as in the investigation of hydrogen sorption in these nanostructures [6]. After hydrostatic compression $D_{9.5}$ and D_{20} structures are maintained at 77 and 300 K for 20 ps under the conditions of NVT ensemble with a Nose-Hoover thermostat.

3. Results and discussion

To compare the dehydrogenation process of the $D_{9.5}$ and D_{20}, structures annealed at 77 and 300 K for 20 ps, the change in the relative volumetric capacity of hydrogen $\rho_V / \rho_{V\,0}$ (where $\rho_{V\,0}$ is the initial volumetric capacity of H_2) that leave the carbon structure was studied. Note that atomic hydrogen was placed in the initial structure of CG, but as a result of temperature exposure, most of the hydrogen was formed by H_2 molecules. Therefore, no chemical sorption of hydrogen by GFs is observed. However, the investigated carbon structure has good properties for the physical sorption of hydrogen, which is shown in this work.

In figure 2(a), the $\rho_V/\rho_{V0}(t)$ curves of hydrogenated $D_{9.5}$ and D_{20} at $\varepsilon = 0.0$ and 0.77 are shown. It can be seen that up to 10 ps (dashed line in figure 2(a)), the diameter of GFs does not affect the sorption capacity of hydrogen at 77 K. However, for $t > 10$ ps, $\rho_V/\rho_{V0}(t)$ of $D_{9.5}$ continues to decrease monotonically, while for D_{20} it is practically unchanged. The volumetric capacity of hydrogenated D_{20} at 20 ps is 10% (15%) higher at $\varepsilon = 0.77$ (0.0) than $D_{9.5}$. Hydrostatic compression also affects the growth of the sorption capacity of CG [6], and it does not depend on the GFs diameter. Figure 2(b) shows the D_{20} (*A*, *B*) and $D_{9.5}$ (*C*, *D*) structures after annealing at 77 K for 20 ps. In figure 2(b) *B* and *D*, it can be seen that H_2 is deposited on the inside of the GFs due to Van-der-Waals forces, regardless of its diameter and degree of hydrostatic compression. However, the compressive deformation of the CG does not lead to the opening of the GFs, which affects the growth of the sorption capacity. The observed decrease in the volumetric capacity of hydrogen $D_{9.5}$ in comparison with D_{20} is due to lower specific surface area of $D_{9.5}$ along with relatively high porosity.

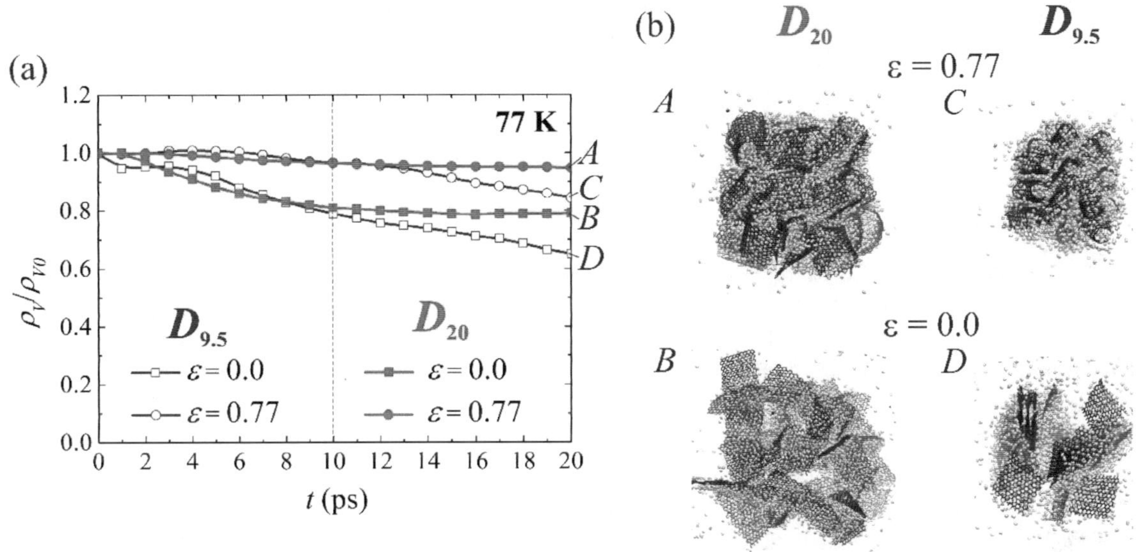

Figure 2. (a) Relative volumetric capacity of hydrogenated $D_{9.5}$ and D_{20} at 77 K for $\varepsilon = 0.0$ and 0.77 as the function of time. (b) Snapshots of the D_{20} (*A*, *B*) and $D_{9.5}$ (*C*, *D*) structure for different values of strain after annealing at 77 K for 20 ps.

An increase in temperature has a negative effect on the sorption capacity of carbon structures (see figure 3(a) and (b)). A sharp decrease in $\rho_V/\rho_{V0}(t)$ at 300 K is observed, regardless of the GFs diameter and the degree of the compression of the structures. However, the dehydrogenation of $D_{9.5}$ occurs somewhat faster than that of D_{20}. Presumably, this is because the structure of $D_{9.5}$ is highly porous, which leads to the break of bonds between GFs (which can be observed in figure 3(b) *C* and *D*). Note that the change in the volume hydrogen capacity after reaching the critical points *A*, *B*, and *C*, *D* in figure 3(a) occurs in different ways. In compressed D_{20}, molecular hydrogen leaves the structure in large clusters (figure 3(a) *A*), which leads to an intense decrease in ρ_V/ρ_{V0}. Therefore, for $t > 3$ ps, ρ_V/ρ_{V0} of the unstrained D_{20} is higher than that of the deformed one. A similar effect is not observed in the $D_{9.5}$ structure and the release of hydrogen in this case occurs more uniformly. The rate of dehydrogenation of the unstrained structure is higher up to the critical points *C* and *D* in figure 3(b), where the amount of remaining hydrogen in the $D_{9.5}$ structure reaches the level of 0.16 and practically does not change during further annealing at 300 K. The deposition of hydrogen molecules on GFs during annealing at 300 K is not observed, since thermal fluttering destroys weak Van-der-Waals forces between hydrogen and carbon atoms, this is the main reason for the dehydrogenation of CG.

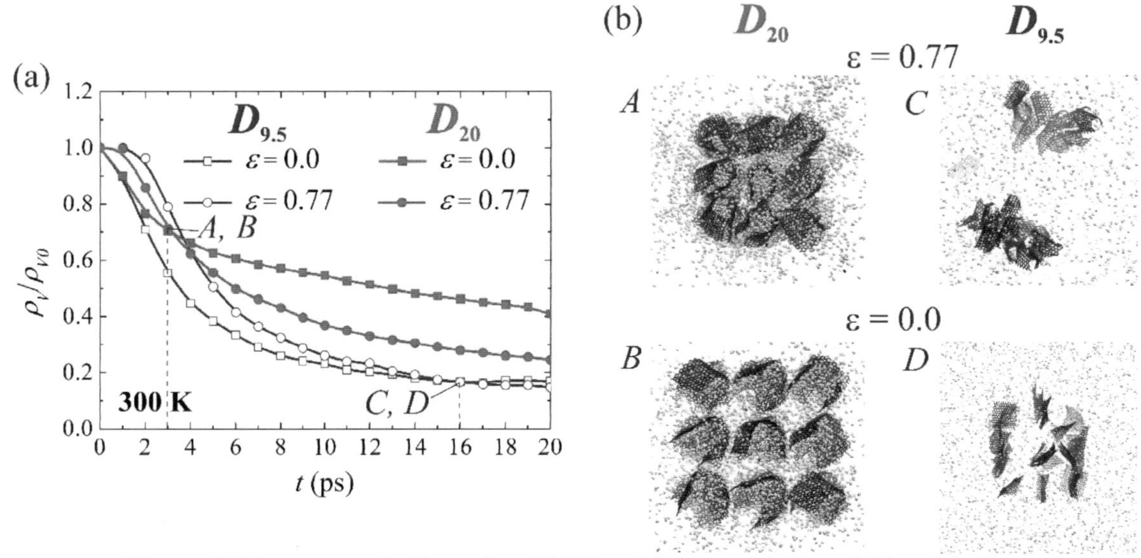

Figure 3. The same as in figure 2 at 300 K and for 3 ps (*A*, *B*) and 16 ps (*C*, *D*).

Table 1 summarises the data on the number of hydrogen atoms N_{out} (%) released from the structures of $D_{9.5}$ and D_{20} at different strains (ε = 0.0, 0.77) after temperature exposure at 77 and 300 K for 20 ps. It can be seen that hydrostatic compression leads to a decrease in the number of hydrogen released from the structure at 77 K. Thus, after annealing for 20 ps, about 5% of H atoms moved out the structure of the D_{20} at ε = 0.77, which is almost 4 times less than N_{out} of the undeformed D_{20}. However, raising the annealing temperature to 300 K significantly increases the dehydrogenation of structures. The smaller is the diameter of the GFs and the greater is the degree of its compression, the more hydrogen will be released from it. This is primarily due to the destruction of the Van-der-Waals forces between atoms due to thermal fluctuations.

Table 1. Number of hydrogen atoms N_{out} (%) moved out from the CGs structures at ε = 0.0 and ε = 0.77 after annealing at 77 and 300 K for 20 ps.

CG type	77 K		300 K	
	ε = 0.0	ε = 0.77	ε = 0.0	ε = 0.77
$D_{9.5}$	56.63	43.75	88.76	89.38
D_{20}	20.97	**5.26**	59.12	75.50

4. Conclusions

Using molecular dynamics simulation it is found that the sorption capacity of crumpled graphene depends on the geometric size (diameter of graphene flakes) of its structural elements. Larger the specific surface area of the crumpled graphene flakes (i.e., the larger the radius of the GFs), the higher its physical sorption. The volumetric capacity of hydrogen D_{20} at ε = 0.77 and holding at 77 K was 10% higher than that of $D_{9.5}$ under the same conditions. Hydrostatic compression of the CG structure also leads to a significant increase in the physical sorption of hydrogen. Thus, the number of hydrogen atoms released from the deformed D_{20} structure (ε = 0.77) after annealing at 77 K for 20 ps is 4 times lower than that of the undeformed D_{20} under the same conditions.

The data obtained significantly expand the understanding of the application of carbon structures for hydrogen storage and transportation. By varying the size of the structural elements of crumpled graphene and its temperature-deformed state, one can increase the sorption capacity of such structures. And by increasing the temperature from 77 to 300 K, molecular hydrogen can be released for further use.

Acknowledgments
The work was supported by thc State Assignment of IMSP RAS.

References

[1] Safina L, Baimova J, Krylova K, Murzaev R and Mulyukov R 2020 *Letters on Materials* **10** 351–60
[2] Dmitriev S V 2016 *Letters on Materials* **6** 86–9
[3] Zhang L, Zhang F, Yang X, Long G, Wu Y, Zhang T, Leng K, Huang Y, Ma Y, Yu A and Chen Y 2013 *Scientific Reports* **3** 1408
[4] Baimova J A, Liu B, Dmitriev S V and Zhou K 2015 *Journal of Physics* D **48** 095302
[5] Broom D, Webb C, Fanourgakis G, Froudakis G, Trikalitis P and Hirscher M 2019 *International Journal of Hydrogen Energy* **44** 7768–79
[6] Krylova K A, Baimova J A, Lobzenko I P and Rudskoy A I 2020 *Physica* B **583** 412020
[7] Ghosh A, Subrahmanyam K S, Krishna K S, Datta S, Govindaraj A, Pati S K and Rao C N R 2008 *The Journal of Physical Chemistry* C **112** 15704–07
[8] Krylova K, Baimova J and Mulyukov R 2019 *Letters on Materials* **9** 81–5
[9] Ritschel M, Uhlemann M, Gutfleisch O, Leonhardt A, Graff A, Täschner C and Fink J 2002 *Applied Physics Letters* **80** 2985–7
[10] Katin K P, Prudkovskiy V S and Maslov M M 2017 *Physics Letters* A **381** 2686–90
[11] Krasnov P O, Shkaberina G S, Kuzubov A A and Kovaleva E A 2017 *Applied Surface Science* **416** 766–71
[12] Pei Q X, Sha Z D and Zhang Y W 2011 *Carbon* **49** 4752–59
[13] Lobzenko I, Baimova J and Krylova K 2020 *Chemical Physics* **530** 110608
[14] Stuart S J, Tutein A B and Harrison J A 2000 *The Journal of Chemical Physics* **112** 6472–86
[15] Baimova J A, Rysaeva L K, Liu B, Dmitriev S V and Zhou K 2015 *Physica status solidi (b)* **252** 1502–7
[16] Baimova Y A, Murzaev R T and Dmitriev S V 2014 *Physics of the Solid State* **56** 2010–16

IOP Conf. Series: Materials Science and Engineering 1008 (2020) 012052 doi:10.1088/1757-899X/1008/1/012052

Molecular dynamics simulation of diffusion in Mg-Al system under pressure

P V Polyakova[1] and J A Baimova[1,2]
[1]Bashkir State University, 32 Validy St., Ufa 450076, Russia
[2]Institute for Metals Superplasticity Problems of the Russian Academy of Sciences, 39 Khalturin St., Ufa 450001, Russia

E-mail: P. V. Polyakova polina.polyakowa@yandex.ru

Abstract. The Mg-Al composite material possesses a large potential value in practical application due to its excellent properties. Molecular dynamics with the embedded atomic method potentials is applied to study aluminium-magnesium (Al-Mg) interface bonding during deformation. Study of fabrication techniques to obtain composites with improved mechanical properties, careful investigation of phase composition, dynamics and kinetics are of high importance. The loading scheme used in the present work is the simplification of the scenario, experimentally observed previously to obtain Al/Cu composites. It is shown that shear strain has a crucial role in the diffusion process. The results indicated that the symmetrical diffusion took place in the Mg-Al interface during deformation. Tensile tests showed that fracture took place in the Mg part of the final composite sample, which means that the interlayer region where the mixing of Mg and Al atoms observed is much stronger than the pure Mg part.

1. Introduction

High-pressure torsion (HPT) is an effective technique to manufacture nanostructured materials with improved mechanical properties [1- 4]. Recently, this method was used to obtain metal-matrix composites based on strongly dissimilar materials consolidated during deformation treatment, like carbon-metal composite [5, 6], ceramic Al/Al_2O_3 composite [7, 8], metal-metal composites [9, 10].

Different initial elements can be used to obtain mixed composite structures like powders or discs. For example, previously, HPT followed by annealing was successfully used to obtain Al-Cu *in-situ* composites from stacked dicks [9-11]. Magnesium and Mg alloys are increasingly used in various industrial applications, for example, for aerospace or automotive industry [12, 13]. A combination of Mg with the other metals can improve the properties of the final structure. Aluminium can prevent the corrosion process of Mg alloy and possess better properties for both Mg and Al [14-16]. Mg-Al alloys are promising for advanced engineering applications.

It was previously shown that temperature and pressure can affect the efficiency of diffusion on the Al and Mg interphase [17, 18]. Thus, studying the diffusion behavior in Al/Mg system is of high importance. To understand the atomic mixing between Al and Mg, the distribution of Al and Mg should be predicted in order to control the formation of intermetallic compounds. X-ray photoelectron spectroscopy or the electron probe micro-analyzer can be used to experimentally determine these distributions however, it is still very hard to get a close view locally in the zones of interests [19].

Molecular dynamics (MD) simulation is an effective tool for studying different processes on atomistic level [20-25]. Different binary systems were successfully studied by MD with the realistic interatomic potential. In the present work, MD simulation is applied in order to study atomic diffusion behavior at the bonding interface in the Al/Mg sample during compression combined with shear strain. Mechanical strength of the obtained composite is studied under uniaxial tension.

2. Simulation details

The atomic structure of a bimetallic AlMg interface is presented in figure 1. A bilayer model of Mg (HCP lattice, red atoms) and Al (FCC lattice, blue atoms) is assembled from two unrelaxed perfect crystals with lattice parameters of $a_{Mg} = 3.203$ Å and $a_{Al} = 4.05$ Å. Magnesium crystal has 31 periodicity-lengths along x and y, $L_x = L_y \approx 100$ Å, and 15 periodicity-lengths along z, $L_z \approx 50$ Å; Al crystal has 24 periodicity-lengths along x and y, $L_x = L_y \approx 100$ Å, and 12 periodicity-lengths along z, $L_z \approx 50$ Å. The interlayer distance between two crystals calculated as $(a_{Mg} + a_{Al})/2 = 3.6$ Å. The periodic boundary conditions are applied along x, y and z directions. The structure is initially relaxed at 300 K to find structural configuration with minimum potential energy.

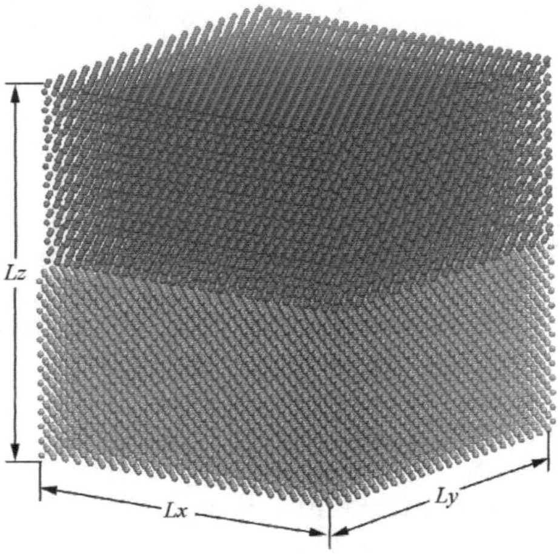

Figure 1. Simulation cell. Al atoms are shown by blue, Mg atoms are shown by red.

The simulations are carried out by molecular dynamics using the LAMMPS free simulation package. Embedded atom method (EAM) interatomic potential for Mg and Al is used [26]. For temperature control, the Nose-Hoover thermostat is applied. Verlet algorithm to integrate the Newtonian equation of motion with an integration time-step of 2 fs is used.

To study the process of fabrication of Mg-Al composite and phase transformation in the system, pressure along z-axis ε_{zz} combined with shear strain ε_{xy}. Strain rate is equal to $\varepsilon_{zz} = 0.00005$. Previously it was shown that this strain rate is small enough to obtain some phase transformations and not increase considerably the simulation time [27].

3. Results and discussion

3.1. Effect of loading

To visualize the evolution of diffusion over time, the representative atomic configurations are drowned out from the data obtained after the deformation of the material. Mg-Al interface is analysed at different strains and two types of loading conditions. The snapshots of interdiffusion processes of Mg and Al atoms at 300 K are displayed in figure 2, where Mg block in the model is shifted to the right by about 100 Å for a clearance. As can be seen, a small number of interdiffusion atoms are observed after pure compression along z-axis even after $\varepsilon_{zz} = 0.16$. Comparatively, when additional shear strain is applied, the number of diffusing atoms considerably increases. It can be seen, that number of Mg atoms diffusing into Al matrix is almost equal to the number of Al atoms diffusing into Mg matrix, which can be connected with the equality of the atomic radius of Al and Mg atoms (r_{Mg}=1.45 Å and r_{Al}=1.43 Å). In addition, the lack of difference in the melting point for Mg (923 K) and Al (933.5 K) is worth considering, which means the bonds in both metals are similarly stronger to fraction.

It is found that key role in the process of atom mixing playing shear strain, because after pure compression (figure 2c,d) even after ε_{zz} = 0.16 only several atoms diffusing near the interlayer region. Thus, further only the case of combined loading would be considered.

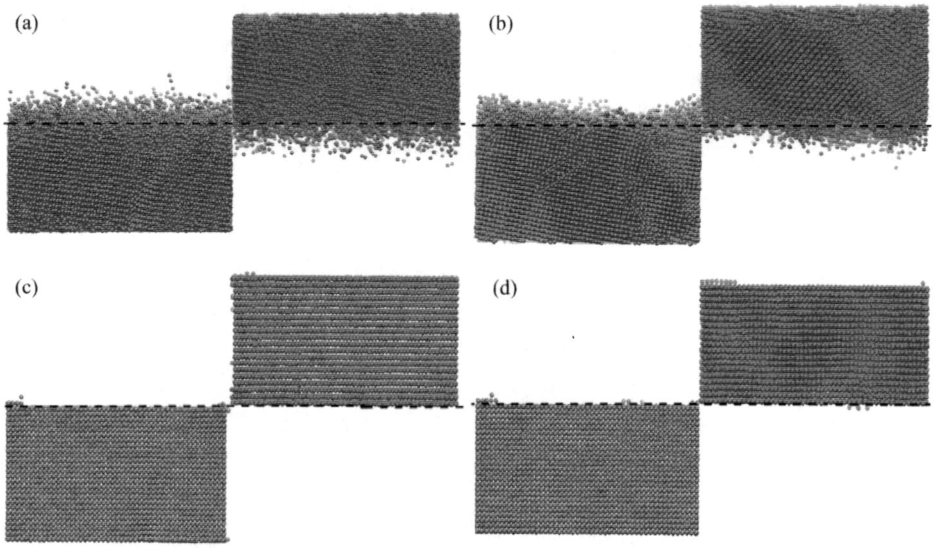

Figure 2. Snapshots of the mutual diffusion of Mg and Al atoms at 300 K for two types of loading: (a,b) uniaxial pressure ε_{zz} combined with shear strain ε_{xy} and (c,d) uniaxial pressure ε_{zz}. Snapshots after ε_{zz} = 0.04 (a,c) and after ε_{zz} = 0.16 (b,d). The red and blue balls represent Mg and Al atoms, respectively. The black dot line represents the initial position of Mg-Al interface.

Figure 3 shows the concentration distributions of Mg and Al atoms along z direction. The area on both sides of the interface where considered to calculate the atomic composition. The thickness of the layer where the diffusion of atoms during deformation took place is 42 Å. As it can be seen, after ε_{zz} = 0.04 thickness of the diffusion layer is equal to 29 Å, while after ε_{zz} = 0.16 equal to 38 Å. This figure also confirms that the number of Mg atoms diffusing into Al matrix is almost equal to the number of Al atoms diffusing into Mg matrix. However, the boundary of the mixing region in Mg part of the sample moves faster than in Al part: shift for 7 Å is observed for Mg part and only for 2 Å for Al part. It can be explained by a more porous HCP lattice of Mg.

3.2. Uniaxial tension

To study mechanical properties of the obtained composite structure, uniaxial tension along z-axis is applied after compression ε_{zz} = 0.16. In figure 4, the stress-strain curve is presented together with the snapshots of the structure close to the critical point. The process initiates with elastic deformation of structures (until about ε_{zz} = 0.02). During plastic deformation, several drops in the stress-strain curve can be seen which is connected with defect structure transformation.

Fraction takes place at ε_{zz} = 0.22 and σ_{zz} = 11 GPa in the Mg block of the composite. As it is known, Mg lattice is more prone to form defects than Al lattice. Analysis of the dislocation structure shows that most of the dislocations nucleated in Mg part of the sample.

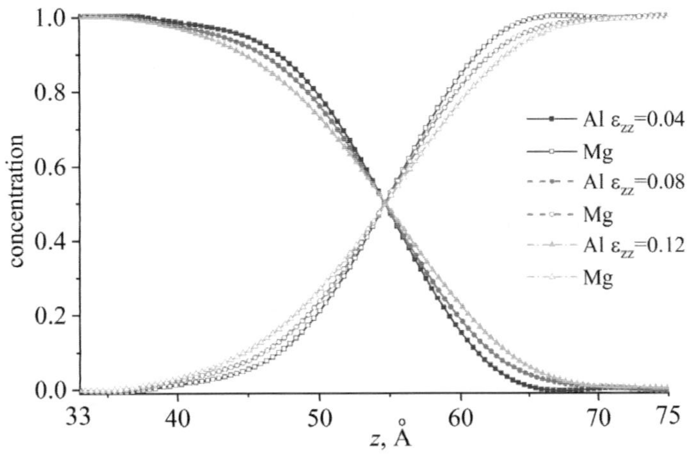

Figure 3. The concentration of Mg and Al atoms along the z-direction after uniaxial pressure combined with shear at (a) $\varepsilon_{zz} = 0.04$ (blue line), (b) $\varepsilon_{zz} = 0.08$ (red line), (c) $\varepsilon_{zz} = 0.16$ (green line).

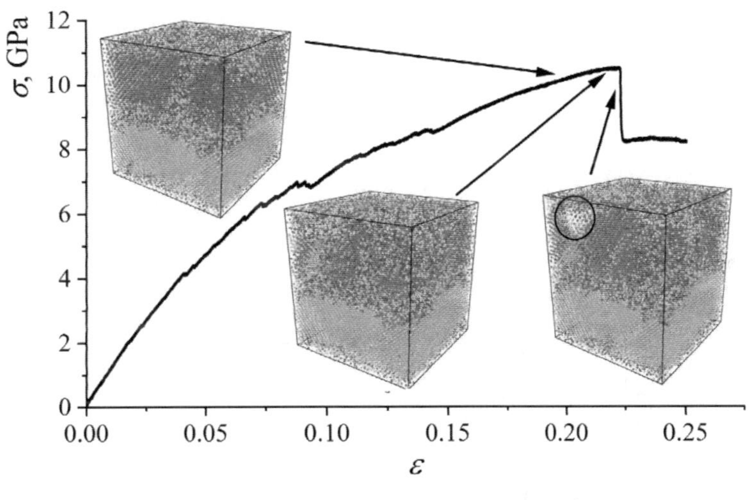

Figure 4. Stress-strain curve for Al-Mg with the snapshots of the structure close to the critical point. Snapshots are shown in accordance with the common neighbour analysis. The green colour is for FCC lattice, red - for HCP lattice, blue - for BCC lattice, and grey - for unknown.

4. Conclusions

In the present paper, the process of formation of Al/Mg composite from two initially separated crystals by severe plastic deformation is analysed on an atomistic level by molecular dynamics simulation. The proposed model is the simplification of scenario, experimentally observed previously in [10, 11, 9, 28]. In the HPT experiments, samples are polycrystals, while in the present model single crystal is considered. However, even such a simple model, allows understanding of mechanisms of diffusion bonding between Mg and Al.

It is found that uniaxial compression combined with shear deformation is an effective way to obtain the composite structure. Shear strain plays the most important role and considerably accelerates atoms mixing. As the strain increases, the number of diffusing atoms and the thickness of the interface region grows rapidly. The symmetrical diffusion bonding took place during the deformation of the initial sample. Further analysis shows that Al atoms diffuse into Mg block easier than Mg atoms diffuse into Al block.

However, promising results are obtained, additional studies are required to underspend the effect of different factors on the diffusion process in Al/Mg binary system. For example, effect of mutual orientation near the interface, because their effect was previously shown for other binary systems.

Acknowledgments
Simulation results were obtained by J.A.B. and supported by Russian Science Foundation grant 18–12-00440. Work of P.V.P. supported by the program of fundamental researches of Government Academy of Sciences of IMSP RAS.

References

[1] Nazarov A A, Mulyukov R R 2002 *Handbook of Nanoscience, Engineering, and Technology* (CRC Press, Boca Raton) chap 22, pp 22–1–22–41
[2] Fu J, Ding H, Huang Y, Pereira P H R, Zhang W and Langdon T G 2015 *Lett. Mater.* **5** 281–6
[3] Mulyukov R R, Nazarov A and Imayev R 2018 *Lett. Mater.* **8** 510–6
[4] Khalikova G R, Korznikova G F and Trifonov V G 2017 *Lett. Mater.* **7** 3–7
[5] Asgharzadeh H, Faraghi H and Kim H S 2017 *Acta Metallurgica Sinica* **30** 973–82
[6] Korznikova G, Czeppe T, Khalikova G, Gunderov D, Korznikova E, Litynska-Dobrzynska L and Szlezynger M 2020 *Mater. Charact.* **161** 110122
[7] Ashida M, Horita Z, Kita T and Kato A 2012 *Mater. Trans.* **53** 13–6
[8] Castro M M, Pereira P H R, Isaac A, Figueiredo R B and Langdon T G 2019 *J. Alloy. Compd.* **780** 422–7
[9] Danilenko V, Sergeev S, Baimova J, Korznikova G, Nazarov K, Khisamov R K, Glezer A and Mulyukov R 2019 *Mater. Lett.* **236** 51–5
[10] Danilenko V N, Korznikova G F, Zhilyaev A P, Sergeev S N, Khalikova G R, Khisamov R K, Nazarov K S, Kiekkuzhina L U and Mulyukov R R 2018 *IOP Conf. Ser.: Mater. Sci. Eng.* **447** 012021
[11] Korznikova G F, Mulyukov R R, Zhilyaev A M, Danilenko V N, Khisamov R K, Nazarov K S, Sergeyev S N, Khalikova G R and Kabirov R R 2018 *AIP Conf. Proc.* **2053** 030028
[12] Castro M, Pereira P H, Figueiredo R and Langdon T 2019 *Lett. Mater.* **9** 541–5
[13] Farbaniec L, Williams C, Kecskes L, et al. 2016 *Int. J. Impact Eng.* **98** 34–41
[14] Lee Z 2004 *Scripta Mater.* **51** 209–14
[15] Lee K, Kim J, Jo Y, Lee S, Heo J, Chang Y and Lee Y 2013 *Mater. Charact.* **75** 138–49
[16] Hajjari E, Divandari M, Razavi S H, Emami S M, Homma T and Kamado S 2011 *J. Mat. Sci.* **46** 6491–99
[17] Minamino Y, Yamane T, Shimomura A, Shimada M, Koizumi M, Ogawa N, Takahashi J and Kimura H 1983 *J. Mat. Sci.* **18** 2679–87
[18] Brennan S, Warren A P, Coffey K R, Kulkarni N, Todd P, Kilmov M and Sohn Y 2012 *J. Phase Equil. Diff.* **33** 121–5
[19] Liu Y, Chen Y and Yang C 2015 *AIP Adv.* **5** 087147
[20] Luo M, Liang L, Lang L, Xiao S, Hu W and Deng H 2018 *Comp. Mater. Sci.* **141** 293–301
[21] Zakharov P V, Poletaev G, Starostenkov M and Cherednichenko A 2017 *Lett. Mater.* **7** 296–302
[22] Poletaev G M, Zorya I V and Starostenkov M D 2018 *J. Micromech. Molec. Phys.* **03** 1850001
[23] Nazarova A A, Dmitriev S V, Baimova Y A, Mulyukov R R and Nazarov A A 2011 *Phys. Metals Metallogr.* **111** 513–9
[24] Baimova J A and Dmitriev S V 2011 *Comp. Mater. Sci.* **50** 1414–17
[25] Baimova Y A, Dmitriev S V, Kuranova N N, Mulyukov R R, Pushin A V and Pushin V G 2018 *Phys. Metals Metallogr.* **119** 589–97
[26] Liu X Y and Adams J 1998 *Acta Mater.* **46** 3467–76

[27] Polyakova P V, Nazarov K S, Khisamov R K and Baimova J A 2020 *J. Phys.: Conf. Ser.* **1435** 012065
[28] Han J K, Lee H J, il Jang J, Kawasaki M and Langdon T G 2017 *Mat. Sci. Eng.* A **684** 318–27

Model calculation of the phase diagram of Cu-Te

A R Kurbangulov[1], L V Tsygankova[1], E V Tikhonov[2], N N Bikkulova[1], D I Safargaliyev[1], G R Nigmatullina[1] and G R Akmanova[3]

[1] Sterlitamak Branch of Bashkir State University, 47a Lenin St, Sterlitamak, 453103 Russia
[3] Northwestern Polytechnical University, 127 West Youyi Road, Xi'an Shaanxi, 710072, China
[3] Bashkir State University, 32 Validy St., Ufa, 450076, Russia

E-mail: bickulova@mail.ru

Abstract. The paper presents the results of model calculations of the phase diagram of the Cu-Te system. A comparative analysis of the experimentally obtained data with theoretical calculations for the obtained model compounds is presented.

Due to the fact that the structure of copper telluride in the non-superionic state has not been unambiguously established, we carried out model computer calculations of the phase diagram of the Cu -Te system using the USPEX program [1], refined the structures of copper tellurides of stoichiometric and non-stoichiometric compositions. It was assumed that such a model calculation would make it possible to obtain a theoretical structural model, which will then be used to calculate the band structure and phonon spectrum.

A comparative analysis of the data in the superionic and non-superionic states could provide a key to understanding the nature of superionic conductivity and determine the criteria by which one could predict the presence of superionic properties in the compounds under study. The essence of the search for the most stable structure comes down to calculating the state of matter that has the lowest energy. The energy in this case depends on the electromagnetic interaction of the nuclei and electrons of the atoms that make up the crystal under study. It can be estimated using quantum mechanical calculations based on the simplified Schrödinger equation. Thus, the USPEX algorithm uses the density functional theory, which was developed in the second half of the last century. Its main purpose is to simplify calculations of the electronic structure of molecules and crystals. The theory makes it possible to replace the many-electron wave function with an electron density, while remaining formally accurate (but in fact the approximations turn out to be inevitable) [1−11]. In practice, this leads to a decrease in the complexity of calculations and, as a consequence, the time that will be spent on them. Thus, quantum mechanical calculations are combined with the evolutionary algorithm in USPEX [1]. The calculation was carried out in cooperation with the laboratory of A.R. Oganov, the purpose of these calculations is to test this method of computer modeling of phase diagrams under critical conditions in order to compare the obtained calculations with experimental phase diagrams. The search for such structures in the Cu-Te system showed that at absolute zero and a pressure of 0.1 GPa, the compositions Cu7Te4, Cu3Te2, Cu5Te4, and Cu5Te4 are stable. The experimental phase diagram under normal conditions contains phases of such compositions.

The structures and parameters of stable phases obtained in the calculation are presented in the table.

IOP Conf. Series: Materials Science and Engineering 1008 (2020) 012053 doi:10.1088/1757-899X/1008/1/012053

Table1. Cell structures and parameters at different pressures.

№	Formula	Cell parameters					
		a, Å	*b*, Å	*c*, Å	α, °	β, °	γ, °
1	$Cu_{1.4}Te$	3.98000	3.98000	6.12000	90.0000	90.0000	90.0000
2	Cu_3Te_2	3.86605	4.47305	7.09674	76.5671	75.7852	64.5499
3	Cu_5Te_4	7.29166	4.00469	7.40392	90.0000	112.9008	90.0000
4	Cu_7Te_4	8.59486	7.31671	4.23495	89.8054	85.7704	88.5299

The results obtained can be used to simulate the evolution of the crystal structure with increasing temperature. To clarify the stability of the theoretically obtained phases, the band structure and phonon spectra were calculated (Figs. 1–12).

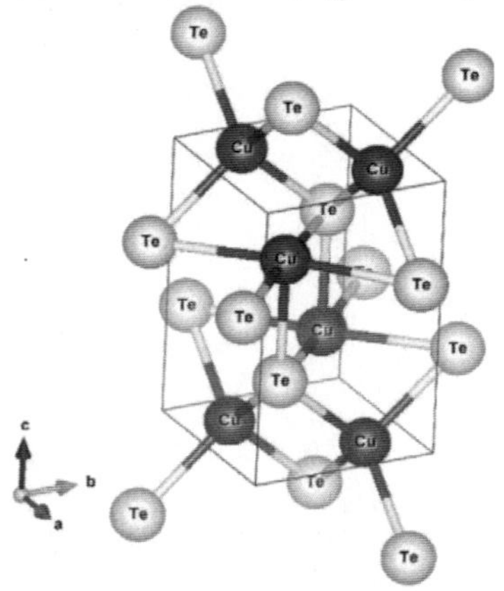

Figure 1. Structure $Cu_{1.4}Te$.

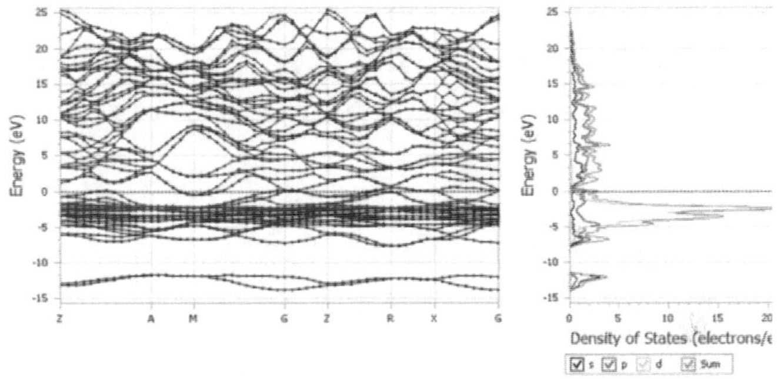

Figure 2. Zone structure $Cu_{1.4}Te$.

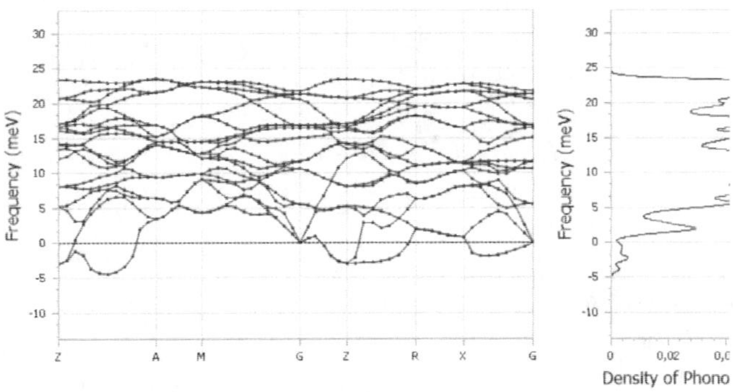

Figure 3. Phonon spectrum $Cu_{1.4}Te$.

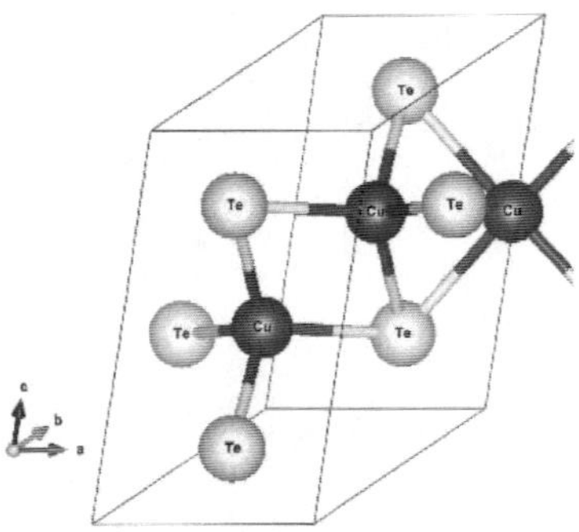

Figure 4. Structure Cu_3Te_2.

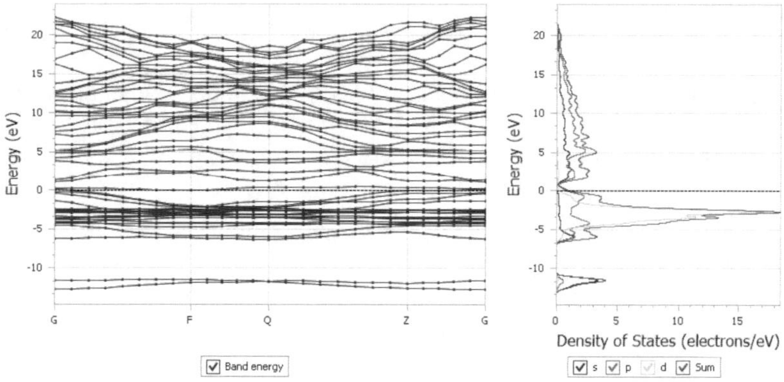

Figure 5. Zone structure Cu_3Te_2.

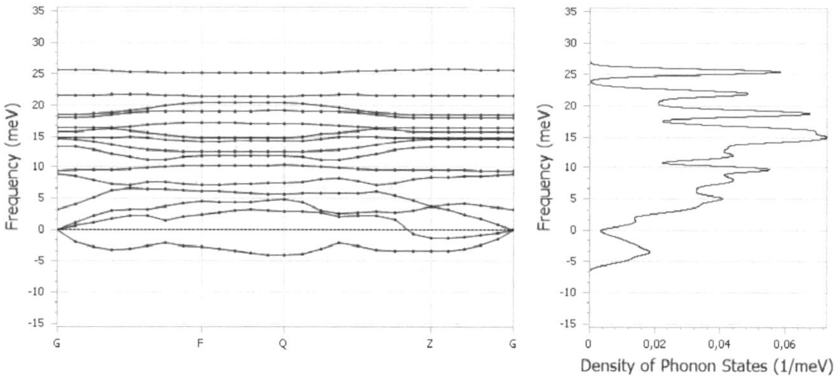

Figure 6. Phonon spectrum Cu_3Te_2.

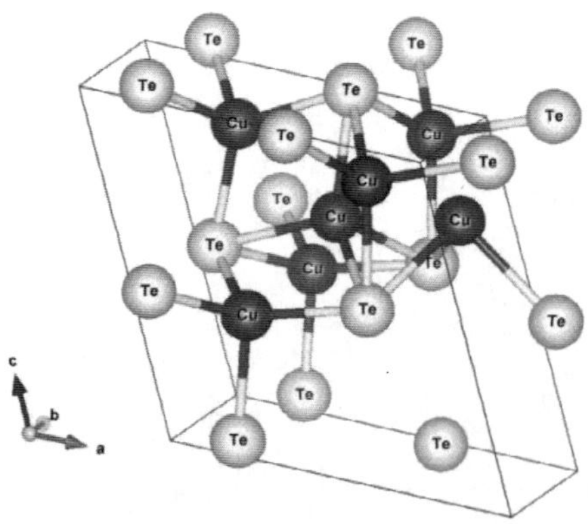

Figure 7. Structure Cu_5Te_4.

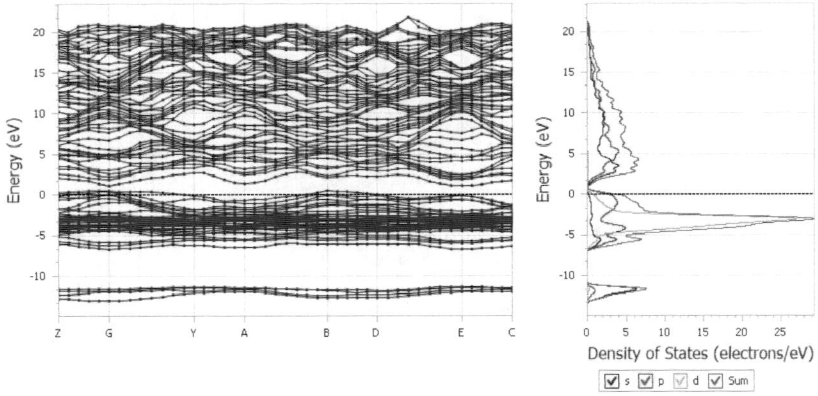

Figure 8. Zone structure Cu_5Te_4.

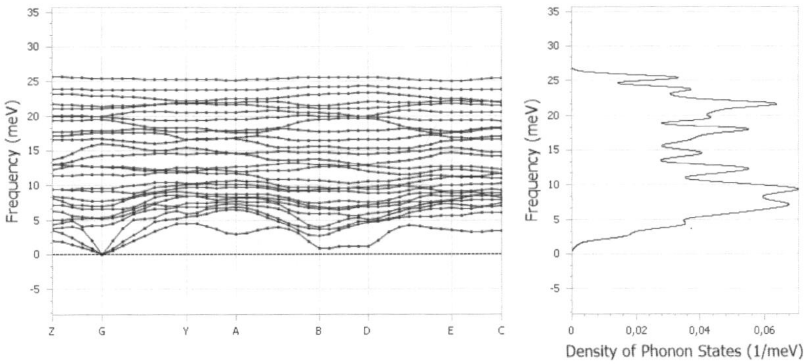

Figure 9. Phonon spectrum Cu_5Te_4.

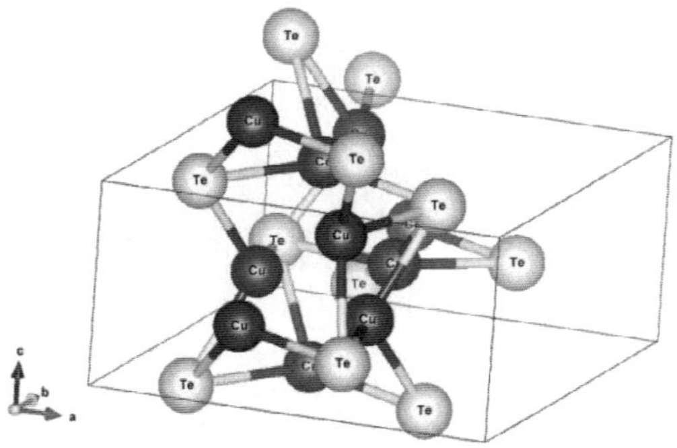

Figure 10. Structure Cu_7Te_4.

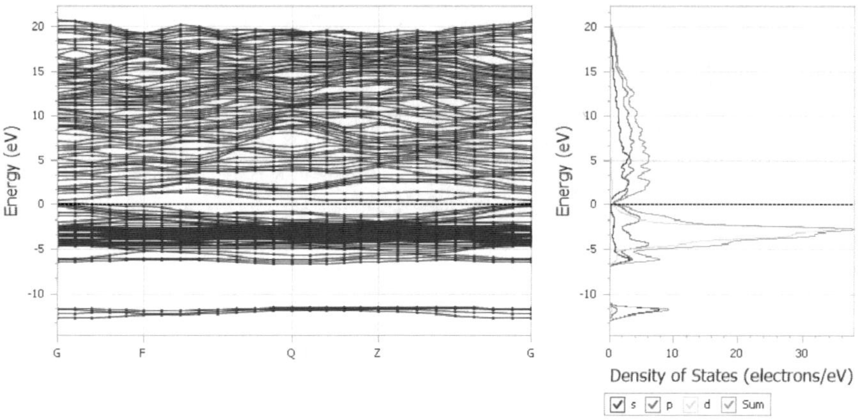

Figure 11. Zone structure Cu_7Te_4.

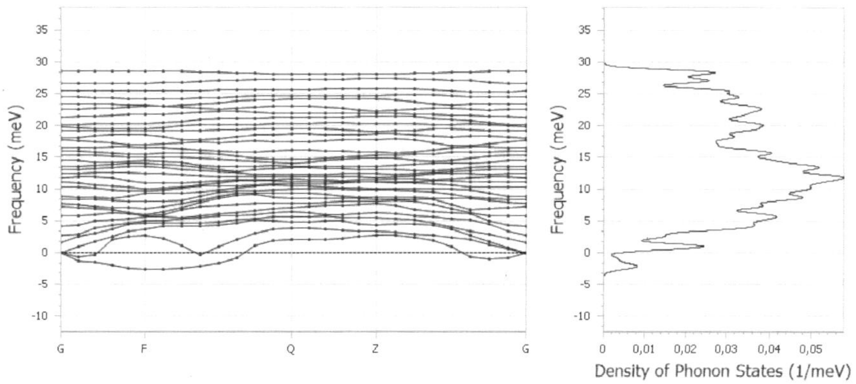

Figure 12. Phonon spectrum Cu_7Te_4.

The band structure and phonon spectrum of copper tellurides were calculated within the framework of the electron density functional theory using the pseudopotential method in the plane wave basis, implemented in the Quantum Espresso software package [7]. In the calculation, we used ultrasoft pseudopotentials for copper, for telluride the norm-preserving pseudopotentials, which were generated by this program [10]. The plane wave cutoff energy was 85–100 Ry. In calculations by this method, the valence electrons of the atoms are taken into account. Automatic selection of reciprocal lattice points (k-points) was used using the Monkhorst-Pack method on an $8 \times 8 \times 8$ grid.

Theoretical calculations predict phase instability due to the presence of imaginary phonon frequencies at the boundary of the Brillouin zone. At the same time, the phonon dispersion of the phases is stable throughout the Brillouin zone. The region around the G point in the structure of the phonon spectrum of the phases shows imaginary frequencies (shown as negative). This indicates the dynamic instability of the systems and assumes that when the ions are displaced in accordance with the wave vector at the point G, a restoring force does not appear in the system, which brings the system to an equilibrium position; the displacement also leads to a decrease in the total energy. At room temperature, the predicted structures exist and are ionic conductors. These compounds are characterized by the anisotropic nature of vibrations and, apparently, the harmonic approximation used in the calculations leads to the appearance of negative frequencies in the phonon spectrum. This fact can be used as one of the possible criteria for assessing and predicting the presence of ionic conductivity in compounds.

The study was carried out with the financial support of the Russian Foundation for Basic Research within the framework of scientific project No. 19-32-80007.

References

[1] USPEX (Universal Structure Predictor: Evolutionary Xtallography) [Electronic resource]. – Access mode: https://uspex-team.org
[2] Brandt N B and Kulbachinsky V A 2007 *Quasiparticles in Condensed Matter Physics* (Moscow: FIZMATLIT) p 632 [in Russian]
[3] Bulat L P, Ivanov A A, Osvensky V B, Pshenay-Suverin D A and Sorokin A I 2017 *Physics of the Solid State* **59** 2071–6.
[4] Chopra K and Das S 1983 *Thin-Film Solar Cells* (Boston: Springer)
[5] Gorbachev V V 1980 *Semiconductor Connections* $A_2^I B^{IV}$ (Moscow: Metallurgy) p 132 [in Russian]
[6] Kim H, Ballikaya S, Chi H, Ahn J P, Ahn K, Uher C, Ctirad U and Kaviany M 2015 *Acta Materialia* **86** 247–53
[7] Rafii-Tabar H, Sutton A P 1991 Long-range Finnis-Sinclair potentials for fcc metallic alloys Philosophical Magazine Letters **63** 217–24
[8] Agekyan V F and Serov A Y 2018 *Exciton Spectrum in a Semiconductor Crystal. Study Guide* (Saint Petersburg State University) p 12 [in Russian]
[9] Quantum-ESPRESSO [Electronic resource]. – Access mode: http://www.quantum-espresso.org
[10] Pseudopotentials Database [Electronic resource]. – Access mode: http://www.pwscf.org
[11] Gimaldinova M, Kochaev A and Maslov M 2020 *Lett. Mater.* **10** 283–7

IOP Conf. Series: Materials Science and Engineering 1008 (2020) 012054 doi:10.1088/1757-899X/1008/1/012054

Molecular dynamics simulation of crumpled graphene filled with Ni nanoparticles

L R Safina[1], R T Murzaev[2] and K A Krylova[2,3]
[1]Ufa State Petroleum Technological University, 1 Kosmonavtov St., Ufa 450062, Russia
[2]Institute for Metals Superplasticity Problems of the Russian Academy of Sciences, 39 Khalturin st., Ufa 450001, Russia
[3]Bashkir State University, 32 Validy St., Ufa 450076, Russia

E-mail: saflia@mail.ru

Abstract. Graphene-nickel composites are novel structures with improved properties intensively studied in the last decades. In the present work, molecular dynamics simulation is used to study the process of formation of a composite obtained from well-known crumpled graphene mixed with Ni nanoparticles on the atomistic level. To obtain the composite from initially separated structural units, hydrostatic compression at elevated temperatures is applied. As it is found, the composite structure can be obtained only at high temperatures in $1200 < T < 1500$ K range. Numerical tensile tests showed that the higher is the temperature during hydrostatic compression, the smaller is the pore size at the final stage of tension. Compression at high temperature leads to the formation of new chemical bonds between graphene flakes, supports the spread of Ni atoms inside the graphene structure and results in a formation of an excellent composite structure.

1. Introduction

Crumpled graphene is a structure which presents a mixture of graphene flakes (GFs) of different sizes and shapes, crumpled in a different way and connected by van-der-Waals weak interactions. In recent decades it has been considered promising for such applications as supercapacitors [1, 2, 3], hydrogen storage [4, 5], the fabrication of new composite materials [6, 7, 8], to name a few. The success in this field is also connected with the fact that a simple and green approach to obtain crumpled graphene was recently proposed [2].

Recent experiments on the production of graphene-metal composites revealed that such new structures can be obtained by quite simple methods and show much better mechanical and physical properties [9, 10, 11]. However, the experiments were unable to reveal in detail the atomistic mechanisms that govern the composite formation, which can only be realized through molecular dynamics (MD) simulations. This method was previously used for studying different properties of crumpled graphene [12, 13, 4, 5, 14].

In the present work, the fabrication of a graphene-nickel composite by hydrostatic compression is studied by molecular dynamics simulation. This work is a continuation of previous works by the Authors, where the mutual interaction of carbon polymorphs with Ni nanoparticles and also the hydrostatic compression of crumpled graphene filled with nanoparticles were studied [15, 8, 16].

2. Simulation details

The initial structure of crumpled graphene with Ni nanoparticles (NPs) is created by a simple combination of one initial unit to a 3D structure. GF obtained from a 2.5 nm long carbon nanotube (CNT) CNT(15,15): a small part of CNT is cut along the CNT axis. After that, the cut nanotube N_C = 252 is filled with the Ni nanoparticles N_{NP47}. The diameter of Ni nanoparticles is N_{NP47} = 7.2 Å. Then, the GFs with Ni nanoparticles are randomly rotated, and the structural unit is repeated four times along three directions with the total number of atoms in the system of 19136 with N_C = 16128, N_{Ni} = 3008. To avoid overlapping, the GFs are placed far from each other. Thus, the initial relaxation is not applied since the large pores will not allow the GFs to interact. However, those pores will disappear during further compression. It should be mentioned, that an increase in the size of the computation cell two times does not lead to significant changes in the results. The initial structure of crumpled graphene filled with NPs is shown as a projection on the *xy* (a), *zx* (b) plane in figure 1.

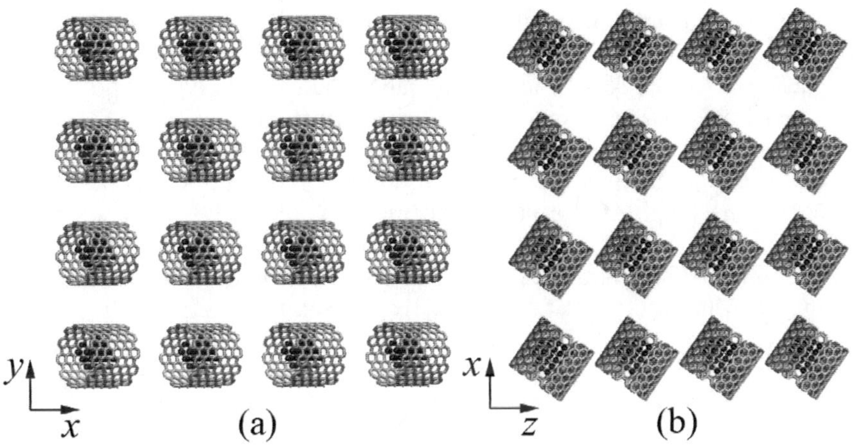

Figure 1. Initial structure of crumpled graphene filled with Ni_{47} nanoparticles as the projection on (a) *xy* and (b) *zx*.

All simulations have been performed employing the MD package large-scale atomic/molecular massively parallel simulator package (LAMMPS). The potential of the Ni- graphene (U_{system}) system can be defined as the sum of three potential energies of carbon-carbon (U_{C-C}), carbon-Ni (U_{C-Ni}) and Ni-Ni (U_{Ni-Ni}) interactions, correspondingly:

$$U_{system} = U_{C-C} + U_{C-Ni} + U_{Ni-Ni} \tag{1}$$

The adaptive intermolecular reactive empirical bond order potential (AIREBO) [17] is used to describe the interatomic interactions between carbon atoms, which include both covalent bonds in the basal plane of graphene and van-der-Waals interactions between GFs. The AREBO potential has been successfully used to study the deformation behaviour of various 3D carbon systems [13, 14], hydrocarbon structures, hydrogen storage in graphene [18, 5], to name a few.

The second and third terms of Eq. (1) are calculated using the Morse potential

$$U_{Ni-C}(r) = D_e[(1 - e^{-\beta(r-Re)})^2 - 1], \tag{2}$$

where D_e is the binding energy, R_e – distance for potential energy minimum and β – potential parameter. The parameters of the Morse potential for describing the interaction of nickel and carbon atoms (table 1) were obtained by the ab-initio method [19, 20, 21].

Table 1. Parameters of the Morse potential.

	De, eV	Re, Å	β, 1/Å	Ref.
Ni-Ni	0.4205	2.78	1.4199	[22]
C-Ni	0.433	2.316	3.244	[19, 20]

The Morse potential was previously successfully used for the investigation of discrete breathers in two- and three-dimensional crystals [23], studying the martensitic transformation [24], point defects in fcc metals [25], to name a few. Numerous MD simulations of Ni nanoparticles are conducted using parameterization by the embedded atom method for the interatomic potential that reproduces reasonably well the properties of Ni clusters. Although the Morse potential is a simplification of a real system, it can provide results of physical relevance.

The other approach is to study the system by one complex potential. Recently, the parameters for C-Ni were proposed for calculation with the potential of reactive force fields (ReaxFF) [21] using the example of graphene rolls wrapped around a metal nanoparticle. Such type of potential can be used to simulate carbon-based nanomaterials, including other elements, such as H, O, N, S, and some metals, such as Ni, Ag, Au, etc. ReaxFF makes it possible to take into account the contributions to the energy of two-, three- and four-particle interactions, which describe covalent bonds and depend on the bond orders calculated by the empirical formula; take into account the valence of atoms; as well as electrostatic and dispersive contributions. All this determines the high accuracy of calculations. However, previously it was shown that qualitatively results obtained by Morse and ReaxFF are similar and for simplicity, the Morse potential can be effectively used for investigation of a complex system, while ReaxFF is more suitable for studying chemical reactions in a small systems [15, 26].

A constant temperature is maintained by a Nose-Hoover thermostat. The equations of motion for the atoms were integrated numerically using the fourth-order Verlet method with a time step of 0.1 fs.

To fabricate a Ni-graphene composite, the hydrostatic compression ($\varepsilon_{xx} = \varepsilon_{yy} = \varepsilon_{zz} = \varepsilon$) is applied to the computational cell. The parameter s is increased at a given strain rate. The hydrostatic pressure $p = (\sigma_{xx} + \sigma_{yy} + \sigma_{zz})/3$, structure density and volume, potential energy and some other characteristics are calculated during deformation.

After compression, the structure and mechanical properties of Ni-graphene composite are analyzed. To study the mechanical behavior, uniaxial tension is applied to the structure, obtained after compression.

3. Results and discussion

To form a strong composite material, the pressure is applied to the initial structure at temperatures $T = (0, 1000, 1200, 1500)$ K, which should activate the formation of new chemical bonds between graphene flakes. It should be noted that this temperature range is rather far from the melting point of graphene (about 5000 K) [27, 28, 29]. In addition, at that temperature nickel nanoparticles would be melted, since the melting point of the nanoparticle is about 1360 K [15]. This will lead to a better distribution of the Ni atoms in the composite structure. Temperatures above 1500 K are not considered because at these temperatures such small clusters of Ni atoms can evaporate.

As it is found previously [15, 8, 16, 26], at zero temperature hydrostatic compression does not lead to the formation of a composite structure. Crumpled graphene filled with Ni nanoparticles can be compressed to high densities of about 6 g/cm^3, however, no chemical bonds appear during compression. After uniaxial tension of the compressed structure, the structural elements return to the initial state - separated Ni nanoparticles covered with graphene flakes. At 0 K, Ni NPs preserve the spherical shape, Ni atoms cannot move between different flakes and covalent bonds in the basal plane of GFs cannot be broken at zero temperature. It should be also mentioned, that at room temperature the formation of a composite is also impossible. Due to this fact only the results for temperature higher than 1000 K are presented in this work.

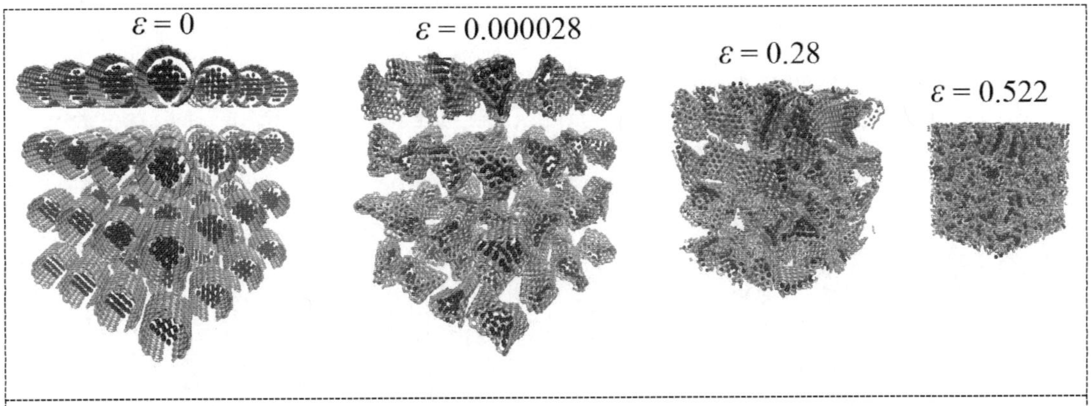

Figure 2. Snapshots of the structure of crumpled graphene filled with Ni_{47} nanoparticles during compression at 1200 K. Colours as in figure 1.

As an example, figure 2 shows the structural changes under hydrostatic compression at $T = 1200$ K. As can be seen, with an increase in temperature, single graphene flakes change their shape. Some flakes are considerably crumpled which simplified the composite formation. Structural units can rotate and fit each other much better than at low temperatures. Ni atoms can get separated from NP and move to the neighboring flakes or take a place between graphene flakes, which leads to the better mixing in the structure. However, some NPs at 1000 K preserve its initial shape which means that this temperature is also not enough for the transition to the composite state.

In figure 3, the stress-strain curves at uniaxial tension of the composites obtained by hydrostatic compression at elevated temperatures are presented. It should be noted, that hydrostatic tension is conducted at 0 K. As it is found, an increase in temperature positively affects the process of composite formation. At 1500 K, the maximum value of the applied pressure is observed. This can be explained by the fact that at this temperature a sufficient number of additional covalent bonds appear and a greater value of stress is required to break them.

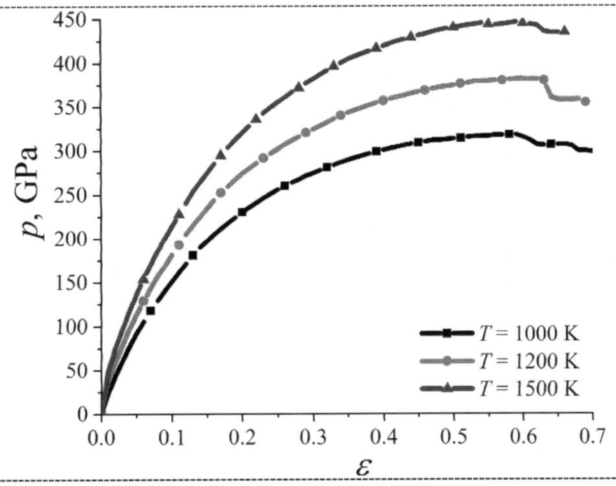

Figure 3. Pressure-strain curves for structures with Ni_{47} nanoparticles under tension. Structure was initially hydrostatically compressed at T = 1000; 1200; 1500 K.

In figure 4, snapshots of the composite structure obtained by hydrostatic compression at 1000, 1200, and 1500 K are presented at different tension strains. The structures withstand high applied tensile pressures about $\varepsilon_{xx} = 0.7$. It can be seen that an increase in temperature leads to better mixing of the elements. At $\varepsilon_{xx} = 0.5$ (T = 1000 K), the first pores begin to appear (figure 4a), and with the strain increase, the size of the pores increases, which is followed by further fracture. As for the structures compressed at 1200 and 1500 K (figure 4b,c), the pores appear a little later. It should also be noted, that the higher the temperature during compression of the structures, the smaller the pore size at the final stage. It can be concluded, that the temperature increase promotes the formation of new chemical bonds. Nearest Neighbour analysis showed that most of the carbon atoms change their hybridisation from sp^2 to sp^3.

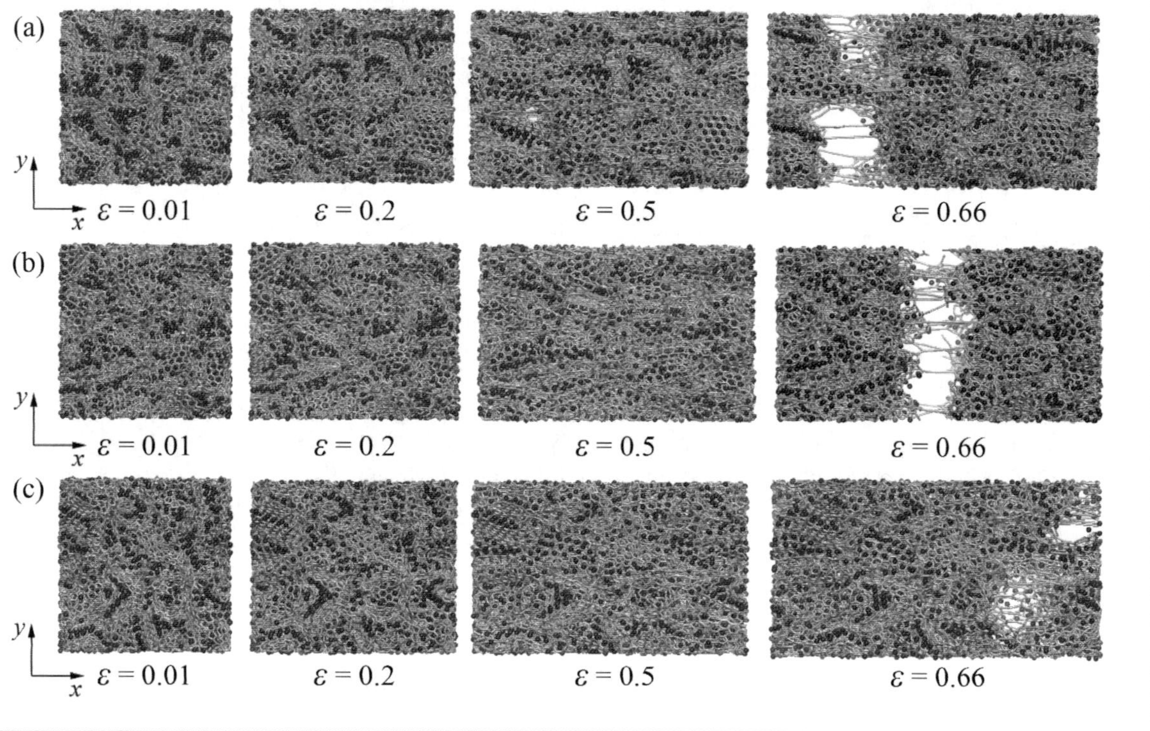

Figure 4. Snapshots of the structure of composite obtained by hydrostatic compression at (a) 1000 K; (b) 1200 K; (c) 1500 K in projection onto the *xy* plane during stretching. Colors as in figure 1.

4. Conclusions
In summary, molecular dynamics simulation is used to study at the atomistic level the process of formation of a composite structure obtained by hydrostatic compression. From the obtained results it is found that crumpled graphene can be used to obtain a graphene-based composite filled with Ni nanoparticles. Only heating to high temperatures close to 1500 K can affect the transformation of graphene flakes and the formation of covalent bonds between neighboring structural elements. If the nanoparticles inside crumpled graphene are melted, it can ease the process of mixing of Ni and C atoms and lead to the formation of a composite.

Acknowledgments
Work of K.K.A. and S.L.R. supported by the grant of Russian Science Foundation (No. 20-72-10112). Work of M.R.T. supported by the program of fundamental researches of Government Academy of Sciences of IMSP RAS.

References

[1] Tang Z, Li X, Sun T, Shen S, Huixin X and Yang J 2018 *Micropor. Mesopor. Mat.* **272** 40–3
[2] Zhang L, Zhang F, Yang X, Long G, Wu Y, Zhang T, Leng K, Huang Y, Ma Y, Yu A and Chen Y 2013 *Sci. Rep.* **3** 1408
[3] Jo E H, Choi J H, Park S R, Lee C M, Chang H and Jang H D 2016 *Electrochimica Acta* **222** 58–63
[4] Krylova K, Baimova J and Mulyukov R 2019 *Lett. Mater.* **9** 81–5
[5] Krylova K A and Safina L R 2020 *J. Phys. Conf. Ser.* **1435** 012064
[6] Hu Z, Tong G, Lin D, Nian Q, Shao J, Hu Y, Saeib M, Jin S and Cheng G J 2016 *J. Mater. Process Tech.* **231** 143–50
[7] Jiang J, He X, Du J, Pang X, Yang H and Wei Z 2018 *Mater. Lett.* **220** 178–81
[8] Safina L R and Krylova K A 2020 *J. Phys. Conf. Ser.* **1435** 012067
[9] Kurapova O, Smirnov I, Solovyeva E, Archakov I and Konakov V 2020 *Lett. Mater.* **10** 164–9
[10] Khisamov R, Nazarov K, Trinh P, Nazarov A, Phuong D and Mulyukov R 2019 *Lett. Mater.* **9** 566–70
[11] Huang H, Tang X, Chen F, Liu J, Sun X and Ji L 2018 *J. Nucl. Mater.* **510** 1–9
[12] Zhang C, Lu C, Pei L, Li J and Wang R 2019 *Applied Nano Materials* **3** 496-505
[13] Baimova J A, Liu B, Dmitriev S V and Zhou K 2015 *J. Phys.* D **48** 095302
[14] Baimova J A, Liu B, Dmitriev S V, Srikanth N and Zhou K 2014 : *Phys. Chem. Chem. Phys*. **16** 19505
[15] Safina L R, Baimova J A and Mulyukov R R 2019 *Mech. Adv. Mater. Mod. Process* **5** 1
[16] Safina L L and Baimova J A 2020 *Micro Nano Lett.* **15** 176–80
[17] Stuart S J, Tutein A B and Harrison J A 2000 *J. Chem. Phys*. **112** 6472–86
[18] Krylova K, Baimova J, Murzaev R and Mulyukov R 2019 *Phys. Lett.* A **383** 1583–88
[19] Katin K P, Prudkovskiy V S and Maslov M M 2018 *Micro Nano Lett.* **13** 160–4
[20] Galashev A Y, Katin K P and Maslov M M 2019 *Phys. Lett. A* **383** 252–8
[21] Bejagam K K, Singh S and Deshmukh S A 2018 *Carbon* **134** 43–52
[22] Girifalco L A and Weizer V G 1959 *Phys. Rev*. **114** 687–90
[23] Dmitriev S V, Khadeeva L Z, Pshenichnyuk A I and Medvedev N N 2010 *Phys. Solid State*. **52** 1499–505
[24] Dmitriev S, Kashchenko M, Baimova J, Babicheva R, Gunderov D and Pushin V 2017 *Lett. Mater*. **7** 442–6
[25] Poletaev G M, Zorya I V and Starostenkov M D 2018 *J. Micromech. Mol. Phys.* **03** 1850001
[26] Safina L, Baimova J, Krylova K, Murzaev R and Mulyukov R 2020 *Lett. Mater*. **10** 351–60
[27] Ganz E, Ganz A B, Yang L M and Dornfeld M 2017 Phys. Chem. *Chem. Phys*. **19** 3756–62
[28] Los J H, Zakharchenko K V, Katsnelson M I and Fasolino A 2015 *Phys. Rev*. B **91** 045415
[29] Orekhov N D and Stegailov V V 2015 *J. Phys. Conf. Ser.* **653** 012090

IOP Conf. Series: Materials Science and Engineering 1008 (2020) 012055 doi:10.1088/1757-899X/1008/1/012055

Structural and phase transformations in austenitic chromium-nickel steels during nanostructuring frictional treatment

P A Skorynina[1], A V Makarov[2], E G Volkova[2], and A L Osintseva[1]

[1]Institute of Engineering Science, Ural Branch of the Russian Academy of Sciences, 34 Komsomolskaya St., Ekaterinburg, 620049, Russia
[2]M.N. Miheev Institute of Metal Physics, Ural Branch of the Russian Academy of Sciences, 18 S. Kovalevskoy St., Ekaterinburg, 620108, Russia

E-mail: polina.skorynina@mail.ru

Abstract. The parameters of the structure and phase composition of austenitic structural steels subjected to frictional treatment with a sliding indenter were studied by X-ray diffraction and transmission electron microscopy. It has been found that the strengthening of deformation-stable austenitic steel was due to the formation of finely divided austenite, with the intensification of the grain-boundary and dislocation strengthening mechanisms. Additional strengthening of metastable austenitic steel was due to the formation in the surface layer of about 70% of the volume fraction of nanocrystalline strain-induced martensite. It is proposed to use nanostructuring frictional pretreatment as a promising method for hardening austenitic structural steels before the subsequent application of superhard, dead-hard, and heat-resistant nanocomposite coatings.

1. Introduction

Austenitic chromium-nickel steels are popular structural materials in the chemical, oil and gas producing, and medical industries, in thermal and nuclear power engineering. To increase the lifetime of stainless steels and to use them under special conditions (high temperatures, corrosive and abrasive environments, high mechanical loads), special processing of the surface layer is required. One of the effective state-of-the-art techniques for increasing operationally important characteristics of metal products is the formation of nanocomposite coatings on their surfaces by physical vapor deposition (PVD) [1, 2]. Such coatings are highly strong, wear-resistant, and thermally stable [3]. However, an essential disadvantage of nanocomposite coatings is their high brittleness, which manifests itself under significant deformation of the base during operation of coated products [4]. Therefore, dead-hard, (HV ~40 GPa), superhard (HV ~60 GPa), and heat-resistant thin-film coatings are mainly applied onto a high-strength structural steel substrate. The strengthening of the surface layer of relatively low-strength structural steels before applying nanocomposite thin-film coatings onto their surfaces considerably expands the area of application of such coatings. This is particularly relevant for soft ductile austenitic steels.

Frictional treatment with a sliding indenter is an effective technique for strengthening thermally nonhardenable chromium-nickel austenitic steels with different contents of nickel (~10 and 14 wt%), which stabilizes austenite to strain-induced $\gamma \rightarrow \alpha'$ transformation [5-7]. Besides, the frictional treatment of austenitic steels prone to contact seizure produces high-quality surfaces with low roughness and without material discontinuities [6, 8]. When coatings are applied, an important role is played by the strength characteristics and the quality of the substrate surface [9, 10]; therefore, frictional treatment can be considered a promising method of surface preparation of chromium-nickel austenitic steels for subsequent deposition of nanocomposite coatings. The structural-phase state of the surface layer of the substrate material is known to be able to affect significantly the mechanical and adhesive properties of coatings [10]. Therefore, it is relevant to study the features of the phase transformations and structural changes that may occur in the surface layers of austenitic steels during frictional treatment [11], i.e. at the stage of preparation for coating application.

Thus, the aim of this paper is to study the effect of frictional treatment on the structure and phase composition of austenitic chromium-nickel steels with different stabilities of austenite to strain-induced transformation. This will create a scientific basis for the formation of thin-film heat-resistant, superhard, and dead-hard nanocomposite coatings on friction-hardened austenitic steel surfaces.

2. Material and experimental procedure

The corrosion-resistant austenitic steels 12Cr18Ni10Ti (0.10 C, 17.72 Cr, 10.04 Ni, 0.63 Ti, 1.33 Mn, 0.57 Si, 0.227 Mo, 0.064 Co, 0.014 Nb, 0.057 Cu, 0.031 P, 0.014 S, the rest Fe, wt%) and 03Cr16Ni14Mo3Ti (0.03 C, 15.69 Cr, 14.17 Ni, 1.17 Ti, 0.25 Mn, 0.64 Si, 2.67 Mo, 0.03 Co, 0.004 Nb, 0.11 Cu, 0.030 P, 0.008 S, 0.043 V, the rest Fe, wt%) were studied. Samples of the 12Cr18Ni10Ti steel sized 98×38×8.6 mm were cut from a rolled sheet and water-quenched from 1050 °C. Samples of the 03Cr16Ni14Mo3Ti steel sized 36×36×6.5 mm were cut from round stock and water-quenched from 1100 °C.

The frictional treatment of austenitic steels was performed with a hemispherical synthetic diamond indenter, with the hemisphere radius R=3.0 mm, sliding in a noncorrosive argon environment. The 12Cr18Ni10Ti steel was processed with the number of scans n=11 and the load on the indenter P=392 N; the 03Cr16Ni14Mo3Ti steel was friction loaded at n=7 and P=294 N. The smaller load and number of scans in the frictional treatment of the 03Cr16Ni14Mo3Ti steel are explained by the development of seizure on its surface when the load on the indenter is P=392 N.

All the samples were cut on a Fanuc Robocut α-0iE electroerosion wire cutting machine. The structure was examined on a JEOL JEM-200CX transmission electron microscope with the use of unilateral mechanical and electrolytic thinning of the workpieces (foils). The phase composition and the integral breadth B of the line (111)γ was determined on a Shimadzu XRD-7000 X-ray diffractometer in CrKα radiation.

3. Experimental results and discussion

X-ray diffraction phase analysis has shown that, during frictional treatment, strain-induced γ→α' transformation actively develops in the surface layer of the 12Cr18Ni10Ti steel (figure 1 a, b). In the diffraction pattern, side by side with the austenite line (111)γ, there is the line (110)α, which corresponds to strain-induced α'-martensite (figure 1 b). The intensity of the peaks of martensite X-ray lines is higher than that of austenite ones, this being indicative of the predominance of the martensitic phase in the structure of the 12Cr18Ni10Ti austenitic steel after frictional treatment. The quantity of α'-martensite that has formed in the surface layer of the steel reaches 70 vol%, and this testifies to the high metastability of the 12Cr18Ni10Ti austenitic steel to strain-induced martensitic phase transformation.

According to the data of X-ray analysis (figure 1 c, d), after frictional treatment, in the severely deformed thin layer of the 03Cr16Ni14Mo3Ti steel, the austenitic structure remains practically intact. In the diffraction pattern shown in figure 1 d there are X-ray line peaks of austenite (111)γ and (200)γ and a very low intensity peak of the (110)α line, which corresponds to strain-induced α'-martensite. The low intensity of the peak of the martensite lines testifies to a low martensite content in the surface layer of the friction-treated 03Cr16Ni14Mo3Ti steel. The calculation has shown that the fraction of strain-induced martensite is at most 2 vol%. Thus, the 03Cr16Ni14Mo3Ti steel has a very high deformation stability, even under conditions of extremely severe plastic deformation implemented during frictional treatment.

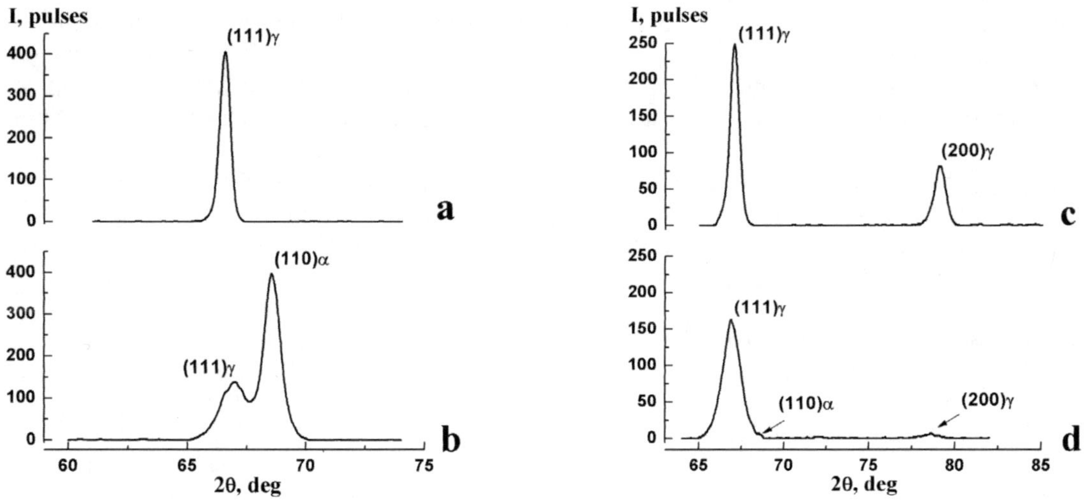

Figure 1. X-ray diffraction patterns of the surfaces of the 12Cr18Ni10Ti (a, b) and 03Cr16Ni14Mo3Ti (c, d) steels, as-quenched (a, c) and friction-treated (b, d).

As is known, the presence of strain-induced martensite may reduce the corrosive characteristics of austenitic steels [12]. However, the application of a nanocomposite coating onto a friction-treated 12Cr18Ni10Ti steel substrate can enhance the corrosion-resistance of the surface layer [13]. The decrease in corrosion resistance can be avoided if the 03Cr16Ni14Mo3Ti steel is used as the substrate material, which was shown above to be highly stable to strain-induced transformation.

The analysis of X-ray diffraction patterns shows that the frictional treatment of the austenitic steels under study increases the integral breadth B of the (111)γ austenite line. For the metastable 12Cr18Ni10Ti steel, this increase amounts to 73 min (from B=13 min for the as-quenched state to B=86 min for the friction-hardened one). For the deformation-stable 03Cr16Ni14Mo3Ti steel, the increase in the austenite X-ray line breadth is 43 min (from B=27 min for the as-quenched state to B=70 min after frictional treatment). This increase in the breadth of the (111)γ X-ray lines is caused, first of all, by increased dislocation density in the austenite and increased microdistortions of the fcc austenite lattice [14].

In the initial as-quenched state, the structure of the 12Cr18Ni10Ti and 03Cr16Ni14Mo3Ti steels is represented by coarse (~2 μm) polyhedral austenite grains with a developed dislocation substructure [6, 7]. In the structure of the 12Cr18Ni10Ti steel there also are individual inclusions of titanium carbide TiC [6]. Transmission electron microscopy on unilaterally thinned thin foils demonstrates (figure 2) that frictional treatment causes intensive dispersion of the structure of the thin (~5 μm) surface layer of the austenitic steels under study. In the electron microdiffraction pattern for the 12Cr18Ni10T steel structure (figure 2 a) annular reflections of austenite (111)γ and martensite (110)α are visible, this being consistent with the data of the X-ray diffraction phase analysis of this steel (figure 1 a, b). The dark-field images in the reflections under study indicate the formation of predominantly nanocrystalline (with crystallites sized below 100 nm), and submicrocrystalline structures of austenite and strain-induced martensite (figure 2 b, c) in the surface layer of the metastable steel under frictional treatment. In the surface layer of the deformation-stable 03Cr16Ni14Mo3Ti austenitic steel, frictional treatment results only in fragmented submicrocrystalline and nanocrystalline austenitic structures, as evidenced by the presence of annular reflections of austenite(111)γ and (200)γ in the electron microdiffraction pattern (figure 2 d).

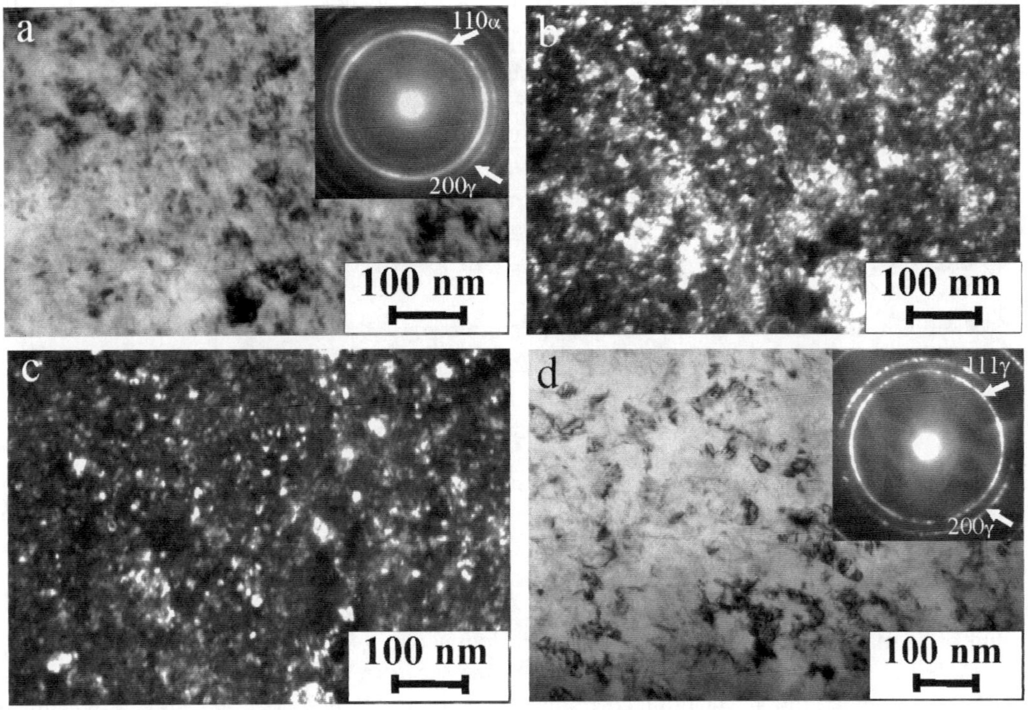

Figure 2. The structure of the surface layer of the friction-treated 12Cr18Ni10Ti (a–c) and 03Cr16Ni14Mo3Ti (d) austenitic steel samples: bright-field images and microdiffractions (a, d), dark-field images in the (110)α (b) and (200)γ reflections (c).

It was reported in [6, 7, 15] that nanostructuring frictional treatment increases the microhardness of quenched austenitic steels from 220 to 710 HV 0.025 (by a factor of 3.2) for the 12Cr18Ni10Ti steel and from 270 to 720 HV 0.025 (by a factor of 2.7) for the 03Cr16Ni14Mo3Ti steel. Consequently, after frictional treatment, the surface microhardness values of these steels are about the same for the two steels (710-720 HV 0.025) despite a lower load applied to the indenter and a lower number of scans in the frictional treatment of the deformation-stable steel than in the treatment of the 12Cr18Ni10Ti steel. Thus, nanostructuring frictional treatment provides intensive strain-hardening of both austenitic steels, independently of the different phase compositions formed on their surfaces (figures 1, 2), and it can be recommended as a preparatory strengthening technique for subsequent application of high-strength nanocomposite coatings onto relatively soft austenitic steels.

The more than 3-fold hardening of the 12Cr18Ni10Ti steel after frictional treatment is due to the formation of a severely dispersed (up to the nano- and submicrocrystalline state) martensite/austenite structure and the appearance of 70 vol% of strain-induced martensite as a result of strain-induced phase transformation. This is evidenced by experimental data presented in figure 1 a, b and figure 2 a, b, c. It follows from figure 1 c, d and figure 2 d that the increase in the microhardness of the 03Cr16Ni14Mo3Ti steel under frictional treatment stems from the severe refinement of austenite grains and the corresponding intensification of the grain-boundary strengthening mechanism. The above-mentioned severe increase in the integral breadth of the austenite X-ray line in both steels under frictional treatment testifies to a sharp increase in the dislocation density during the formation of the nano- and submicrocrystalline structures of the surface layer.

Thus, the main contribution to the hardening of the 03Cr16Ni14Mo3Ti steel under frictional treatment is made by the dislocation and grain-boundary strengthening mechanisms, which become more active during the formation of nano- and submicrocrystalline austenite. In the case of the frictional treatment of the 12Cr18Ni10Ti steel, side by side with the above-mentioned strengthening mechanisms, an additional contribution to strengthening is made by the formation of finely divided strain-induced martensite in the structure of the thin surface layer.

The effectiveness of the formation of heat-resistant, superhard, and dead-hard coatings on an austenitic steel substrate and the service properties of thin-film coatings can be largely determined by the structural-phase state of the surface layer, which has been formed by strengthening frictional pretreatment. Particularly, the studies have shown that, depending on the chemical composition of austenitic steels, frictional treatment can form a substrate with both austenitic and mixed martensite/austenite structures. This will enable one to study the effect the type of the lattice of the nanostructured steel substrate has on the adhesion of deposited thin-film coatings.

4. Conclusion

Frictional treatment has been proposed to be used as a preparatory strengthening technique for subsequent application of high-strength thin-film coatings. The development of structural and phase transformations in chromium-nickel austenitic steels under frictional treatment has been comparatively analyzed. It has been shown that, depending on the chemical composition of austenitic steels, frictional treatment can form a nanostructured substrate both with mixed martensite/austenite structures (in the metastable 12Cr18Ni10Ti steel) and with a single-phase austenitic structure (in the deformation-stable 03Cr16Ni14Mo3Ti steel). Effective surface hardening by frictional treatment (from 220-270 to 710-720 HV 0.025) is implemented in the 03Cr16Ni14Mo3Ti steel due to the grain-boundary and dislocation strengthening of austenite; in the 12Cr18Ni10Ti steel, an additional contribution is made by the progress of the strain-induced $\gamma \rightarrow \alpha'$ transformation. The intensification of the studied mechanisms of strengthening promotes the formation nano- and submicrocrystalline structures of austenite and strain-induced α'-martensite in the thin surface layers of austenitic steels under frictional treatment.

The results obtained will serve as an important scientific basis for substantiating the new comprehensive technique for strengthening structural austenitic steels, which consists in frictional treatment followed by deposition of heat-resistant, superhard, and dead-hard nanocomposite coatings.

Acknowledgments

The work was funded by RFBR and BRFBR, project number 20-58-00057 Bel_a, in the part of the studying the structure and phase composition. The methodology of the frictional treatment was developed within the state assignment for IMP UB RAS, theme “Structure” No. AAAA-A18-118020190116-6, and the assignment for IES UB RAS, theme No. AAAA-A18-118020790148-1. The studies were performed at the collaborative access center “Testing center of nanotechnology and advanced materials” of IMP UB RAS and collaborative access center “Plastometriya” of IES UB RAS.

References

[1] Bobzin K, Brögelmann T, Kalscheuer C, Stahl K, Lohner T and Yilmaz M 2019 *Trib. Int.* **140** 105817
[2] Xu X, Su F and Li Z 2019 *Wear* **430-431** 67
[3] Ravi N, Markandeya R and Joshi S V 2016 *Mater. Today: Proc.* **3** 3002
[4] Malygin G A 2007 *Phys. Solid State* **49** 1013
[5] Makarov A V, Skorynina P A, Osintseva A L, Yurovskikh A S and Savrai R A 2015 *Obrab. Met. (Met. Work. Mat. Sci.)* **69** 80
[6] Makarov A V, Skorynina P A., Yurovskikh A S and Osintseva A L 2017 *Phys. Met. Metallogr.* **118** 1225
[7] Makarov A V, Skorynina P A, Volkova E G and Osintseva A L 2020 *Met. Sci. Heat Treat.* **61** 764
[8] Savrai R A, Makarov A V, Malygina I Yu, Rogovaya S A and Osintseva A L 2017 *Diagnostics, Resource and Mechanics of Materials and Structures* **5** 43

IOP Conf. Series: Materials Science and Engineering 1008 (2020) 012055 doi:10.1088/1757-899X/1008/1/012055

[9] Podgornik B, Leskovšek V, Tehovnik F and Burja J 2015 *Surf. Coat. Technol.* **261** 253

[10] Podgornik B, Sedlaček M, Čekada M, Jacobson S and Zajec B 2015 *Surf. Coat. Technol.* **277** 144

[11] Makarov A V, Savrai R A, Skorynina P A and Volkova E G 2020 *Met. Sci. Heat Treat.* **62** 61

[12] Emurlaev K I, Bataev I A, Lazurenko D V, Burov V G, Ivanov I V and Emurlaeva Yu Yu 2020 *Mater. Today: Proc.* **25** 424

[13] Jannat S, Rashtchi H, Atapour M, Golozar M A, Elmkhah H and Zhiani M 2019 *J. Power Sources* **435** 226818

[14] Skorynina P A, Makarov A V, Men'shakov A I and Osintseva A L 2019 *Obrab. Met. (Met. Work. Mat. Sci.)* **21** 97

[15] Makarov A V and Korshunov L G 2019 *Phys. Met. Metallogr.* **120** 303

IOP Conf. Series: Materials Science and Engineering 1008 (2020) 012056 doi:10.1088/1757-899X/1008/1/012056

Rheological behavior of a Cu-Cr alloy subjected to severe plastic deformation

R N Asfandiyarov[1,2], D A Aksenov[1,2], G I Raab[1] and M P Baryshnikov[3]

[1]Ufa State Aviation Technical University, 12 K. Marx str., Ufa 450008 Russia
[2]Ufa Federal Research Center RAS, 71 pr. Oktyabrya, Ufa 450054 Russia
[3]Nosov Magnitogorsk State Technical University, 38 Lenin str, Magnitogorsk 455000 Russia

E-mail: a.r.n@list.ru

Abstract. The choice of rate conditions of deformation is one of the key factors that determine the mode of thermomechanical processing of metallic materials. The aim of this work is to study the rheological behavior as well as the structural response of the Cu-0.6Cr alloy under conditions of large deformations. For this, physical modeling of upsetting of specimens on a Gleeble 3500 installation at deformation rates of 3, 30, and 300 mm/s, structural characterization of the obtained specimens and computer simulation of the equal-channel angular pressing (ECAP) process in the Deform 3D software package were carried out. It was found that with an increase in the rate of deformation, the value of deformation heating increased. If at the rate of ECAP 30 mm/s and at the initial temperature of 20°C, the temperature of the main volume of a round billet with a diameter of 20 mm reached 80-90°C, at 300 mm/s increased up to 150°C. It is shown that with an increase in the deformation rate up to 300 mm/s the necessary pressing forces increase by 1.5 times, and, accordingly, the pressures during pressing and the load on the tool increase.

1. Introduction

Chromium bronze considered in this study is a classic representative of the group of low-alloy heat-strengthened bronzes for electrical applications. The scope of its application is quite extensive, but one of the most promising areas of its use is high-speed rail transport. This material meets modern requirements, is widespread and relatively inexpensive. One of the modern methods of obtaining contact wires is the method of pressing according to the Conform scheme [1, 2]. However, this process, due to the relatively large volume of the deformation zone, is carried out at high temperatures, which reduces the characteristics of finished products. A consistent development of this approach is the ECAP-Conform method, which allows obtaining higher characteristics of finished products [3,4] including contact wires. It should be noted that the main role in the kinetics of structural and phase transformations during this process belongs to shear deformations. Therefore, the temperature regime is important during deformation of low-alloy heat-strengthened bronzes. However, direct control and, accordingly, the temperature correction in the deformation zone, is difficult to implement due to the design features of the equipment. Existing work on the analysis of deformation heating during ECAP is based either on direct observations using thermometric instruments, mainly in areas close to the deformation zone, or on mathematical calculations performed using tabular values and coefficients found in the databases of the software systems used for modeling [5-7].

In this work, to study the rheological behavior of the Cu-0.6Cr alloy during ECAP at different rates of deformation, a combination of physical and finite element modeling was used.

2. Material and experimental procedures

As the material for the study, we selected the low-alloyed alloy Cu-0.6Cr (wt.%). The initial state of the alloy was obtained by high-temperature treatment at 1000°C for 1 hour followed by quenching in water that resulted in a supersaturated solid solution of chromium. In the initial state, the alloy had electrical conductivity equal to 35% IACS, average grain size of 130±12 μm and ultimate strength of 270±10 MPa.

Physical modeling was carried out using the Gleeble 3500 complex. Nine cylindrical samples were prepared for physical modeling with the diameter of 10 mm and length of 14 mm. Temperature control was carried out using a K type thermocouple welded to the sample by means of contact welding. After the preparation of the samples, their upsetting was carried out up to the strain value of $e = 1.1$, which corresponded to the accumulated strain during one cycle of ECAP [8]. In the process of upsetting, the values of the temperature of the samples, the deformation force, and stresses during deformation were recorded.

Computer modeling was carried out in the Deform 3D software package. Based on the data obtained using the Gleeble facility, a database was prepared for the finite element computer simulation of the upsetting process and then, after verification, for the ECAP process. The heat transfer coefficient of the tool with the workpiece was taken equal to 11 N/s/mm/°C (11000 $W/m^2 \cdot °C$). For the workpiece, a finite element mesh consisting of 15,000 tetrahedra was generated. Since a volumetric deformation pattern with high contact stresses was simulated, the Siebel friction factor was used. Based on experimental and literature data [9, 10], the Siebel friction factor for a steel-bronze pair was taken to be 0.1. The impermeability condition was set on the contact surfaces of the rig. The number of simulation steps is 300-1000. The strain rates were equal to 3, 30, and 300 mm / s. As a result of the simulation of the ECAP process of chrome bronze with different rates of deformation, data on deformation heating, power characteristics of the process, and the stress-strain state of the workpiece during deformation were obtained and studied.

3. Results

3.1. Microstructural analysis and properties after physical modeling

Characterization of the microstructure of samples after upsetting (figure 1) indicates a significant refinement of structural elements down to the range of nanometers. One can see that in all cases a predominantly elongated structure is formed. The average transverse size of the structural components, depending on the rate of deformation, is about 155 ± 10 nm at 3 mm/s, 215 ± 13 nm at 30 mm/s, and 220 ± 12 nm at 300 mm/s. At the same time, at the deformation rate of 3 mm / s, dislocation clusters are observed mainly in the grain boundary regions, while at higher deformation rates of 30 and 300

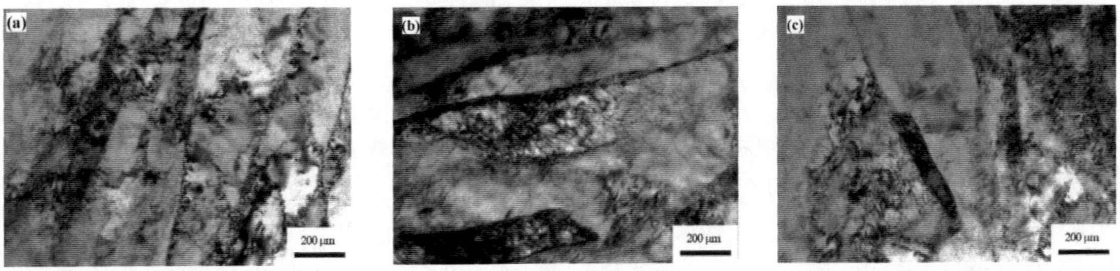

Figure 1. Microstructure of Cu-0.6Cr alloy after upsetting: a - 3 mm/s; b- 30 mm/s, c- 300 mm/s.

mm/s, dislocation networks and walls are observed in the grain body. That is, at a lower velocity, the redistribution of dislocations and their migration to the boundaries of grains and subgrains has time to occur.

In the course of the physical experiments, it was found that with an increase in the rate of deformation, stresses increase, while the electrical conductivity of the alloy also increases, which indirectly indicates an intensification of the process of deformation-stimulated decomposition of a supersaturated solid solution with an increase in the rate of deformation (Table 1).

Table 1. Physical and mechanical characteristics.

Deformation rate, mm/s	Stress, MPa	δ, %IACS
3	485±15	61
30	520±15	64
300	690±20	70

3.2. Computer simulation results

3.2.1. Deformation heating. Analysis of the data shows that at the deformation rate of 3 mm/s (figure 2a), the temperature field in the workpiece is fairly uniform, so the difference between the maximum and minimum temperatures in the deformation zone is about 20°C. The maximum value of deformation heating is 50-55°C. At the deformation rate of 30 mm/s (figure 2b), the temperature field of the workpiece is nonuniform, so the difference between the maximum and minimum temperatures in the deformation zone is about 30-35°C. The maximum value of deformation heating is 95-100°C. Data analysis shows that at the deformation rate of 300 mm/s (figure 2c), the temperature field of the workpiece is extremely inhomogeneous, so the difference between the maximum and minimum temperatures in the deformation zone reaches up to 130°C, depending on the initial temperature. The maximum value of deformation heating is about 230°C.

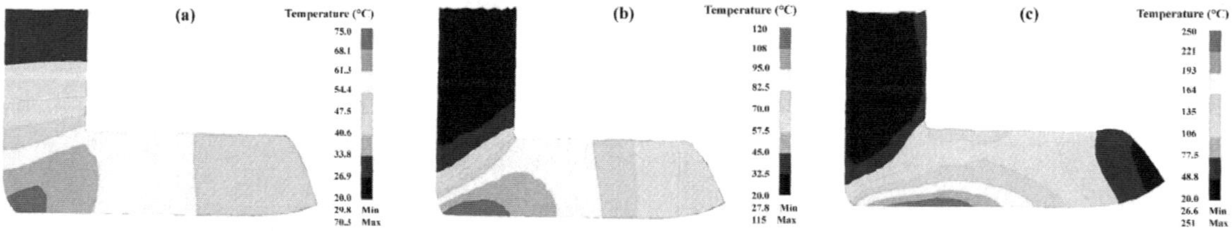

Figure 2. Change in temperature fields in the process of ECAP at speeds: a- 3 mm / s; b- 30 mm / s, c- 300 mm / s.

Thus, under conditions of severe plastic deformation by the ECAP method, intense deformation heating occurs. The deformation rate has a significant effect on heating, so at the initial room temperature of 20°C with an increase in the deformation rate from 3 to 300 mm, deformation heating increases from 50 to 230°C.

3.2.2. *Stress-strain state.* Figures 3 a, b show the distribution of the accumulated strain in the longitudinal section of the workpiece. The distribution is non-uniform, the maximum values are observed in the peripheral areas and are equal to e ~ 2.5. This can have a positive effect on the strength and wear resistance characteristics of the finished product in the form of a contact wire (in the area of the friction pair of the contact wire). The minimum values in the central region are of the order of e ~ 1.4.

The stress state was estimated from the average stress in the deformation zone. According to the analysis of the data obtained in the deformation zone, regardless of the deformation rate, compressive stresses prevail, which in turn contributes to the production of defect-free products. Figures 4 c, d show the distribution of average stresses at the deformation rate of 30 mm/s. One can see that the stress state is inhomogeneous, the maximum values of compressive stresses reach 230 MPa.

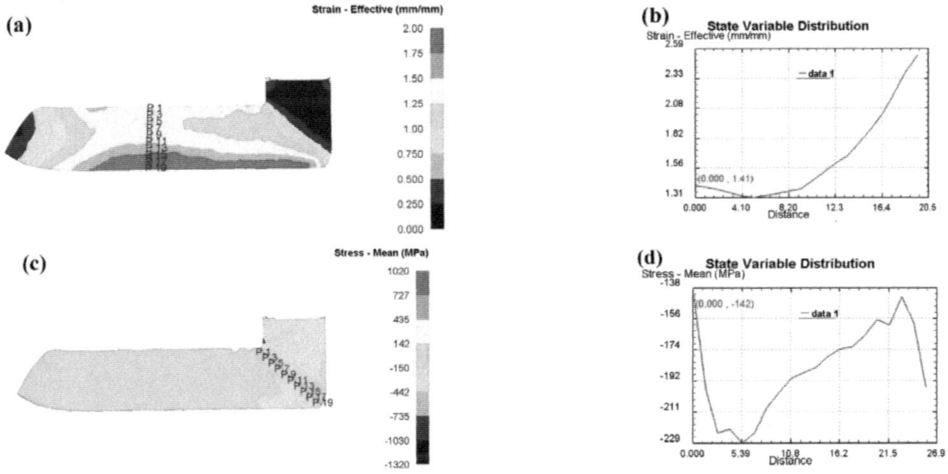

Figure 3. Stress-strain state of the workpiece during ECAP at an initial temperature of 20 °C and a deformation rate of 30 mm/s: a, b - accumulated strain distribution; c, d - distribution of average stresses

According to the simulation, as the rate of deformation increases, the required deformation force increases. For the deformation rate of 3 mm/s, the required force is about 210 kN, for 30 mm/s and 300 mm/s it amounts 220 kN and 280 kN, respectively.

Conclusions

Thus, in the course of computer modeling, a comprehensive analysis of the stress-strain state of the workpieces during processing by the ECAP method under various rate conditions was carried out.

It was found that during severe plastic deformation by the ECAP method, intense deformation heating occurs. The deformation rate has a significant effect on heating, so at the initial room temperature of 20°C with an increase in the deformation rate from 3 to 300 mm/s, the deformation heating increases from 50 to 230 °C. An increase in the rate of deformation also leads to an increase in the required force. At 20°C, an increase in the deformation rate from 3 to 300 mm/s leads to an increase in the force from 210 to 280 kN.

Based on the data obtained during the physical experiment, it was found that with an increase in the deformation rate, not only an increase in stresses and the required force will occur, but also the decomposition of a supersaturated solid solution will occur more intensively, which makes it possible to increase the electrical conductivity of the alloy to 70% IACS. The strain of e = 1.1 ensures the formation of a grain-subgrain type structure with the size of structural elements down to 150-220 nm.

Acknowledgments

The study was supported by the Russian Science Foundation (project №19-19-00432).

References

[1] Barcellona A, Cannizzaro L and Riccobono R 2000 *Proc. of the 33rd Int. MATADOR Conf.* 353-358
[2] Babcock Wire Equipment Ltd http://www.bwe.co.uk
[3] Xu C, Schroeder S, Berbon P B and Langdon T G 2010 *Acta Mater.* **58** 4 1379-86

[4] Raab G I, Valiev R, Gunderov D, Lowe T C, Misra A and Zhu T 2008 *Mater. Sci. Forum* **584-586** 80-5
[5] Karamia J S 2016 *Mater. Res.* **19** 640-7
[6] Kim H S 2001 *Mater. Trans.* **42**(3) 536-8
[7] Pei Q X, Hu B H, Lu C and Wang Y Y 2013 *Scripta Mater.* **49** 303-8
[8] Utyashev F Z and Raab G I 2013 *Deformation Methods for the Fabrication and Processing of Ultrafine-Grained and Nanostructured Materials* (Ufa, Gilem) (in Russian)
[9] Loginiv Yu N 2006 *Copper and Wrought Copper Alloys* (Ekaterinburg, USTU) (in Russian)
[10] Male A T 1964 *J. Inst. Metal.* **93** 38-46

Contribution of structural parameters to the strength and electrical conductivity of Cu-0.5%Cr alloy subjected to ECAP and cold rolling

E A Sarkeeva and I V Alexandrov

Ufa State Aviation Technical University, 12 K. Marx, 450000 Ufa, Russia

E-mail: sarkeeva.e@inbox.ru

Abstract. The effect of cold rolling with a reduction rate of 95% at room temperature and subsequent aging on the evolution of the microstructure, strength properties and electrical conductivity of the Cu-0.5Cr alloy (wt.%), which is widely used in the electrical industry in the form of strips and ribbons, has been studied. The cold rolling and aging of the alloy was carried out in coarse-grained and ultrafine-grained states after equal-channel angular pressing. The regularities found were used to assess the contribution of various structural parameters to the level of strength properties as well as the electrical conductivity of the alloy in the obtained states.

1. Introduction

Chrome bronzes are precipitation-hardening alloys and have an optimal combination of physical, mechanical and operational properties after thermal and thermomechanical treatment. In recent years, a new approach has been developed that makes it possible to achieve a significant increase in the strength of alloys of the Cu-Cr system while maintaining high electrical conductivity and ensuring the creation of new generation conductors [1, 2]. This approach is based on replacing traditional machining with methods of severe plastic deformation (SPD). SPD, in particular, equal-channel angular pressing (ECAP), makes it possible to control the mechanisms of hardening and electrical conductivity by refinement of the copper matrix to submicron sizes. The subsequent breakdown of solid solution is accompanied by the formation of nanosized particles of secondary phases, which makes an additional contribution to the strengthening and increase in the electrical conductivity of the formed ultrafine-grained (UFG) states. ECAP makes it possible to obtain bulk billets, as a rule, in the form of rods, however, the issue of obtaining thin strips from ultrafine-grained materials for subsequent use, including in electrical engineering [3], is urgent. Flat rolling of copper, titanium, etc. makes it possible to give massive UFG blanks obtained by ECAP the form of strips [4-6]. In this case, a combined deformation scheme is realized. Changing the deformation mode from ECAP to flat rolling affects the mechanisms of microstructure evolution and physical and mechanical properties. To further improve the functional properties of UFG alloys of the Cu-Cr system, it is necessary to understand the physical nature of increasing the strength and electrical conductivity of these materials.

This paper presents the results of a study on the microstructure and physico-mechanical properties of the Cu-0.5Cr alloy subjected to ECAP, subsequent cold rolling and aging. The regularities obtained were used to assess the contribution of various structural parameters to the level of strength properties, as well as the electrical conductivity of the alloy in the obtained states.

2. Method of experimental research

In this work, studies were carried out on an electrical alloy Cu-0.5Cr (wt/%). At the first stage, the samples were exposed at a temperature of 1000 °C for 1 hour, followed by quenching in water to create a supersaturated solid solution of alloying elements in a copper matrix. Thus the initial coarse crystalline state (the IS state) was formed. ECAP was carried out in a die with an internal angle ψ=90° along route B_C (turning the workpiece 90° clockwise along the longitudinal axis after each pass) at room temperature at a speed of 0.2 mm/s. The number of passes is 8 (the ECAP8 state). Subsequently, cold rolling (CR) was carried out at room temperature with a total reduction ratio of 95% (the ECAP8+CR state). In one pass during rolling, the reduction in thickness was approximately 5%. Between the passes, billets were cooled in air to minimize the effect of deformation heating on the microstructure. The rolling direction was parallel to the direction of extrusion. The equivalent deformation in one pass of ECAP was 1.15, and after rolling by 95% - 3.5. Aging was carried out in air in an oven SUOL 45 at 425 °C for 40 minutes. At all stages of the study, the microstructure was investigated by optical, scanning and transmission electron microscopy (TEM). Studies of the microstructure and properties were carried out in the rolling plane. X-ray structural analysis (XRD) of the samples was performed by diffractometer Rigaku Ultima IV (Bragg-Brentano geometry) using $Cu_{K\alpha}$ monochromatic radiation at the reflected beam (parabolic graphite monochromator). The measurements were conducted along the area of 2×2 mm² situated in the center of the samples. The values of the lattice parameter a, the size of the coherent scattering regions (CSR) D_{XRD}, and the level of elastic microdistortions $\langle\varepsilon^2\rangle^{\frac{1}{2}}$ were calculated by the Rietveld refinement method using the MAUD program. The dislocation density $\rho_\perp$ was estimated by the formula (1)

$$\rho_\perp = \frac{2\sqrt{3}\langle\varepsilon^2\rangle^{\frac{1}{2}}}{D_{XRD}b}, \quad (1)$$

where b is Burgers vector.

Tensile tests were performed at a strain rate of 10^{-3} s^{-1}. Samples were cut parallel to the direction of deformation from the central part of the ECAP blanks and sheets after rolling. Uniaxial tensile tests were carried using small specimen on an Instron 8801 tester at room temperature. Conductivity studies were carried out at a temperature T = 23°C by the eddy current method. The measurement error was ± 2%.

3. Results and discussion

The results of detailed studies of the microstructure and properties of the Cu-0.5Cr alloy obtained by optical scanning and transmission electron microscopy, as well as by XRD, are presented in table 1.

In the initial state, the lattice parameter is 3.6184 Å, which indicates the maximum dissolution of alloying elements in the copper matrix among all the studied states. In all states after deformation (CR, ECAP8, ECAP8+CR), the lattice parameter decreases, which indicates a decrease in the concentration of alloying elements in the solid solution. These results are in agreement with the results obtained by TEM. In these states a small number of dispersed particles with a size of about 6 nm are observed in the microstructure, which are distributed inhomogeneously. Subsequent aging leads to a further decrease in the lattice parameter, which approaches the lattice parameter of pure copper 3.6150 Å.

Cold rolling of the coarse-grained state leads to an increase in the dislocation density up to $2.54 \cdot 10^{14}$ m^{-2}. Conducting ECAP8 increases the dislocation density to values of $1.24 \cdot 10^{14}$ m^{-2}. In this case, the subsequent rolling of ECAP8 increases the dislocation density to $2.19 \cdot 10^{14}$ m^{-2}, which is slightly lower than in the case of the CR state. Probably, this is due to the formation of a mixed microstructure with two types of grains/subgrains characterized by two mean sizes after cold rolling of the UFG state obtained by ECAP. The first type of grains/subgrains are small with an average size of 320 ± 20 nm and a high dislocation density. The second type of the observed grains, the fraction of which is 0.25 of the total volume, is characterized by large dislocation-free grains with an average size of 620 ± 30 nm. Most likely, they are created by mechanisms of dynamic recovery, leading to annihilation of dislocations, which reduces the total density of dislocations. Also, in the ECAP8+CR state there is a significantly higher proportion of high-angle boundaries, which is consistent with the results obtained for pure copper [8, 9].

Cold rolling of the initial state increased the strength up to 490 MPa. As a result of ECAP, the strength reached 470 MPa. The highest strength 605 MPa was demonstrated by samples after complex deformation ECAP8+CR (table 1). Strength growth is accompanied by a slight change in electrical conductivity. In IS it was 37% IACS, after CR - 39% IACS, after ECAP8 - 36% IACS and ECAP8+CR - 37% IACS. Post-deformation aging leads to a 2-fold increase of electrical conductivity for all states. In this case, the strength for the rolled CG state and the ECAP8 state change within the error limits. In the rolled UFG state, it decreases up to 470 MPa.

Table 1. Microstructure parameters, mechanical properties and electrical conductivity.

State	D_{av}, nm [7]	d_{av}, nm [7]	L_{av}, nm [7]	a, Å	$\langle\varepsilon^2\rangle^{\frac{1}{2}}$, %	D_{XRD}, nm	$\rho_\perp \cdot 10^{14}$, m^{-2}	σ_{UTS}, MPa	σ_{YS}, MPa	ε, %	IACS, %
IS	260000	-	-	3,6184(4)	0.015	188	0.011	160	60	42	37
CR	190±10	≈6	-	3,6163(4)	0.12	62	2.54	490	480	8	39
ECAP8	250±15	≈6	-	3,6159(4)	0.11	120	1.24	475	450	17	36
ECAP8 +CR	160±10	≈6	-	3.6159(4)	0.10	59	2.19	605	565	13	37
CR+A	270±30	8	170	3,6153(5)	0.02	104	0.23	495	480	20	74
ECAP8 +A	300±20	10	160	3,6154(4)	0.02	117	0.19	460	430	24	72
ECAP8 +CR+ A	320±20 620±30	12	120	3,6152(4)	0.07	125	0.77	470	430	20	74

The strength characteristics of metals and alloys are determined by a number of mechanisms. Based on the obtained microstructural parameters, the main strengthening mechanisms were assessed. The superposition of each of the strengthening mechanisms with the lattice friction stress is often linearly additive, i.e. the yield strength of a strain-hardened material is the sum:

$$\sigma_{0.2} = \sigma_0 + \Delta\sigma_{ss} + \Delta\sigma_{dis} + \Delta\sigma_{GB} + \Delta\sigma_{ph} \tag{2}$$

where σ_0 – strengthing due to friction in the crystal lattice, $\Delta\sigma_{ss}$ an increase in the yield strength due to solid solution hardening, $\Delta\sigma_{dis}$ an increase in the yield strength due to dislocation (strain) hardening, $\Delta\sigma_{GB}$ the grain boundary strengthening associated with a decrease in the grain size, and $\Delta\sigma_{ph}$ increase in the yield strength due to precipitation hardening [6, 10, 11].

Strength due to friction in the crystal lattice (σ_0) is determined by the Paierls-Nabarro stress, which for copper is equal to $\sigma_0 = 25$ МПа [10].

The variety of possible mechanisms of solid solution hardening greatly complicates the exact quantitative description of these processes. To calculate the solid-solution hardening [11], we used the expression (3):

$$\Delta\sigma_{SS} = \frac{MG\varepsilon_S^{3/2}c^{1/2}}{760}, \tag{3}$$

where M is the orientation factor, for FCC crystals M=3.06 [12], G shear modulus of the matrix, G = 42.1 ГПа [13], c the concentration of the dissolved element, and ε_S dimensional mismatch parameter. The solubility of Cr in the copper matrix at 1000 °C is 0.4 wt.% (0.49 at.%) [3] In general, the effect of solid solution hardening in the alloy is small and amounts to 0.045 Pa. Therefore, it can be considered ineffective.

Dislocation hardening caused by an increase in the density of dislocations can be described by the formula proposed by Taylor [14-15]

$$\Delta\sigma_{dis} = \alpha MGb\sqrt{\rho_\perp}, \tag{4}$$

where α is a coefficient depending on the nature of the interaction of dislocations during strain hardening, which can vary from 0.089 to 0.5, and b= 0.256 nm the magnitude of Burgers vector [10].

Dispersion strengthening is of a particular interest. The interaction of dislocations with particles can be realized through various mechanisms depending on the distribution density of particles, their size, the difference between the shear modulus of the particle and the matrix, and the type of interphase boundaries. After deformation, the contribution of dispersed particles can be neglected, since their distribution is nonuniform, and also because the density of precipitates is rather low. After aging, the size of the strengthening particles was 8-12 nm (table 1), then, according to [16], the strengthening proceeds according to the Orowan mechanism:

$$\Delta\sigma_{DS} = QM\frac{Gb}{2\pi(L-d)}\Phi\ln(\frac{L-d}{4b}), \tag{5}$$

where Q is a parameter equal to 0.81-0.85, that takes into account the uneven distribution of particles in the matrix, Φ a factor taking into account the type of dislocation, $\Phi = 1$ in the case of a screw dislocation, $\Phi = (1-\vartheta)^{-1}$ – for edge dislocation, $\Phi = 0.5(1+\frac{1}{1-\vartheta}) = 1.25$ – for mixed deployment, L the average distance between particles, and d the average particle size.

The contribution of grain-boundary strengthening was estimated as

$$\Delta\sigma_{GB} = kd^{-\frac{1}{2}}, \tag{6}$$

where k is the coefficient of grain boundary strengthening, which characterizes the material and the state of the boundaries. In this work, the value of the coefficient k is chosen as 0.21 for IS and 0.11 for UFG states, which corresponds to the calculations presented in [13]. Since after aging of the rolled UFG state large grains are observed, the proportion of which is 0.25 of the total volume, while the UFG microstructure is preserved, the contribution of grain boundaries to grain boundary strengthening was estimated due to the equivalent average grain size d_{eq}, calculated as follows:

$$d_{eq} = (\sum\frac{f_i d_i^2}{\sum f_j d_j^3})^{-1}, \tag{7}$$

where f_i is the proportion of grains with an average size d_i

The calculation results are shown in figure 1.

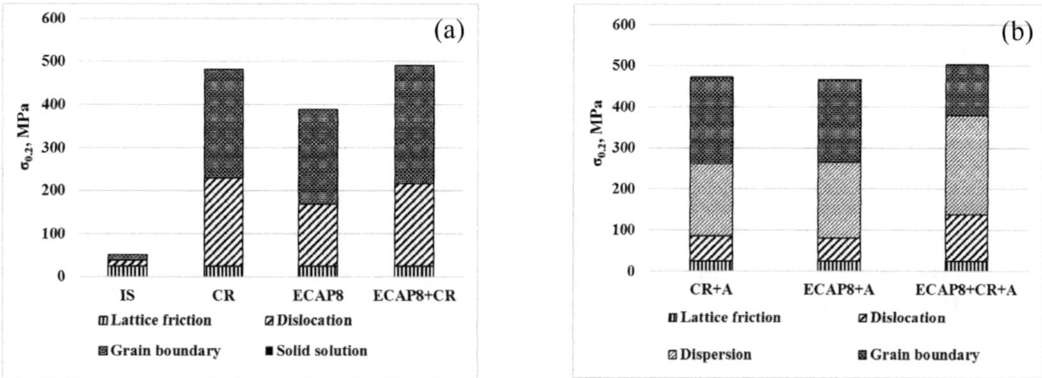

Figure 1. Contributions to the strengthening of various mechanisms.

Figure 1 shows that the dominant mechanism of strengthening after deformation in all states is grain boundary strengthening by the Hall-Petch mechanism. The dislocation mechanism ranks second in terms of contributions. It should be noted that in a rolled coarse-grained sample (CR), the dislocation strengthening numerically approaches the grain-boundary strengthening. In specimens after ECAP8 and ECAP8+CR, grain-boundary strengthening dominates over dispersion strengthening.

Aging leads to a partial decrease in the contributions of dislocation and grain-boundary hardening, due to the processes occurring during the recovery of the dislocation structure. The separation of nanosized precipitates rich in Cr leads to an increase in dispersion strengthening. The strength of the obtained states after aging is the result of a combination of grain boundary strengthening and precipitation strengthening.

The contribution of various structural parameters to the level of electrical resistance was estimated based on the Matthiessen - Flemming rule:

$$\rho = \rho_0 + C_v\Delta\rho_v + L_{dis}\Delta\rho_{dis} + S_{GB}\Delta\rho_{GB} + \sum_i C_{sol}^i\Delta\rho_{sol} \quad (8)$$

where ρ_0 is the electrical resistivity of pure copper. Since the resistivity measurements were carried out at room temperature, we will consider the electrical resistivity of pure Cu at room temperature ($\rho_0 = 1.72 \cdot 10^{-6}\ \Omega \cdot \text{cm}$ [3]). Here, $\Delta\rho_v$, $\Delta\rho_{dis}, \Delta\rho_{GB}, \Delta\rho_{sol}$ are additions that take into account the contribution, respectively, of vacancies, dislocations, grain boundaries, and dissolved elements in the copper matrix. C_v is the concentration of vacancies, L_{dis} dislocation density, S_{GB} fraction of grain boundaries per unit volume and C_{sol}^i concentration of solute *i* in the matrix.

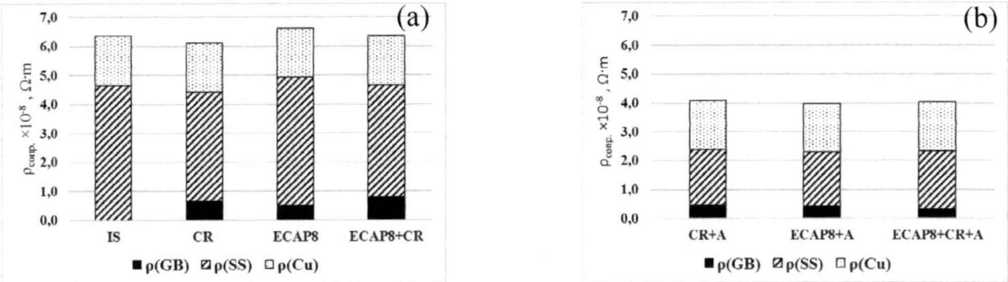

Figure 2. Contributions electrical resistivity of various mechanisms.

In this work, we did not measure the concentration of vacancies. According to [16, 17], the maximum concentration of vacancies in copper subjected to severe plastic deformation reaches 10^{-5} wt.%. However, it is important to note that even if such a high concentration of vacancies is achieved during the deformation of the alloy, this should lead to an increase in the specific electrical resistance by only 0.0013 μΩ· cm, i.e. several orders of magnitude lower than ρ_0. Therefore, the influence of vacancies on electrical conductivity can be neglected.

The deformation of IS leads to a significant increase in the dislocation density, while the electrical conductivity remains practically unchanged (table 1). It is known that the calculated values of the increase in the electrical resistivity per unit density of chaotically distributed dislocations in pure copper are equal to $\Delta\rho_{dis} = 1.9 \cdot 10^{-19}$ μΩ·m² [18]. The maximum value of the dislocation density is attained in the rolled CG state, which is $2.54 \cdot 10^{14}$ m^{-2}. It follows from these data that the contribution of dislocations to the electrical resistivity of the Cu-0.5Cr alloy at room temperature does not exceed 0.28% relative to the electrical resistivity of pure copper. Consequently, the dislocation mechanism has no significant effect on the electrical conductivity, which is consistent with the results obtained in [19].

The grain boundary contribution is much more significant. A decrease in the grain/subgrain size and an increase in the length of the boundaries lead to additional scattering of conduction electrons. The estimate of this contribution is $\Delta\rho_{GB} = 2.1 \cdot 10^{-16}$ Ω·m [18].

According to the literature data, the effect of the solid solution on the electrical resistance is quite large. This is due to the distortion of the lattice of the copper matrix, in connection with which the scattering of conduction electrons increases. The electrical properties of a solid solution are also related to the chemical interaction of components [3]. However, it is rather problematic to determine the exact concentration of alloying elements in the copper matrix required for calculations. Taking all other contributions to the electrical resistance to be insignificant, one can calculate the contribution of the solid solution as the difference between the experimental values and calculated values. The contribution of the solid solution to the scattering is maximum in IS, when a supersaturated solid solution is formed and the lattice is maximally distorted. After deformation, in all considered states, the lattice parameter decreases. In the structure, a small amount of a dispersed phase is observed, which is distributed inhomogeneously. However, the values of electrical conductivity remain practically unchanged. X-ray diffraction data indirectly indicate that the decomposition of the supersaturated solid solution proceeds is more complete after aging. Aging also leads to the development of recovery processes, which increases the size of grains/subgrains and, accordingly, reduces the grain boundary contribution to electrical resistance. As a result, the electrical resistance is significantly reduced.

Thus, the change in the electrical resistance in this alloy is controlled by the composition of the solid solution. An estimate of the change in electrical resistance suggests that the dominance of this mechanism is so strong that it exceeds the contributions of dislocation and grain-boundary scattering.

4. Conclusions

As a result of studies on the relationship between microstructural parameters and a combination of strength and electrical conductivity in different states of the Cu-0.5Cr alloy, the following was established:

- an increase in strength (almost 3-4 times) as a result of deformation realized by cold rolling and ECAP is mainly due to dislocation and grain-boundary contributions (in the rolled initial state these contributions are equal to each other, in the rolled UFG state the grain-boundary contribution prevails);
- aging leads to the predominance of the dispersion contribution to the strength of deformed states;
- the electrical conductivity of the deformed states is low and its low level is due to the solid solution contribution;
- aging leads to a significant (approximately twofold) increase in electrical conductivity for all deformed states, due to the decomposition of the solid solution, while the contributions of dislocation and grain-boundary electrons scattering are small.

Acknowledgments

This work was supported by the Ministry of Education of the Russian Federation, project 0838-2020-0006 Fundamental study of new principles for the creation of promising electromechanical energy converters with characteristics above the world level, with increased efficiency and minimum specific indicators, using new highly efficient electrotechnical materials. Authors are grateful to the personnel of the research and technology Joint Research Center, 'Nanotech', Ufa State Aviation Technical University for their assistance with instrumental analysis.

5. References

[1] Vinogradov A, Ishida N, Kitagawa K and Kopylov V 2005 *Acta Mater.* **53** 2181
[2] Purcek G, Yanar H, Saray O, Karaman I and Maier H 2014 *Wear* **311** 149
[3] Osintsev O E, Phyodorov V N 2004 *Copper and Copper Alloys, Russian and Foreign Brands* (Moscow: Mashinostroenie) p. 215
[4] Ranjbar Bahadori Sh, Dehghani K and Bakhshandeh F 2013 *Mater. Sci. & Eng. A* **583** 36–42
[5] Jia Qiang Tao, Yao Jiang, Ying Liu, Yu Wei Liu, Jia Jie Meng, Jing Tao Wang 2020 *Mater. Sci. Eng. A* **770** 138556
[6] Stolyarov V V, Zhu Y T, Alexandrov I V, Lowe T C and Valiev R Z 2003 *Mater. Sci. Eng. A* **343** 43-50
[7] Sarkeeva E A, Raab G I, Wei W and Alexandrov I V 2019 *IOP Conf. Series: Mater. Sci. Eng.* **672** 012056
[8] Stepanov N D, Kuznetsov A V, Salishchev G A, Raab G I and Valiev R Z 2012 *Mater. Sci. Eng. A* **554** 105–115
[9] Ranjbar Bahadori Sh, Dehghani K and Bakhshandeh F 2001 *Mater. Sci. & Eng. A* **588** 260–264
[10] Zhilyaev A P, Shakhova I, Moroova A, Belyakov A and Kaibyshev R 2016 *Mater. Sci. Eng. A* **654** 131-142
[11] Kahn R W and Haazen P 1987 *Physical metallurgy* (Moscow: Metallurgy) p. 663
[12] Hull D and Bacon D J 2011 *Introduction to Dislocations* (Amsterdam, Elsevier) p. 258
[13] Nohrin A V, Chuvildeev V N, Smirnova E S, Pirozhnikova O E, Melehin N V, Lopatin U G, Schavleva A V and Kopylov V I 2007 *Mechanical Properties of Nano- and Microcrystalline Metals* (Nizhny Novgorod, Nizhny Novgorod University) 46 p.
[14] Liu J 1963 *Trans. Metall. Soc. AIME* **227** 239

IOP Conf. Series: Materials Science and Engineering 1008 (2020) 012057 doi:10.1088/1757-899X/1008/1/012057

[15] Hughes D A and Hansen N 2000 *Acta Mater.* **48** 2985–3004
[16] Tamás Ungár, Erhard Schafler and Jenö Gubicza 2009 *Bulk Nanostructured Material* (Germany: GmbH & Co)
[17] Petegema S V, Torre F D, Segersa D and Swygenhoven H V 2003 *Scripta Mater.* **48** 17
[18] Karolik A S, Luhvich A A, 1994 *Phys.: Condens. Mater* **6** 873-886
[19] Chembarisova R G 2020 *Technical Physics* **65** 593–601

Crystallogeometry and energy of planar superstructure defects of ternary alloys

A R Khalikov[1], V T Nguen[1], V A Valitov[2], S V Dmitriev[2,3]

[1] Ufa State Aviation Technical University, 450077, 12 Karl Marx, Ufa, Russia
[2] Institute for Metals Superplasticity Problems of RAS, 450001, 39 Khalturin St., Ufa, Russia
[3] Ufa State Petroleum Technological University, 450064, 1 Kosmonavtov St., Ufa, Russia

E-mail: khalikov.albert.r@gmail.com

Abstract. Planar superstructure defects have a great influence on the mechanical and functional properties of ternary Heusler alloys; however, their complete crystallogeometric and energy analysis is lacking in the literature. In this work, Heusler alloys of stoichiometric composition X_2YZ of the $L2_1$ superstructure are analysed. An expression is given for finding the planes of occurrence of all possible planar superstructure defects. It is established that among them there are only shear and thermal antiphase boundaries. The results obtained are useful for analyzing slip systems of the considered family of alloys and the configuration of magnetic domain walls.

1. Introduction

Among ternary alloys, a special place is occupied by the Heusler family of alloys [1, 2], which are usually defined as ternary intermetallic compounds formed with the stoichiometric composition X_2YZ with the $L2_1$ structure, although these also include alloys with a large number of components and with a different symmetry [3].

Most magnetic shape memory alloys belong to the Heusler family of alloys. These are alloys based on Ni (and somewhat on the basis of Co) having an austenitic cubic phase with an $L2_1$ superstructure at high temperatures and a low-symmetry phase at low temperatures, which arises as a result of martensitic transformation [4]. Martensitic transformations are responsible for the manifestation of the mechano-caloric effect [4], which consists in cooling the sample subjected to uniaxial deformation (elastocaloric effect) or volumetric deformation (barocaloric effect). Heusler alloys exhibit Seebeck and Peltier thermoelectric effects [5-7]. Some rare-earth Heusler alloys exhibit a very strong magnetocaloric effect, which creates real preconditions for the development of magnetic refrigeration devices on their basis [8]. Plastic deformation leads to grain refinement in Heusler alloys, which noticeably affects their strength and physical properties.

The physical and mechanical properties of Heusler alloys are greatly influenced by planar superstructure defects. Despite the important role of planar superstructure defects in the formation of the physical and mechanical properties of Heusler alloys, there is no systematic information in the literature on the classification and energy characteristics of such defects. At the same time, there is a significant theoretical foundation, well tested on two-component ordered alloys [9-17], which makes it possible to carry out such a study for three-component Heusler alloys. Some information on the structure and energetics of various planar defects in alloys with the $L2_1$ superstructure can be found in [10-15]. It seems important to give a detailed description of the coordination of atoms and possible planar superstructure defects, as well as to calculate the energy of these defects in the hard-sphere model (without taking into account atomic relaxation).

2. Crystallogeometry and energy of planar superstructure defects

The superstructure $L2_1$will be represented both on the basis of a cubic translation cell with 16 atoms, and on the basis of a primitive translation cell of minimum volume with 4 atoms. Let us describe the packing of atoms in ternary Heusler alloys of stoichiometric composition X_2YZ with the superstructure $L2_1$. As shown in figure 1, this superstructure is a union of four fcc lattices, two of which are occupied by atoms of type *X*, one by atoms of type *Y*, and one by atoms of type *Z*. Each fcc lattice is a set of four simple cubic lattices, that is, the superstructure $L2_1$ can be decomposed into 16 simple cubic gratings. Let the cubic translational cell shown in figure 1 in the Cartesian coordinate system xyz, has a lattice parameter a.

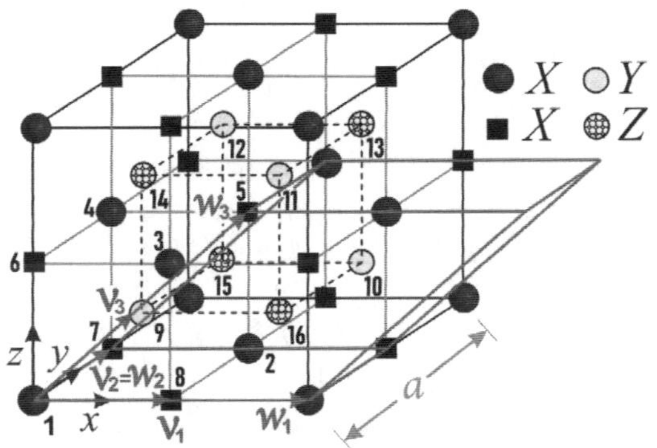

Figure 1. Cubic and primitive translation cells of the $L2_1$ superstructure based on translation vectors w_i, $i = 1,2,3$ (highlighted in red). The cell contains two *X*-type atoms, one *Y* and one *Z*-type atom. The primitive cell volume is 4 times less than the cubic cell volume. The bcc lattice is based on translation vectors v_i, $i = 1,2,3$.

To calculate the energy of any planar superstructural defect, we delaminate the superstructure into two-dimensional monoatomic packings Ω parallel to the orientation of the defect (*h*, *k*, *l*):

$$\tilde{Q} = \cup_{m=1}^{4} \cup_{\xi=-\infty}^{+\infty} (P + \boldsymbol{\theta}_m + \xi \boldsymbol{u}_3)_{s_m} \quad (1)$$

where *P* is a two-dimensional lattice parallel to the plane of the defect, based on vectors $\boldsymbol{u}_1$, $\boldsymbol{u}_2$, $\boldsymbol{\theta}_m$ are translation vectors, $\boldsymbol{u}_3$ is a vector not lying in the plane of vectors $\boldsymbol{u}_1$, $\boldsymbol{u}_2$. Note that the vectors $\boldsymbol{u}_i$ generate the lattice $\tilde{L}$ in the expression. The choice of vectors $\boldsymbol{u}_i$ determines the plane of the planar defect.

A planar superstructure defect is formed by joining along the plane (*h*, *k*, *l*) of two half-crystals

$$\tilde{Q}_- = \cup_{m=1}^{4} \cup_{\xi=-\infty}^{-1} (P + \boldsymbol{\theta}_m + \xi \boldsymbol{u}_3)_{s_m},$$
$$\hat{Q}_+ = \cup_{m=1}^{4} \cup_{\xi=0}^{+\infty} (P + \boldsymbol{\theta}_m + \xi \boldsymbol{u}_3)_{s_{\alpha m}}, \quad (2)$$

which differ in the arrangement of types of atoms along the sublattices, and these arrangements are energetically equivalent. In (2), the subscripts "-" and "+" mark the lower and upper half-crystals. The defect energy per unit area is calculated according to

$$E = E(\tilde{Q}_- \leftrightarrow \hat{Q}_+) - \frac{E}{2}(\tilde{Q}_- \leftrightarrow \tilde{Q}_+) - \frac{E}{2}(\hat{Q}_- \leftrightarrow \hat{Q}_+) \quad (3)$$

where the first term on the right is the energy released when two different domains join, and the other two are the surface energy of these domains, per unit area. The first of the three energies is expressed by the following sum:

$$E(\tilde{Q}_- \leftrightarrow \hat{Q}_+) = \frac{1}{|\boldsymbol{u}_1 \times \boldsymbol{u}_2|} \sum_{m=1}^{4} \sum_{n=1}^{4} \sum_{\eta_3=-\infty}^{-1} \sum_{\eta_3=0}^{\infty} \sum_{\eta_1,\eta_2=-\infty}^{+\infty} \varphi_{s_m s_{\alpha n}}(|\boldsymbol{r}|), \quad (4)$$

where $|\boldsymbol{r}| = \boldsymbol{r}_{s_m} - \boldsymbol{r}_{s_{\alpha n}} = (m\boldsymbol{\theta}_m + \eta_3 \boldsymbol{u}_3) - (n\boldsymbol{\theta}_n + \eta_1 \boldsymbol{u}_1 + \eta_2 \boldsymbol{u}_2 + \eta_3 \boldsymbol{u}_3)$, and the multiplier before the sum normalizes the energy per unit area.

The energies $E(\tilde{Q}_- \leftrightarrow \tilde{Q}_+)$ and $E(\hat{Q}_- \leftrightarrow \hat{Q}_+)$ are calculated in the same way, but the argument of the sum will be $\varphi_{S_m S_n}(|\boldsymbol{r}|)$ and $\varphi_{S_{\alpha_m} S_{\alpha_n}}(|\boldsymbol{r}|)$, respectively.

Let us consider superstructure defects with different bedding planes, starting with the (001) plane. As seen from figure 1, the bundle into monoatomic two-dimensional packings parallel to the (001) plane is determined by the vectors

$$\boldsymbol{u}_1 = (4,0,0),\ \boldsymbol{u}_2 = (0,2,0),\ \boldsymbol{u}_3 = (2,2,2). \tag{5}$$

The shift vectors will be:

$$\boldsymbol{\theta}_1 = (0,0,0),\ \ \boldsymbol{\theta}_2 = (2,0,0),\ \ \boldsymbol{\theta}_3 = (1,1,1),\ \ \boldsymbol{\theta}_4 = (3,1,1). \tag{6}$$

Substituting (5) and (6) into (3) taking into account (4), we obtain the results that are presented in terms of ordering energies $\omega_{S_m S_n}$ and energy parameters $\Delta_{S_m S_n}$. So, for three vectors of antiphase we have the following expressions for the energies of antiphase boundaries.

1. Case $S_m = XXYZ,\quad S_k = XXZY,$

$$E = \frac{2}{a^2}\left(2\omega_{YZ}(R_2) - 8\omega_{YZ}(R_3) + 8\omega_{YZ}(R_5) - 4\omega_{YZ}(R_6) + 24\omega_{YZ}(R_8)\right).$$

2. Case $S_m = XXYZ,\quad S_k = YZXX.$

$$E = \frac{2}{a^2}\begin{pmatrix} 4\omega_{XY}(R_1) + 4\omega_{XZ}(R_1) - 4\omega_{YZ}(R_1) - 4\Delta_{XY}(R_1) - 4\Delta_{XZ}(R_1) \\ -2\omega_{XY}(R_2) - 2\omega_{XZ}(R_2) + 2\omega_{YZ}(R_2) \\ -8\omega_{XY}(R_3) - 8\omega_{XZ}(R_3) \\ 20\omega_{XY}(R_4) + 20\omega_{XZ}(R_4) - 16\omega_{YZ}(R_4) - 12\Delta_{XY}(R_4) - 12\Delta_{XZ}(R_4) \\ -8\omega_{XY}(R_5) - 8\omega_{XZ}(R_5) + 8\omega_{YZ}(R_5) \\ -4\omega_{XY}(R_6) - 4\omega_{XZ}(R_6) \\ 28\omega_{XY}(R_7) + 28\omega_{XZ}(R_7) - 20\omega_{YZ}(R_7) - 12\Delta_{XY}(R_7) - 12\Delta_{XZ}(R_7) \\ -24\omega_{XY}(R_8) - 24\omega_{XZ}(R_8) + 24\omega_{YZ}(R_8) \end{pmatrix}.$$

3. Case $S_m = XXYZ,\ S_k = ZYXX.$

Here the result is the same as in case 2.

Note that the energy of the first of the three defects is expressed in terms of the ordering energies $\omega_{S_m S_n}$, indicating that this is a shear antiphase boundary that does not violate the stoichiometry of the alloy. At the same time, the energies of the second and third defects are not expressed only in terms of $\omega_{S_m S_n}$, since these are thermal antiphase boundaries, leading to a local violation of stoichiometry near the boundary. This result is fully consistent with the results of the analysis of the planes of occurrence of the SAFB, where it was shown that in the plane (001) only one antiphase vector, $\boldsymbol{t}_1$, gives the SAFB, and the other two vectors do not give such a defect.

Defects in the plane (011). Using the method described above, we get the following results.

1. Case $S_m = XXYZ,\quad S_k = XXZY,$

$$E = \frac{2}{a^2}\left(4\omega_{YZ}(R_2) - 14\omega_{YZ}(R_3) + 12\omega_{YZ}(R_5) - 8\omega_{YZ}(R_6) + 40\omega_{YZ}(R_8)\right)$$

2. Case $S_m = XXYZ,\quad S_k = YZXX,$

$$E = \frac{2}{a^2}\begin{pmatrix} 6\omega_{XY}(R_1) + 6\omega_{XZ}(R_1) - 3\omega_{YZ}(R_1) \\ -4\omega_{XY}(R_2) - 4\omega_{XZ}(R_2) + 4\omega_{YZ}(R_2) \\ -14\omega_{XY}(R_3) - 14\omega_{XZ}(R_3) \\ +30\omega_{XY}(R_4) + 30\omega_{XZ}(R_4) - 15\omega_{YZ}(R_4) \\ -12\omega_{XY}(R_5) - 12\omega_{XZ}(R_5) + 12\omega_{YZ}(R_5) \\ -8\omega_{XY}(R_6) - 8\omega_{XZ}(R_6) \\ +38\omega_{XY}(R_7) + 38\omega_{XZ}(R_7) - 19\omega_{YZ}(R_7) \\ -40\omega_{XY}(R_8) - 40\omega_{XZ}(R_8) + 40\omega_{YZ}(R_8) \end{pmatrix}$$

3. Case $S_m = XXYZ,\quad S_k = ZYXX.$

Here the result is the same as in case 2.

It is worth noting that the defect energy for all three antiphase vectors is expressed through the ordering energies $\omega_{S_m S_k}$, that is, in all three cases we have shear antiphase boundaries that do not violate the stoichiometry of the alloy. These results are consistent with the analysis of possible planes of occurrence of the shear antiphase boundaries, where it was shown that in the plane (011) all three vectors of antiphase, $\boldsymbol{t}_i$, give the shear antiphase boundaries.

3. Conclusions

A crystallogeometrical analysis of X_2YZ stoichiometric ternary Heusler alloys with $L2_1$ superstructure is presented. An expression is given for the sublimation energy of an alloy in terms of the energies of pair interactions, from which it follows that on the first coordination sphere, only the *XY* and *XZ* bonds contribute to the sublimation energy. In the second sphere, the contribution from the *XX* and *YZ* bonds appears. The *YY* and *ZZ* bonds contribute only starting from the 3rd coordination sphere.

All energetically equivalent but geometrically different representations of the $L2_1$ superstructure are found. It was found that all structures follow from shifts by one of the lattice vectors, while point symmetry transformations do not give new geometrically different representations of the $L2_1$ superstructure. On this basis, it was concluded that there are no planar superstructure defects in the $L2_1$ superstructure that differ from shear and thermal antiphase boundaries. Methodology developed in this study can be applied to the lattices of any dimension, in particular, to the novel 2D materials [18-22].

Acknowledgments
V.A. Valitov expresses gratitude for the financial support from the Russian Science Foundation, Grant No. 18-19-00685. This work was supported by the State Assignment of IMSP RAS No. AAAA-A17-117041310220-8.

References

[1] Heusler F, Starck W and Haupt E 1903 *Verh. DPG* **5** 220
[2] Heusler F 1903 *Verh. DPG* **5** 219
[3] Yousuf S and Gupta D C 2018 *J. All. Compd.* **766** 241
[4] Manosa L and Planes A 2017 *Advanced Materials* **29** 1603607
[5] Singh S, Zeeshan M, Singh U, Brink J and Kandpal H C 2020 *J. Phys. Cond. Matter* **32** 055505
[6] Sofi S A and Gupta D C 2020 *Physica* B **577** 411792
[7] Hinterleitner B et al 2019 *Nature* **576** 85
[8] Modak R, Raja M M and Srinivasan A 2018 *J. Magn. Magn. Mater.* **448** 146
[9] Starostenkov M D, Gorlov N V and Tsaregorodtsev A I 1986 *Phys. Met. Metallogr.* **62** 1
[10] Baranov M, Starostenkov M D and Nikiforov A G 1997 *Metallofizika i Noveishie Tekhnologii* **19** 47 [in Russian]
[11] Starostenkov M D, Dmitriev S V and Starostenkova O Kh 1997 *J. Struct. Chemistry* **38** 930
[12] Dmitriev S V, Wusatowska-Sarnek AM and Starostenkov M D 1998 *Acta Crystallogr.* A **54** 430
[13] Dmitriev S V, Starostenkov M D and Zhdanov A N 1995 *Fundamentals of Crystal-Geometric Analysis of Defects in Metals and Alloys* (Barnaul, AltSTU Publishing House) 256 p [in Russian]
[14] Khalikov A R, Sharapov E A, Korznikova E A, Potekaev A I, Starostenkov M D, Galieva E V and Dmitriev S V 2019 *Russ. Phys. J.* **62** 691
[15] Khalikov A R and Iskandarov A M 2013 *Russ. Phys. J.* **55** 1470
[16] Pavlukhina O O, Sokolovskiy V V, Buchelnikov V D and Zagrebin M A 2019 *Lett. Mater.* **9** 395
[17] Zagrebin M, Derevyanko S, Sokolovskiy V and Buchelnikov V 2018 *Lett. Mater.* **8** 21
[18] Barani E, Korznikova E A, Chetverikov A P, Zhou K and Dmitriev S V 2017 *Phys. Lett.* A **381** 3553
[19] Yoon T, Kang S, Yeob Kang T and Kim T-S 2019 *Comp. Model. Eng. Sci.* **120** 351–61
[20] Shcherbinin S A, Zhou K, Dmitriev S V, Korznikova E A, Davletshin A R and Kistanov A A 2020 *J. Phys. Chem.* C **124** 10235
[21] Babicheva R I, Dahanayaka M, Liu B, Korznikova E A, Dmitriev S V, Wu M S and Zhou K 2020 *Mater. Sci. Eng.* B **259** 114569
[22] Kistanov A A, Khadiullin S K, Dmitriev S V and Korznikova E A 2019 *ChemPhysChem* **20** 575

Influence of rolling modes and the level of accumulated deformation on the mechanical and plastic properties of stainless austenitic steel AISI 304L

A G Raab[1], G I Raab[1], S V Dobatkin[2,3], A A Tokar[2,3], P La[4]

[1]Institute of Physics of Advanced Materials, Ufa State Aviation Technical University, 12 K. Marx st., Ufa 450008 Russia
[2]Baikov Institute of Metallurgy and Materials Science RAS, 49 Leninskiy pr., Moscow 119334 Russia
[3]National University of Science and Technology "MISIS", 4 Leninskiy pr., Moscow 119049 Russia
[4]State Key Laboratory of Advanced Processing and Recycling of Nonferrous Metals, Lanzhou University of Technology, Lanzhou, China

E-mail: agraab@mail.ru

Abstract. The paper presents the results of studies of rolling modes (direct, reverse, cross ones) at 500°C for samples of stainless steel AISI 304L and their influence on mechanical and plastic properties of the material. The stress-strain state was investigated using a numerical simulation program and the deformation parameters of processing were established. The mechanical properties were studied by microhardness measuring and tensile testing methods. The results of the experiment were analyzed and discussed, and conclusions were drawn. The regularities of the effect of strength parameters depending on the modes of processing and accumulated deformation during rolling have been established.

1. Introduction

Stainless steels are widely used in industry, and improvement of their properties is an important task for increasing the reliability of products obtained from them. Recently, to improve the properties of austenitic stainless steels, methods of severe plastic deformation (SPD) have been used, which ensure the accumulation of large degrees of deformation. In this case, it is possible to form a structure with a grain size of down to 1 μm or less with a large proportion or prevalence of boundaries with high-angle misorientation [1-3].

However, these methods have a number of limitations and are not technologically advanced for the production of sheet blanks. Therefore, in the study of sheet materials, including stainless steel, rolling in a wide temperature range and with the accumulation of high degrees of deformation is quite widely used to improve the mechanical properties. In addition, austenitic stainless steels demonstrate efficient grain refinement and hardening during cold and warm working [4-6]. It should be noted that grain refinement in austenitic steels is also associated with deformation twinning and martensitic transformation. A decrease in the rolling temperature promotes significant strengthening due to an increase in the dislocation density and the development of deformation twinning [5, 7].

In this regard, the purpose of this paper is to study the effect of rolling modes on the hardening intensity and plastic properties of austenitic stainless steel AISI 304. Direct, reverse and cross rolling at 500°C, strain rates of 5 s^{-1} with a relative accumulated deformation of 50 and 70% were used as the modes of rolling. It is assumed that a change in the rolling direction may affect the nature of hardening due to the non-monotonous effect.

2. Material and experimental procedure

AISI-304L steel was used as a material for the study after holding for 1 hour at a temperature of 1050°C and cooling in water. The chemical composition of the steel is shown in table 1. The samples had the shape of a disc with a diameter of 60 mm and a height of 15 mm.

Rolling was carried out on a six-roller strip mill HANKOK M-TECH INDUSTRIES CO LTD with a roll diameter of 70 mm, the total relative deformation was 50 and 70% (which is equal to the true strain e = 0.8 and e = 1.1, respectively). Deformation by rolling was carried out in 15 passes, the relative reduction per pass was 5%. During direct rolling, the samples were rolled in the same direction every pass, while during the inverse and cross ones the sample was rotated to 180° and 90° around the normal direction before each subsequent pass, respectively. The workpieces were preheated in a Nabertherm N321/13 muffle furnace to 500°C before each pass. Samples for measuring the microhardness were cut out transversely to the direction of the final rolling of the sample on an ARTA120 electro-erosion machine.

The Vickers hardness (HV) was measured on a Buehler micro-hardness tester with a load of 100 g and a holding time of 15 sec. The number of measurements for each sample was at least 15. Mechanical tensile testing was carried out at room temperature in accordance with GOST 1497-84, on an Instron 5982 universal testing machine at a crosshead speed of 1 mm / min.

Table 1. Chemical composition of AISI 304L (wt%)

C	Si	Mn	Ni	Cr	P	S
0.024	0.516	1.063	7.64	17.04	0.022	0.032

The stress-strain state for each processing mode was investigated by the finite element method in the Deform 3D software package. The rheological characteristics of steel AISI 304L were taken from the database of the software product.

3. Results and discussion

3.1. Computer simulation

As a result of computer simulation in the software product DEFORM 3D, maps and graphs of the strain intensity distribution in the cross-section of the samples received in the studied rolling modes were obtained (figure 1).

The analysis of the obtained images and graphs shows that the deformation during rolling in the studied modes is not uniformly distributed across the section. The surface layers of the workpieces accumulate higher strain values compared to the central region. This indicates that with a relatively small reduction per pass (5%), at the early stages of processing, the surface areas of the samples are deformed first of all. Moreover, after the first cycle of processing, the maximum value of the surface layers is e=1.1, and the central area it is practically equal to zero. With a further increase in the number of passes up to three, an increase in the accumulated strain occurs in the central region, and with cross-rolling - more intensively.

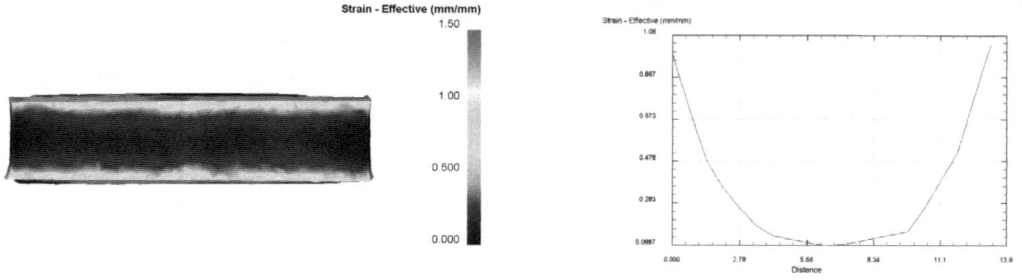

(a)

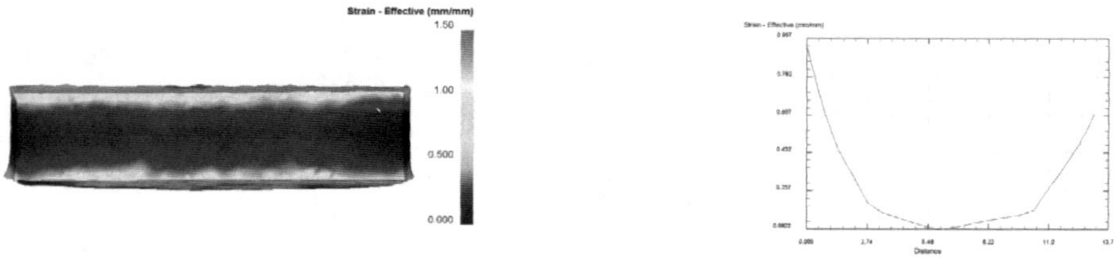

(b)

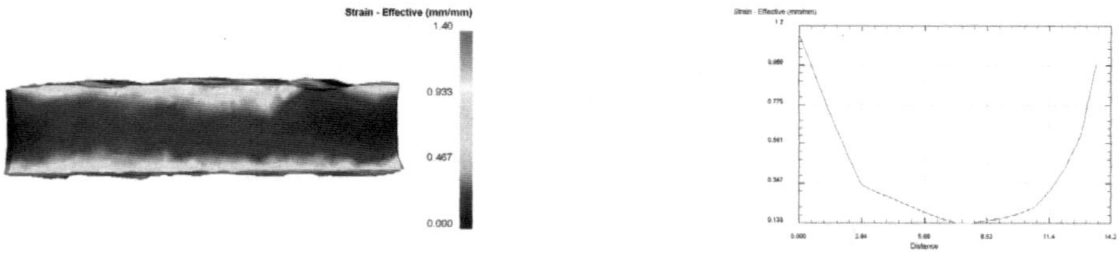

(c)

Figure 1. Maps of the deformed state and graphs of the distribution of accumulated deformation in the cross-section of the sample during direct (a), reverse (b) and cross (c) rolling of AISI-304L steel after three passes.

3.2. Experimental study

Rolling of blanks in three modes showed that after deformation at 500°C, the samples did not exhibit mechanical destruction and the formation of surface defects or cracks with a relative accumulated deformation of up to 70%. The samples obtained during rolling are shown in figure 2.

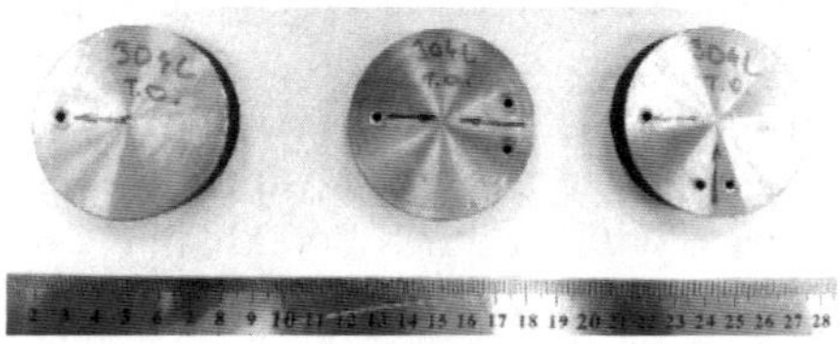

(a)

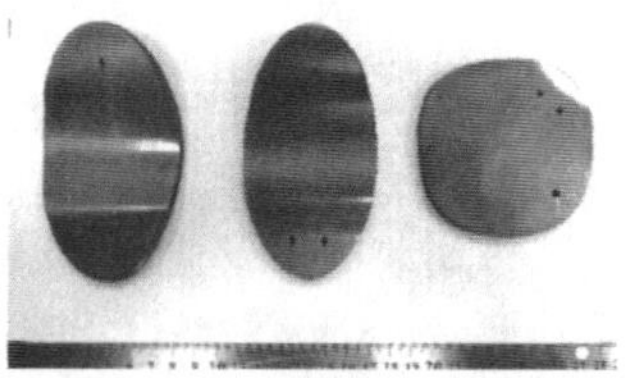

(b)

Figure 2. Initial samples for experimental studies (a), samples obtained by direct, reverse and cross-rolling at a temperature of 500°C and a relative accumulated deformation of 70%.

After direct and reverse rolling, the samples are elongated in the rolling direction and acquire a wavy shape, i.e. lose flatness, and after cross-rolling the sample has a symmetrical shape and looks like a flat plate.

3.3. Study of mechanical properties

The results of studies on the mechanical properties of the specimens in tension after all processing modes are presented in figure 3, and the results of microhardness measurements are given in figure 4. The obtained data show that during rolling there is an increase in ultimate strength and yield stress, and a higher degree of deformation of 70% provides higher values of these characteristics. In this case, the maximum value of the tensile strength of 972 MPa and the yield stress of 938 MPa is observed when using reverse rolling, and the minimum values of the tensile strength of 937 MPa and the yield stress of 911 MPa are observed after cross-rolling. However, while the difference in mechanical characteristics between the modes of rolling is insignificant (only about 35MPa), the elongation of the samples after reverse rolling and cross-rolling differs almost twice.

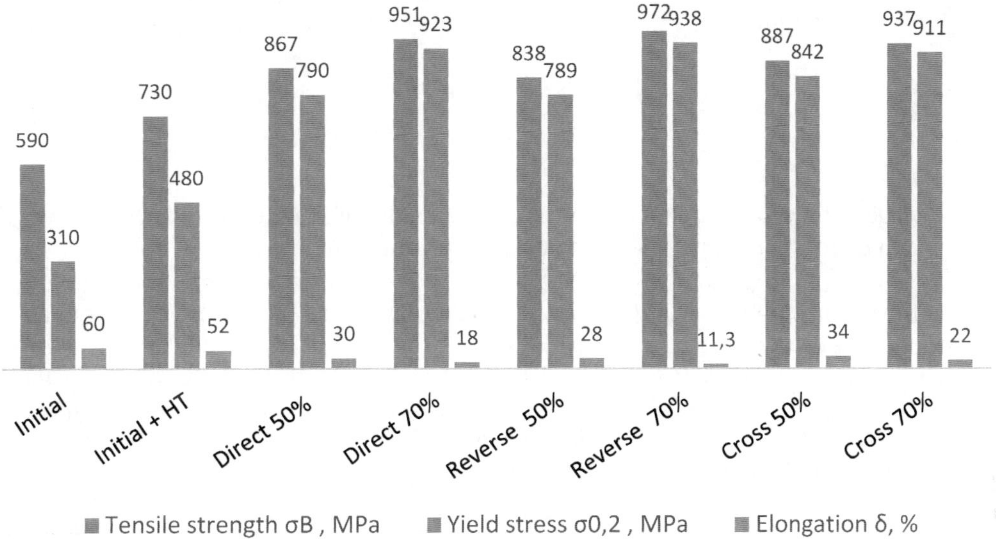

Figure 3. Tensile data for AISI 304L steel specimens after various treatments.

Measurements of the microhardness HV after three rolling modes presented in figure 4 were performed along the central zone of the sample cross section.

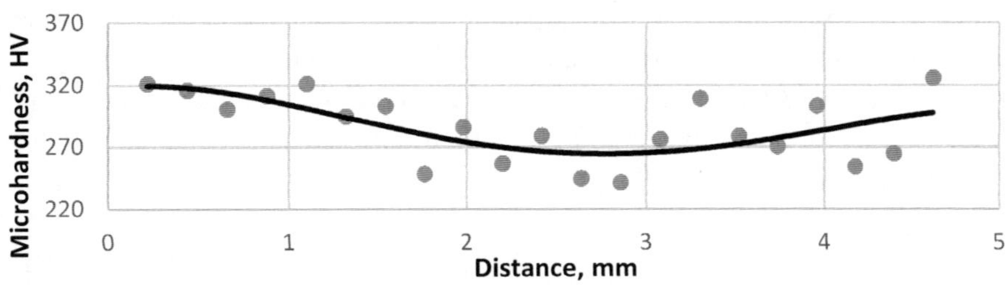

(a)

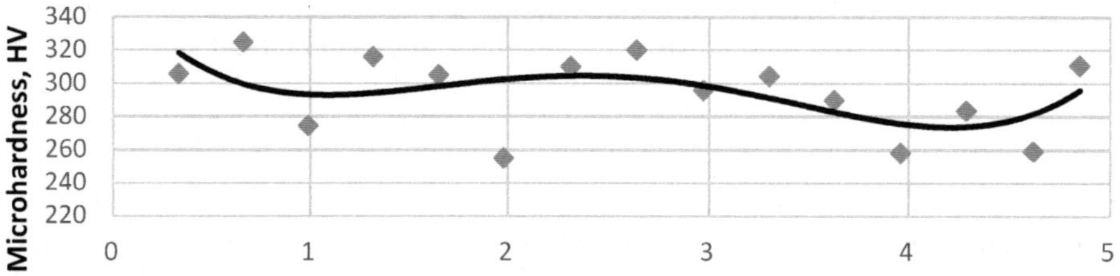

(b)

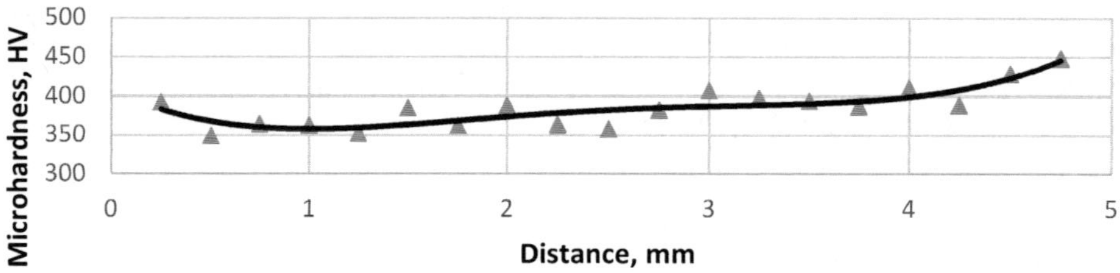

(c)

Figure 4. Change in the microhardness of the cross section of the samples after direct (a) reverse (b) and cross rolling (c) when reaching the maximum deformation of 70%

Analysis of the results on microhardness measurements show that in the case of direct and reverse rolling, a nonuniform distribution of microhardness over the section is observed. Thus, during direct rolling, the surface layers are hardened more intensively up to 320HV, and the difference with the middle layers is approximately 60HV. In reverse rolling, the microhardness distribution is more uniform, and the difference between the middle and surface layers is about 20HV. The maximum microhardness values are shown for the samples after cross rolling - 400 ± 40HV.

4. Conclusions

The influence of rolling modes (direct, reverse, cross) at 500°C and a degree of relative deformation of 50 and 70% on the strength and plasticity of properties has been studied. Thus, an increase of deformation from 50 to 70% allows to increase the mechanical properties by an average of 12% and 30% as compared to the original heat-treated sample. It was found that the use of cross-rolling leads to the formation of higher plastic properties in comparison with other rolling modes, for example, 2 times compared to reverse rolling, with similar strength characteristics. Cross-rolling also provides the most homogeneous and 1.25 times higher microhardness values in comparison with other modes. The study of the deformed state by numerical modeling shows that a high fractional deformation of 5% per pass leads to the formation of an inhomogeneous deformation field with higher deformation values in the surface layers. This is most pronounced in direct rolling.

Acknowledgements

The work was performed in the framework of a joint Russian-Chinese research project funded by the Russian Foundation for Basic Research (project No. 19-53-53022) and the National Natural Science Foundation of China (51911530119).

References

[1] Valiev R Z, Zhilyaev A P and Langdon T G 2014 *Bulk Nanostructured Materials: Fundamentals and Applications* (Hoboken, New Jersey: Wiley)

[2] Dobatkin S V, Rybal'chenko O V and Raab G I 2007 *Mater. Sci. Eng.* A **463** 41–5
[3] Dobatkin S V, Kaputkina L M, Rybalchenko O V and Komlev V S 2012 *Metals* **5** 28-37
[4] Stolyarov V V, Valiev R Z and Zhu Y T 2006 *Appl. Phys. Lett.* **88** 041905
[5] Shakhova I, Dudko V, Belyakov A, Tsuzaki K and Kaibyshev R 2012 *Mater. Sci. Eng.* A **545** 176–86
[6] Dobatkin S V and Sauvage X 2009 *Bulk Nanostructured Materials* (Wiley- WCH) pp. 571-603
[7] Belyakov A, Odnobokova M, Kipelova A, Tsuzaki K and Kaibyshev R 2014 *IOP Conf. Ser. Mater. Sci. Eng.* **63** 012156

Features of the formation of a diffusion layer and the mechanism of diffusion during boronizing of steels

A.M. Guryev[1,2], S. G. Ivanov[2], M. A. Guryev[2], S. Mei [1], and B.D. Lygdenov[1,3]

[1]Wuhan Textile University, Wuhan, People's Republic of China (PRC)
[2]Polzunov Altai State Technical University, Barnaul, Russia
[3]East Siberian State University of Technology and Management, Ulan-Ude, Russia

E-mail: gurievam@mail.ru

Abstract. The paper presents the results of studies on the formation of a diffusion layer during the boriding of steel with a carbon content of 0.38 wt%. The study of diffusion coatings by SEM and TEM methods allows one to more clearly determine the mechanisms of diffusion of boron atoms and the formation of diffusion boride coatings. Diffusion of boron in the processes of diffusion boriding of iron-carbon alloys proceeds by a mixed mechanism: the delivery of active boron atoms to the diffusion front proceeds along the grain and phase boundaries. The reaction mechanism is realized by a gradual substitution of carbon with boron in iron carbides. Due to the continuous diffusion of boron, the reaction boundary is constantly moving in the direction from the surface to the core of the saturated material. In this case, as the diffusion proceeds, the concentration of carbon atoms under the boride layer constantly increases. This provides resistance to diffusion of boron atoms by the atomic mechanism and provokes diffusion by the reaction mechanism.

1. Introduction

The most common cause of failure of machine parts and tools is the wear of their working surface. It prevails in more than 75% of cases. The problem of increasing the wear resistance of the tool is now becoming increasingly relevant. In this regard, various methods of surface hardening, including diffusion, are widely demanded. This trend is due to the problem of saving alloying metals (manganese, chromium, cobalt, tungsten, vanadium, molybdenum, niobium, etc.) and the transition to the consumption of economically alloyed steels. The use of diffusion hardening and protective coatings can significantly reduce the cost of production and improve the performance of the created products. In some cases, chemical and thermal treatment is the only possible means of obtaining the required performance properties not only of the surface, but also of the product as a whole.

Boriding occupies a special place among the processes of chemical and thermal treatment. Increased interest in the process of boriding is caused by the possibility of obtaining mono - or multiphase boride layers in the surface area of the processed products, characterized by a unique complex of physicochemical properties and its acceptability for processing a wide range of structural and tool steels and alloys. Boron-based coatings are the leaders in improving properties such as wear resistance, heat resistance and hardness. However, the processes of diffusion of boron and related elements in the diffusion hardening of the surface of steels have not been sufficiently studied. At the same time, while the zone of continuous borides is still being studied, as a rule, insufficient attention is paid to the transition zone Although some studies show that the structure and physical and mechanical properties of the transition zone have a significant impact on the performance of the entire coating.

An important place is occupied by the question of the mechanism of boron diffusion in the boriding of steels and alloys and the formation of diffusion coatings. It has not been fully resolved and is still debatable [1-14]. Some authors adhere to the reaction mechanism of formation of boride coatings directly as a result of the chemical reaction between iron and boron. Other authors claim that first a solid solution of boron in iron is formed, and after reaching the limit concentrations by a chemical reaction between iron and boron borides are formed. There is no clear understanding of both diffusion mechanisms and methods of process control in order to obtain coatings of a given composition and with the required properties, and this leads to the appearance of technological recommendations that differ 2 to 50 times in labor and energy costs for diffusion boriding. Therefore, the study of the process of formation of the diffusion layer and the mechanism of boron diffusion in the boriding of steels is an important task.

2. Methods

In this paper, we study the process of formation of boride coatings on carbon steel with a carbon content of 0.38 wt. %. Saturation was carried out in an airtight container of a powder mixture based on boron carbide [15] at a temperature of 920°C for 2.5 hours [16-18]. The microstructure of the boride layer and the transition zone was studied on transverse microspheres using optical and scanning electron microscopy. To study the methods of transmission electron microscopy, templets 0.2 mm thick were cut using an electroerosive machine. The cutting mode was chosen in such a way that it did not introduce additional deformation and, therefore, did not affect the structure of the sample. Foils were polished electrolytically. The composition of the electrolyte was as follows: saturated solution of phosphoric acid, chrome anhydride. The temperature of the electrolyte during the preparation of foils was 30-50°C. The phase composition was characterized by diffraction patterns obtained by the methods of x-ray diffraction analysis and diffraction electron microscopy.

3. Results and discussion

The studies show that boron diffuses along the grain and phase boundaries [1–3]. At the same time, due to the limited solubility of boron in iron, saturated solid solutions of boron in iron are formed relatively quickly. Along with this, boron alloys the iron carbides by substitution reactions, replacing part of the carbon in cementite. Carbon displaced from cementite together with boron creates additional volumes of boronizing cementite $Fe_3(C, B)$ at some distance from the reaction front of iron boride formation. This can explain the significant volumes of perlite in the transition zone. Often the content of perlite in the transition zone reaches 100% when bordering steels with a carbon content of more than 0.4 wt % (figure 1). The structure of the transition zone is presented in more detail in figure 2.

As can be seen from figures 1 and 2, the content of perlite in the transition zone reaches 95-98% by volume. At the same time, the thickness of the transition zone and, accordingly, the volume occupied by it, significantly exceed the thickness and volume of the boride zone (4-6 times). Even taking into account the fact that from the zone of the boride layer carbon is displaced almost completely, the formation of such a large amount of cementite with the participation of only the amount of carbon displaced by boron is impossible.

The data of energy dispersion analysis, as well as diffraction patterns and electron diffraction patterns obtained by SEM methods confirm the doping of cementite transition zone. The content of boron in cementite ranges to 0.8 ($Fe_3(C_{0.2}, B_{0.8})$) in the upper part of the transition zone. The minimum concentration of boron in boronizing cementite corresponds to the gross formula $Fe_3(C_{0.8}, B_{0.2})$ and this compound is predominantly in the lower part of the transition zone, at a depth of about 800 μm from the surface. Figure 2 shows that the cementite formed as a result of diffusion boriding has a thinner structure compared to the cementite that existed initially and underwent boriding.

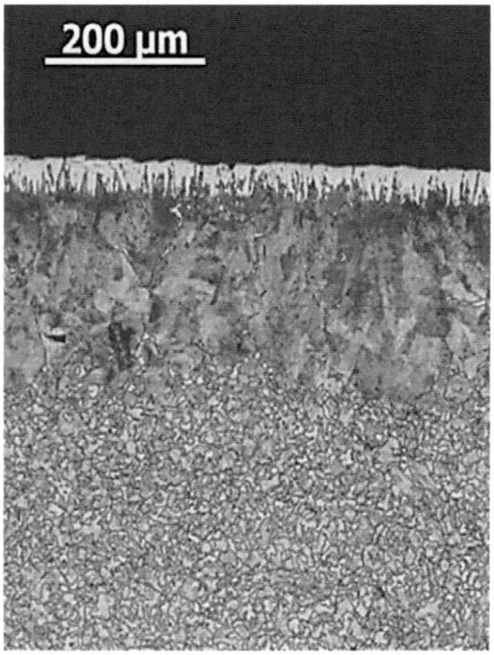

Figure 1. Microstructure of boride coating on the steel with 0.38 wt. % carbon content.

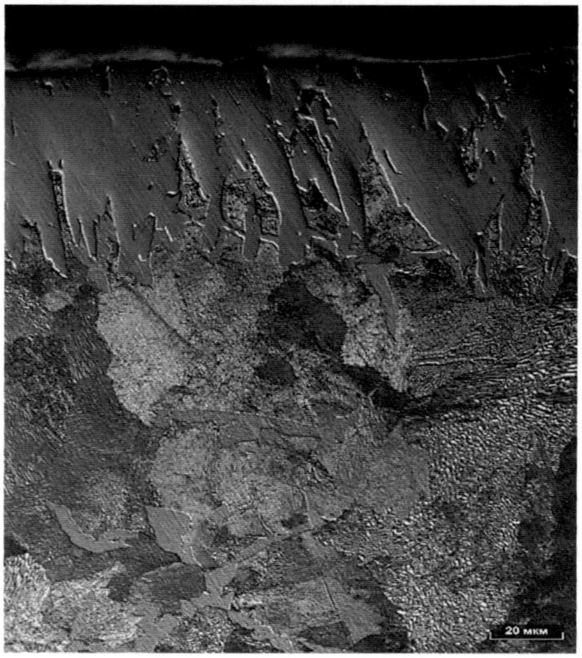

Figure 2. The microstructure of the transition layer on the steel with 0.38 wt. % carbon content.

The corresponding microstructures obtained by SEM methods are shown in figures 3 and 4. The microstructure of boronizing cementite, shown in Figure 4, is represented by fragments of plates destroyed by boron diffusion. At the same time, on the side faces of the plate fragments, the nuclei of boride phases of different composition having a globular shape and a size of about 20-50 nm are noticeable.

Boron diffusion predominantly occurs along the grain and phase boundaries, as well as along substructural units (subgrains), and at some distance from the reaction front, the formation of particles of carboborides $Fe_{23}(C, B)_6$ and $Fe_3(C, B)$) occurs. As the boron content in carboborides increases, their chemical composition approaches the boron cementite Fe_3B. This compound is unstable at diffusion boriding temperatures and decomposes according to the reaction $Fe_3B = Fe + Fe_2B$. As boron further diffuses along the interface between the boride needle and the iron matrix, the phase composition is balanced and passes into the Fe_2B compound. This argument is supported by darker areas on cementite plates in figure 4.

It should be noted that, along with the carboboride mechanism of the formation of iron hemiboride, its formation also occurs as a result of a direct reaction between iron and boron at the border of the boride needle, and the direct reaction of the formation of iron hemiboride Fe_2B at the initial stage is predominant. As the process of boron diffusion develops deeper into the iron matrix and carbon is displaced by boron, the concentration of carbon atoms increases at the border of the boride layer. As the concentration of carbon atoms approaches approximately 0.8 wt. %, the reactions of direct formation of borides by direct interaction of boron and iron and the reaction of obtaining borides through the decomposition of borated cementite Fe_3B have approximately equal proportions. This assumption is indirectly confirmed by the fact that at the initial stages of the boriding process, no destroyed cementite plates are observed in the transition zone. Whereas the formation of the pearlite structure under the boride layer is observed practically from the very beginning of diffusion boriding.

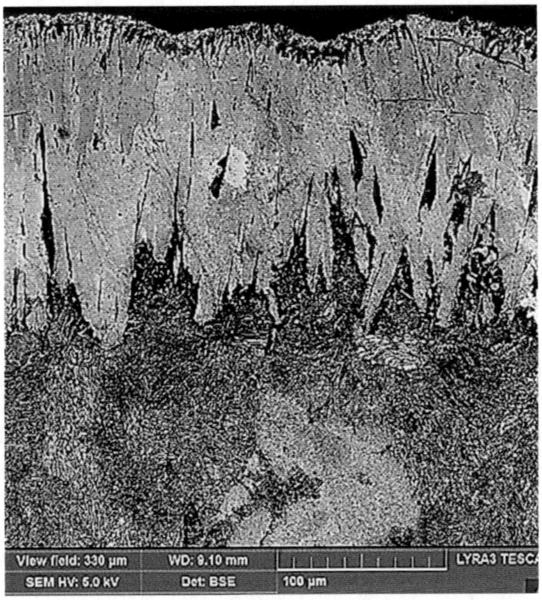

Figure 3. Microstructure of diffusion boride layer and transition zone obtained by SEM methods, general microstructure, BSE.

Figure 4. Microstructure of transition zone obtained by SEM methods, microstructure of boronizing cementite plates, BSE.

Directly in the transition zone itself, under the influence of diffusing boron, a partial dissolution of cementite occurs, as a result of which it takes the form of fragmented plates. Fragmentation and destruction of cementite plates in borated pearlite probably occurs as a result of boron diffusion, which is accompanied by phase transformations with a change in volume. When analyzing the microstructures shown, it is obvious that boron atoms prefer to diffuse over areas depleted by carbon (subsequently, it is ferrite after cooling). At the same time, the grain boundaries and ferrite grains are enriched with boron, while the destruction of cementite plates in perlite has not yet occurred.

4. Conclusion

The following conclusions can be made from the results of the present study.

1. Boron diffusion in the processes of diffusion boriding of iron-carbon alloys proceeds by a mixed mechanism: the delivery of active boron atoms to the diffusion front proceeds along the grain and phase boundaries. The reaction mechanism is realized by the gradual substitution of carbon with boron in iron carbides.
2. A significant contribution is made by the mechanism of the reaction of the formation of iron borides, which is implemented according to a two-stage scheme as follows: $Fe_3C + B + Fe \rightarrow Fe_3(C,B)$, and the boron content in borating cementite ranges from 20 to 80 at. %. When adding new amounts of boron atoms, the boron content in cementite reaches 100 at. % and boric cementite is formed, which, in turn, decomposes as a result of peritectic reactions to FeB, Fe_2B. Carbon displaced during boriding is removed from the reaction front into the depth of the transition zone and there, together with boron, forms additional amounts of iron carboborides $Fe_3(C,B)$ and $Fe_{23}(C, B)_6$.
3. Iron atoms, reacting at the border of the nuclei of boride phases directly with boron diffusing along the borders of these nuclei, form additional volumes of iron hemiboride Fe_2B, and subsequently monoboride FeB.
4. Due to the continuous diffusion of boron, the reaction boundary is constantly moving in the direction from the surface to the core of the saturated material. In this case, as evolution the diffusion processes in time, the concentration of diffusing atoms of carbon in the boundary layer are constantly increasing, providing resistance to the processes of physical diffusion of active atoms of boron.

References

[1] Sudarshan T S, Khor K A and Jeandin M 1999 *Review of Surface Modification Technologies XII* (ASM International, Am. Techn. Publ. Ltd)
[2] Bindal C and Ucisik A H 1999 *Surf. Coat. Technol.* **122** 208
[3] Voroshnin L G and Lyakhovich L S 1978 *Boriding Steels* (Moscow, Metallurgiya) (in Russian)
[4] Kuz'ma Yu B 1983 *Crystal Chemistry of Borides* (Kiev, Vishcha Shk) (in Russian)
[5] Campos I, Oseguera J, Figueroa U, Garcı J A, Bautista O and Kelemenis G 2003 *Mater. Sci. Eng.* **A 352** 261
[6] Campos I, Ramirez G, Figueroa U, Martinez J and Morales O 2007 *Appl. Surf. Sci.* **253** 3469
[7] Kayali Y, Gunes I and Ulu S 2012 *Vacuum* **86** 1428
[8] Sinha A K 1991 *ASM Handbook. J. Heat Treat.* **4** 437
[9] Mao M X, Ge S H, Chen Z H, Chen G L, Zhang C L, Zhang Y D, Hines WA and Budnick J I 1994 *Magn. Mater.* **138** 301
[10] Li F and Li Z 2014 *J. Alloy. Compd.* **587** 267
[11] Van Rompacy T, Hari Kumar K C and Wollants P 2002 *J. Alloy. Compd.* **334** 173
[12] Filonenko N Yu, Beryoza O Yu and Bezrukava O G 2013 *PAST* **87** 168
[13] Ivanov S G, Guriev A M, Starostenkov M D, Ivanova T G and Levchenko A A 2014 *Russian Physics Journal* **57** 266
[14] Ivanov S G, Garmaeva I A, Guriev M A, Guriev A M and Starostenkov M D 2015 *Lecture Notes in Mechanical Engineering* **22** 49
[15] Garmaeva I A, Guriev A M, Ivanova T G, Guriev M A and Ivanov S G 2016 *Letters on Materials* **6**(4) 262 (in Russian)
[16] Gur'ev A M, Ivanov S G, Gur'ev M A, Chernikh E V and Ivanova T G 2015 *Steel in Translation* **45**(8) 555
[17] Mei S, Dong Y, Guriev A M, Zhang W, Liu C, Ivanov S G and Chernykh E V 2016 *4th Annual International Conference on Material Science and Engineering (ICMSE) AER-Advances in Engineering Research* **101** 308
[18] Guriev A M, Ivanov S G, Guriev M A and Chernykh E V, Mei S 2019 *IOP Conference Series-Materials Science and Engineering* **479** 012077

Influence of composition nonstoichiometry and ordering effects on the properties of ferroelectric ceramics based on lead magnoniobate

A R Lebedinskaya[1] and A G Rudskaya[2]

[1]Academy of Architecture and Arts, South Federal University, 39 Budennovskiy Av., Rostov-on-Don, 344002, Russia
[2]The Department of Physics SFU, 5 Zorge st., Rostov-on-Don, 344090, Russia

E-mail : lebed1989@rambler.ru

Abstract. The effects of violation of the stoichiometry of lead magnoniobate in the preparation of compositions $(1\text{-}x)PbMg_{1/3}Nb_{2/3}O_3 - xPbMg_{1/2}Nb_{1/2}O_{2,75}$ with a step of 0.1 and $0 \leq x \leq 1$ have been studied by X-ray diffraction methods. Differences in the unit cell parameters of the ferroelectric ceramic samples of the composition under study were found. It was shown that the effect of ordering of different types of atoms in the cells, leading to the formation of a superstructure and a doubling of the lattice period, was not observed in any of the samples. . Measurements of the dielectric constant of the synthesized samples and the tangent of the dielectric loss angle showed a violation of the Curie-Weiss law, which indicates the diffusion of the phase transition.

1. Introduction

The development of production and research of nanocrystalline materials and structures [1-5] is primarily associated with the potentially unique prospects of their use in electronic engineering. However, the realization of this perspective requires a deep understanding of the essence of physical phenomena and careful development of the physical foundations of the corresponding technological processes for their production and purposeful control of their properties [6-8]. The possibility of using nanocrystalline materials and structures in various technological processes makes it necessary to purposefully control their functional properties. The solution of these problems requires the development of the physical foundations of the corresponding technological processes for obtaining ceramic materials with desired functional properties.

Relaxor ferroelectrics [1,8], discovered more than half a century ago, are of great interest to researchers today as highly promising functional materials in the form of single crystals, ceramics, and thin films due to the unique set of their properties.

Among oxide perovskites of complex composition $A_3B'B''_2O_9$, lead magnoniobate $PbMg_{1/3}Nb_{2/3}O_3$ (PMN), which is still a traditional model object for studying relaxor properties, is of great interest to researchers. In lead magnoniobate, the superstructure of the period doubling of the perovskite cell was found in regions up to 30 Å by the methods of electron, X-ray and neutron diffraction [6-8].

Of course, single crystals, which, as a rule, have high characteristics, are very expensive and unique in preparation, and their use is often limited due to the appearance of defects during their processing. The undoubted advantage of ceramic materials in comparison with single-crystal materials is their availability, practicality, and the identity of the functional properties observed in them during the transition to the nanocrystalline state in comparison with single crystals.

In the synthesis of PMN-based solid solutions, the main problem is associated with the presence of the pyrochlore phase simultaneously with the perovskite phase in the samples. This phase is piezoelectrically inactive and its presence leads to low values of the piezoelectric parameters of ceramic elements.

The $PbMg_{1/2}Nb_{1/2}O_{2.75}$ component was chosen because ordering in the B-sublattice of differently charged *Mg* and *Nb* atoms with doubling of the periods of the perovskite cell is possible [6,8,9].

2. Materials and methods

2.1. Fabrication of samples and X-ray structural studies

Ferro-ceramic samples were synthesized from a mixture of *PbO*, *MgO*, and Nb_2O_5 oxides selected in appropriate proportions corresponding to the compositions $(1-x)PbMg_{1/3}Nb_{2/3}O_3 - xPbMg_{1/2}Nb_{1/2}O_{2.75}$, where x varied from 0 to 1 with a step of 0.1. The synthesis of the samples was carried out according to the usual ceramic technology for 3 hours at a temperature of 890°C. It should be emphasized that both the synthesis and sintering of the samples were carried out under almost completely identical conditions, which excludes the possibility of the influence of changes in the conditions of the technological process on the properties of the ceramic [7,8].

After each stage of the synthesis, X-ray structural studies were carried out on an URS-50 IM X-ray diffractometer using Cuk_α - radiation with $Cuk_{\beta 1}$ filtration by a nickel filter. Lines with $h^2 + k^2 + l^2 = 14$ *(321)* and *24 (422)* were recorded, the measurement accuracy was 0.003 Å in the first case and 0.001 Å in the second.

2.2. Investigation of dielectric properties of samples

Measurements of the dielectric constant of the obtained samples and the tangent of the dielectric loss angle were carried out at the frequency of the measuring field f = 1 kHz in the range of temperatures from 25 to 150°C with the heating rate of the samples $v = 1$ °C / min. The temperature dependences of the electrical capacity (C) and electrical conductivity (G) of the samples were measured. The calculation of the required values was carried out according to the formulas:

$$\varepsilon = \frac{4 \cdot \pi \cdot h}{S} \cdot C \quad (1)$$

$$tg\delta = \frac{G}{\omega \cdot C} \quad (2)$$

where ε is the dielectric constant, $tg\delta$- tangent of the dielectric loss angle, h and S- thickness and surface area of the sample, C and G - measured electrical capacity and electrical conductivity of the sample, $\omega = 2\pi f$, where $f = 1$ kHz.

3. Experimental results and discussion

It was found from the diffraction patterns that, for the selected synthesis and sintering mode, compounds of a perovskite-like type and ideally cubic cell are formed in all compositions. The indexing of the X-ray diffraction patterns showed differences in the cell parameters of the studied compositions with different values of x. The obtained results on the cell parameters in the compositions of nonstoichiometric PMN for two types of diffraction reflections of the type (321) and (422) are presented in table 1.

For compositions with x changing from 0 to 0.5, in addition to the perovskite phase, reflections of impurity phases, pyrochlore and two modifications of Nb_2O_5, were recorded on the diffraction patterns. From the relative intensities of the lines, it was found that the impurities are concentrated mainly on the surfaces of the samples.

Table 1. Cell parameters (a_{per}) in nonstoichiometric PMN compositions for two types of diffraction reflections, type (321) and (422).

Compositions (1-x)$PbMg_{1/3}Nb_{2/3}O_3$- – x$PbMg_{1/2}Nb_{1/2}O_{2,75}$	a_{per}(321), Å	a_{per}(422), Å
x = 0.0	4.041(3)	4.044(1)
x = 0.1	4.045(3)	4.047(1)
x = 0.2	4.054(3)	4.049(1)
x = 0.3	4.048(3)	4.048(1)
x = 0.4	4.054(3)	4.049(1)
x = 0.5	4.054(3)	4.050(1)
x = 0.6	4.064(3)	4.052(1)
x = 0.7	4.059(3)	4.049(1)
x = 0.8	4.052(3)	4.049(1)
x = 0.9	4.048(3)	4.047(1)
x = 1.0	4.048(3)	4.047(1)

A careful examination of the obtained diffraction patterns of the samples did not reveal any superstructural reflections for all the studied sample compositions, which indicates the absence of the effect of ordering atoms of different types in the B-sublattice leading to a doubling of the unit cell parameters.

Table 2. Values of the dielectric constant (ε) and tangent of the dielectric loss angle(tgδ) at temperatures T = 25°C and 50°C.

Compositions (1-x)$PbMg_{1/3}Nb_{2/3}O_3$- – x$PbMg_{1/2}Nb_{1/2}O_{2,75}$	T=25 °C		T=50 °C	
	ε	tgδ	ε	tgδ
x = 0.0	6406	0.2055	4404	0.0342
x = 0.1	4449	0.0676	3483	0.0238
x = 0.2	3854	0.1329	3030	0.1160
x = 0.3	4777	0.0512	3650	0.0441
x = 0.4	1948	0.1113	2059	0.2595
x = 0.5	10535	0.1842	9210	0.1885
x = 0.6	7237	0.2938	6769	0.3446
x = 0.7	7631	0.0881	5949	0.0786
x = 0.8	8267	0.0719	6233	0.0516
x = 0.9	6624	0.1777	5580	0.1951
x = 1.0	4460	0.0635	3536	0.0497

As is known, lead magnoniobate belongs to the class of ferroelectric relaxors [3,4,10-13]. For all samples, it was found that the Curie-Weiss law is not fulfilled, which indicates a smearing of the phase transition. It is generally accepted that in the case of ferroelectrics with a diffuse phase transition, instead of the usual Curie-Weiss law $1/\varepsilon = (T - T_o)/C_w$, one should use the formula $1/\varepsilon = A + B\cdot(T-T_o)^m$, where A and B are some constants, T_o Curie temperature, and m takes on different values from 1.5 to 2.0. For lead magnoniobate, it is usually assumed that $m = 2$. From the temperature dependences ε (T) at temperatures from 25 to 150°C, the values of the Curie temperature, of constants A and B were obtained by calculation. The values of these parameters are given in tables 2 and 3.

Table 3. The values of the A and B constants and Curie temperature for different compositions.

Compositions $(1\text{-}x)PbMg_{1/3}Nb_{2/3}O_3$- – $xPbMg_{1/2}Nb_{1/2}O_{2,75}$	A, $\times 10^{-4}$	B, $\times 10^{-8}$	T_o, °C
x = 0.0	0.376355	1.523466	-58.26
x = 0.1	1.964185	3.096179	-3.68
x = 0.2	0.612425	3.028753	-48.53
x = 0.3	1.870444	3.515499	+0.98
x = 0.4	2.544109	7.887078	-15.78
x = 0.5	0.935968	2.134559	+22.07
x = 0.6	1.384200	2.972646	+31.70
x = 0.7	1.236226	2.657120	+8.40
x = 0.8	1.069574	2.442676	+2.32
x = 0.9	1.433186	2.951184	+12.31
x = 1.0	2.102110	3.865964	+6.50

The calculated Curie temperature turned out to be significantly different for different sample compositions, which once again confirms the dependence of the ceramic properties on stoichiometry [14].

4. Conclusions

Analysis of the currently available experimental data on the study of the effect of deviations of the oxygen content from stoichiometric one on the physical properties of samples with the perovskite structure does not give an accurate picture of the observed changes in the structure and properties. An experimental study of the synthesis of nonstoichiometric compositions $(1\text{-}x)PbMg_{1/3}Nb_{2/3}O_3$ – $xPbMg_{1/2}Nb_{1/2}O_{2,75}$ ($0 \leq x \leq 1$, $\Delta x = 0.1$) shows that the formation of the perovskite phase most likely corresponds to the effects characteristic of nanocrystalline materials. In the course of the study, it has been found that the properties of the obtained ferroelectric ceramics based on lead magnoniobate substantially depend on the stoichiometry. In ceramics, a structure with a predetermined ratio of magnesium and niobium is not realized; some of these substances contribute the formation of the perovskite structure, and some are released in the form of impurity phases. A diffuse phase transition was observed in the synthesized samples.

Thus, the studies of the nature and conditions of the formation of the structural and electromagnetic properties of ceramic samples with substitutions of different-valence magnesium and niobium ions in the B-sublattice and deviations from stoichiometry in oxygen are promising both in terms of determining the relationship and mutual influence of structural parameters on physical properties, and in connection with the urgent need to create promising functional materials and control their characteristics.

References

[1] Pandya S, Wilbur J, Kim J, Gao R, Dasgupta A, Dames C and Martin L W 2018 *Nature Mater.* **17**(5) 432

[2] Kolodiazhnyi T, Padchasri J and Yimnirun R 2018 *J. Eur. Ceram. Soc.* **38**(4) 1517
[3] Smolensky G A, Bokov V A, Isupov V A, Kraynik N N, Pasynkov N and Shur M S 1971 *Ferroelectrics and Antiferroelectrics* (Leningrad, Nauka) 476 p. [in Russian]
[4] Blinc R 2011 *Advanced Ferroelectricity* (Oxford, New York, Oxford University Press) 271 p.
[5] Bokov A and Ye Z-G 2012 *J. Adv. Diel.* **2**(2) 1241010
[6] Davies P K and Akbas M A 2000 *J. Phys. Chem. Sol.* **61** 159
[7] Kupriyanov M F and Fesenko E G 1967 *News of the USSR Academy of Sciences. Physical Series* **31**(7) 1086 [in Russian]
[8] Surovyak Z, Chekay D, Gomez M J M, Kofanova N B, Kuprina Yu A, Chebanova E V and Kupriyanov M F 2002 *Izvestiya RAN. Physical Series* **66** **(**6) 866 [in Russian]
[9] Hwang J, Rao R R, Giordano L, Katayama Y, Yu Y and Shao-Horn Y 2017 *Science* **358** 751
[10] Vakhrushev S B, Kvjatkovskiy B E, Naberezhnov A A, Okuneva N M and Tolerverg B P 1989 *Ferroelectrics* **90** 173
[11] Cowley R A, Gvasaliya S N, Lushnikov S G, Roessli B and Rotaru G M 2011 *Adv. Phys.* **60**(2) 229
[12] Verbaere A, Piffard Y, Ye Z-G and Husson E 1992 *Mat. Res. Bull.* **27** 1227
[13] Lebedinskaya A R and Kasparova N G 2019 *Adv. Mater.* **224** 267
[14] Gusev A I 2007 *Nonstoichiometry, Disorder, Short-Range and Long-Range Order in a Solid* (Moscow, Fizmatlit) 856 p. [in Russian]

Superplastic behavior of a nickel-base superalloy containing topologically close-packed phases

S K Mukhtarov[1], A A Ganeev[1], R V Shakhov[1], V M Imayev[1] and A V Logunov[2]

[1] Institute for Metals Superplasticity Problems of Russian Academy of Sciences, 39 Stepan Khalturin, Ufa, 450001, Russia
[2] PJSC "UEC-Saturn", 163 Lenin ave., Rybinsk, 152903, Russia

E-mail: shamilm@imsp.ru

Abstract. Superplastic behaviour of a polycrystalline nickel-base superalloy Ni-47(Cr,Co,W,Al,Ta,Hf)-0.2(C,B) (wt.%) in the fine-grained condition obtained via thermomechanical treatment (TMT) has been studied. TMT transformed a coarse-grained as-cast structure into a homogeneous fine-grained one. Both the initial as-cast and the thermomechanically treated conditions contained not only the matrix γ phase, the hardening γ' phase and carbides but also topologically close-packed (TCP) phases. In spite of the presence of the TCP phases, the superalloy in the fine-grained condition demonstrated superplastic properties (δ>300% and low flow stresses) at subsolvus temperatures. Superplastic deformation led to some refinement and alignment of the TCP phase particles along the tensile axis.

1. Introduction

Development of advanced gas turbine engines is dependent on developing novel heat resistant materials, such as nickel-base superalloys, capable of operating at higher temperatures and loadings. In regard to the nickel-base superalloys, high alloying with the γ' forming elements and the substitution elements for strengthening the matrix γ phase remain relevant [1]. However long-term operation of such superalloys (typically single-crystals or produced via directional crystallization) can lead to the formation of unfavorable topologically close-packed (TCP) phases [2-7]. At present, there are practically no data on the deformability of nickel-base superalloys containing TCP phases. The present work was aimed to study the microstructure and superplastic properties of a fine-grained nickel-base superalloy heavily alloyed with Co and W and containing TCP phases.

2. Material and methods

The nominal composition of the superalloy was Ni-47(Cr,Co,W,Al,Ta,Hf)-0.2(C,B) (wt.%). The as-cast ingot with a size of Ø45 × 270 mm was manufactured and supplied by STC "Technologies of special metallurgy" Ltd, MISiS (Russia). The actual composition of the ingot was very close to its nominal composition. The γ' solvus temperature (T_s) was determined via quenching experiments and shown to be equal to $T_s \approx 1185$°C.

The as-cast superalloy was subjected to homogenization heat treatment followed by slow cooling to cause the coarsening of the γ' phase and to improve the hot workability. After that, thermomechanical treatment (TMT) developed in our previous work [5] was carried out for small workpieces of the superalloy. The TMT consisted of unidirectional two-stage hot forging in the temperature range of

1100-1150°C with intermediate recrystalliation annealing. The forging procedure was carried out with the strain rate of $\acute{\varepsilon} = 10^{-3}$-$10^{-2}$ s^{-1}, the total strain value during forging was $e \approx 1.9$.

The thermomechanically treated condition of the superalloy was used for tensile testing. To do it, the flat samples with a gauge section of 10×3×2 mm^3 were cut from the fine-grained workpieces using spark cutting. All surfaces of the samples were finely ground before testing. Tensile tests were carried out at T = 1075-1150°C with an initial strain rate of $\acute{\varepsilon} \approx 8.3\times10^{-4}$ s^{-1}. The tensile tests were performed in air. Elongation to rupture δ, the true stress value σ_{30} corresponding to 30% of elongation were determined and the true stress-elongation curves σ-e were drawn. The true stress was calculated taking into account the graduate narrowing of the sample cross section during tensile testing. The strain rate sensitivity coefficient of the flow stress m was determined by the strain rate switching method.

Microstructure examination was carried out on a scanning electron microscope (SEM) in the back-scattering (BSE) or secondary electron (SE) mode. To identify the elemental composition of the bright phases precipitated in the superalloy (TCP phases), energy dispersive X-ray analysis (EDS) was applied. Electron backscatter diffraction (EBSD) analysis was performed with a scan-step size of 0.1 μm to characterize the thermomechanically treated condition. EBSD analysis was conducted using the CHANNEL 5 processing software. The grain boundaries having misorientation angle less than 2° were not considered taking into account the measurement accuracy. The grain boundaries with misorientation angle more than 15° were assumed as high-angle ones. The specimen surfaces before studying by SEM were mechanically ground and polished.

3. Results and discussion

3.1. *Effect of processing on the microstructure*

Figure 1 a, b represents the microstructural images of the as-cast superalloy. One can see that the superalloy contains not only the γ and γ′(Ni_3Al) phases and carbides (rounded bright particles) but also TCP phases located predominantly along γ grain boundaries. The γ grain size was $d_\gamma \sim 100$ μm. The γ′ phase in the as-cast condition was mainly presented by precipitates with a size of 0.2-0.25 μm (figure 1b). The volume fraction of the γ′ phase was defined to be about 60%. The volume fraction of carbides was not more than 1-2% due to the small amount of carbon. The TCP phases were observed along the γ grain boundaries in the form of elongated and equiaxed particles of 0.5-20 μm in size; its volume fraction was about 2.5%. Homogenization annealing led to precipitation of the TCP phases, which were uniformly precipitated as lamellar and equiaxed particles. The volume fraction of the TCP phases increased up to about 5% (figure 1c). The EDS analysis of the TCP phases revealed that they were enriched with W, Ta and Hf.

The TMT led to a uniform development of recrystallization processes in the forged workpieces that in turn led to the formation of a fine-grained microstructure with the mean γ grain size of $d_\gamma \approx 5$ μm (figure 1d) [5]. The primary γ′ phase not dissolved during TMT had a size of 1-2 μm and occupied about 20 vol.% (figure 1d). The secondary γ′ particles with a size of about 0.25 μm precipitated within γ grains during cooling after TMT [5]. The lamellar TCP phases after TMT were partially globularized and aligned parallel to the material flow (figure 1d).

Figure 2 shows the EBSD orientation map and corresponding misorientation angle distribution for grain boundaries obtained for the thermomechanically treated condition. One can see that mostly high-angle boundaries were obtained after TMT showing that recrystallization processes occurred extensively throughout the forged workpiece. A small peak observed near misorientation of 60° corresponded to twin boundaries that were formed immediately after completing the forging processing (figure 2b).

3.2. *Mechanical behaviour*

Figure 3 shows the results of tensile tests of samples in the fine-grained condition obtained via TMT. High elongations (δ = 300-380%), low and nearly constant flow stresses (σ = 12-18 MPa) and an increased strain rate sensitivity coefficient of the flow stress (m > 0.3) typical of superplasticity were achieved at T = 1100-1150°C and $\acute{\varepsilon} = 8.3\times10^{-4}$ s^{-1}. The highest elongations (δ = 350-380%) were reached at 1100 and 1125°C. Tensile testing at 1075°C gave much lower elongation and higher flow stresses. Apparently, this tensile test did not correspond to the superplastic temperature-strain rate conditions and, therefore, elongation was much smaller than 300% (figure 3).

Microstructure examination of the gauge sections of samples strained in tensiion at 1100-1150°C showed that γ grain growth occurred after superplastic deformation. For instance, after tensile straining at 1100°C the mean γ grain size grew up to d ≈ 7 μm (figure 4). The grain growth is typical of superplasticity of fine-grained alloys and results from the grain boundary sliding, which occurs in the course of the superplastic deformation. The size of the prime γ' phase after superplastic elongation was found to be the same as prior to tensile testing and did not exceed 2 μm. Secondary γ' phase precipitated after superplastic deformation and had also near the same size as before tensile testing. The carbides and particles of the TCP phases were slightly refined and aligned along the tensile axis. The length of lamellar TCP precipitates decreased after superplastic deformation and did not exceed 15 μm. Fragmentation of the lamellar TCP phase particles promoted the formation of voids near the fracture zone of the superplastically strained samples. Most likely, this provoked premature failure and limited the superplastic elongation. Nevertheless, in spite of the presence of the TCP phases, the fine-grained condition provided superplastic properties of the superalloy at subsolvus temperatures (T = 1100-1150°C).

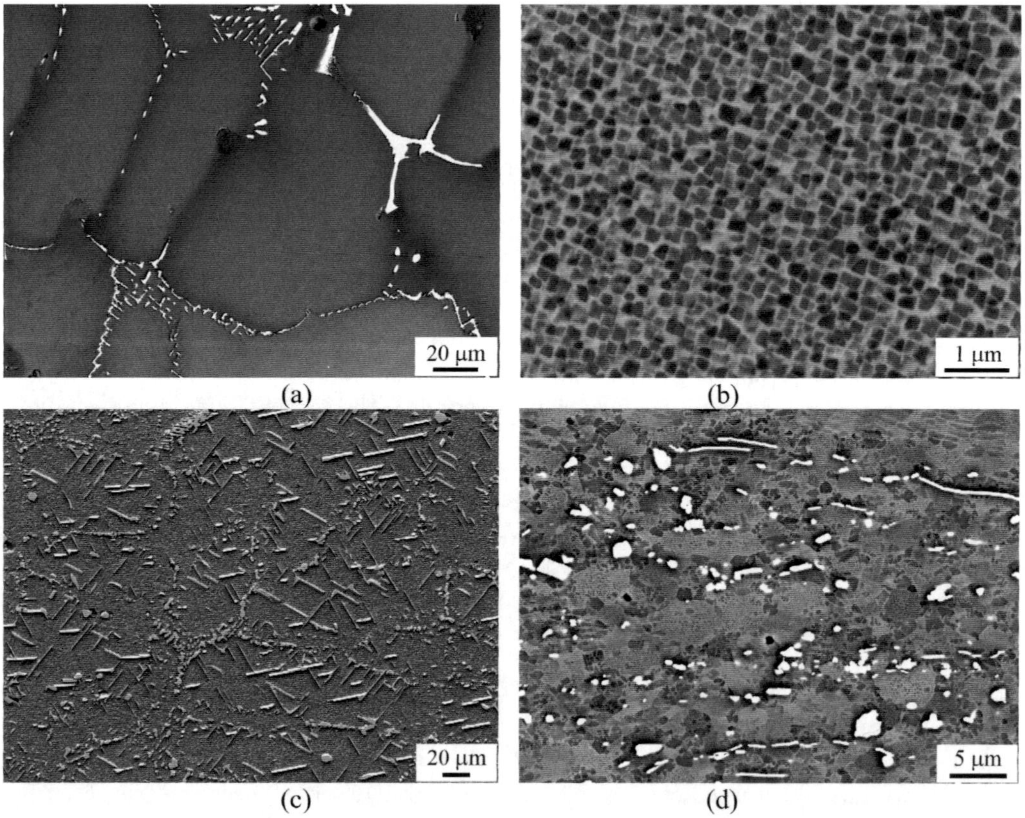

Figure 1. The microstructure images of the superalloy: (a, b) the as-cast condition (BSE), (c) the cast condition subjected to homogenization heat treatment (SE), the condition obtained after TMT (BSE).

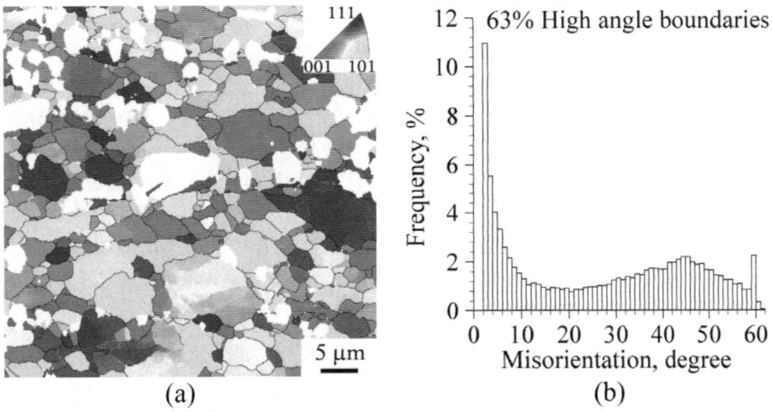

Figure 2. Normal-direction EBSD (inverse-pole-figure) map (a), and the corresponding misorientation-angle distribution for grain boundaries (b) obtained for the superalloy after TMT. The forging axis is vertical, high- and low-angle grain boundaries are indicated by black and white lines, respectively.

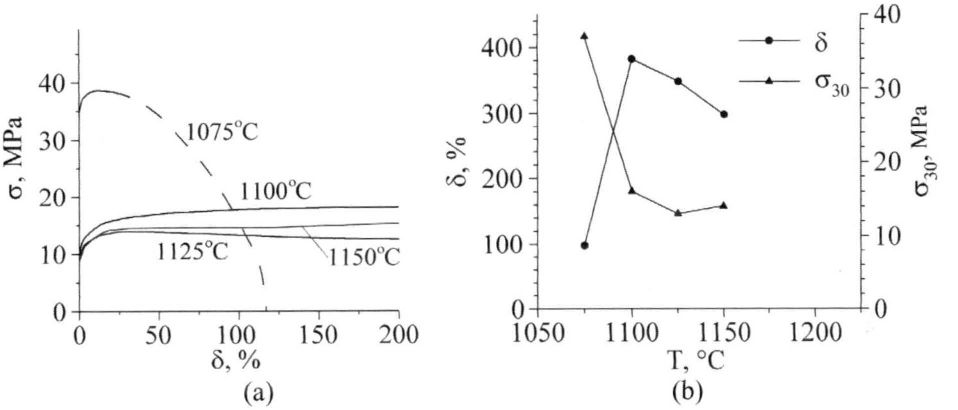

Figure 3. The temperature dependences of (a) the true stress vs. elongation and (b) the total elongation and the flow stress (σ_{30}) obtained as a result of tensile tests at T = 1075-1150°C ($\dot{\varepsilon} = 8.3\times10^{-4}$ s^{-1}).

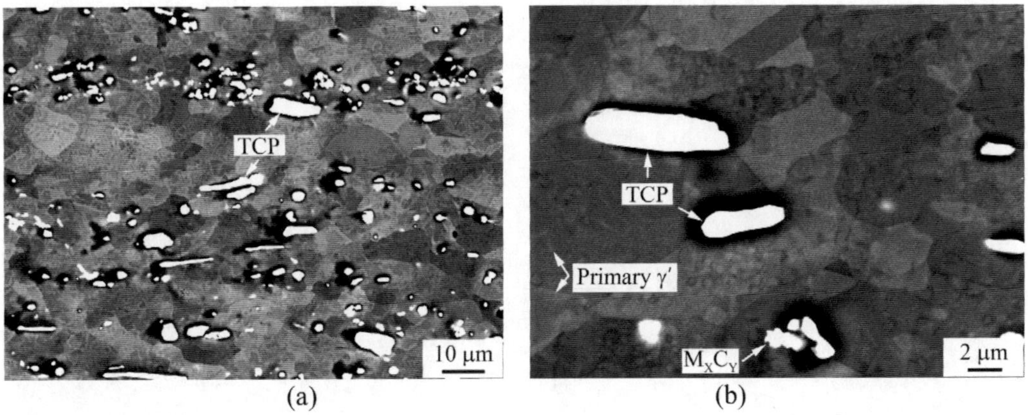

Figure 4. The BSE images obtained from the gauge area near the fracture zone of the superalloy sample after tensile testing at T = 1100°C ($\dot{\varepsilon} = 8.3\times10^{-4}$ s^{-1}). The primary γ' phase, TCP phases and carbides are arrowed.

4. Conclusions

The heavily alloyed nickel-base superalloy containing TCP phases was subjected to homogenization heat treatment and thermomechanical treatment, which led to a uniform occurrence of recrystallization processes and formation of a fine-grained structure in the forged workpieces. In spite of the presence of the TCP phases, the superalloy in the fine-grained condition exhibited superplastic behavior at subsolvus temperatures. In the temperature range of 1100-1150°C at strain rate of $\dot{\varepsilon} \approx 8.3\times10^{-4}$ s^{-1}, the fine-grained samples of the superalloy showed high elongations ($\delta = 300$-380%), low and nearly constant flow stresses ($\sigma = 12$-18 MPa) and increased values of the strain rate sensitivity coefficient ($m > 0.3$). After superplastic deformation the TCP phase particles were refined and aligned along the tensile axis.

Acknowledgments

The present work was funded by the Russian Foundation for Basic Research (Research Project No 18-08-00997-a). The work was performed using the facilities of the shared services center «Structural and Physical-Mechanical Studies of Materials» at the Institute for Metals Superplasticity Problems of Russian Academy of Sciences.

References

[1] Reed R C 2006 *The Superalloys: Fundamentals and Applications* (Cambridge University Press)
[2] Rae C M F, Karunaratne M S A, Small C J, Broomfield R W, Jonesand C N and Reed R C 2000 *Proc. of the 9th Int. Symp. Superalloys 2000* (Warrendale, Pa., TMS) p 767
[3] Pyczak F, Neumeier S and Göken M 2010 *Mater. Sci. Eng.* A **527** 7939
[4] Matuszewski K, Rettig R, Matysiak H, Peng Z, Povstugar I, Choi P, Müller J, Raabe D, Spiecker E, Kurzydłowski K J and Singer R F 2015 *Acta Mater.* **95** 274
[5] Shakhov R V, Ganeev A A, Mukhtarov Sh Kh and Logunov A V 2018 *Lett. Mater.* **8** (4) 494
[6] Sugui T, Minggang W, Tang L, Benjiang Q and Jun X 2010 *Mater. Sci. Eng.* A **527** 5444
[7] Darolia R, Lahrman D F, Field R D and Sisson R 1988 *Proc. of the 6th Int. Symp. Superalloys 1988* (Warrendale, Pa.,) p 255

IOP Conf. Series: Materials Science and Engineering 1008 (2020) 012063 doi:10.1088/1757-899X/1008/1/012063

Behaviour of carbon nanotube bundle under quasistatic and dynamic transverse compression

L Kh Rysaeva[1], R T Murzaev[1], A A Kudreyko[2], E A Korznikova[1,3] and S V Dmitriev[1,4]

[1] Institute for Metals Superplasticity Problems of RAS, 450001, 39 Khalturin St., Ufa, Russia
[2] Department of Medical Physics and Informatics, Bashkir State Medical University, 47 Zaki Validi st., 47, 450000 Ufa, Russia
[3] Ufa State Aviation Technical University, 450077, 12 Karl Marx, Ufa, Russia
[4] Institute of Molecule and Crystal Physics, Ufa Federal Research Centre of the Russian Academy of Sciences, 71 October Avenue, 450054 Ufa, Russia

E-mail: lesya813rys@gmail.com

Abstract. Carbon nanotubes (CNTs) have a unique combination of physical and mechanical properties, which makes them attractive for a number of applications. Investigation of the mechanical response of a CNT bundle under conditions of plane strain (uniaxial and biaxial lateral compression), since it can be used as an elastic damper for vibration and shock protection. In this work, using a chain model with a reduced number of degrees of freedom, the behaviour of a CNT bundle under uniaxial lateral compression is investigated. Quasistatic loading is used to calculate the elastic constants of the bundle. Dynamic loading reveals the propagation of shock waves of two types: the faster one propagates with the speed of longitudinal sound and causes insignificant deformation of the CNT cross sections, while the second one propagates approximately three times slower and leads to the collapse of the CNTs.

1. Introduction

Carbon nanotubes (CNTs) are attracted to each other by relatively weak van der Waals forces and can create molecular crystals, also called CNT bundles [1-3]. CNT bundles can be obtained by various techniques [4,5]. They have excellent mechanical properties, since individual CNT has a very high tensile strength in the range from 11 to 63 GPa, tensile Young's modulus ranging from 1.0 to 1.3 TPa and a high strain to fracture of about 10% [6,7]. In addition, CNTs are flexible, lightweight, good thermal and electrical conductors; therefore, they can be offered for a number of applications [8, 9]. The most important mechanical applications of CNTs are high-strength ropes [2, 10], fibers [11, 12], composites with a polymer and metal matrix [12, 13], solid lubricants [14], shock protecting structures [15], etc. Computer simulation studies contribute to a better understanding of the physical and mechanical properties of CNT bundles. Transformation of a vertically aligned into a horizontally aligned CNT forest by applying pressure has been studied in [16,17]. The shell model has been used to describe an ensemble of CNTs of different morphology [18]. The applicability of the beam, plate and shell models to the analysis of the mechanical properties of nanomaterials has been analyzed in [19,20]. CNTs having diameter is above a threshold value can exist either in circular or collapsed state [21-23]. For the study of mechanical response of CNT bundles the nonlinear coarse-grained potentials has been developed [24]. In order to reduce the number of degrees of freedom considered in simulation of some sp^2-carbon nanostructures, the chain model has been developed [25]. This model was successfully applied to modeling of structure and properties of carbon nanoribbon folds and scrolls [25-29] and surface ripplocations [30]. In the works [31,32], the chain model was extended to the study of CNTs under lateral compression in plane strain conditions. Damping properties of CNT bundle were described in [33]. Mechanical properties and phase transitions in CNT bundle under lateral compression were analyzed in [34,35]. Twisted graphene nanoribbons show enhanced resistance to axial compression [36,37]. Solitary waves [30,38] and shock waves [39] in nanomaterials have also been analyzed, although dynamic problems remain much less studied.

So far, the chain model was used for the analysis of static loading of CNT bundles [31-35], and in this work, we also consider dynamical loading (shock compression).

2. Simulation details and numerical results

2.1. Simulation setup

The CNT bundle consists from single walled CNTs of the same diameter with a zigzag orientation. The *z*-axis of the Cartesian coordinate system is directed along the axes of CNTs. The bundle cross section is parallel to the *xy*-plane. The CNT cross sections form a triangular lattice with the close packed direction aligned with the *x*-axis [see figure 1(a)]. The computational cell includes 10×12 (8×200) nanotubes for static (dynamic) problem. The boundary conditions are periodic. The CNT bundle is treated under the plain strain condition. Each CNT cross section is represented by 30 carbon atoms, and each atom represents a rigid row of atoms parallel to the CNT axis (*z*-axis). Each atom has two degrees of freedom, the *x*- and *y*-components of the displacement vector.

Quasistatic compression is performed along the *x*-axis ($\varepsilon_{xx}<0$, $\varepsilon_{yy}=0$) and dynamic loading along the y-axis ($\varepsilon_{xx}=0$, $\varepsilon_{yy}<0$). Quasistatic compression is done in a stepwise manner with the increment $\Delta\varepsilon_{xx}$=-0.005 followed by relaxation. Absolute value of the volumetric strain $|\theta|=|\varepsilon_{xx}+\varepsilon_{yy}|/2=\varepsilon_{xx}/2$ is used as the measure of quasistatic strain. Shock compression loading is performed by moving the upper row of CNTs down and fixing positions of atoms in the bottom row of CNTs. The upper row moves with the velocity V=-1000m/s, which is the speed of a bullet.

2.2. Quasistatic loading

We firstly present the simulation results for quasistatic loading. In figure 1 CNT structure at different levels of uniaxial compression along the *x*-axis are shown: (a) $|\theta|=0$, (b) $|\theta|=0.08$ and (c) $|\theta|=0.3$. In (b) the undeformed structure has translational symmetry with single CNT in a primitive translational cell. Uniaxial compression results in elliptization of CNT cross sections keeping the crystallinity of the structure. At $|\theta|=0.07$ a phase transition with period doubling in two directions occurs. It can be seen that in (b) the CNT bundle has translational symmetry with four CNTs in a primitive translational cell. At $|\theta|=0.12$ another phase transition takes place when CNTs start to collapse. With increasing compressive strain the fraction of collapsed CNTs gradually increases and in (c) at $|\theta|=0.3$ almost all CNTs are collapsed.

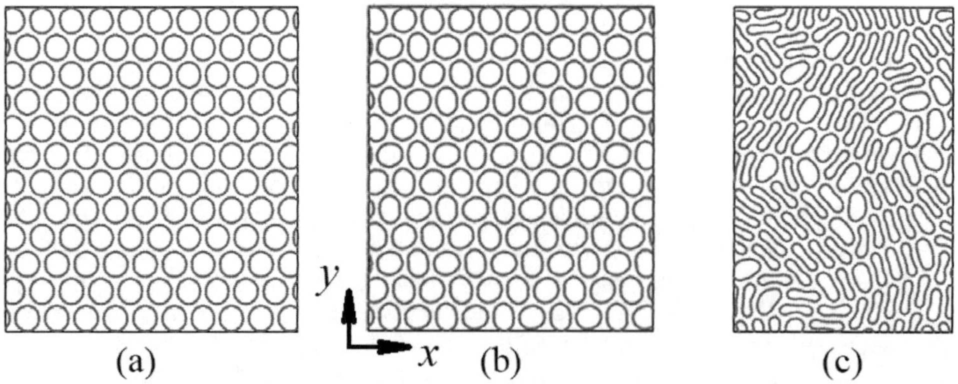

Figure 1. CNT bundle structure evolution during uniaxial compression along the *x*-axis: (a) initial state at $|\theta|=0$, (b) period doubling for the primitive translational cell at $|\theta|=0.08$ and (c) almost all CNTs are collapsed at $|\theta|=0.3$.

In figure 2(a) the stress-strain curves are presented. It can be seen that in the range $0<|\theta|<0.07$ stress components grow rapidly with strain and compressive stress σ_{xx} is greater than σ_{yy}. We do not show the shear stress because it is close to zero. At $|\theta|=0.07$ stress components show a sharp change in the slope and in the range $0.07<|\theta|<0.12$ they vary linearly with strain. At $|\theta|=0.12$ a drop of stress components takes place due to the collapse of some CNTs.

We then calculate elastic constants (Poisson's ratios ν_{xy} and ν_{yx}, Young's moduli E_{xx} and E_{xy} and shear modulus G) for the CNT bundle as the functions of volumetric strain. The Poisson's ratios are shown in figure 2(b), while the Young's moduli in figure 2(c). Interestingly, Poisson's ratios in the range $0<|\theta|<0.07$ are close to unity and they approach this value with increasing compressive strain. Even more interesting is that in the range $0.07<|\theta|<0.12$ ν_{yx} is negative and ν_{xy} remains positive with the value close to 0.3. This means that in this range of volumetric strain CNT bundle is a partial auxetic. Young's moduli and shear modulus grow linearly with compressive strain in the range $0<|\theta|<0.07$. The shear modulus exhibits a considerable increase in the range $0.07<|\theta|<0.12$, while Young's moduli a smaller in this range of strain.

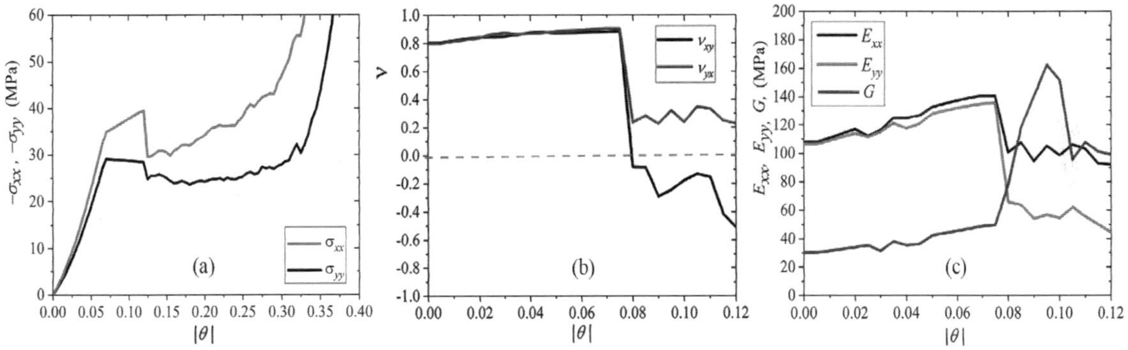

Figure 2. (a) Stress-strain curves. Change in (b) Poisson's ratios and (c) in shear modulus and Young's moduli with compressive volumetric strain.

2.3. *Dynamic loading*

We now turn to the presentation of the results of dynamical loading of CNT bundle, which are given in figure 3. In (a) one can see the vertical displacements of the centres of gravity of CNTs at t_1=10 ps (black line), t_2=20 ps (red line) and t_3=30 ps (blue line). Note that the horizontal rows of CNTs are numbered by the index n. In (b), as the function of n, the ellipticity of CNTs is shown as the ratio of the minimal to maximal diameters. In (c) and (d) the averaged energies of valence bonds and valence angles, respectively are shown as the functions of n. Comparison of the results shown in (a) and (b) suggests that propagation of the faster wave results in a weak elliptization of CNTs with D_{min}/D_{max}=0.9. The second slow wave leads to collapse of CNTs with D_{min}/D_{max}=0.2. As it follows from (c), energies of valence bonds are approximately same in the collapsed and non-collapsed CNTs. On the other hand, the energies of the valence angle [see in (d)], as expected, are much higher in the collapsed CNTs as compared to CNTs with small ellipticity.

3. Conclusions

Molecular dynamics simulations of quasistatic and dynamic lateral compression of CNT bundle in plane strain state were performed. Quasistatic compression along the close packed direction of the bundle (along the x-axis) has revealed that in the crystalline structural state with single CNT in the primitive translational cell Poisson's ratios are close to unity and approach this value with increasing compressive strain. It should be noted that most elastic bodies have Poisson's ratio close to 0.3, and the value of Poisson's ratio, close to unity, is extremely large. This means that the CNT bundle exhibits unusual elastic properties. When CNT bundle has crystalline structure with four CNTs in the primitive translational cell, one of the two Poisson's ratios becomes negative meaning that the bungle is a partial auxetic. Shock compression of CNT bundle in the direction normal to the close packed direction (along the y-axis) has revealed propagation of two waves moving with different velocities. The faster wave caused small elliptization of CNTs, while the following slow wave causes their collapse. Results presented in this work demonstrate that elastic bodies composed of highly deformable units, such as CNT cross sections, can demonstrate unusual mechanical properties. In future works, the chain model can be extended for 2D materials other than graphene [41].

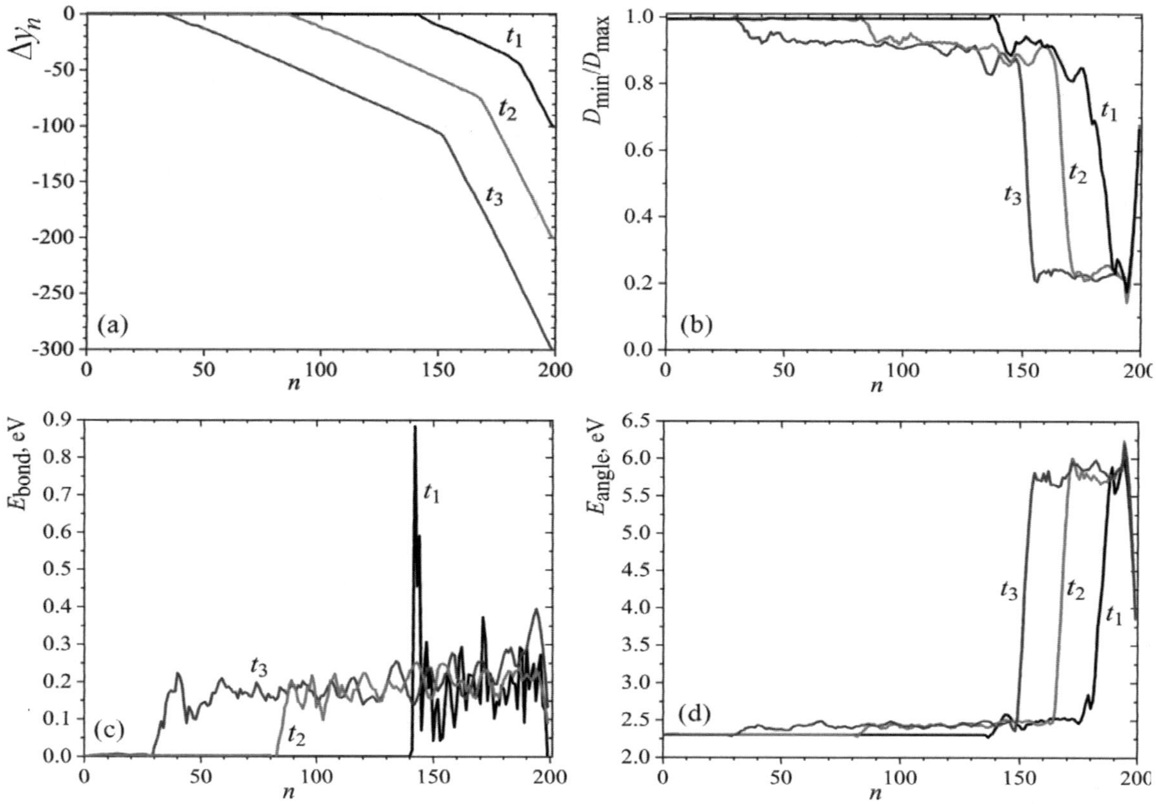

Figure 3. Displacements of the centers of mass of nanotubes under compression (a), minimal to maximal diameter ratio (b), energy valence bonds (c) and angles (d) at loading speed of 1000 m/c. t_1, t_2 and t_3 correspond to the simulation time 10 ps - black curve, 20 ps - red curve and 30 ps -blue curve.

Acknowledgments
This work was supported by the State Assignment of IMSP RAS No. AAAA-A17-117041310220-8.

References

[1] Tersoff J and Ruoff R S 1994 *Phys. Rev. Lett.* **73** 676
[2] Thess A, Lee R, Nikolaev P, Dai H, Petit P, Robert J, Xu C, Lee Y H, Kim S G, Rinzler A G, Colbert D T, Scuseria G E, Tomanek D, Fischer J E and Smalley RE 1996 *Science* **273** 483
[3] Saether E, Frankland S J V and Pipes R B 2003 *Compos. Sci. Technol.* **63** 1543
[4] Rakov E G 2013 *Russ. Chem. Rev.* **82** 538
[5] Bedewy M, Meshot E R, Guo H, Verploegen E A, Lu W and Hart A J 2009 *J. Phys. Chem.* C **113** 20576
[6] Shenderova O A, Zhirnov V V and Brenner D W 2002 *Crit. Rev. Sol. State* **27** 227-356
[7] Yu M-F 2004 *J. Eng. Mater. Technol.* **126** 271-278
[8] Truong T K, Lee Y and Suh D 2016 *Current Applied Physics* **16** 1250-1258
[9] Yao X, Hawkins S C and Falzon B G 2018 *Carbon* **136** 130-138
[10] Yu M-F, Files B S, Arepalli S and Ruoff R S 2000 *Phys. Rev. Lett.* **84** 5552-5555

[11] Bai Y, Zhang R, Ye X, Zhu Z, Xie H, Shen B, Cai D, Liu B, Zhang C, Jia Z, Zhang S, Li X and Wei F 2018 *Nature Nanotechnology* **13** 589-595
[12] Dang Z-M, Yuan J-K, Zha J-W, Zhou T, Li S-T and Hu G-H 2012 *Prog. Mater. Sci.* **57** 660
[13] Bakshi S R, Lahiri D and Agarwal A 2010 *Int. Mater. Rev.* **55** 41
[14] Reinert L, Lasserre F, Gachot C, Grützmacher P, Maclucas, Souza N, Mücklich F and Suarez S 2017 *Sci. Rep.* **7** 42873
[15] Randjbaran E, Majid D L, Zahari R, Sultan M T H and Mazlan N 2020 *Facta Universitatis Ser.: Mech. Eng.* **18** 229
[16] Wittmaack B K, Volkov A N and Zhigilei L V 2019 *Carbon* **143** 587
[17] Wittmaack B K, Volkov A N and Zhigilei L V 2018 *Compos. Sci. Technol.* **166** 66
[18] Yakobson B I, Brabec C J and Bernholc J 1996 *Phys. Rev. Lett.* **76** 2511
[19] Rafii-Tabar H, Ghavanloo E and Fazelzadeh S A 2016 *Phys. Rep.* **638** 1
[20] Harik V M 2001 *Solid State Commun.* **120** 331
[21] Impellizzeri A, Briddon P and Ewels C P 2019 *Phys. Rev.* B **100** 115410
[22] Chopra N G, Benedict L X, Crespi V H, Cohen M L, Louie S G and Zettl A 1995 *Nature* **377** 135
[23] Chang T 2008 *Phys. Rev. Lett.* **101** 175501
[24] Ji J, Zhao J and Guo W 2019 *J. Mech. Phys. Solids* **128** 79
[25] Savin A V, Korznikova E A and Dmitriev S V 2015 *Phys. Rev.* B **92** 035412
[26] Savin A V, Korznikova E A and Dmitriev S V 2015 *Phys. Solid State* **57** 2348
[27] Savin A V, Korznikova E A, Lobzenko I P, Baimova Y A and Dmitriev S V 2016 *Phys. Solid State* **58** 1278
[28] Savin A V, Korznikova E A, Dmitriev S V and Soboleva E G 2017 *Comp. Mater. Sci.* **135** 99
[29] Savin A V and Mazo M A 2019 *Adv. Struct. Mat.* **94** 241
[30] Savin A V, Korznikova E A and Dmitriev S V 2019 *Phys. Rev.* B **99** 235411
[31] Korznikova E A, Rysaeva L K, Savin A V, Soboleva E G, Ekomasov E G, Ilgamov M A and Dmitriev S V 2019 *Materials* **12** 3951
[32] Abdullina D U, Korznikova E A, Dubinko V I, Laptev D V, Kudreyko A A, Soboleva EG, Dmitriev S V and Zhou K 2020 *Computation* **8** 27
[33] Rysaeva L K, Korznikova E A, Murzaev R T, Abdullina D U, Kudreyko A A, Baimova J A, Lisovenko D S and Dmitriev S V 2020 *Facta Universitatis Ser.: Mech. Eng.* **18** 1
[34] Abdullina D U, Kudreyko A A, Korznikova E A, Shepelev I A and Dmitriev S V 2020 *Progress in Biomedical Optics and Imaging - Proceedings of SPIE* **11459** 1145911
[35] Babicheva R I, Dmitriev S V, Korznikova E A and Zhou K 2019 *J. Exp. Theor. Phys.* **129** 66
[36] Savin A V, Korznikova E A, Krivtsov A M and Dmitriev S V 2020 *Eur. J. Mech.* A. **80** 103920
[37] Savin A V, Korznikova E A and Dmitriev S V 2019 *Mech. Mater.* **137** 103123
[38] Korznikova E A and Dmitriev S V 2014 *J. Phys.* D **47** 34
[39] Shepelev I A, Chetverikov A P, Dmitriev S V and Korznikova E A 2020 *Comp. Mater. Sci.* **177** 109549
[40] Hosseini-Hashemi S, Sepahi-Boroujeni A and Sepahi-Boroujeni S 2018 *Appl. Surf. Sci.* **437** 366
[41] Romanov A E, Rozhkov M A and Kolesnikova A L 2018 *Lett. Mater.* **8** 384
[42] Barani E, Korznikova E A, Chetverikov A P, Zhou K and Dmitriev S V 2017 *Phys. Lett.* A **381** 3553
[43] Shcherbinin S A, Zhou K, Dmitriev S V, Korznikova E A, Davletshin A R and Kistanov A A 2020 *J. Phys. Chem.* C **124** 10235
[44] Kistanov A A, Khadiullin S K, Dmitriev S V, Korznikova E A 2019 *Chem. Phys. Lett.* **728**, 53
[45] Babicheva R I, Dahanayaka M, Liu B, Korznikova E A, Dmitriev S V, Wu M S and Zhou K 2020 *Mater. Sci. Eng.* B **259** 114569
[46] Kistanov A A, Khadiullin S K, Dmitriev S V and Korznikova E A 2019 *Chem. Phys. Chem.* **20** 575

IOP Conf. Series: Materials Science and Engineering 1008 (2020) 012064 doi:10.1088/1757-899X/1008/1/012064

Influence of structural state on corrosion behaviour of bearing steel 110C-18Cr-Mo

M M Abramova[1], E R Tuktamysheva[1], Yu B Saburova[1] and N A Enikeev[1]

[1] Ufa State Aviation Technical University, K. Marx str. 12, Ufa, 450008, Russia
E-mail: abramovamm@yandex.ru

Abstract. The paper report on the studies of the effect of severe plastic deformation on the structure and properties, including the corrosion behavior, of 110C-18Cr-Mo bearing steel. Various structural states were formed by the methods of severe plastic deformation (SPD) and subsequent heat treatment. It is shown that SPD and subsequent heat treatment lead to the formation of a homogeneous structure with a uniform distribution of carbides. In this case, heat treatment leads to a reduction of the volume fraction of carbides and an increase in the microhardness of the samples. Steel subjected to SPD demonstrates a decrease in the corrosion rate in one-molar solutions of salt and sodium hydroxide. Subsequent heat treatment leads to an additional decrease in the corrosion rate.

1. Introduction

Bearing steels are similar to instrumental ones by the composition and properties, and their final properties are achieved by heat treatment, namely, by quenching followed by tempering. This treatment provides a high level of strength and hardness required for operation of the bearings. However, the performance properties are determined not only, and in some cases even not so much by the level of hardness, but by the uniformity of hardness distribution in the bulk of the material. In its turn, the homogeneity of properties is determined by the homogeneity of the microstructure, as applied to bearing steels, primarily by the uniformity of the distribution of the carbide phase [1]. In addition, corrosion resistance is also a structurally sensitive property. In the case of bearing steels, the main problem is a significant heterogeneity of carbide, which leads to a decrease in wear resistance and premature failure. To solve this problem, severe plastic deformation (SPD) by equal channel angular pressing (ECAP) was used. In the last decade, it has been shown that this method can be successfully used not only for the refinement of the grain structure of the matrix phase, but also for the redistribution of particles of the second phases [2]. Hiasch et al. [3] showed that the microstructure of steel could strongly influence its behavior during electrochemical dissolution. The dissolution of different phases (ferrite and cementite) occurs in different ways, because different phases have different electrode potentials, which leads to the formation of galvanic pairs [4, 5]. Qu et al. [6] showed that ferrite and bainite formed a galvanic pair in a solution of NaCl ferrite – bainite dual-phase steel, and corrosion proceeded mainly along the boundary of these phases and depended on the volume fraction of each phase. A number of studies of the corrosion behavior of ferrite-pearlite and ferrite-bainitic steels have been carried out [3-7]. On the other hand, not only different phases, but also SPD and size refinement of structural elements can affect the corrosion behavior. In work [8], the effect of SPD on austenitic or low- carbon steels considered. There is practically no information on studies of hard-to-deform high-carbon steels.

2. Materials and methods

Structural bearing steel 110X18M of standard chemical composition (table 1) was chosen as the material for the study.

Table 1 Chemical composition of steel 110C-18Cr-M

Steel	C	Cr	Mo	Si	Mn
110C-18Cr-M	1,10÷1,20	16,5÷18,0	0,50÷0,80	0,53÷0,93	0,50÷1,00

SPD was carried out by the ECAP method. The workpieces were in the shape of cylinders 10 mm in diameter and 60 mm in length. The deformation was carried out at a temperature of 600°C, the angle between the channels was 120°, the number of passes n = 5. Standard heat treatment (HT) was used prior to SPD: quenching in water from a temperature of 1050°C, holding for 1 hour and subsequent low tempering at 170°C (1 hour). Microstructure analysis was carried out by scanning electron microscopy on a JSM-6490LV microscope. The structure was revealed by etching in a solution of 3% nitric acid and 97% ethyl alcohol. Microhardness was measured using a Duramin device at a load of 100 g for 10 seconds. Corrosion behavior was investigated by potentiodynamic method in 3% sodium chloride and sodium hydroxide 1M, which are the most common media for this steel.

Polarization curves were recorded on a PG12-100 programmable potentiostat-galvanostat up to an anode potential of + 0.5V. Based on the obtained polarization curves, the corrosion currents were calculated by the three-point method and by constructing the Evans curves.

3. Results and discussion

The microstructure in the initial state is a granular pearlite with an average carbide size of 2.9 ± 0.3 μm and 15.6 ± 0.8 μm, the volume fraction of carbide particles is 51 ± 2% (figure 1a). It should be noted that the structure is characterized by a heterogeneity of the sizes of the carbide phase. Very large carbides up to several tens of micrometers in size are observed, and the distribution is bimodal. Such heterogeneity of the carbide structure leads to a decrease in contact fatigue [1]. After SPD, the average size of carbides was 5.1 ± 0.5 μm, with a volume fraction of 25 ± 1% (figure 1b). Compared to the initial state, the carbides are refined, the size heterogeneity remains, but the distribution of carbides over the ferrite matrix becomes more uniform.

The structure of the bearing steel after quenching and tempering is highly dispersed. Heat treatment of steel is carried out to eliminate carbide inhomogeneity, tempering martensite, retained austenite and undissolved cementite are present in the structure, the amount of large carbides is noticeably reduced (figure 1c, d). The average size of carbides is 2.5 ± 1.5 μm and the volume fraction is 19 ± 1% for the initial state and 4.3 ± 3.0 μm, the volume fraction of carbides is 16 ± 1% for the state after SPD and HT.

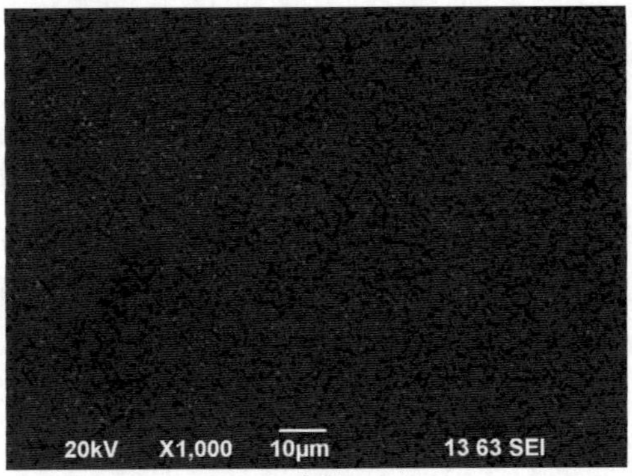

(a)

(b)

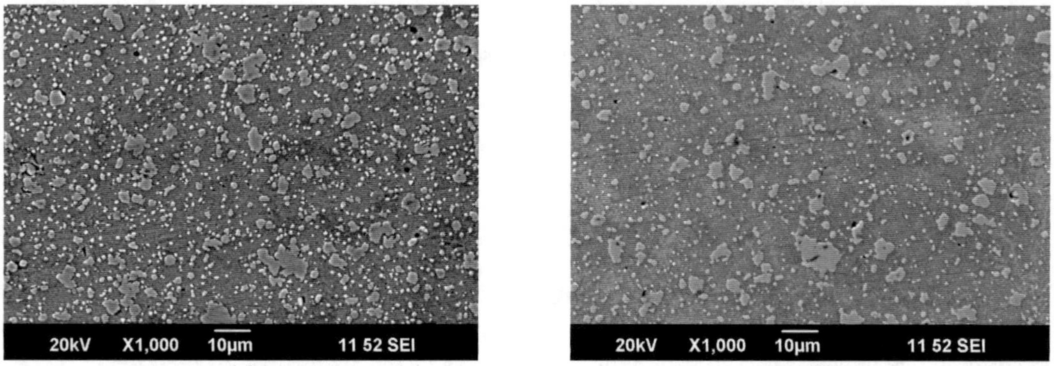

(c) (d)

Figure 1. Microstructure of steel in various states: a - initial, b - after SPD, c - initial + standard heat treatment, d - SPD + standard heat treatment

The microhardness of the initial state is 287 ± 6 HV, after ECAP the microhardness increased to 420 ± 10 HV, in the sample after heat treatment, the microhardness increased by 2.5 times compared to the initial state and amounted 709 ± 23 HV, the highest microhardness is observed in the sample subjected to ECAP and subsequent heat treatment - 829 ± 35 HV. The distribution of microhardness over the diameter of the samples is uniform.

Studies of corrosion behavior have shown that dissolution for samples after heat treatment begins at higher positive potentials, i.e. dissolution is influenced to a greater extent by heat treatment, after which the volume fraction of carbides in the structure decreases (figure 2). The original sample is more corrosive, which is explained by a heterogeneity of the sizes of the carbide phase and their greater number and size.

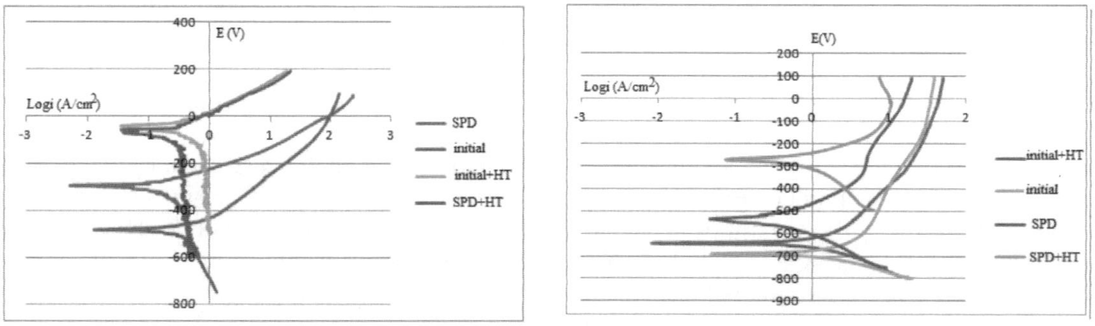

(a) (b)

Figure 2. Polarization semilogarithmic curves in solutions: a - sodium chloride 3%, b - sodium hydroxide 1M.

Steel after SPD is more corrosion-resistant in sodium hydroxide compared to coarse-grained state, the components of steel dissolve in sodium hydroxide, chromium can dissolve, molybdenum also reacts with hydroxide, and since the structure after ECAP is more homogeneous, dissolution occurs at lower rates compared to with a coarse-grained state.

Heat treatment promotes a decrease in corrosion rates both for specimens obtained by SPD by the ECAP method and for the initial coarse-grained specimen. However, the effect is more pronounced for the samples obtained by ECAP.

4. Conclusions

Thus, the use of SPD by the ECAP processing of bearing steel 110C-18Cr-Mo followed by heat treatment leads to a decrease in the volume fraction of carbides, the amount of large carbides, which, firstly, were refined during deformation, and, secondly, partially dissolved as a result of heat treatment, significantly decreases.

The study of corrosion resistance by the potentiodynamic method showed that samples with a more homogeneous structure, obtained as a result of a combined processing of ECAP with subsequent heat treatment, exhibited the highest corrosion resistance.

Acknowledgments
This work was supported by the Grant of the President of the Russian Federation MK-1458.2020.8.

References

[1] Nasedkina Ya I, Karavaeva M V and Kaibyshev O A 2012 *Bulletin of USATU* **16** 145-8 [in Russian]
[2] Karavaeva M V, Nureyev S K, Zaripov N G, Ganeev A V and Valiev R Z 2012 *Metal Sci. Heat Treat. Met.* **4** 3-7.
[3] Hiasch T, Mittemerijer E J and Schultze J W 2002 *Mater. Corros.* **53** 740-55
[4] Panda B, Balasubramaniam R and Dwivedi G 2008 *Corros. Sci.* **50** 1684-92
[5] Katiyar P K, Misra S and Mondal K 2018 *J. Mater. Eng. Perform.* **27** 1753-62
[6] Qu S, Pang X, Wang Y and Gao K 2013, *Corros. Sci.* **75** 67-77
[7] Guo J, Yang S, Shang C, Wang Y, and He X 2008 *Corros. Sci.* **51** 242-51
[8] Miyamoto H 2016 *Mater. Trans.* **57** 559-72

Microstructure and phase composition of Al-Nb metal-matrix composites produced by shear deformation under pressure

V N Danilenko, L U Kiekkuzhina, K S Nazarov and R R Mulyukov

Institute for Metals Superplasticity Problems of Russian Academy of Sciences, 39 Khalturin st., Ufa 450001, Russia

E-mail: vdan@anrb.ru

Abstract. Al and Nb billets with a coarse grain structure in the form of thin discs were stacked in the order of Al-Nb-Al and then deformed by shear under pressure. After the deformation the samples were annealed at 500°C for 30 min. The formation of an intermetallic compound Al_3Nb was revealed by using X-ray diffraction and energy-dispersive spectroscopy. After post-deformation annealing, the amount of this compound increased. At the same time, the study of microhardness demonstrates a decrease of the value of microhardness after post-deformation annealing.

1. Introduction

The development of new materials is one of the main tasks of materials science, which ensures the progress in the modern technology. As such materials, composites can significantly improve the properties of products for various purposes and therefore play an important role in the development of modern industry.

One of the methods of solid-phase preparation of composites is shear deformation under pressure [1-3]. Such a method of obtaining composite materials combined with post-deformation annealing makes it possible to form a reliable connection of strengthening phase with the matrix.

Recently, the results of studies on the processing of *in-situ* metal-matrix composites have been reported for such binary systems as Al-Nb [1], Al-Mg [2], and Al-Cu [4, 5]. Of a particular interest is the Al-Nb binary system investigated in [1], where materials with a grain size of less than 0.5 μm were used as billets.

A study on the effect of coarse-grained structure of billets on the microstructure and phase composition of an Al-Nb metal matrix composite obtained by shear deformation under pressure and post-deformation annealing was the aim of this work.

2. Experimental

Al (purity 99.1%) and Nb (purity 99.7%) billets with a coarse-grain structure in the form of thin disks were stacked in the order of Al-Nb-Al in a weight ratio of 22.5% Al to 77.5% Nb. The stacks of thin billets were placed between flat Bridgman anvils and processed under 5 GPa for 5 turns at a rate of 1 rpm at room temperature. Then the obtained samples were subjected to post-deformation annealing at 500°C for 30 minutes.

The phase composition of the samples was studied by X-ray diffraction (XRD) analysis using DRON 4-07 diffractometer with graphite crystal monochromator on the diffracted ray using Cu - Kα radiation.

IOP Conf. Series: Materials Science and Engineering 1008 (2020) 012065 doi:10.1088/1757-899X/1008/1/012065

Energy dispersive spectroscopy (EDS) and microstructure studies were carried out on a VEGA 3 SBH scanning electron microscope with a backscattered electron detector. The microhardness of the samples was determined by Vickers method using an Axiovert 100A optical microscope with an MNT-10 hardness measuring device. The static load applied to the diamond tip for 10 seconds was 100g.

3. Results and discussion

The results of X-ray diffraction analysis of the fabricated composite are shown in figure 1. After deformation, the diffraction maxima of aluminum and niobium and a diffraction maximum from the (002) plane of the intermetallic compound Al_3Nb were found on the diffraction pattern (figure 1a). Post-deformation annealing at 500°C led to the broadening of the maximum at $2\theta = 38.40$, which can be associated with the appearance of a diffraction maximum from the (112) plane of the intermetallic compound Al_3Nb (figure 1 b). The observed displacements of the diffraction maxima from Al and Nb after post-deformation annealing can be associated with the formation of a solid solution of niobium in aluminum and aluminum in niobium.

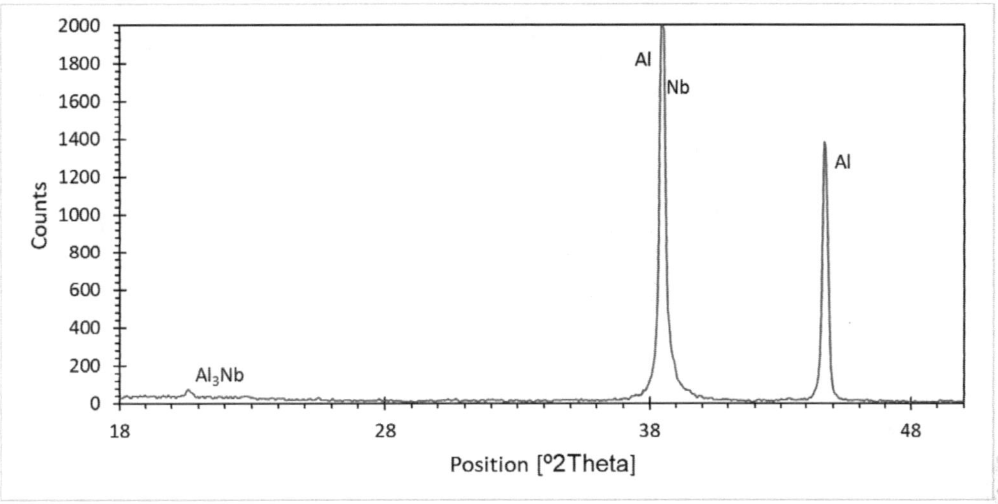

a)

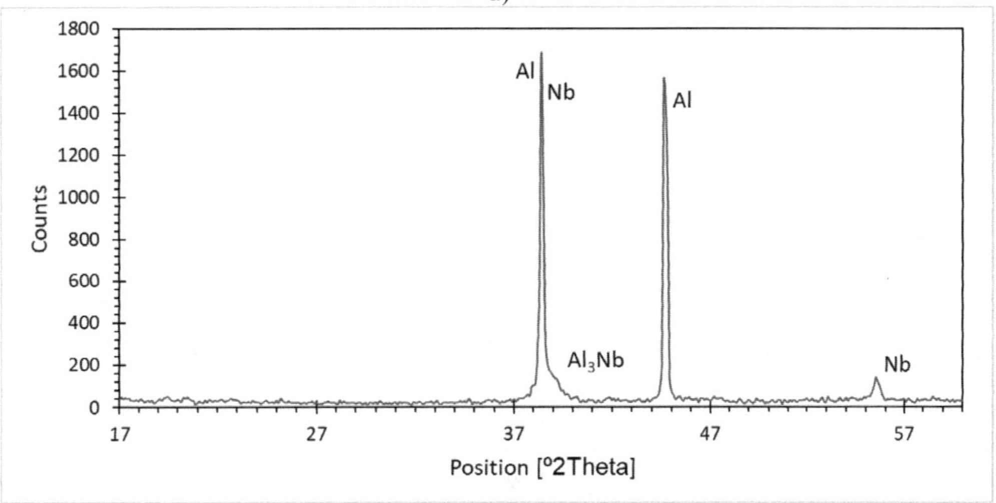

b)

Figure 1. X-ray diffraction profiles of the fabricated Al-Nb composite after shear deformation under pressure (a), after post-deformation annealing at 500°C for 30 minutes (b).

Energy dispersive analysis carried out after deformation of a three-layer stack of Al-Nb-Al showed that mixing of niobium and aluminium was not observed in the central region but it occurs at the edge of the sample (figure 2, table 1). Low mixing of the metals in the centre of the fabricated composite is associated with the inhomogeneity of the shear deformation along the radius of the sample.

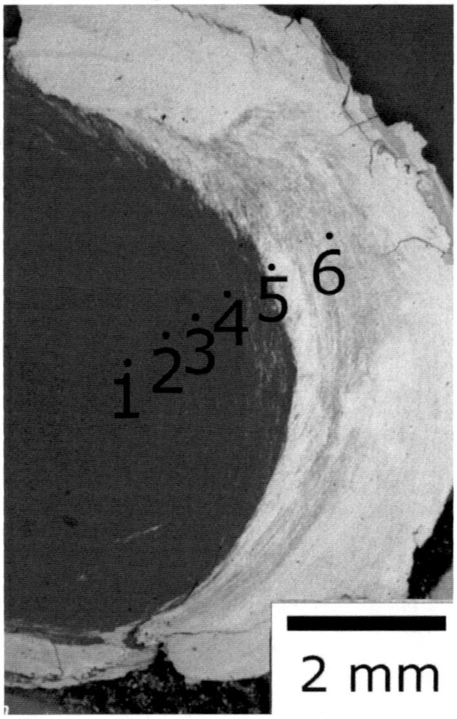

Figure 2. BSE image of the Al-Nb composite at a small magnification.

Table 1. Fractions of Al and Nb at the points of composite sample indicated in figure 2.

Point number	Al	Nb
1	100	0
2	100	0
3	100	0
4	100	0
5	25,2	74,8
6	26,2	73,8

Taking into account the X-ray diffraction analysis data (figure 1a), it can be assumed that the formation of the intermetallic phase begins in the process of cold plastic deformation.

The results of microhardness measurements on a fabricated composite after post-deformation annealing and aluminum after deformation for 5 turns are shown in figure 4. The microhardness of the composite is half that of deformed aluminium in the centre area, but more on the edge. It may be due to the formation of intermetallic compounds during deformation and subsequent annealing.

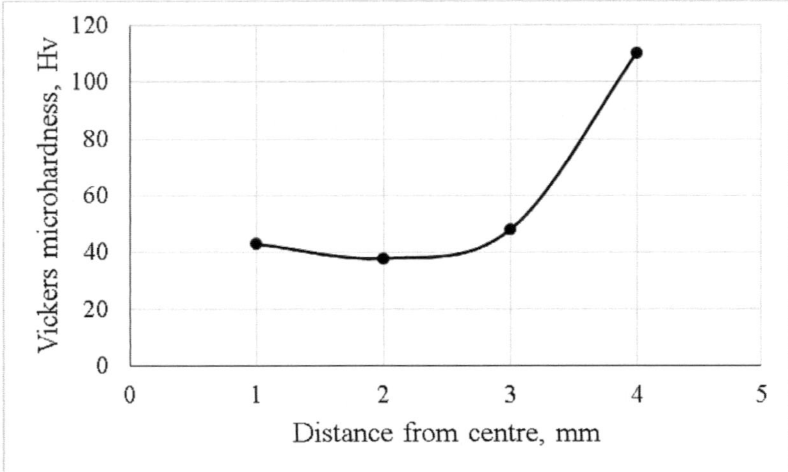

Figure 4. Vickers hardness variation for the Al-Nb composite after post-deformation annealing at 500°C for 30 minutes.

A similar phenomenon was observed in [1], where a similar experiment was carried out, but the initial Al and Nb had an ultrafine-grained structure. The microhardness of the Al-Nb composite after post-deformation annealing at 500°C for 30 minutes decreased [1]. According to X-ray diffraction data, the formation of intermetallic compounds in [1] after post-deformation annealing proceeds more intensively than in this work. In the cases of using billets with ultrafine-grained structure, a high-intensity diffraction maximum of an intermetallic compound is observed in comparison with the case of using blanks with a coarse-grained structure. Therefore, the structure of the starting materials influences the intensity of the phase formation process in the Al-Nb binary system.

This behaviour of microhardness is associated with phase formation processes both upon deformation and upon annealing. This phenomenon requires a further study.

4. Summary

The data of X-ray diffraction, energy dispersive spectroscopy and the results of microhardness measurements suggest that the formation of an intermetallic phase in a stack of thin coarse-grained aluminium and niobium billets, stacked in the order Al-Nb-Al begins during shear deformation under pressure via solid-phase reaction. Post-deformation annealing continues the phase formation process. During shear deformation under pressure of a stack of thin aluminium and niobium billets and post-deformation annealing in a solid solution of aluminium in niobium and niobium in aluminium, inclusions of the intermetallic compound Al_3Nb are formed and grow, i.e. the *in-situ* metal-matrix composite is formed. The intensity of the phase formation process depends on the structure of the initial billets. The number and rate of intermetallic inclusions Al_3Nb formation increases when using initial billets with an ultrafine-grained structure.

Acknowledgements

The present work was accomplished according to the state assignment of IMSP RAS and supported by the Russian Science Foundation (Grant No. 18-12-00440). Experimental studies were carried out on the facilities of shared services center of the Institute for Metals Superplasticity Problems of Russian Academy of Sciences «Structural and Physical-Mechanical Studies of Materials».

References

[1] Danilenko V N, Galushev S N and Mulyukov R R 2009 *Perspektivnuye Materialy* **7** 94 (in Russian)
[2] Ahn B, Zhilyaev A, Kawasaki M and Langdon TG 2015 *Mater. Sci. Eng. A* **635** 109

IOP Conf. Series: Materials Science and Engineering 1008 (2020) 012065 doi:10.1088/1757-899X/1008/1/012065

[3]. Castro M M, Wolf W, Isaac A, Kawasaki M and Figueiredo R B 2019 *Lett. Mater.* **9** 546
[4] Oh-ishi K, Edalati K, Kim H S, Hono K and Horita Z 2013 *Acta Mater.* **61(9)** 3482
[5] Danilenko V N, Korznikova G F, Zhilyaev A P, Sergeev S N, Khalikova G R, Khisamov R Kh, Nazarov K S, Kiekkuzhina L U and Mulyukov R R 2018 *IOP Conf. Ser.: Mater. Sci. Eng.* **447** 012021

Peierls-Nabarro potential for kinks in nonlinear chains

Yu V Bebikhov[1] and S V Dmitriev[2,3]

[1] North-Eastern Federal University, Polytechnic Institute (branch) in Mirny, 678170 Mirny, Sakha (Yakutia), Russia
[2] Institute for Metals Superplasticity Problems of RAS, 450001, Khalturin St. 39, Ufa, Russia
[3] Ufa State Petroleum Technological University, Kosmonavtov Str., 1, Ufa, 450062, Republic of Bashkortostan, Russia

E-mail: bebikhov.yura@mail.ru; dmitriev.sergey.v@gmail.com

Abstract. The aim of this work is to review various discrete models supporting topological solitons, in which the Peierls-Nabarro potential can be significantly lowered or even reduced to zero. These theoretical results are discussed in relation to the Peierls stresses for dislocations in a variety of crystals. Derivation of the discrete models free of the Peierls-Nabarro potential has been done by a number of authors with the use of analytical calculations. Peierls stresses for dislocations in crystals described in the literature have been estimated within the framework of molecular dynamics and *ab initio* simulations. These theoretical results are discussed in connection with the variability of the Peierls stress in different crystals.

1. Introduction

Localized nonlinear excitations in the form of topological solitons play an important role in many areas of physics and are very often studied in discrete media [1-5]. For example, in solid state physics, they describe domain walls, dislocations, and crowdions in crystals. In all these applications, the mobility of topological solitons is an important problem, given that the discreteness of the medium destroys the translational invariance of the continuum. In continuum systems topological solitons can move radiating no energy, while in discrete media, typically, static solitons cannot be placed arbitrarily with respect to the lattice because the lattice creates for the solitons a potential energy profile (called Peierls-Nabarro potential) with minima for particular high-symmetry configurations. Motion of the solitons in discrete media is possible only if they get enough energy to leave the minimum of the Peierls-Nabarro potential or, in other words, to overcome this potential.

From a physical point of view, the mobility of topological solitons can be analysed from two different points of view. On the one hand, one can estimate the minimum external force required to set a standing soliton in motion, and, on the other, determine the deceleration rate of a moving soliton in the absence of external forces. In the first case, the mobility is higher when the minimum external force is lower, and in the second case, it is higher when the deceleration rate is smaller.

In the theory of dislocations the mobility of dislocations is estimated by the Peierls stress – the value of share stress τ that is needed to overcome the Peierls-Nabarro potential [6]. In a recent study [7], the Peierls stresses have been estimated by a discretized Peierls-Nabarro model for a variety of crystals based on the results of *ab initio* simulations and compared to the experimentally measured values reported in [8]. Remarkably, the Peierls stress can differ noticeably for crystals belonging to the same group. For instance, for copper the ratio τ_P/G, where τ_P is the Peierls stress and G is the shear modulus, is five times smaller than for gold (note that both Cu and Au are metals with fcc lattice). Strength of interatomic bonds in crystals correlates with such parameters as melting temperature and heat of fusion. Both Cu and Au have very close melting temperatures of 1357 and 1337 K, respectively, and close values of heat of fusion, 13.3 and 12.6 kJ/mol, respectively. A natural question arises: why the height of the Peierls-Nabarro potential for two metals with the same crystal lattice and nearly same interatomic binding energy can be so different?

In order to address this question, we turn to the discussion of the Peierls-Nabarro potential in various nonlinear chains (simplified one-dimensional models of crystals). Note that the Frenkel-Kontorova chain was the first theoretical model used to study dislocations [1,2].

The aim of this article is to demonstrate that discreteness of media not always produces the Peierls-Nabarro potential, which is important for the discussion of dislocation mobility and mechanisms of plastic deformation of metals [24-31].

2. Exceptional discretizations of the φ^4 model

Existence of the integrable lattices [9], where topological solitons propagate freely with any velocity below the speed of sound, suggests that discreteness of media does not exclude mobility of solitons. However, the number of known integrable lattices is very limited. On the other hand, several wide classes of non-integrable discretizations of the Klein-Gordon field have been derived and analyzed, where static kinks do not experience Peierls-Nabarro potential and the kink can be placed anywhere with respect to the lattice [10-23]. The first such model has been derived by Speight and Ward [10-12]. Their model conserves total energy but does not conserve momentum. Then Kevrekidis has derived a new class of models free of the static Peierls-Nabarro potential that conserve momentum but not energy [13]. Rather general approach to derivation of translationally invariant discrete kinks, based on the discretized first integral approach, has been offered in [18,21]. Such discretizations were named as exceptional in the work [18].

Let us give one example of derivation of an exceptional discretization of Klein-Gordon equation based on the discretized first integral approach [21].

The continuum Klein-Gordon model has the Hamiltonian

$$H = \frac{1}{2}\int_{-\infty}^{\infty}[\varphi_t^2 + \varphi_x^2 + 2V(\varphi)]dx, \tag{1}$$

where $\varphi(x,t)$ is the unknown field and $V(\varphi)$ is the given on-site potential. The corresponding equation of motion has the form

$$\varphi_{tt} = \varphi_{xx} - V'(\varphi) \equiv D(\varphi(x,t)). \tag{2}$$

Equation (2) will be discretized with respect to the variable $x_n=hn$, where $n=0, \pm1, \pm2, ...,$ and $h>0$ is the lattice spacing. Let us write out the traditional discretization of equation (2),

$$\ddot{\varphi}_n = (1/h^2)(\varphi_{n-1} - 2\varphi_n + \varphi_{n+1}) - V'(\varphi_n), \tag{3}$$

in which kinks do experience the Peierls-Nabarro potential. A method for constructing discrete models with kinks free of Peierls-Nabarro potential, based on the use of a discretized first integral, was proposed in [21]. Following this method, consider the first integral of the static Eq. (2), $\varphi_x^2 - V(\varphi) + C = 0$, where C is the constant of integration. This first integral can be taken in a modified form

$$u(x) \equiv \varphi_x - \sqrt{2V(\varphi) - C} = 0. \tag{4}$$

Next, we rewrite the Hamiltonian Eq. (1) in terms of $u(x)$,

$$H = \frac{1}{2}\int_{-\infty}^{\infty}\left\{\varphi_t^2 + [u(x)]^2 + 2\varphi_x\sqrt{2V(\varphi) - C}\right\}dx. \tag{5}$$

The first integral Eq. (4) can be discretized as follows

$$u(\varphi_{n-1}, \varphi_n) \equiv \frac{\varphi_n - \varphi_{n-1}}{h} - \sqrt{2V(\varphi_{n-1}, \varphi_n) - C} = 0 \tag{6}$$

where we assume that in the continuum limit ($h \to 0$) $V(\varphi_{n-1}, \varphi_n) \to V(\varphi)$. Using Eq. (6), we obtain a discrete version of the Hamiltonian of Eq. (1)

$$H=\frac{1}{2}\sum_n\left\{\begin{array}{l}\dot{\varphi}_n^2+\left[u(h,\varphi_{n-1},\varphi_n)\right]^2\\+2\dfrac{\varphi_n-\varphi_{n-1}}{h}\sqrt{2V(\varphi_{n-1},\varphi_n)-C}\end{array}\right\} \tag{7}$$

If the potential is discretized as suggested in [10], namely,

$$\sqrt{2V(\varphi_{n-1},\varphi_n)-C}=\frac{G(\varphi_n)-G(\varphi_{n-1})}{\varphi_n-\varphi_{n-1}},\quad \text{where}\quad G'(\varphi)=\sqrt{2V(\varphi)-C}, \tag{8}$$

then the second term in curly brackets in Eq. (7) will vanish as a result of telescopic summation. Under these conditions, the Hamiltonian takes the form

$$H=\frac{1}{2}\sum_n\left\{\dot{\varphi}_n^2+\left[u(h,\varphi_{n-1},\varphi_n)\right]^2\right\}.$$

Further, using Eq. (8), we rewrite Eq. (6) in the form

$$u(h,\varphi_{n-1},\varphi_n)\equiv\frac{\varphi_n-\varphi_{n-1}}{h}-\frac{G(\varphi_n)-G(\varphi_{n-1})}{\varphi_n-\varphi_{n-1}}. \tag{9}$$

Then the equation of motion is written in the form

$$\ddot{\varphi}_n=-u(h,\varphi_{n-1},\varphi_n)\frac{\partial}{\partial\varphi_n}u(h,\varphi_{n-1},\varphi_n)-u(h,\varphi_n,\varphi_{n+1})\frac{\partial}{\partial\varphi_n}u(h,\varphi_n,\varphi_{n+1}). \tag{10}$$

Equilibrium static solutions of this model can be found from the two-point map $u(h, \varphi_n, \varphi_{n-1})=0$, where the function $u(h, \varphi_n, \varphi_{n-1})$ is defined by Eq. (9). Such solutions can be constructed iteratively, starting from any admissible value φ_{n-1} or φ_n, and therefore the Peierls-Nabarro potential is absent.

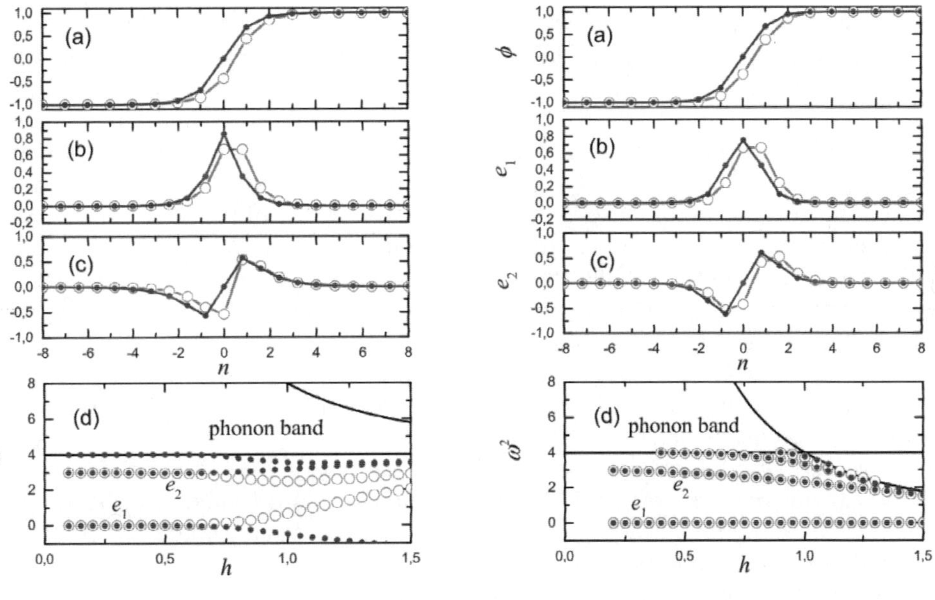

Figure 1. Left: Standard discretization of the φ^4 field Eq. (3): (a) on-site (dots) and inter-site (circles) kinks, (b) Goldstone mode localized on the kink, (c) kink's internal vibrational mode for h=0.8. (d) Frequencies of the kink's internal vibrational modes as the functions of the lattice spacing h. Solid lines give the upper and lower edges of the spectrum of vacuum. In all panels the results for the on-site (inter-site) kink are given by dots (circles). Right: Same as in figure 1, but for the exceptional discretization of the φ^4 field Eq. (10).

3. Results and discussion

Let us discuss some properties of the kinks in the exceptional discretization Eq. (10) in comparison with the standard discretization Eq. (3) taking as an example the φ^4 model with the on-site potential

$$V(\varphi)=\frac{1}{2}\left(1-\varphi^2\right)^2. \tag{11}$$

This is a two-well potential with the minima at $\varphi=\pm1$ supporting the kink solution interpolating between these two minima.

Static kink solution for the standard discrete model can be found numerically with the use, for example, of the conjugate gradient method. There exist two static kink solutions, one centered on a lattice site and another one centered on the middle of a bond. One of such configurations has greater energy and it is unstable, while another one has minimal energy and it is stable. Actually difference between energies of these two high-symmetry states defines the height of the Peierls-Nabarro potential.

Exact static kink solution for the exceptional discretization, as mentioned above, can be found iteratively from the two-point map $u(h, \varphi_n, \varphi_{n-1})=0$, where the function $u(h, \varphi_n, \varphi_{n-1})$ is defined by Eq. (9), starting from any admissible value $-1<\varphi_{n-1}<1$ or $-1<\varphi_n<1$. The fact that the initial value of φ_n can be chosen arbitrarily implies that the static kink can be placed arbitrarily with respect to the lattice, and all such configurations have precisely the same energy.

Most of the modes are the extended phonon modes with frequencies belonging to the spectrum of vacuum solution $\varphi=\pm1$. A few modes have frequencies below the phonon band, those are the modes localized on the kink.

In figure 1, panels at left, for the standard discretization Eq. (3) with the discreteness parameter h=0.8, we present (a) the profiles of the static on-site and inter-site kinks, (b,c) the two eigenmodes localized on the kink and (d) the eigenfrequencies as the functions of the discreteness parameter h. The eigenmodes form the phonon band with the borders shown by the black solid lines, with the additional frequencies of the modes localized on the kink. Results for the on-site (inter-site) kink are shown by dots (circles). In figure 2, panels at right, the same is given for the exceptional discretization Eq. (10).

Comparison of the left and right panels of figure 1 reveals the difference between the standard and exceptional discretizations of the φ^4 field. The kink profiles [panels (a)] and the eigenmodes in the two models [panels (b) and (c)] are very similar. On the other hand, vibrational spectrums of the two models differ noticeably. In figure 1, left panels, the width of the phonon band decreases monotonically with increasing degree of discreteness h. In figure 2, right panels, the upper and lower edges of the phonon band cross at h=1. For this particular value of h the model becomes purely anharmonic. The most important difference is the variation with h of the lowest frequency mode e_1. In figure 1(d), left panels, one can see that for h<0.5 frequency of the mode e_1 is nearly zero. For larger values of h the frequency of the stable inter-site kink configuration increases with h, while frequency of the unstable on-site configuration becomes imaginary ($\omega^2<0$).

In figure 2(d), right panels, frequency of the e_1 mode for both on-site and inter-site kinks is precisely zero for any h. Actually, it is zero not only for the two symmetric but also for any asymmetric kink configuration. This reflects the absence of the Peierls-Nabarro potential in the exceptionally discretized model.

The difference in the behavior of the kinks in the standard and exceptional discretizations can be revealed by applying external force f. The results of numerical simulation of the kink dynamics under external force f are presented in figure 2 (a,b) for the standard discretization Eq. (3) and in (c) for the exceptional discretization Eq. (10).

In the case of standard discretization, inter-site kink is initially at rest and at t=0 the external force starts to grow with time linearly, as shown in (a). In (b), the kink kinetic energy is given as the function of time. It can be seen that before the force f reaches certain value, the kink is at rest since it is trapped by the Peierls-Nabarro potential. For h=0.7 the depth of the Peierls-Nabarro potential is smaller than in the case of h=0.8, that if why, in the latter case the kink starts to move at a greater value of the external force.

In the case of exceptional discretization at t=0 constant in time force f is turned on. Three values of the force are considered, $f=10^{-5}$, 10^{-4} and 10^{-3}. Kink kinetic energy as the function of time is shown in figure 2(c). It can be seen that the kink starts to move immediately after the force is turned on even though the values of the external force are very small as compared to the force applied to the kink in the standard discrete model, see figure 2(a). Note that the dotted line in figure 2(c) shows the slope equal to 2 (log-log scale is used here). Kink kinetic energy increases quadratically in time, which means that the kinks under constant external force demonstrate uniformly accelerated motion.

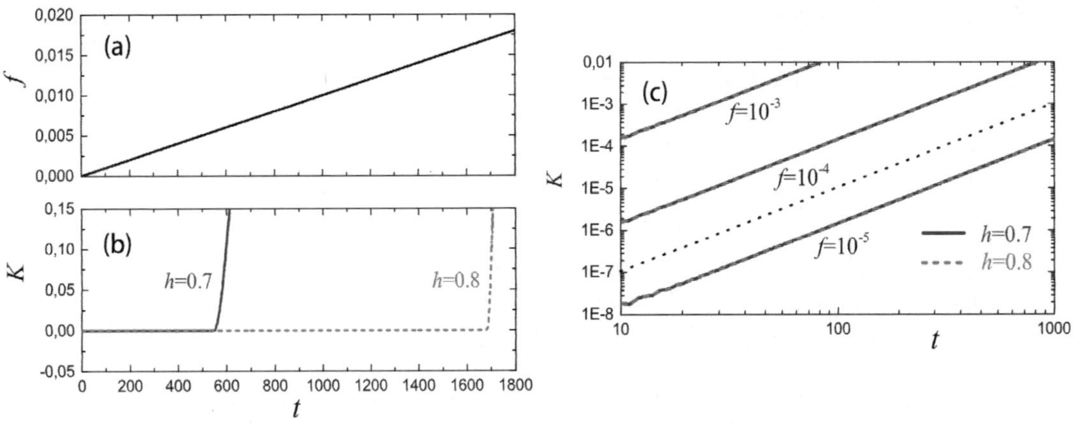

Figure 2. (a,b) Results for the standard discretization of the φ^4 field Eq. (3). (c) Results for the exceptional discretization Eq. (10). (a) External force as a function of time. (b) Kink kinetic energy as the function of time. Results for h=0.7 (blue solid line) and h=0.8 (red dashed line). (c) Kink kinetic energy as a function of time for constant in time external force f, as specified for each curve. The discreteness parameter is h= 0.7 (blue solid line) and h= 0.8 (red dashed line). The dotted line shows the slope equal to 2.

4. Conclusions

Theoretical models presented in Section. 2 suggest that the discreteness of media does not necessarily produce a Peierls-Nabarro potential for topological solitons. Several wide classes of non-integrable discretizations of the Klein-Gordon field have been offered [10-23] in addition to a few integrable ones [9]. In such exceptional discretizations kinks are accelerated by any small external force. In terms of the dislocation theory, the Peierls stress is exactly zero for kinks in such models. The reason for vanishing of the Peierls-Nabarro potential in exceptionally discrete models is the many-body interactions. Indeed, in the classical discrete model Eq. (3), the particles interact via pairwise bonds, while in the discretization Eq. (10), one has three-body interactions.

Our main conclusion in the light of the theoretical results reviewed here is that in a discrete media with many-body interactions Peierls-Nabarro potential can be small, and indeed, the Peierls stress can differ noticeably for similar crystals [7,8]. This conclusion can be of interest in the theory of dislocations.

A final note is that discrete breathers and other localized phenomena can also experience the Peierls-Nabarro potential [32-40].

Acknowledgments
This work was supported by the State Assignment of IMSP RAS No. AAAA-A17-117041310220-8.

Reference

[1] *The Sine-Gordon Model and its Applications: From Pendula and Josephson Junctions to Gravity and High-Energy Physics* 2014 ed. J Cuevas-Maraver, P G Kevrekidis and F Williams (Heidelberg: Springer-Verlag)

[2] Braun O M and Kivshar Yu S 2004 *The Frenkel-Kontorova Model: Concepts, Methods, and Applications* (Berlin: Springer-Verlag)

[3] Archilla J F R, Kosevich Yu A, Jimenez N, Sanchez-Morcillo V J and Garcia-Raffi L M 2015 *Phys. Rev.* E **91** 022912

[4] Dmitriev S V, Medvedev N N, Chetverikov A P, Zhou K and Velarde M G 2017 *Phys. Status Solidi RRL* **11** 1700298

[5] Sergeev K S, Dmitriev S V, Korznikova E A and Chetverikov A P 2018 *Nelineinaya Dinamika* **14** 195
[6] Hirth J P and Lothe J 1982 *Theory of Dislocations*, 2nd ed. (New York: Wiley)
[7] Edagawa K, Kamimura Y, Iskandarov A M, Umeno Y and Takeuchi S 2019 *Materialia* **5** 100218
[8] Kamimura Y, Edagawa K and Takeuchi S 2013 *Acta Mater.* **61** 294
[9] Orfanidis S J 1978 *Phys. Rev.* D **18** 3822
[10] Speight J M and Ward R S 1994 *Nonlinearity* **7** 475
[11] Speight J M 1997 *Nonlinearity* **10** 1615
[12] Speight J M 1999 *Nonlinearity* **12** 1373
[13] Kevrekidis P G 2003 *Physica* D **183** 68
[14] Bender C M and Tovbis A 1997 *J. Math. Phys.* **38** 3700
[15] Dmitriev S V, Kevrekidis P G and Yoshikawa N 2005 *J. Phys.* A **38** 7617
[16] Roy I, Dmitriev S V, Kevrekidis P G and Saxena A2007 *Phys. Rev.* E **76** 026601
[17] Cooper F, Khare A, Mihaila B and Saxena A 2005 *Phys. Rev.* E **72** 036605
[18] Barashenkov I V, Oxtoby O F and Pelinovsky D E 2005 *Phys. Rev.* E **72** 35602
[19] Dmitriev S V, Kevrekidis P G and Yoshikawa N 2006 *J. Phys.* A **39** 7217
[20] Oxtoby O F, Pelinovsky D E and Barashenkov I V 2006 *Nonlinearity* **19** 217
[21] Dmitriev S V, Kevrekidis P G, Yoshikawa N and Frantzeskakis D J 2006 *Phys. Rev.* E **74** 046609
[22] Speight J M and Zolotaryuk Y 2006 *Nonlinearity* **19** 1365
[23] Dmitriev S V, Kevrekidis P G, Khare A and Saxena A 2007 *J. Phys.* A **40** 6267
[24] Xu S, Su Y, Smith L T W and Beyerlein I J 2020 *J. Mech. Phys. Solids* **141** 104017
[25] Zhang M, Yang S, Zhu B, Liu G, Ding J, Zhan Q, Shang C and Wan F 2020 *Solid State Commun.* **314** 113917
[26] Xu S, Mianroodi J R, Hunter A, Svendsen B and Beyerlein I J 2020 *Int. J. Plasticity* **129** 102689
[27] Shang S-L, Shimanek J, Qin S, Wang Y, Beese A M and Liu Z-K 2020 *Phys. Rev.* B **101** 024102
[28] Jozić S, Dumanić I and Bajić D 2020 *Facta Universitatis, Series: Mechanical Engineering* **18** 13
[29] Zhang P, Bian J-J, Yang C, Zhang J-Y, Liu G, Weiss J and Sun J 2019 *Materialia* **7** 100416
[30] Avtokratova E, Sitdikov O, Latypova O and Markushev M 2020 *Facta Universitatis, Series: Mechanical Engineering* **18** 255
[31] Lam N D 2020 *Current Nanoscience* **16** 86
[32] Semenov A S, Murzaev R T, Bebikhov Y V, Kudreyko A A and Dmitriev S V 2020 *Lett. Mater.* **10** 185
[33] Murzaev R T, Babicheva R I, Zhou K, Korznikova E A, Fomin S Y, Dubinko V I and Dmitriev S V 2016 *Eur. Phys. J.* B **89** 168
[34] Zakharov P V, Korznikova E A, Dmitriev S V, Ekomasov E G and Zhou K 2019 *Surf. Sci.* **679** 1
[35] Korznikova E A, Bachurin D V, Fomin S Y, Chetverikov A P and Dmitriev S V 2017 *Eur. Phys. J.* B **90** 1-8
[36] Barani E, Korznikova E A, Chetverikov A P, Zhou K and Dmitriev S V 2017 *Phys. Lett.* A **381** 3553
[37] Marjaneh A M, Saadatmand D, Evazzade I, Babicheva R I, Soboleva E, Srikanth N, Zhou K, Korznikova E and Dmitriev S 2018 *Phys. Rev.* E **98** 023003
[38] Dmitriev S, Korznikova E and Chetverikov A 2018 *J. Exp. Theor. Phys.* **126** 347
[39] Babicheva R I, Evazzade I, Korznikova E A, Shepelev I A, Zhou K and Dmitriev S V 2019 *Comp. Mater. Sci.* **163** 248
[40] Shepelev I A, Chetverikov A P, Dmitriev S V and Korznikova E A 2020 *Comp. Mater. Sci.* **177** 109549

Dynamics of supersonic *N*-crowdions in bcc tungsten

R T Murzaev[1], E A Korznikova[1,2] and S V Dmitriev[1,2]

[1] Institute for Metals Superplasticity Problems of Russian Academy of Sciences, 39 Khalturin St., 450001 Ufa, Russia
[2] Institute of Molecule and Crystal Physics, Ufa Federal Research Centre of the Russian Academy of Sciences, 71 October Avenue, 450054 Ufa, Russia

E-mail: murzaevrt@gmail.com

Abstract. An interstitial atom placed in a close-packed atomic row of a crystal is called a crowdion. Such defects are highly mobile; they can move along the close-packed row at subsonic and even supersonic speeds, transferring mass and energy. In this paper, we consider the initial conditions for excitation of *N*-crowdions, in which *N*-atoms move simultaneously with high speed. The most efficient methods for initiating supersonic *N*-crowdions with N=2 and 4 are demonstrated using molecular dynamics modelling for bcc tungsten.

1. Introduction

Migration of point defects in metals contributes to the mass and energy transfer during plastic deformation, heat treatment, irradiation, etc. Interstitial atoms, as compared to vacancies, have a higher energy but much lower mobility [1-4], with the exception of crowdions – interstitials embedded in a close-packed atomic rows [5], that are highly mobile. Crowdions can propagate along its close-packed atomic row transferring mass of an atom with subsonic or even supersonic speed [6-10].

Moving crowdions are difficult to identify experimentally, that is why, molecular dynamics is a very popular and reliable method of their investigation. This method proved to be an effective approach in analysis of surface voidions [11], discrete breathers [12,13] and recently discovered 2D nanomaterials [14-17]. Molecular dynamics is a suitable tool for the study of crowdions because they have nanometer size, corresponding to several interatomic distances, and they travel fast, so that the computational cell and the simulation time can be relatively small. Since crowdions move in a close packed atomic rows, their motion can be considered as quasi one-dimensional and can be described by the chain models, such as the Frenkel-Kontorova model [18,19]. In the frames of this model it was shown that crowdions initiated by a molecule impact on the surface have smaller energy threshold required for crowdion initiation in comparison to the initiation with single atoms [19]. This effect has been confirmed for Pt crystal, where the crowdion was found to emerge at lower energies and propagate longer distances, when it was excited by a molecule impact rather than by single atom impact [20]. The results of the work [21-23] show that clusters larger than three to four self interstitial atoms in Cu and Fe can demonstrate one-dimensional motion. Moving subsonic crowdions can bear a vibrational mode with the frequency lying well above the phonon spectrum [9].

In molecular dynamics simulations, crowdion dynamics is determined by the initial conditions. In this work, we study the effect of initial conditions used for crowdion initiation in the crystal structure of bcc W. Note that supersonic crowdions in bcc tungsten are poorly studied.

2. Simulation setup

A three-dimensional bcc lattice of tungsten is considered, the equilibrium interatomic distance of which is x_0 = 3.16 Å. As shown in figure 1, the x axis is directed along the close-packed atomic rows parallel to the [111] direction. For modelling, the EAM interatomic potential for tungsten is used, which is included in the molecular dynamics software package LAMMPS. The size of the computational cell is 543.7×51.3×44.4 Å and it included 80,000 atoms. Periodic boundary conditions were used along all three axes.

To launch a crowdion, the following initial conditions are used: the equal initial velocity V_x is given to two or four neighbouring atoms in a close packed row parallel to the x-axis, as shown in figure 1 (a) and (b), respectively. The initial displacements and velocities of all other atoms in the structure are equal to zero. Thus, the total energy of the system is equal to the initial kinetic energy of excited atoms, $E=NMV_x^2/2$, where N is the number of excited atoms and M is the atom mass. As a result of application of such initial conditions for sufficiently large V_x, supersonic 2- or 4-crowdions are formed. In our simulations we take different values of V_x and the dynamics of supersonic 2- and 4-crowdions.

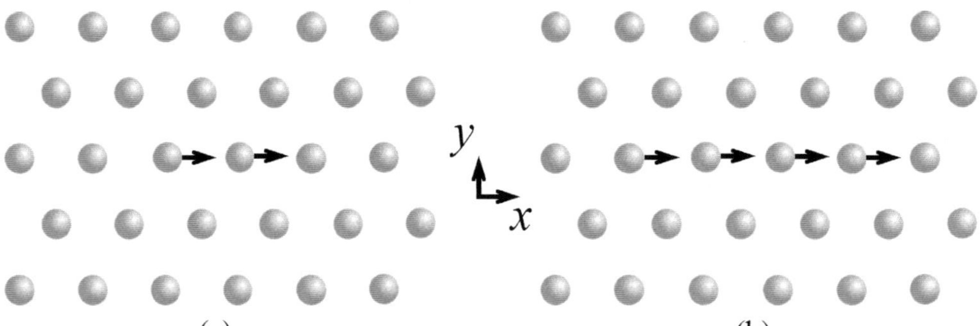

Figure 1. Initial conditions used for excitations of supersonic (a) 2- and (b) 4-crowdions. Equal initial velocity V_x is given to two or four neighbouring atoms in a close packed row parallel to the x-axis.

3. Simulation results

In figure 2, as the function of time, we plot normalized displacements of atoms in the close-packed atomic row where the supersonic N-crowdion moves. The left column is for 2-crowdion and the right column for 4-crowdion. Initial velocities of the atoms are from V_x=40 to 70 Å/ps, as indicated for each of the four rows. The displacements of the atoms along x-axis, dx, are normalized to the equilibrium interatomic distance, x_0.

In the first row of figure 2 the results for V_x=40 Å/ps are presented. For the 2-crowdion (left column) the two first atoms loose their initial kinetic energy at dx/x_0=0.8, so that they do not reach the position of the neighboring atom moving with the high speed and they reach this positin dfifting with a small velocity. The supersonic 2-crowdion rapidly transforms into a subsonic one after travelling about a dozen of interatomic distances. The 4-crowdion (right panel) lounched with V_x=40 Å/ps has two times greater initial energy. The atoms almost reach the position of the neighboring atoms and supersonic motion is observed during the whole simulation run (1 ps). Note that the supersonic 2-crowdion carries one interstitial atom, while supersonic 4-crowdion carries 2 interstitials. One can say that the 4-crowdion transports mass more efficiently because with the double energy it carries double mass over longer distance.

In the second row of figure 2 the results for V_x=50 Å/ps are plotted. For the 2-crowdion the atoms loose their initial momentum still a bit before reaching the positions of the neighboring atoms. On the other hand, for the 4-crowdion one can see that the atoms stop nearly at the positions of the nearest atoms, i.e., at dx/x_0=1. In the third row of figure 2 we give the results for V_x=60 Å/ps. This initial velocity is sufficient for the 2-crowdion to shift the atoms to the positions of the neighboring atoms. For the 4-crowdions the atoms overshoot these positions. Finally, for V_x=70 Å/ps (see the last row of figure 2) the initial velocity is big enough for both 2- and 4-crowdions to shift the atoms to the positions beyond the energy minimu.

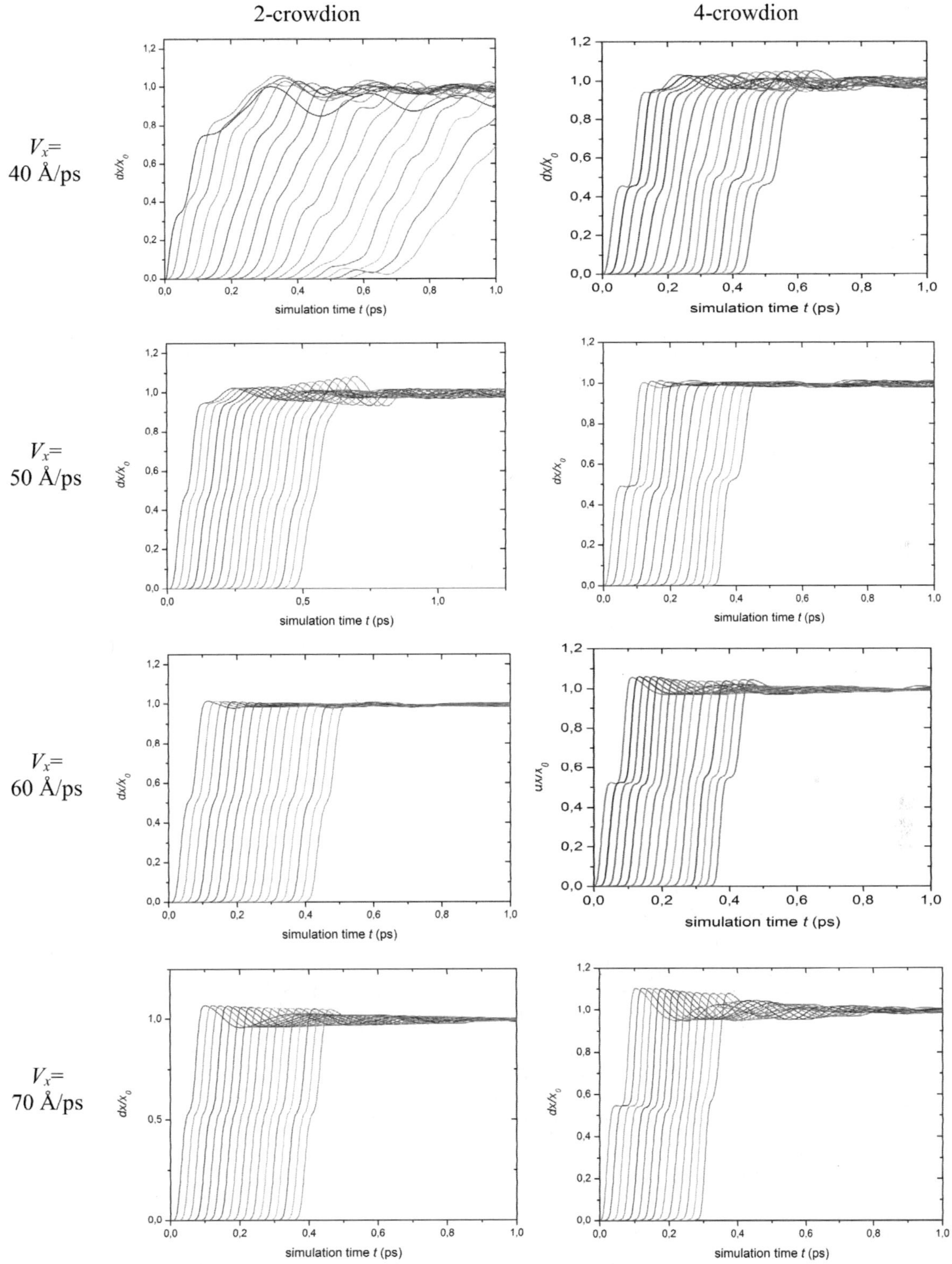

Figure 2. Normalized displacements of atoms, in the close-packed atomic row where the supersonic *N*-crowdion moves, as the functions of time. The left column is for 2-crowdion and the right column for 4-crowdion. Initial velocities of the atoms are from V_x=40 to 70 Å/ps, as indicated for each of the four rows.

IOP Conf. Series: Materials Science and Engineering 1008 (2020) 012067 doi:10.1088/1757-899X/1008/1/012067

4. Conclusions

According to the results of our simulations, the supersonic 4-crowdion transports mass more efficiently than the supersonic 2-crowdion because with the double initial energy it carries double mass over longer distance. The initial velocity of V_x=60 Å/ps is sufficient for the 2-crowdion to shift the atoms to the positions of neighbouring atoms. For the 4-crowdion, V_x=50 Å/ps is sufficient.

Acknowledgments
This work was supported by the State Assignment of IMSP RAS No. AAAA-A17-117041310220-8.

Reference

[1] Nordlund K 2019 *J. Nucl. Mater.* **520** 273
[2] Bjorkas C, Nordlund K and Dudarev S 2009 *Nucl. Instrum. Meth. B* **267** 3204
[3] Sand A E, Nordlund K and Dudarev S 2014 *J. Nucl. Mater.* **455** 207
[4] Dudarev S L, Mason D R, Tarleton E, Ma P-W and Sand A E *Nucl. Fusion* **58** 126002
[5] Paneth H R 1950 *Phys. Rev.* **80** 708
[6] Chetverikov A, Shepelev I, Korznikova E, Kistanov A, Dmitriev S and Velarde M G 2017 *Compu. Condens. Matter* **13** 59
[7] Korznikova E, Shepelev I, Chetverikov A, Dmitriev S, Fomin S Y and Zhou K 2018 *J. Exp. Theor. Phys.* **127** 1009
[8] Korznikova E, Shepelev I, Chetverikov A, Fomin S Y and Dmitriev S 2018 *IOP Conf. Series: Mater. Sci. Eng* **447** 012030
[9] Shepelev I, Korznikova E A, Bachurin D, Semenov A S, Chetverikov A P and Dmitriev S V 2020 *Phys. Lett. A* **384** 126032
[10] Dmitriev S V, Korznikova E A and Chetverikov A P 2018 *J. Exp. Theor. Phys.* **126** 347
[11] Mazilova T, Sadanov E, Voyevodin V, Ksenofontov V and Mikhailovskij I 2018 *Surf. Sci.* **669** 10
[12] Zakharov P V, Korznikova E A, Dmitriev S V, Ekomasov E G and Zhou K 2019 *Surf. Sci.* **679** 1
[13] Murzaev R T, Babicheva R I, Zhou K, Korznikova E A, Fomin S Y, Dubinko V I and Dmitriev S V 2016 *Eur. Phys. J. B* **89** 168
[14] Liu B, Bai L, Korznikova E A, Dmitriev S V, Law A W-K and Zhou K 2017 *J. Phys. Chem. C* **121** 13876
[15] Savin A V, Korznikova E A and Dmitriev S V 2015 *Phys. Rev. B* **92** 035412
[16] Bai L, Srikanth N, Korznikova E A, Baimova J A, Dmitriev S V and Zhou K *Wear* **372** 12
[17] Korznikova E A, Shcherbinin S A, Ryabov D S, Chechin G M, Ekomasov E G, Barani E, Zhou K and Dmitriev S V 2019 *Phys. Status Solidi B* **256** 1800061
[18] Dudarev S 2003 *Philos. Mag.* **83** 3577
[19] Marjaneh A M, Saadatmand D, Evazzade I, Babicheva R I, Soboleva E G, Srikanth N, Zhou K, Korznikova E A and Dmitriev S V 2018 *Phys. Rev. E* **98** 023003
[20] Babicheva R I, Evazzade I, Korznikova E A, Shepelev I A, Zhou K and Dmitriev S V 2019 *Comp. Mater. Sci.* **163** 248
[21] Osetsky Y N, Bacon D, Serra A, Singh B and Golubov S 2003 *Philos. Mag.* **83** 61
[22] Korznikova E A, Sunagatova I R, Bayazitov A M, Semenov A S and Dmitirev S V 2019 *Lett. Mater.* **9** 386
[23] Zorya I V, Poletaev G M, Rakitin R Y, Ilyina M A and Starostenkov M D 2019 *Lett. Mater.* **9** 207

Interaction of supersonic 2-crowdions in fcc platinum

A M Bayazitov[1], E A Korznikova[1,2,3], A S Semenov[4], D I Borisov[5] and S V Dmitriev[1,2]

[1] Institute of Molecule and Crystal Physics, the Ufa Federal Research Centre of the Russian Academy of Sciences, 71 October Avenue, 450054 Ufa, Russia
[2] Institute for Metals Superplasticity Problems of the Russian Academy of Sciences, 39 Khalturin St., 450001 Ufa, Russia
[3] Ufa State Aviation Technical University, 12 K. Marx St., 450008 Ufa, Russia
[4] North-Eastern Federal University, Polytechnic Institute (branch) in Mirny, 678170 Mirny, Sakha (Yakutia), Russia
[5] Institute of Mathematics with Computing Centre, the Ufa Federal Research Centre of RAS, 112 Chernyshevsky St., 450008 Ufa, Russia

E-mail: ayrseptember@mail.ru

Abstract. In metals subjected to ion implantation or irradiation, mass transfer by supersonic crowdion clusters takes place. The interaction between supersonic 2-crowdions moving in parallel close-packed atomic rows has been studied only in 2D Morse lattice and here it is analysed with the use of molecular dynamics simulations in fcc platinum.

1. Introduction

It is widely accepted that point defect play an important role in the diffusion in crystal lattices thus contributing to the structure evolution upon the external impacts [1]. Among different types of zero-dimension defects present in the material, one can distinguish vacancies [2] or voidions [3] and interstitials in the form of atoms embedded in an octahedral or tetrahedral pore [4] or crowdions [5–9]. Crowdions are much more mobile than conventional interstitial defects, however having higher formation enthalpy. These facts allow to conclude that crowdions play an important role in mass transfer in crystals during extreme impacts. Recently a phenomenon of supersonic crowdion has been extended to supersonic *N*-crowdion where several (*N*) atoms move at a high speed simultaneously in a close packed row and this type of motion was found to be much more effective than the single crowdion propagation [10].

Features of supersonic crowdion motion have been described in different types of crystals [5,6,11-13]. However the scenarios of crowdion interaction remain uncovered at the moment. The aim of the present work is to analyze peculiarities of interaction of supersonic 2-crowdions in fcc platinum crystal by means of molecular dynamics simulations.

2. Simulation details

Since crowdion motion is studied at the atomic scale, molecular dynamics is an effective tool for this study. Previously this approach, based in the solving of Newtonian equations of motion, has recommended itself as a powerful method allowing to analyze nonlinear dynamics of crystal lattices

[14–20], phase transformations [21], evolution of atomic structure and mechanical properties upon external loading [22–25] and many other features.

At the initial stage, the fcc lattice of Pt was simulated with the lattice parameter a = 3.920 Å and the interatomic distance $d=a/\sqrt{2}$=2.772 Å. The x and y axes of the Cartesian coordinate system are oriented along the close-packed directions {110} and {$\bar{1}$10}, and the z axis is oriented along the {001} direction (see Fig. 1). The computational cell includes 30, 8 and 8 cubic translational cells in the x, y and z direction, respectively, and each cell contains 4 atoms. The total number of atoms is 15360. Periodic boundary conditions are used. Integration time step is 0,2 fs.

We have studied interaction between three supersonic 2-crowdions. Single 2-crowdion is excited by giving same initial velocity to two neighbouring atoms in a close packed row along this row (along the x-axis). The initial velocities of all other atoms and the initial displacements of all atoms in the computational cell are equal to zero. Thermal vibrations were not introduced, i.e. the simulation was carried out at a temperature of 0 K. Thus, the total energy of the system is equal to the kinetic energy of excited atoms at t=0. The atomic mass of Pt is 195.084 amu. The simulation was carried out using the LAMMPS software package and the many-body interatomic potential [26].

The following configurations of three 2-crowdions were investigated: (i) vertical parallel to the xz plane, (ii) horizontal parallel to the xy plane, and (iii) diagonal (see figure 2).

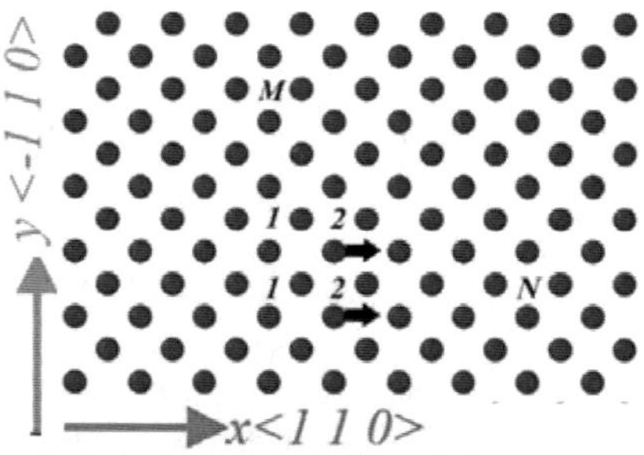

Figure 1. Atoms of the fcc Pt crystal projected onto the xy plane, where the axes of the Cartesian coordinate system x, y and z are directed along the crystallographic directions <110>, <-110> and <001>, respectively.

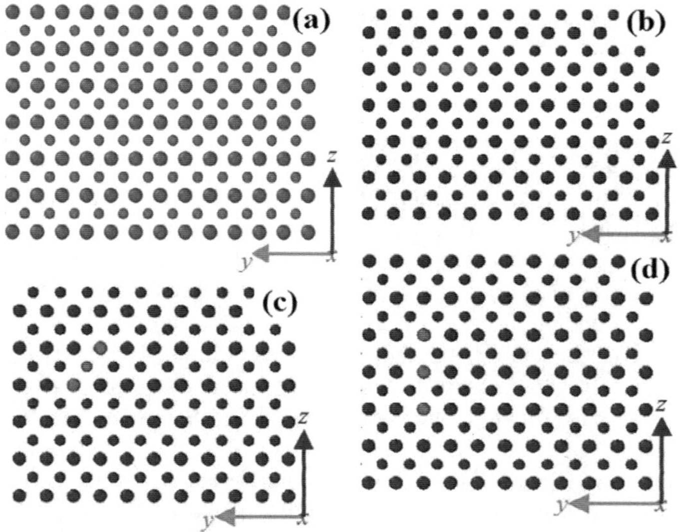

Figure 2. In (a), the atoms located on the front plane are highlighted in blue, while red atoms are on the second plane. Three investigated configurations of clusters of supersonic 2-crowdions: (b) horizontal, (c) diagonal and (d) vertical.

3. Simulation results

In the course of molecular dynamics modeling of three supersonic 2-crowdions with an initial velocity of 60 Å/ps, it was revealed that the vertical configuration shows itself to be the most stable in comparison with the horizontal and diagonal configurations (see figure 3). The degree of instability depends on the distance between close-packed rows, where 2-crowdions are launched. In the case of the horizontal and diagonal configuration, the distance $S_d = S_h = d$ = 2.772 Å is less than in the case of the vertical configuration $S_v = a = 3.92$ Å. Therefore, in horizontal and diagonal configurations the interaction of adjacent 2-crowdions is expected to be greater than in the vertical one. The diagonal configuration has a violation of symmetry, which is an additional factor affecting the dynamics of this cluster of 2-crowdions.

It should be noted that, with a diagonal configuration, the appearance of an additional crowdion is observed at a 450 time step, which is explained by the recombination of atoms in the cone-shaped tail of 2-crowdion clusters.

Figure 4 demonstrates atomic displacements as the functions of time for the atoms in the central close-packed row. The displacements satisfy the self-focusing condition [10] in all three cases. At the same time, it can be observed that the displacements in the horizontal and diagonal configurations intersect. This indicates a tendency to the formation of an additional vacancy, which also indicates an unstable dynamics of the three supersonic 2-crowdion in horizontal and diagonal configurations.

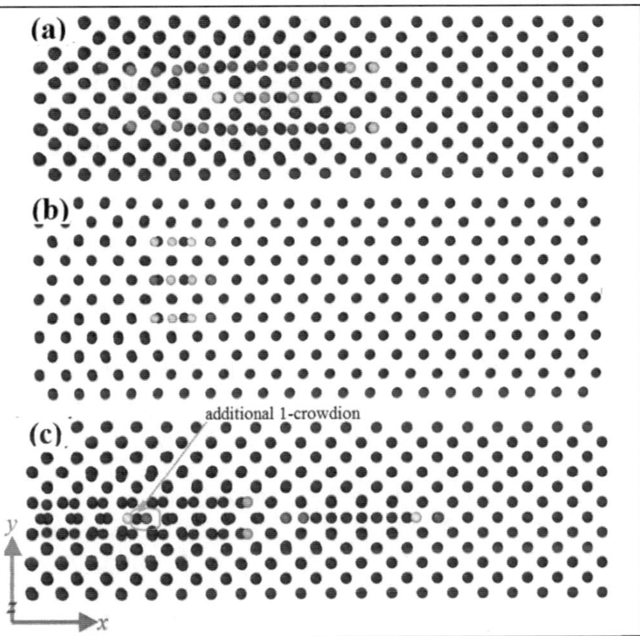

Figure 3. Dynamics of the studied configurations of clusters of supersonic 2-crowdions: (a) horizontal, (b) vertical and (c) diagonal configuration.

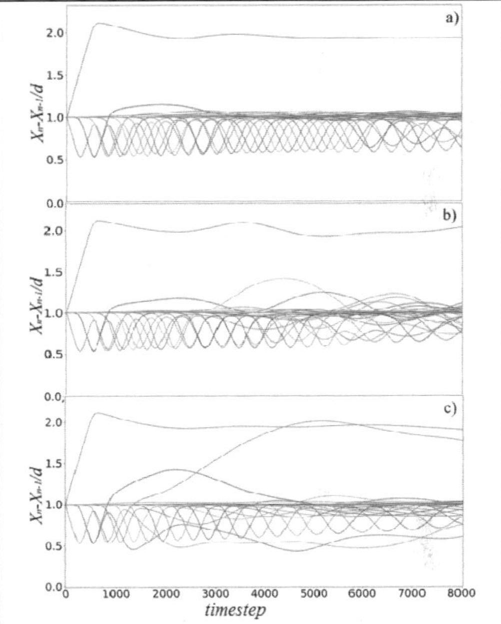

Figure 4. Displacement of atoms as a function of time in the central close-packed row for (a) vertical, (b) horizontal and (c) diagonal configuration.

4. Conclusions

Molecular dynamics modeling was carried out for three supersonic 2-crowdions moving in parallel close-packed rows in the crystal lattice of platinum, using a many-body interatomic potential. A stable vertical configuration was found, in which the three 2-crowdions propagate with minimal disturbances. In this case, the dynamics of the 2-crowdions practically does not differ from how they move without interacting with each other. In the case of a horizontal configuration, there is a stronger interaction of the compound 2-crowdions with each other and a difference in the dynamics of movement. The movement is unstable, however, the three 2-crowdions still propagate along the close packed directions. When three 2-crowdions are launched in a diagonal configuration, the two outer 2-crowdions are converted into 1-crowdions, at which an additional 1-crowdion appears in the central close-packed direction. Obtained results contribute to the fundamental understanding of mass transfer mechanisms in crystals in non-equilibrium conditions.

Acknowledgement
E.A.K. thanks for the financial support the Council of the President of the Russian Federation for state support of young Russian scientists, grant No. MD-3639.2019.2 (writing the manuscript). This work was supported by the State Assignment of IMSP RAS No. AAAA-A17-117041310220-8.

References
[1] Lam N Q, Rothman S J and Sizmanns R 1974 *Radiation Effects* **23** 53–9
[2] Korznikova E A, Mironov S Yu, Korznikov A V, Zhilyaev A P and Langdon T G 2012 *Mater. Sci. Eng.* A **556** 437–45
[3] Mazilova T I, Sadanov E V, Voyevodin V N, Ksenofontov V A and Mikhailovskij I M 2018 *Surf. Sci.* **669** 10–5
[4] Yang Z Q, Ma S Y, Hu Q M, Ye H Q and Chisholm M F 2017 *Mater. Charact.* **128** 226
[5] Shepelev I A, Bachurin D V, Korznikova E A and Dmitriev S V 2020 *Eur. Phys. J.* B **93** 167
[6] Shepelev I A, Dmitriev S V, Kudreyko A A, Velarde M G and Korznikova E A 2020 *Chaos, Solitons and Fractals* **140** 110217
[7] Shepelev I A, Korznikova E A, Bachurin D V, Semenov A S, Chetverikov A P and Dmitriev S V 2020 *Phys. Lett.* A **384** 126032
[8] Chetverikov A P, Shepelev I A, Korznikova E A, Kistanov A A, Dmitriev S V and Velarde M G 2017 *Comput. Condens. Matter* **13** 59–64
[9] Dmitriev S V, Korznikova E A and Chetverikov A P 2018 *J. Exp. Theor. Phys.* **126** 347–52
[10] Dmitriev S V, Medvedev N N, Chetverikov A P, Zhou K and Velarde M G 2017 *Phys. Status Solidi RRL* **11** 1700298
[11] Babicheva R I, Evazzade I, Korznikova E A, Shepelev I A, Zhou K and Dmitriev S V 2019 *Comp. Mater. Sci.* **163** 248
[12] Moradi Marjaneh A, Saadatmand D, Evazzade I, Babicheva R I, Soboleva E G, Srikanth N, Zhou K, Korznikova E A and Dmitriev S V 2018 *Phys. Rev.* E **98** 023003
[13] Korznikova E A, Sunagatova I R, Bayazitov A M, Semenov A S and Dmitriev S V 2019 *Lett. Mater.* **9** 386
[14] Evazzade I, Lobzenko I P, Saadatmand D, Korznikova E A, Zhou K, Liu B and Dmitriev S V 2018 *Nanotechnology* **29** 215704
[15] Korznikova E A, Shcherbinin S A, Ryabov D S, Chechin G M, Ekomasov E G, Barani E, Zhou K and Dmitriev S V 2019 *Phys. Status Solidi* B **256** 1800061
[16] Murzaev R T, Bachurin D V, Korznikova E A and Dmitriev S V 2017 *Phys. Lett.* A **381** 1003
[17] Barani E, Korznikova E A, Chetverikov A P, Zhou K and Dmitriev S V 2017 *Phys. Lett.* A **381** 3553
[18] Korznikova E A, Fomin S Y, Soboleva E G and Dmitriev S V 2016 *JETP Letters* **103** 277–81
[19] Zakharov P V, Korznikova E A, Dmitriev S V, Ekomasov E G and Zhou K 2019 *Surface Science* **679** 1–5
[20] Shepelev I A, Chetverikov A P, Dmitriev S V and Korznikova E A 2020 *Comp. Mater. Sci.* **177** 109549
[21] Chen H-Y and Tsou N-T 2019 *Computer Modeling in Engineering & Sciences* **120** 319
[22] Tsukanov A and Psakhie S 2016 *Facta Universitatis, Series: Mechanical Engineering* **14** 269
[23] Bhui A S, Singh G, Sidhu S S and Bains P S 2018 *Facta Universitatis, Series: Mechanical Engineering* **16** 337
[24] Lee W-J, Lo Y-C, Yang A, Chen K and Chen N-Y 2019 *Computer Modeling in Engineering & Sciences* **120** 293
[25] Babicheva R I, Dahanayaka M, Liu B, Korznikova E A, Dmitriev S V, Wu M S and Zhou K 2020 *Mater. Sci. Eng.* B **259** 114569
[26] Plimpton S 1995 *J. Comput. Phys.* **117** 1-19

IOP Conf. Series: Materials Science and Engineering 1008 (2020) 012069 doi:10.1088/1757-899X/1008/1/012069

Behavior of the mechanical system composed of highly deformable structural elements

D U Abdullina[1], L Kh Rysaeva[2], E A Korznikova[2,3] and S V Dmitriev[2,3]

[1] Ufa State Aviation Technical University, 450077, Karl Marx, 12, Ufa, Russia
[2] Institute for Metals Superplasticity Problems of RAS, 450001, Khalturin St. 39, Ufa, Russia
[3] Institute of Molecule and Crystal Physics, Ufa Federal Research Centre of the Russian Academy of Sciences, 71 October Avenue, 450054 Ufa, Russia

E-mail: dina.abdullina25@gmail.com

Abstract. Carbon nanotubes (CNTs) are attractive for many applications because they possess a unique combination of mechanical and physical properties. Horizontally aligned CNT bundles under lateral compression behave as an elastic body with highly deformable elements, since their cross sections can collapse. Mechanical properties of such systems is poorly investigated in spite of the fact that they are promising for vibration and shock protection. Here we use a chain model with a reduced number of degrees of freedom in order to study the behaviour of a CNT bundle under uniaxial and biaxial lateral compression. Stress-strain curves are obtained and evolution of the CNT bundle structure is analyzed.

1. Introduction

Carbon nanotubes (CNTs) attract each other by non-valence interactions forming CNT bundles [1-3] that can be obtained by various experimental methods [4,5]. Individual CNTs have excellent mechanical properties, such as tensile strength from 11 to 63 GPa, Young's modulus during axial tension from 1.0 to 1.3 TPa and strain to fracture exceeding 10% [6,7]. Moreover, CNTs are lightweight, flexible, bendable, they have good thermal and electrical conductivities, that is why they can be applied in various nanotechnologies [8, 9]. CNT bundles are used to produce high-strength ropes [2, 10], fibers [11, 12], polymer and metal matrix composites [12, 13], solid lubricants [14], projectile protectors [15], etc.

Physical and mechanical properties of CNT bundles have been extensively studied with the use of computer simulation methods. Vertically aligned CNTs can be transformed into horizontally aligned forest by pressing [16,17]. The thin shell theory has been used to describe CNT ensembles of different structure [18]. In the works [19,20], the applicability of the beam, plate and shell theories to the analysis of the mechanical properties of nanomaterials has been analyzed. Importantly for the present study, CNTs having diameter above a threshold value, have two equilibrium configurations, circular and collapsed [21-23]. Mechanical properties of CNT bundles can be efficiently studied using the nonlinear coarse-grained potential [24]. In order to considerably reduce the number of degrees of freedom, the chain model has been developed for simulation of some sp^2-carbon nanostructures [25]. In particular, this model was successfully used to study structure and properties of various conformations of carbon nanoribbons [25-29] and surface ripplocations [30]. The chain model was extended to the study of CNTs under lateral plane strain compression in [31,32]. CNT bundle as an elastic damper was analyzed in [33]. Laterally compressed CNT bundle undergoes phase transitions [34,35]. Buckling critical load for axially compressed graphene nanoribbons can be increased by twisting about long axis [36,37]. Propagation of solitary waves and shock waves in carbon and other nanomaterials have been studied in [30,38,39].

In this work, we demonstrate the application of the chain model to the analysis of the structure evolution of CNT bundle under uniaxial and biaxial compression. In Sec. 2 we describe the model and in Sec. 3 present simulation results. Sec. 4 concludes this work.

2. Simulation method

Presented in figure 1 is a schematic representation of the computational model used in this study. The zigzag nanotubes of the same diameter are aligned with the z-axis of the Cartesian coordinate system and they create a triangular lattice in cross-section. CNTs are numbered by the two indices, $i = 1, ..., I$ and $j = 1, ..., J$ ($I=J=2$ in the figure). The atoms move on the (x, y) plane and each atom stands for a rigid row of atoms normal to the (x, y) plane. Atoms within a CNT are numbered anticlockwise by the index $n = 1, ..., N$, as shown in figure 1 for the tube with $i=j=1$. Total number of carbon atoms in the computational cell is $I \times J \times N$. Positions of the atoms are specified by the radius-vectors $r_{ijn} = (x_{ijn}, y_{ijn})$. We use periodic boundary conditions along directions of translation.

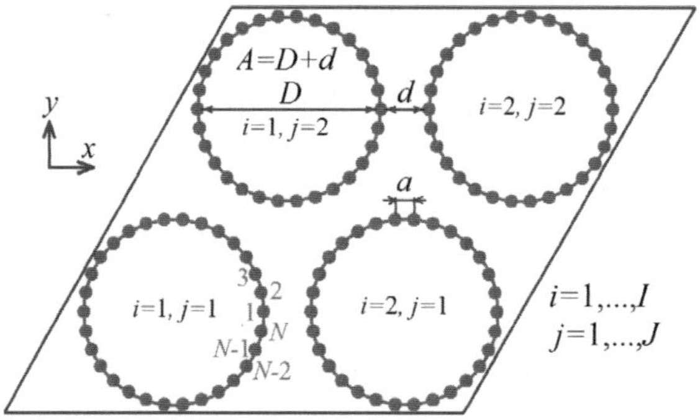

Figure 1. The computational cell that includes $I \times J$ CNTs ($I = J = 2$ in this case) numbered by the indices $i = 1, ..., I$ and $j = 1, ..., J$, with the atoms within a CNT numbered by the index $n = 1, ..., N$ anticlockwise. Atoms have two degrees of freedom moving on the (x, y) plane. Each atom stands for a row of atoms parallel to z-axis, which moves as a rigid body. CNT diameter is D, distance between walls of neighbouring CNTs is d, and the distance between atoms in the CNT wall is a.

The equilibrium interatomic distance in graphene is $\rho = 1.418$ Å. In the zigzag CNT, the distance between neighbouring atomic rows parallel to the z-axis is $a = \rho(3^{1/2})/2 = 1.228$ Å. Diameter of CNT is $D = a/\sin(\pi/N)$, where N is the number of atoms in the cross section of CNT. If d is the distance between neighbouring CNT walls, then the distance between centres of neighbouring CNTs is $A=D+d$. In our simulations we take $N = 30$, then $D = 11.75$ Å and $d = 3.088$ Å.

Carbon nanotube bundle in the frames of the chain model can be described by the Hamiltonian

$$H=K+U_B+U_A+U_{\mathrm{vdW}}, \tag{1}$$

where the first term in kinetic energy

$$K = \frac{M}{2}\sum_{i=1}^{I}\sum_{j=1}^{J}\sum_{n=1}^{N}\left(\dot{x}_{ijn}^2 + \dot{y}_{ijn}^2\right), \tag{2}$$

the second term gives the energy of valence bonds

$$U_B = \sum_{i=1}^{I}\sum_{j=1}^{J}\sum_{n=1}^{N} V(|r_{ijn+1} - r_{ijn}),\ \text{where}\ V(r) = \frac{k}{2}(r-a)^2, \tag{3}$$

the third term describes the energy of valence angles

$$U_A = \sum_{i=1}^{I}\sum_{j=1}^{J}\sum_{n=1}^{N} P(\theta_{ijn}),\ \text{where}\ P(\theta) = \varepsilon[cos(\theta)+1], \tag{4}$$

and the fourth term describes the van der Waals interactions

$$U_{VdW} = \sum_{i=1}^{I}\sum_{j=1}^{J}\sum_{n=1}^{N}\sum_{i'=1}^{I}\sum_{j'=1}^{J}\sum_{n'=1}^{N} W\left(|r_{ijn} - r_{i'j'n'}|\right), \tag{5}$$

$$\text{where}\ |n'-n| > 3\ \text{when}\ i = i', j = j'$$

In Eq. (2), M is the mass of carbon atom.

According to Eq. (3), valence bonds are described by the harmonic potential with stiffness constant $k = 25.279$ eV/Å^2.

In Eq. (4), the angle between valence bonds, $r_{ijn} - r_{ijn-1}$ and $r_{ijn+1} - r_{ijn}$, is found from

$$\cos(\theta_{ijn}) = \frac{(r_{ijn}-r_{ijn-1}, r_{ijn+1}-r_{ijn})}{|r_{ijn}-r_{ijn-1}||r_{ijn+1}-r_{ijn}|}. \quad (6)$$

In Eq. (4) the parameter $\varepsilon = 3.50$ eV.

The van der Waals interactions in Eq. (5) are

$$W(r) = \frac{\varepsilon}{6}\left[5\left(\frac{\sigma}{r}\right)^{11} - 11\left(\frac{\sigma}{r}\right)^{5}\right], \quad (7)$$

with $\varepsilon = 0.00166$ eV and $\sigma = 3.61$ Å.

Detailed information on the chain model is given in [32,33].

3. Simulation results

In figure 2, we plot the stress-strain curves obtained for (a) compression along x, (b) compression along y and (c) biaxial compression. Black (blue) line shows stress σ_{xx} (σ_{yy}) as the function of absolute value of volumetric strain $|\theta|=(1/2)|\varepsilon_{xx}+\varepsilon_{yy}|$.

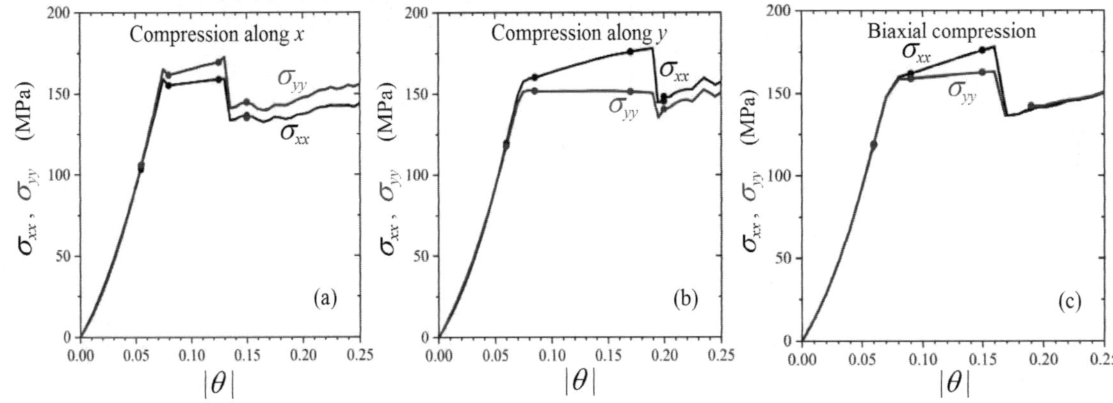

Figure 2. Stress-strain curves for (a) compression along x, (b) compression along y, (c) biaxial compression. Black (blue) line shows stress σ_{xx} (σ_{yy}) as the function of absolute value of volumetric strain $|\theta|=(1/2)|\varepsilon_{xx}+\varepsilon_{yy}|$.

As it can be seen from figure 2, the stress-strain curves can be divided into three regions. For $|\theta|<0.07$, both σ_{xx} and σ_{yy} rapidly increase with strain, and this is the first region. Within the second region, stress components show a very slow increase with strain and even a slow decrease of σ_{yy} can be seen in (b). Note that in (a) $\sigma_{yy}>\sigma_{xx}$, but in (b) and (c) $\sigma_{xx}>\sigma_{yy}$ within the second range. The second range is terminated by a sharp drop of stress components. In the third range, the stress components continue to slowly increase.

In figure 3, structure evolution of the CNT bundle is presented. The top, middle and bottom rows show the results for the compression along x, along y, and biaxial compression, respectively. The absolute values of volumetric strain are given for each panel. The values of strain are shown by dots on the stress-strain curves presented in figure 2.

For all modes of loading the same scenario of structure transformation is observed, reflecting the three regions observed on the stress-strain curves (see figure 2). During the first stage, crystal structure remains unchanged with slightly deformed CNT cross section, see figure 3(a), (e) and (i). In the second stage translational symmetry is still preserved but the primitive translational cells increase in size. In (b) and (c) they contain four CNTs, while in (f) and (g), as well as in (j) and (k), two CNTs. The third stage is characterised by the appearance of collapsed CNTs and the loss of crystal order. Compressive deformation during the third stage results in the increase of the fraction of collapsed CNTs.

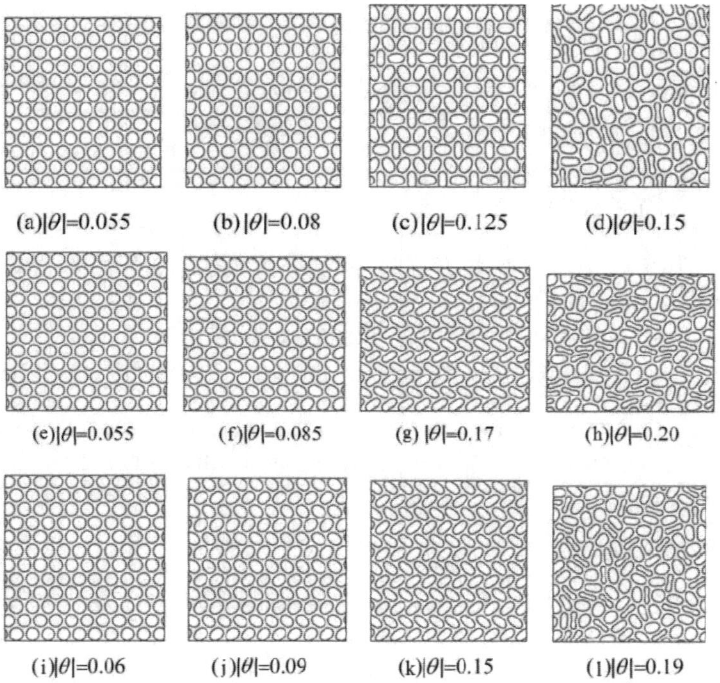

Figure 3. Structure evolution of the CNT bundle under (a-d) compression along *x*, (e-h) compression along *y*, and (i-l) biaxial compression. Absolute value of the volumetric strain is shown for each panel.

4. Conclusions

Molecular dynamics modelling of the lateral compression of CNT bundle was performed with the use of the chain model with greatly reduced number of degrees of freedom under the assumption that the plane strain conditions are satisfied. The efficiency of the used model was demonstrated. Peculiarities of the stress-strain curves were described and linked to the structure transformation of the CNT bundle. Overall, a CNT bundle under lateral compression is an elastic material with unusual properties due to the high deformability of structural units.

In future works, the chain model can be extended for 2D materials other than graphene [40-43].

Acknowledgments

This work was supported by the State Assignment of IMSP RAS No. AAAA-A17-117041310220-8.

Reference

[1] Tersoff J and Ruoff R S 1994 *Phys. Rev. Lett.* **73** 676-679
[2] Thess A, Lee R, Nikolaev P, Dai H, Petit P, Robert J, Xu C, Lee Y H, Kim S G, Rinzler A G, Colbert D T, Scuseria G E, Tomanek D, Fischer J E and Smalley RE 1996 *Science* **273** 483-487
[3] Saether E, Frankland S J V and Pipes R B 2003 *Compos. Sci. Technol.* **63** 1543-1550
[4] Rakov E G 2013 *Russ. Chem. Rev.* **82** 538-566
[5] Bedewy M, Meshot E R, Guo H, Verploegen E A, Lu W and Hart A J 2009 *J. Phys. Chem.* C **113** 20576
[6] Shenderova O A, Zhirnov V V and Brenner D W 2002 *Crit. Rev. Sol. State* **27** 227
[7] Yu M-F 2004 *J. Eng. Mater. Technol.* **126** 271
[8] Truong T K, Lee Y and Suh D 2016 *Curr. Appl. Phys.* **16** 1250
[9] Yao X, Hawkins S C and Falzon B G 2018 *Carbon* **136** 130
[10] Yu M-F, Files B S, Arepalli S and Ruoff R S 2000 *Phys. Rev. Lett.* **84** 5552
[11] Bai Y, Zhang R, Ye X, Zhu Z, Xie H, Shen B, Cai D, Liu B, Zhang C, Jia Z, Zhang S, Li X and Wei F 2018 *Nature Nanotechnology* **13** 589

[12] Dang Z-M, Yuan J-K, Zha J-W, Zhou T, Li S-T and Hu G-H 2012 *Progr. Mater. Sci.* **57** 660
[13] Korznikova G, Czeppe T, Khalikova G, Gunderov D, Korznikova E, Litynska-Dobrzynska L and Szlezynger M 2020 *Mater. Charact.* **161** 110122
[14] Reinert L, Lasserre F, Gachot C, Grützmacher P, Maclucas, Souza N, Mücklich F and Suarez S 2017 *Sci. Rep.* **7** 42873
[15] Randjbaran E, Majid D L, Zahari R, Sultan M T H and Mazlan N 2020 *Facta Universitatis Ser.: Mech. Eng.* **18** 229
[16] Wittmaack B K, Volkov A N and Zhigilei L V 2019 *Carbon* **143** 587
[17] Wittmaack B K, Volkov A N and Zhigilei L V 2018 *Compos. Sci. Technol.* **166** 66
[18] Yakobson B I, Brabec C J and Bernholc J 1996 *Phys. Rev. Lett.* **76** 2511
[19] Rafii-Tabar H, Ghavanloo E and Fazelzadeh S A 2016 *Phys. Rep.* **638** 1
[20] Harik V M 2001 *Solid State Commun.* **120** 331
[21] Impellizzeri A, Briddon P and Ewels C P 2019 *Phys. Rev.* B **100** 115410
[22] Chopra N G, Benedict L X, Crespi V H, Cohen M L, Louie S G and Zettl A 1995 *Nature* **377** 135
[23] Chang T 2008 *Phys. Rev. Lett.* **101** 175501
[24] Ji J, Zhao J and Guo W 2019 *J. Mech. Phys. Solids* **128** 79
[25] Savin A V, Korznikova E A and Dmitriev S V 2015 *Phys. Rev.* B **92** 035412
[26] Savin A V, Korznikova E A and Dmitriev S V 2015 *Phys. Solid State* **57** 2348
[27] Savin A V, Korznikova E A, Lobzenko I P, Baimova Y A and Dmitriev S V 2016 *Phys. Solid State* **58** 1278
[28] Savin A V, Korznikova E A, Dmitriev S V and Soboleva E G 2017 *Comp. Mater. Sci.* **135** 99
[29] Savin A V and Mazo M A 2019 *Adv. Struct. Mat.* **94** 241
[30] Savin A V, Korznikova E A and Dmitriev S V 2019 *Phys. Rev.* B **99** 235411
[31] Korznikova E A, Rysaeva L K, Savin A V, Soboleva E G, Ekomasov E G, Ilgamov M A and Dmitriev S V 2019 *Materials* **12** 3951
[32] Abdullina D U, Korznikova E A, Dubinko V I, Laptev D V, Kudreyko A A, Soboleva EG, Dmitriev S V and Zhou K 2020 *Computation* **8** 27
[33] Rysaeva L K, Korznikova E A, Murzaev R T, Abdullina D U, Kudreyko A A, Baimova J A, Lisovenko D S and Dmitriev S V 2020 *Facta Universitatis Ser.: Mech. Eng.* **18** 1
[34] Abdullina D U, Kudreyko A A, Korznikova E A, Shepelev I A and Dmitriev S V 2020 *Pro. Biomed. Opt. Imag. - Proc. of SPIE* **11459** 1145911
[35] Babicheva R I, Dmitriev S V, Korznikova E A and Zhou K 2019 *J. Exp. Theor. Phys.* **129** 66
[36] Savin A V, Korznikova E A, Krivtsov A M and Dmitriev S V 2020 *Eur. J. Mech.* A. **80** 103920
[37] Savin A V, Korznikova E A and Dmitriev S V 2019 *Mech. Mater.* **137** 103123
[38] Korznikova E A and Dmitriev S V 2014 *J. Phys.* D. **47** 34
[39] Shepelev I A, Chetverikov A P, Dmitriev S V and Korznikova E A 2020 *Comp. Mater. Sci.* **177** 109549
[40] Romanov A E, Rozhkov M A and Kolesnikova A L 2018 *Lett. Mater.* **8** 384
[41] Belenkov M B and Chernov V M 2020 *Lett. Mater.* **10** 254
[42] Baimova J A, Dmitriev S V and Zhou K 2012 *Phys. Stat. Sol.* B **249** 1393
[43] Baimova J A, Liu B, Dmitriev S V and Zhou K 2014 *Physica Status Solidi RRL* **8** 336
[44] Barani E, Korznikova E A, Chetverikov A P, Zhou K and Dmitriev S V 2017 *Phys. Lett.* A **381** 3553
[45] Shcherbinin S A, Zhou K, Dmitriev S V, Korznikova E A, Davletshin A R and Kistanov A A 2020 *J. Phys. Chem.* C **124** 10235
[46] Yoon T, Kang S, Yeob Kang T and Kim T-S 2019 *CMES - Comp. Model. Eng.* **120** 351
[47] Babicheva R I, Dahanayaka M, Liu B, Korznikova E A, Dmitriev S V, Wu M S and Zhou K 2020 *Mater. Sci. Eng.* B **259** 114569
[48] Kistanov A A, Khadiullin S K, Dmitriev S V and Korznikova E A 2019 *Chem. Phys. Chem.* **20** 575

Comparative analysis of fine-grained structure formation in an Al-Zn-Mg-Cu alloy during multidirectional isothermal forging and uniaxial compression

O Sitdikov [1] and T Sakai [2]

[1] Institute for Metals Superplasticity Problems of the Russian Academy of Sciences, 39 Khalturin St., Ufa 450001, Russia
[2] UEC Tokyo (The University of Electro-Communications), Chofu, Tokyo 182-8585, Japan

E-mail: sitdikov.oleg@imsp.ru

Abstract. A comparative study of the mechanical behavior and microstructural evolution in a cast coarse-grained 7475 Al alloy during uniaxial compression and multidirectional forging at 490°C and 10^{-4} s^{-1} was performed. It was shown that high-temperature deformation of the alloy led to grain refinement accompanied by strain softening. At similar processing conditions, the deformation path provided by multidirectional forging was found to be more favorable for grain refinement and strain softening than that of uniaxial compression.

1. Introduction

Producing fine-grain structures upon plastic working of metals and alloys is an important technical problem, the solution of which may have a high commercial potential due to improved technological and service properties of processed semi-finished products [1,2]. Studies carried out in past decades have shown that methods of severe plastic deformation (SPD), such as high-pressure torsion, equal-channel angular pressing, isothermal multidirectional forging (MDF) etc., which ensure the achievement of high strains, can be very effective for grain refinement [1-6]. Among the above SPD methods, MDF is the versatile and affordable material processing technique, because it assumes the use of universal equipment and tooling and allows processing large-sized billets [3-5].

The principle of MDF is based on free forging operations in several consecutive passes with changing the loading direction at each pass: x → y → z → ..., etc. [1,3,5]. Since the billet does not practically change its shape and dimensions, high strains can be accumulated during this processing. At the same time, the use of MDF may be important for grain refinement not only due to achieving high strains, but also ensuring a certain deformation route that may be more favorable for the development of a new fine-grain structure [5] than those in any "monotonic" schemes, such as rolling, extrusion, or compression. Unfortunately, the influence of the MDF processing route on grain refinement is currently not well addressed in the literature and still remains undisclosed. This work was aimed to provide a comparative analysis of the mechanical behavior and structural changes occurring in a high-strength coarse-grained 7475 Al alloy under high-temperature uniaxial compression and MDF. It was found in [5,7] that this alloy demonstrated grain refinement in both processing routes at elevated temperatures. The formation mechanisms of new grains, as well as the effect of the deformation paths on the development of fine-grain structure in this alloy are discussed.

2. Material and methods

The direct-chill cast ingot of the 7475 Al alloy (Al–6.04%Zn–2.46%Mg–1.77%Cu–0.23%Cr–0.16%Zr–0.03%Mn–0.03%Si–0.04%Fe, wt. %) was homogenized at 490°C for 20h. Coarse dendrite-like grains were present after such heat treatment in the initial structure, where some segments of the grain boundaries were alternatively straight and corrugated (figure 1). The sizes of grains were within 1-10 mm in the longitudinal direction and 50-250 μm in the transverse direction. Two types of dispersoids identified as Al_3Zr and Al_3Cr, with the mean sizes of 20 and 100 nm, respectively, were uniformly distributed in the alloy [5,7].

Rectangular samples of about 10×9×6 mm^3 (1.8:1.7:1) were machined from the ingot. The first compression axis was aligned with the ingot axis. Uniaxial compression to the true strain of e=1.9 and MDF to the total strain of e=3.5 were carried out under isothermal conditions at 490°C and a constant strain rate of 3×10^{-4} s^{-1}. Upon MDF, the samples were deformed with the strain per a pass of 0.7. The testing machine was equipped with a special device enabled to automatically maintain a constant strain rate during compression and a water-quenching apparatus to fix the structural changes occurring during deformation. Microstructural analysis in conventional and polarized light was carried out in the central part of the deformed specimens with an Olympus PME-3 optical microscope. The average size and fraction of strain-induced new grains were evaluated by a line-intercept and a point-count techniques, respectively. Electron backscattering diffraction (EBSD) analysis was performed using a Hitachi-3500A scanning electron microscope with OIM Data Analysis software. In the EBSD maps obtained, different colors corresponded to different orientations, while the low- (Θ≥2°), medium- (Θ≥5°) and high-angle (Θ≥15°) misorientations between neighboring grid points were marked by thin white-, narrow- and bold black lines, respectively. To study the deformation relief, the samples with a pre-polished surface that was perpendicular to the z-axis were used. They were loaded along the x axis with e=0.16; and also, sequentially along the x-axis (e=0.16) and after rotating the sample by 90° - along the y-axis (e=0.16) up to the total strain of e=0.32.

3. Experimental results

Typical true stress-strain curves derived from uniaxial compression and MDF are shown in figure 2. It is seen that the σ-e curve plotted for uniaxial compression (dotted line in figure 2) demonstrated a peak of flow stresses just after yielding and subsequent significant material softening followed by a steady state of flow stresses at e of beyond 0.6. On the contrary, the stress-strain curve envelope that was obtained by MDF showed a more significant softening and lower flow stresses achieved at the same effective strains. In this case, the steady state of plastic flow was reached at e of beyond 3 with a much lower stress level. It is also noteworthy in figure 2 that the flow stresses at the onset of each subsequent MDF pass (i.e. at reloading) were approximately equal to those immediately before

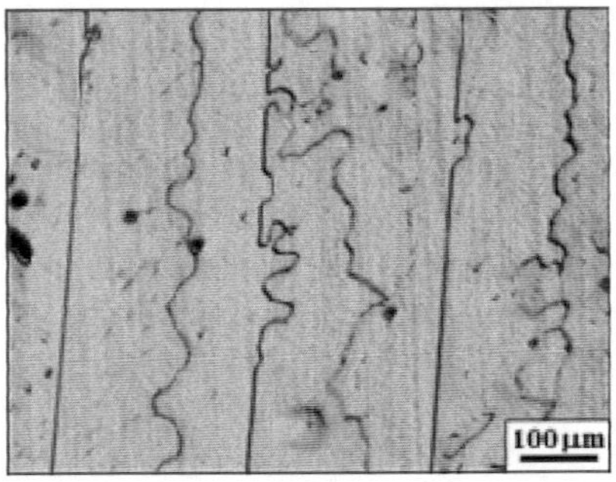

Figure 1. The initial structure of the 7475 Al alloy showing layered dendrite-like grains. The ingot axis is vertical.

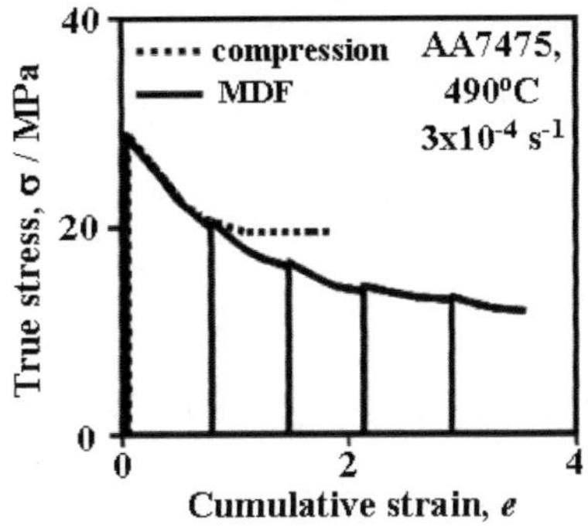

Figure 2. True stress - true strain curves of the 7475 Al alloy plotted for uniaxial compression and MDF.

the material unloading in the preceding compression passes. This suggested that any softening processes associated with static recovery or static recrystallization in the material scarcely occurred upon overheating and soaking the samples at the high temperature between MDF passes. Accordingly, like in the case of uniaxial compression, the structural changes during MDF were mainly affected by plastic deformation (i.e. strain accumulation in each compression pass).

The typical microstructures after (a) uniaxial compression at e=1.9; (b) MDF at e=2.1; and (c) MDF at e=3.5, are shown in figure 3. The dark areas in figure 3 were composed by fine recrystallized grains processed by deformation, while light ones corresponded to the fragments of coarse original grains. It is seen that the deformation under both processing conditions resulted in the development of partially recrystallized bimodal grain structures. In the case of uniaxial compression to the strain of 1.9, new grains with the size of about 6 μm were formed mainly near the original grain boundaries and their volume fraction, V_{rex}, did not exceed 20% in the entire sample (figure 3a). Meanwhile, during MDF to the close value of strain e=2.1, the regions of new grains of 6-7 μm in size were formed more homogeneously and occupied about 60% of the total material volume (figure 3b). With a further increase in the MDF strain to e=3.5, an almost fully recrystallized fine-grained structure evolved (figure 3c). Figure 4 shows strain dependences of V_{rex} for both the material processing routes. It should be noted in figure 4 that changes in the route from uniaxial compression to MDF significantly accelerated the kinetics of grain refinement. At the same time, as noticed above, only a negligibly small effect of the considered change in deformation path on the size of fine recrystallized grains was found, which, apparently, was mainly controlled by the temperature-strain rate conditions [8].

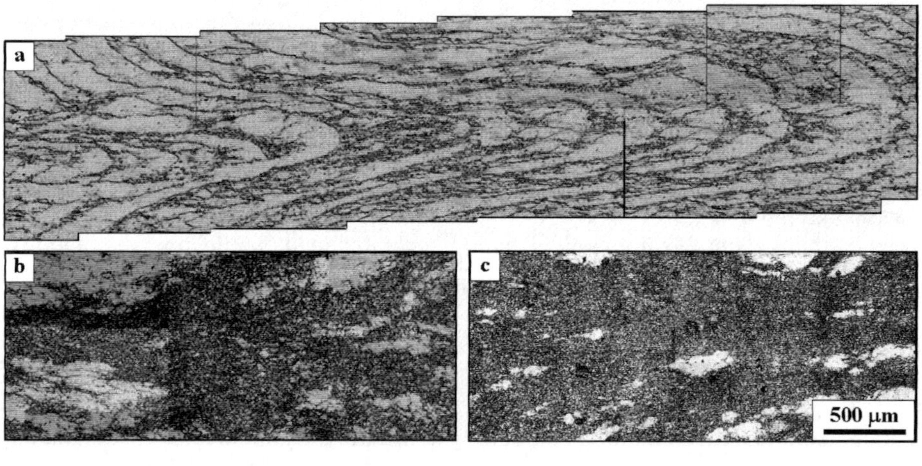

Figure 3. Typical microstructure formed in the 7475 Al alloy after (a) uniaxial compression to e = 1.9; (b) MDF to e = 2.1; and (c) MDF to e = 3.5. The last compression axis is vertical.

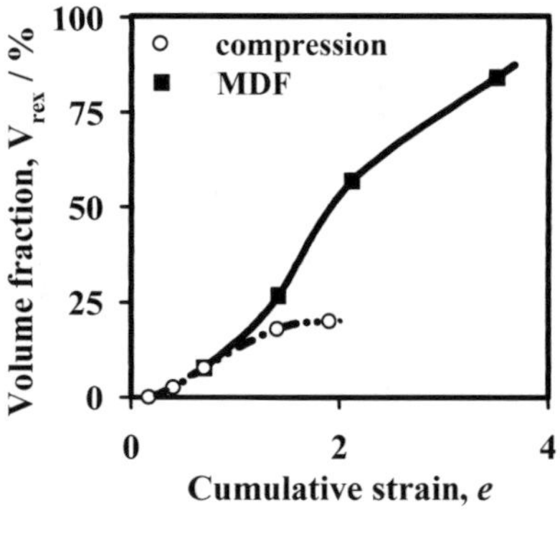

Figure 4. Strain dependences of the volume fractions of fine grains, V_{rex}, developed by uniaxial compression and MDF.

4. Discussion

The results obtained show that both the hot compression and MDF resulted in grain refinement of the alloy accompanied by significant strain softening. However, the implementation of these processing schemes led to different microstructural behaviors of the alloy. Namely, at approximately the same effective strain of about 2, the volume fraction of new grains was stabilized at about 20% under uniaxial compression, while with MDF it reached about 60% and continued to increase with increasing strain (figure 4). The effect of the strain path on the evolution of the microstructure can be explained as follows.

Figure 5 shows a typical deformation relief that developed on the polished surface of samples (a, b) under uniaxial compression to e=0.16; and (c) under two sequential loadings with the same strains of 0.16, when the compression direction was changed by 90° between passes (i.e., up to the total strain of e=0.32). It can be seen that the deformation of the alloy with a coarse-grained layered structure occurred extremely non-uniformly even at the initial stages of the plastic flow due to a high deformation incompatibility of such a structure and caused the intensive formation of arrays of approximately parallel deformation bands that fragmented the original grains. This suggested that the individual regions in each grain bounded by deformation bands were significantly misoriented relative to each other. These bands (the dislocation boundaries with medium-to-high angle misorientation) can also be seen at higher strains on the etched surfaces of samples using polarized optical microscopy [5] and EBSD analysis, as shown in figure 6. Hence, the evolution of deformation bands could play a key role in grain refinement. The process of formation of new grains during high-temperature equal-channel angular pressing and MDF of some Al alloys, related to the development of deformation bands, was discussed in the previous studies [5,9-11]. In particular, it has been pointed out that deformation bands, that were recognized as microshear bands (MSBs) [8,12], subdivided the original grains into separate domains surrounded by medium-angle boundaries. A gradual increase in the number of deformation bands and misorientation angles of their boundaries during deformation led to the transformation of medium-angle boundaries into high-angle ones and the development of a fine-grain structure in accordance to the mechanism of continuous dynamic recrystallization [8].

Analysis of the data [5,7] showed that the above-described structural mechanism associated with the development of MSBs may be responsible for the development of new grains in the present 7475 Al alloy both under uniaxial compression and MDF. However, it is clearly seen using polarized microscopy (figure 6) that during uniaxial compression, the interiors of non-recrystallized grains contained a large number of bands, the boundaries of which remained approximately parallel to each other at all strains and gradually rotated toward the material flow direction. Thus, we can imagine [13] that if the deformation/microshear bands are formed parallel to each other inside a grain (figure 5a,b), their evolution can lead to the formation of mainly two-dimensional flat structures, as schematically shown in figure 7a. With increasing strain under conditions of the constant strain path in the uniaxial

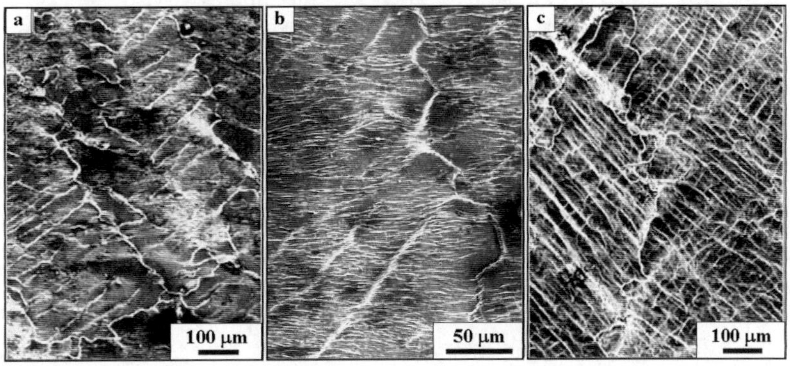

Figure 5. Typical surface features, developed by compression (a, b) along one axis to e=0.16 and (c) sequentially along two axes to total e =0.32.

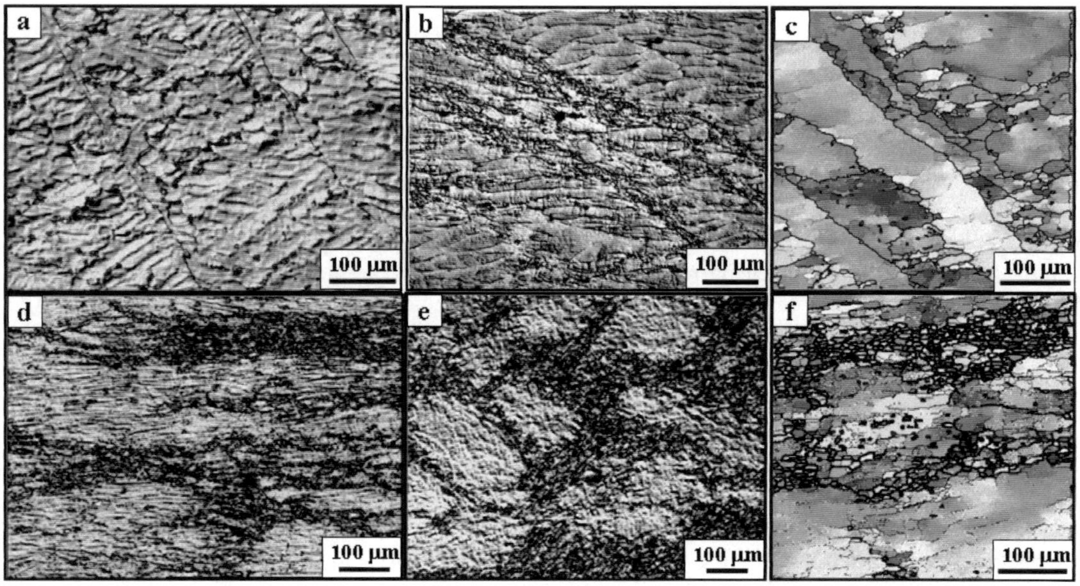

Figure 6. Microstructures formed in the alloy under uniaxial compression to (a) e=0.3; (b, c) e=0.7; (d) e=1.9 and (e, f) MDF to e=2.1: (a, b, d, e) polarized optical microscopy; (c, f) EBSD analysis.

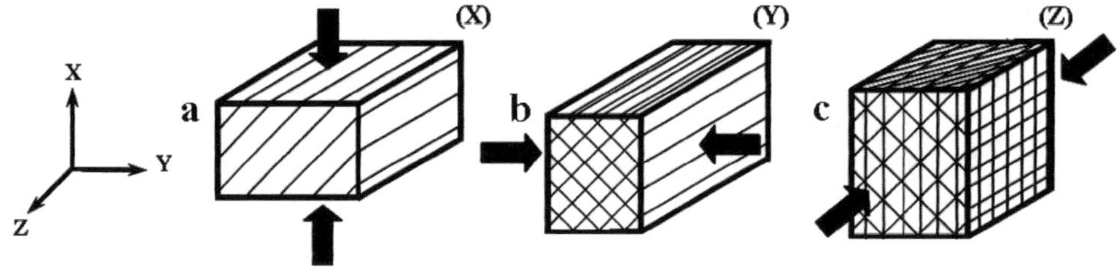

Figure 7. Schematic representation for the formation of the microstructure during deformation: (a) uniaxial compression; (b) two-axial compression; (c) MDF [13].

compression, these structures could tend to orient perpendicular to the compression axis, and their spacing should decrease with increasing the strain. Note that such structures can also be observed at high strains in cold rolling, as well as torsion under high pressure [e.g. 12,14,15]. Therewith, when the present 7475 Al alloy was compressed to e=1.9, the boundaries of MSBs with medium-to-high misorientations remained mostly extended and were interspersed with many low-angle ones. Fragmentation of the initial grains and subsequent formation of the equiaxed fine-grained structures were feasible only near the initial grain boundaries (figure 6c), where the increased stress level and associated local lattice distortions [16] facilitated the development of MSBs in different directions.

On the contrary, the change in the deformation path in MDF, caused by a sequential change in the direction of the loading axis, ensured the development of MSBs in various directions (figure 5c). This is schematically illustrated in figure 7b,c [13]. An increase in the density of the bands and their mutual crossing during deformation led to a continuous fragmentation of initial grains (figure 6e,f). Concurrently, new crystallites were formed in the places fragmented by the boundaries of the bands. Moreover, as can be seen in figure 3b,c, new fine grains were extensively formed not only near the initial high-angle boundaries, but also inside the grains. This provided a more uniform development of the fine-grained structure and significantly increased the volume fraction of new grains at high strains (figure 4). Thus, the change in the deformation path during high-temperature MDF could control structural changes and play a role in grain refinement in the alloy.

5. Conclusions

1. Under hot uniaxial compression conditions, the σ–e curve of the 7475 Al alloy demonstrated a peak of flow stresses just after yielding with subsequent alloy softening. A steady state of plastic flow was reached after a strain of about 0.6. MDF at the same temperature and strain rate led to a more significant decrease in flow stresses, followed by their stabilization at higher strains of beyond 3.

2. Both uniaxial compression and MDF led to the development of a fine-grained structure. New grains of 6-7 μm in size evolved as a result of fragmentation of the original grains upon the formation of deformation bands, followed by transformation of their boundaries to high-angle grain boundaries.

3. The implementation of the above processing methods led to different microstructural behaviors of the alloy. During uniaxial compression, deformation/microshear bands developed mainly parallel to each other within each grain and, with an increase in the strain, were oriented in the direction perpendicular to the applied load. In this case, fragmentation of the original grains followed by grain refinement took place only near the original high-angle boundaries, where development of deformation/microshear bands in various directions was facilitated by higher stress concentrations and larger lattice distortions.

4. Changing the deformation path during MDF ensured the development of the deformation / microshear bands in various directions and led to the continuous fragmentation of original grains. At approximately the same strain of about e=2, the volume fraction of new grains processed by uniaxial compression stabilized at the value of about 20%, while upon MDF, it reached about 60% and proceeded to increase continuously to about 85% with further alloy straining to e=3.5. This suggests that the deformation path provided by MDF controls the structural changes and plays an important role in the grain refinement.

Acknowledgements
The work was performed in accordance to the state assignment of IMSP RAS no. AAAA-A19-119021390107-8. Financial support from the UEC Tokyo should also be acknowledged with gratitude.

References

[1] Humphreys F J, Prangnell P B, Bowen J R, Gholinia A and Harris C 1999 *Phil. Trans. R. Soc. London: A* **357** 1663
[2] Valiev R Z, Islamgaliev R K and Alexandrov I V 2000 *Progr. Mater. Sci.* **45** 103
[3] Imaev R M, Salishchev G A, Senkov O N, Imayev V M, Shagiev M R, Gabdullin N K, Kuznetsov A V and Froes F H 2001 *Mater Sci Eng. A* **300** 263
[4] Mulyukov R R, Imayev R M, Nazarov A A, Imayev M F and Imayev V M 2014 *Superplasticity of Ultrafine Grained Alloys: Experiment, Theory, Technologies* (Moscow: Nauka) p 284 (in Russian)
[5] Sitdikov O, Sakai T, Goloborodko A, Miura H and Kaibyshev R 2005 *Phil. Mag.* **85** 1159
[6] Markushev M V 2011 *Lett. Mater.* **1** 36
[7] Kaibyshev R, Sitdikov O, Goloborodko A and Sakai T 2003 *Mater. Sci. Eng.* **344** 348
[8] Humphreys F J and Hatherly M 2004 *Recrystallization and Related Annealing Phenomena* (Amsterdam: Elsevier) p 628
[9] Mazurina I, Sakai T, Miura H, Sitdikov O and Kaibyshev R 2009 *Mater. Trans.* **50** 101
[10] Sitdikov O, Avtokratova E and Sakai T 2015 *J. Alloys Compd.* **648** 195
[11] Sitdikov O, Avtokratova E and Markushev M 2019 *Materials Characterization* **157** 109885
[12] Hurley P J and Humphreys F J 2003 *Acta Mater.* **51** 1087
[13] Sakai T, Miura H and Yang X 2009 *Mater Sci Eng.: A* **499** 2
[14] Huang X, Winther G, Hansen N, Hebesberger T, Vorhauer A, Pippan R and Zehetbauer M 2003 *Mater. Sci. Forum* **426-432** 2819
[15] Valeev I, Valeeva A, Ilyasov R, Sitdikov O and Markushev M 2019 *Lett. Mater.* **9** 447
[16] Belyakov A, Gao W, Miura H and Sakai T 1998 *Metall. Mat. Trans* A **29** 2957

Methods of controlling mechanical behavior of layered titanium material under impact loading

A A Sarkeeva, A A Kruglov and R Ya Lutfullin

Institute for Metals Superplasticity Problems of Russian Academy of Sciences, 39 Khalturin St., 450001 Ufa, Russia

E-mail: aigul-05@mail.ru

Abstract. The results of studies on the mechanical behavior of a Ti-6Al-4V laminated material obtained by diffusion welding are analyzed. It is shown that the impact behavior of the layered material depends on many factors including the interface defects, the orientation of the notch tip relative to interface, the number and microstructural state of the layers and their crystallographic texture.

1. Introduction

Layered materials (LMs) attract a great attention due to the possibility of significant improvements in their fracture toughness, fatigue and impact characteristics, wear and corrosion resistance and formability [1]. These materials are manufactured by various technological methods based on joining of both the similar and dissimilar metallic materials. Diffusion bonding, one of the joining processes, is carried out at a temperature below the melting temperature of materials under an applied pressure [2, 3]. The process of joint formation and the level of properties obtained by diffusion bonding can be controlled by selecting its parameters.

One of distinctive features of laminated materials in comparison with monolithic ones is the presence of numerous internal interfaces between the layers, which play a significant role in the mechanical behavior. The number of layers, interfacial properties and the type of arrangement of the interfaces relatively to the direction of impact load are critical factors. The properties of joints obtained by diffusion bonding depend on the microstructural characteristics in interfaces, in particular, on the presence of pores. It is worth noting that impact testing is the most sensitive method to assess the quality of joints processed by diffusion bonding [4]. The crack propagation in the layered materials is usually studied in two orientations: a) the crack divider orientation when the crack propagates simultaneously through all the layers; b) the crack arrester orientation when the crack propagates sequentially from one layer to another [5-7].

Another distinctive feature of layered materials is their ability to have a gradient structure. The production of materials, which are simultaneously strong and ductile, has been a permanent challenge for a long time. It is known, for example, that ultra-strong nanocrystalline materials usually have a low ductility [8]. Therefore, the development of materials with a gradient structure is important, since the strength and ductility are two of the most important mechanical properties but often are mutually exclusive. According to the published data [9], the combination of strong and plastic layers inhibits the crack development.

IOP Conf. Series: Materials Science and Engineering 1008 (2020) 012071 doi:10.1088/1757-899X/1008/1/012071

Mechanical properties of many materials depend not only on the structure but also on the texture [10, 11]. Texture effects play a significant role in the technological characteristics of titanium alloys [12].

The possible ways to control the mechanical behavior of diffusion bonded Ti–6Al–4V layered materials are shown in this paper.

2. Experimental

Ti-6Al-4V alloy sheets with microcrystalline (MC) and ultrafine-grained (UFG) structures were produced by VSMPO, Verkhnyaya Salda, Russia and multiple step forging [13] with subsequent isothermal rolling [14], respectively. Three types of layered materials were manufactured by diffusion bonding. The LM of type I consisted of thirteen MC sheets of an approximately 0.8 mm thickness, type II of seven MC sheets of ~ 1.5 mm thickness. The type III laminates were composed of six MC sheets of ~ 1.5 mm thickness and five UFG sheets of ~ 0.35 mm thickness assembled in an alternative sequence. Two methods of assembling sheets in a package were used when fabricating the layered material of types I: in one case, they were stacked relative to each other so that the rolling directions in them coincided, and in the other, they did not coincide, the angle between rolling directions in the neighboring sheets was 90°. Diffusion bonding of MC sheets was conducted at temperature 900°C, MC and UFG sheets at 600°C. The technology of diffusion bonding has been described in detail in Ref. [15].

The impact tests were carried out according to ASTM E23 at room temperatures using the impact testing machine that enables recording dynamic loading diagrams in the "load-displacement" coordinates. Full–sized U-notched Charpy specimens of dimensions 10×10×55 mm^3, each with a 2 mm deep U-notch, were used in the tests. The samples were tested in the crack divider and arrester orientations, as shown schematically in figure 1. In the crack divider orientation (figure 1a), the initial notch/crack tip intersects all the layers of the test sample so that the crack propagates ''sees'' all the layers interfaces simultaneously. In the crack arrester orientation (figure 1b), the initial notch/crack tip ends within an individual layer of the test sample so that the crack front ''sees'' each layer interface sequentially during loading [5]. The division of the total energy of fracture (A) into its components such as the energy of crack initiation (A_i) and the energy of crack propagation (A_{pr}) was performed on the basis of an analysis of the experimental diagrams of dynamic loading [16].

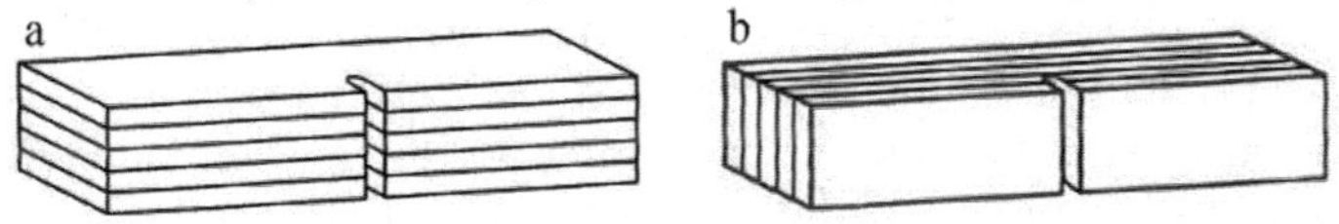

Figure 1. Samples for impact test: (a) crack divider orientation and (b) crack arrester orientation.

Metallographic examinations and fractography analysis were carried out using an optical microscope Nikon L150 and a TESCAN MIRA3 LMU scanning electron microscope.

3. Results and discussion

As previously published in Ref. [17], the as-received Ti-6Al-4V alloy sheets have a prismatic texture (11 $\bar{2}$0) [10 $\bar{1}$0]. The prismatic texture generates a difference in the properties along and transverse the rolling direction in the plane of a sheet. It is in such a texture that the strong anisotropy of fracture toughness is observed [18]. This texture is inherited by a layered material in which the sheets a stacked relative to each other so that the rolling directions in neighboring sheets coincide and thus causes the anisotropy in impact strength. The transverse samples of the 13-layer material in the crack divider orientation had a maximum impact strength value equal to 0.73 MJ/m^2. The layered material in which the sheets are stacked relative to each other so that the rolling directions in neighboring sheets do not coincide has the isotropy of properties due to practically the same distribution of the pole density of the basis in the rolling and transverse directions. The impact strength value of this material is equal to 0.66 MJ/m^2 for the crack divider orientation and 0.56 MJ/m^2 for the crack arrester orientation [15]. The laminated material in the crack divider orientation has higher impact strength due to the increased work of crack propagation. It is worth noting that the level of anisotropy in the layered material also can be decreased by using UFG sheets with the isotropic mechanical properties as intermediate layers [19]. In addition, the application of UFG sheets allows manufacturing a layered material at a lower temperature of diffusion bonding.

The impact strength of the layered material in the crack arrester orientation can be improved by the presence of pores in interfaces which weaken them [19]. The results of studies of the 13-layer material showed that with a relative length of pores in the interfaces increasing from ~1 to 30%, the impact strength value decreases by 1.7 times for the crack divider orientation, and it increases by 14 times for the crack arrester orientation. The quantitative assessment of the fracture characteristics made it possible to clearly establish that the decrease in the impact strength of the laminate in the crack divider orientation is caused by a decrease in the work of crack propagation. Such a significant improvement in the fracture resistance of the layered material in the crack arrester orientation is determined by the need for multiple nucleation of new cracks and the higher dissipation of the fracture energy due to the formation of delamination. It is known that the delamination in layers ahead of the crack tip results in a reduction and redistribution of local stresses [20]. The delamination is formed along the interface between layers, and it is located perpendicularly to the direction of the main crack propagation. Delaminations are usually considered as a principal mechanism of crack arresting in layered materials [20, 21].

The level of properties of the layered material can be controlled by adjusting the number of layers [17]. The test results of the layered material without pores showed that when the crack propagates simultaneously through all layers (the crack divider orientation) the impact strength value is higher for the 7-layer material. In contrast, when the crack propagates sequentially from one layer to another (the crack arrester orientation) the impact strength value is higher for the 13-layer material. The decrease in the impact strength value of the layered material in the crack divider orientation with the number of layers is due to a decrease in both the crack initiation and propagation energies. In contrast, the improvement of impact strength of the layered material in the crack arrester orientation occurs due to the significant increase in the crack propagation energy at an insignificant decrease in the crack initiation energy.

To establish the effect of the gradient structure, a comparative analysis of the results of Charpy impact tests of the layered materials consisting of seven MC sheets and alternating sheets with MC and UFG structures (the latter is called structural composite) was carried out. The impact strength of the structural composite in the crack divider orientation is lower than that of the layered material (table 1). It is important to note that the structural composite is characterized by interfacial pores that facilitate the formation of delamination during fracture. On the other hand, this composite includes the UFG sheets. It is known that the Ti–6Al–4V alloy with UFG structure has a lower impact strength value due to the lower work of crack propagation [22]. The samples of structural composite tested in the crack arrester orientation did not fail as a result of the tests. The impact strength value was estimated by calculating the energy value expended on both the failure of the majority of the layers and the bend of the unfailed layers.

Table 1. The impact strength (KCU), the crack initiation energy (A_i) and the crack propagation energy (A_{pr}) of the layered material (MC/MC) and structural composite (MC/UFG) in the crack divider and arrester orientations.

Material	KCU (MJ/m^2)		A_i (J)		A_{pr} (J)	
	crack divider	*crack arrester*	*crack divider*	*crack arrester*	*crack divider*	*crack arrester*
MC/UFG	0.70	> 1.74	21.1	-	33	-
MC/MC [17]	0.81	0.47	23.1	22.4	39.6	13.8

Figure 2. Macrographs of fractured impact samples of the structural composite: (a) crack divider orientation and (b) crack arrester orientation.

It is important to note that the studied orientations of interfaces relative to the direction of impact load action are characteristic for the real structures from titanium alloys manufactured by the DB/SPF process [23-25].

4. Conclusion

The technology methods of controlling properties of the layered material based on Ti-6Al-4V alloy are shown in the present work. The layered material with isotropic mechanical properties can be manufactured by assembling sheets in a package so that the rolling directions in them do not coincide or by alternation of MC and UFG sheets. The impact strength of the layered material in the crack arrester orientation can be improved by the presence of pores in interfaces contributing to the formation of delaminations. The impact strength of the layered material with an increasing number of layers decreases for the crack divider orientation and increases for the crack arrester orientation. The level of properties in the structural composite can be improved by optimizing the diffusion bonding conditions and the methods of combining layers with different structures.

Acknowledgment

This work was accomplished according to the state assignment of IMSP RAS No. AAAA-A17-117041310221-5.

References

[1] Lesuer D R, Syn C K, Sherby O D, Wadsworth J, Lewandowski J J and Hunt W H 1996 *International Materials Reviews* **41** 169
[2] Kazakov N F 1985 *Diffusion Bonding of Materials* (Moscow: Publishers) 304 p [In Russian]
[3] Dunford D V and Wisbey A 1993 *Materials Research Society Symposium Proceedings* **314** 39
[4] Gómez de Salazar J M, Ureña A and Carrión J G 1996 *Scripta Materialia* **35(4)** 479
[5] Embury J D, Petch N J, Wraith A E and Wright E S 1967 *Transaction of the Metallurgical Society of AIME* **239** 114
[6] He X F, Dong Y H, Li Y H and Wang X M 2018 *International Journal of Fatigue* **106** 1
[7] Cepeda-Jiménez C M, Garcia-Infanta J M, Rozuelo M, Ruano O A and Carreno F 2009 *Scripta Materialia* **61(4)** 407
[8] Valiev R Z, Zhilyaev A P and Langdon T G 2013 *Bulk Nanostructured Materials: Fundamentals and Applications* (Wiley-Blackwell) 440 p
[9] Petuhov A 2005 *Dvigatel* **5** 6 [In Russian]
[10] Miklyaev P G and Fridman Ya B 1986 *Anisotropy of Mechanical Properties of Metals* (Moscow:

Metallurgy) 224 p. [In Russian]
[11] Bache M and Evans W J 2001 *Materials Science and Engineering A* **319** 409
[12] Rubina E B and Betsofen S Ya 1990 *Physics of Metals and Metallography* **4** 191
[13] Kaibyshev O A, Salishev G A, Galeyev R M, Lutfullin R Ya and Valiakhmetov O R 1998 Patent PCT/US97/18642, WO 9817836
[14] Astanin V V and Kaibyshev O A **2008** Patent RU № 2320771
[15] Sarkeeva A A, Kruglov A A, Lutfullin R Ya, Gladkovskiy S V, Zhilyaev A P and Mulyukov R R. 2020 *Composites Part B* **187** 107838
[16] Georgiev M 2007 *Impact Crack-Resisting of Metals* 231 (Sofia: Bulvest 2000) 231 p [In Bulgarian]
[17] Sarkeeva A A 2020 *Letters on Materials* 10(3) 345
[18] Gorynin I V, Chechulin B B 1990 *Titanium in Mechanical Engineering* (Moscow: Mashinostroyeniye) 400p [In Russian]
[19] Ganeeva A A, Kruglov A A and Lutfullin R Ya 2010 *Reviews on Advanced Materials Science* **25** 136
[20] Rohatgi A, Harach D J, Vecchio K S and Harvey K P 2003 *Acta Materialia* **51** 2933
[21] Cepeda-Jiménez C M, Garcia-Infanta J M, Rozuelo M, Ruano O A and Carreno F 2009 *Scripta Materialia* **61(4)** 407
[22] Sarkeeva A A, Lutfullin RYa, Kruglov A A and Astanin V V 2012 *Letters on Materials* **2** 99
[23] Mulyukov R R, Nazarov A A and Imayev R M 2018 *Letters on Materials* **8 (4s)** 510
[24] Beck W, Duong L and Rogall H 2008 *Materialwissenschaft und Werkstofftechnik* **4-5** 293
[25] Kruglov A A, Mulyukov R R, Rudenko O A, Karimova A F and Enikeev F U 2019 *Letters on Materials* **9(4)** 433

Concentration wave method for description of all possible energy states of A_2BC ordered alloy

A R Khalikov[1], V T Nguen[1], V A Valitov[2] and S V Dmitriev[2,3]

[1] Ufa State Aviation Technical University, 450077, 12 Karl Marx, Ufa, Russia
[2] Institute for Metals Superplasticity Problems of RAS, 450001, 39 Khalturin St., Ufa, Russia
[3] Ufa State Petroleum Technological University, 450064, 1 Kosmonavtov St., Ufa, Russia

E-mail: khalikov.albert.r@gmail.com

Abstract. The procedure for describing all possible energy states of an alloy based on the concentration wave method proposed by Khachaturian is employed to describe all possible energy states of ordered A_2BC alloy based on N-dimensional lattice. Within the framework of this method, a complete enumeration of the structures obtained by the superposition of N plane concentration waves with all possible wave vectors is carried out, provided that the given stoichiometry is preserved. For each such superposition, the order parameters on the first I coordination spheres are calculated, thereby determining the point in the I-dimensional order parameter space corresponding to the given structure. For the case of I=2 it is shown that a complete enumeration of all structures generated by one plane concentration wave fills a non-convex figure in the space of two order parameters.

1. Introduction

It seems interesting to solve the problem of describing all possible energy states realized in a structure of a given composition for a given crystal lattice under the assumption that the energies of interatomic interactions can take any values, while taking into account the pair bonds of atoms in the first few coordination spheres. The solution of this problem will make it possible to describe possible superstructures realized in an alloy of a given stoichiometry, for a given topology of the lattice on which the atoms of the alloy are located, and for a given number of coordination spheres of pair interatomic interactions taken into account. An algorithm for solving this problem based on the method of concentration waves has been proposed in [1].

This paper proposes a simplified algorithm for determining the boundaries of variation of the order parameters for the structure of stoichiometry A_2BC, given on an arbitrary N-dimensional lattice, taking into account interatomic interactions in the first I coordination spheres [1]. When solving this problem, only the structural characteristics of the alloy are taken into account, and the values of the energies of interatomic interactions are not involved in consideration [2]. The algorithm is applied for ternary alloys of stoichiometric composition, taking into account the interactions of atoms in the first two coordination spheres (I=2) [3-7]. As examples of ternary alloys, Heusler alloys should be mentioned as materials with a number of applications [8,9].

2. Model formulation

We consider the case of a three-component alloy of the composition A_2BC with the atoms of types A, B, and C located in the sites of a given lattice. Interactions of atoms in the first I coordination spheres are taken into account. The lattice dimension N can be any. The concentrations of atoms of types A, B, and C are equal, respectively, to

$$c_A = \frac{m}{m+n+k}, \quad c_B = \frac{n}{m+n+k}, \quad c_C = \frac{k}{m+n+k}. \tag{1}$$

Here $p_{KL}^{(i)}$ denote the probability that on the i-th coordination sphere of an atom of type K ther is an atom of type L, where $K,L = \{A,B,C\}$. In a three-component alloy, there are the following relationships between nine probabilities, $p_{KL}^{(i)}$ and three concentrations c_A, c_B, c_C:

$$p_{AA}^{(i)} + p_{AB}^{(i)} + p_{AC}^{(i)} = 1, \quad p_{BA}^{(i)} + p_{BB}^{(i)} + p_{BC}^{(i)} = 1, \quad p_{CA}^{(i)} + p_{CB}^{(i)} + p_{CC}^{(i)} = 1, \quad c_A + c_B + c_C = 1, \tag{2}$$

$$c_A\left(p_{AA}^{(i)} - c_A\right) = -\frac{1}{2}\left[c_A\left(p_{AB}^{(i)} + p_{AC}^{(i)}\right) + c_B\left(p_{BA}^{(i)} - 2c_A\right) + c_C\left(p_{CA}^{(i)} - 2c_A\right)\right], \tag{3}$$

$$c_B\left(p_{BB}^{(i)} - c_B\right) = -\frac{1}{2}\left[c_B\left(p_{BA}^{(i)} + p_{BC}^{(i)}\right) + c_A\left(p_{AB}^{(i)} - 2c_B\right) + c_C\left(p_{CB}^{(i)} - 2c_B\right)\right], \tag{4}$$

$$c_C\left(p_{CC}^{(i)} - c_C\right) = -\frac{1}{2}\left[c_C\left(p_{CA}^{(i)} + p_{CB}^{(i)}\right) + c_A\left(p_{AC}^{(i)} - 2c_C\right) + c_B\left(p_{BC}^{(i)} - 2c_C\right)\right]. \tag{5}$$

Let us denote by $\varphi_{KL}^{(i)}$ the binding energy of a pair of atoms of types K and L located at a distance equal to the radius of the i-th coordination sphere.

The potential energy of the structure per atom, taking into account the interaction of atoms in the first I coordination spheres can be written in the form

$$E = \sum_{i=1}^{I} \frac{N_i}{2} \left[\begin{array}{l} c_A\left(p_{AA}^{(i)}\varphi_{AA}^{(i)} + p_{AB}^{(i)}\varphi_{AB}^{(i)} + p_{AC}^{(i)}\varphi_{AC}^{(i)}\right) \\ + c_B\left(p_{BA}^{(i)}\varphi_{BA}^{(i)} + p_{BB}^{(i)}\varphi_{BB}^{(i)} + p_{BC}^{(i)}\varphi_{BC}^{(i)}\right) \\ + c_C\left(p_{CA}^{(i)}\varphi_{CA}^{(i)} + p_{CB}^{(i)}\varphi_{CB}^{(i)} + p_{CC}^{(i)}\varphi_{CC}^{(i)}\right) \end{array} \right]. \tag{6}$$

The energy of the completely disordered state of the structure is determined by expression (6) for $p_{AA}^{(i)} = p_{BA}^{(i)} = p_{CA}^{(i)} = c_A$, $p_{AB}^{(i)} = p_{BB}^{(i)} = p_{CB}^{(i)} = c_B$, $p_{AC}^{(i)} = p_{BC}^{(i)} = p_{CC}^{(i)} = c_C$, which gives

$$E_{disord} = \sum_{i=1}^{I} \frac{N_i}{2}\left[c_A^2\varphi_{AA}^{(i)} + c_B^2\varphi_{BB}^{(i)} + c_C^2\varphi_{CC}^{(i)} + 2c_Ac_B\varphi_{AB}^{(i)} + 2c_Ac_C\varphi_{AC}^{(i)} + 2c_Bc_C\varphi_{BC}^{(i)}\right]. \tag{7}$$

The energy of decomposition into pure components is determined by expression (6) for $p_{AB}^{(i)} = p_{BA}^{(i)} = p_{AC}^{(i)} = p_{CA}^{(i)} = p_{BC}^{(i)} = p_{CB}^{(i)} = 0$ $p_{AA}^{(i)} = 1$, $p_{BB}^{(i)} = 1$, $p_{CC}^{(i)} = 1$, which gives

$$E_{decomp} = \sum_{i=1}^{I} \frac{N_i}{2}\left[c_A\varphi_{AA}^{(i)} + c_B\varphi_{BB}^{(i)} + c_C\varphi_{CC}^{(i)}\right]. \tag{8}$$

Let us choose the energy E_{disord} as a reference point, and characterize the energy of any structure by the difference

$$\Delta E = E - E_{disord} = -\sum_{i=1}^{I} \frac{N_i}{4}\left[\alpha_{AB}^{(i)}\omega_{AB}^{(i)} + \alpha_{AC}^{(i)}\omega_{AC}^{(i)} + \alpha_{BC}^{(i)}\omega_{BC}^{(i)}\right], \tag{9}$$

where the short-range order parameters

$$\begin{aligned} \alpha_{AB}^{(i)} &= c_A p_{AB}^{(i)} + c_B p_{BA}^{(i)} - 2c_Ac_B, \\ \alpha_{AC}^{(i)} &= c_A p_{AC}^{(i)} + c_C p_{CA}^{(i)} - 2c_Ac_C, \\ \alpha_{BC}^{(i)} &= c_B p_{BC}^{(i)} + c_C p_{CB}^{(i)} - 2c_Bc_C, \end{aligned} \tag{10}$$

and the ordering energies

$$\begin{aligned} \omega_{AB}^{(i)} &= \varphi_{AA}^{(i)} + \varphi_{BB}^{(i)} - 2\varphi_{AB}^{(i)}, \\ \omega_{AC}^{(i)} &= \varphi_{AA}^{(i)} + \varphi_{CC}^{(i)} - 2\varphi_{AC}^{(i)}, \end{aligned} \tag{11}$$

$$\omega_{BC}^{(i)} = \varphi_{BB}^{(i)} + \varphi_{CC}^{(i)} - 2\varphi_{BC}^{(i)},$$

are introduced.

As we see from (9), the deviation of the energy of any structure from the energy of a disordered alloy is uniquely determined by the N_i coordination numbers, by the ordering energies $\omega_{KL}^{(i)}$, and by the parameters $\alpha_{KL}^{(i)}$, which are Cowley short-range order parameters [10].

To determine the range of variation of the order parameters α_i, the method of concentration waves [1] is used, adapted for the purposes of this study as follows. A computational cell with superimposed periodic boundary conditions is considered, which, for the three-dimensional lattice (N=3) contains the $N_x \times N_y \times N_z$ atoms. Then the following values are calculated:

$$C_{ijl} = \sin\left(\frac{2\pi i k_x^{(1)}}{N_x} + \frac{2\pi i k_y^{(1)}}{N_y} + \frac{2\pi i k_z^{(1)}}{N_z} + \varepsilon_1\right) + \sin\left(\frac{2\pi i k_x^{(2)}}{N_x} + \frac{2\pi i k_y^{(2)}}{N_y} + \frac{2\pi i k_z^{(2)}}{N_z} + \varepsilon_2\right) + \sin\left(\frac{2\pi i k_x^{(3)}}{N_x} + \frac{2\pi i k_y^{(3)}}{N_y} + \frac{2\pi i k_z^{(3)}}{N_z} + \varepsilon_3\right) + C, \quad (12)$$

where $0 \le i \times N_x$, $0 \le i \times N_y$, $0 \le i \times N_z$. The components of the vectors $\mathbf{k}^{(1)}$, $\mathbf{k}^{(2)}$, $\mathbf{k}^{(3)}$ in the expression (12) have integer values $0 \le k_x^{(1)}, k_x^{(2)}, k_x^{(3)} \le N_x$, $0 \le k_y^{(1)}, k_y^{(2)}, k_y^{(3)} \le N_y$ $0 \le k_z^{(1)}, k_z^{(2)}, k_z^{(3)} \le N_z$. The phase shifts ε_1, ε_2, and ε_3 are introduced in order to avoid the vanishing of the C_{ijl} numbers. The constant C determines the stoichiometry of the structure of the three-component A_2BC alloy and it should be taken as

$$C = \cos[\pi(c_B + \frac{c_c}{2})]. \quad (13)$$

Using the C_{ijl} numbers, the computational cell is filled with atoms of types A, B, and C in such a way that the node with the number (i, j, l) is assigned sort A if $C_{ijl} > 0$ and type B if $C_{ijl} < 0$ and $C_{ijl} \le \cos\left(\pi \frac{c_C}{2}\right) - \cos(\pi \cdot c_A)$. In this case, when filling the computational cell with atoms, the type C is placed if the values $C_{ijl} > 0$ and $C_{ijl} \ge -\left(\cos\left(\pi \frac{c_C}{2}\right)\right) - \cos(\pi \cdot c_A)$.

After filling the cell with atoms, the observance of stoichiometry is checked and, if it is violated, this structure is not considered. For structures with stoichiometry A_2BC, the short-range order parameters $\alpha_{KL}^{(i)}$, $i = 1,...,I$ are calculated and displayed as points in the I-dimensional $(\alpha_1,...,\alpha_I)$ space. This procedure is performed for all possible vectors $\mathbf{k}^{(1)}$, $\mathbf{k}^{(2)}$, $\mathbf{k}^{(3)}$ in (12) which, for sufficiently large N_x, N_y and N_z makes it possible to get an idea of the ranges of possible changes in the order parameters of the $\alpha_{KL}^{(i)}$.

The described algorithm is much more efficient than the method of full enumeration of the arrangement of atoms of types A, B and C over the periodicity cell, since its application is limited by the relatively small dimensions of the cell, N_x, N_y and N_z, and, which does not allow describing some of the possible structures.

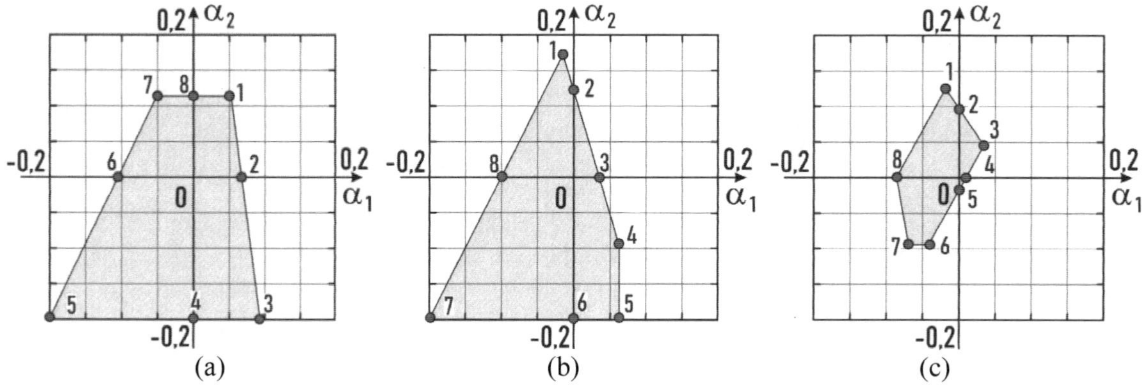

Figure 1. Range of admissible values of short-range order parameters for the A_2BC alloy on the bcc lattice for (a) $\alpha_{AB}^{(i)}$, (b) $\alpha_{AC}^{(i)}$ and (c) $\alpha_{BC}^{(i)}$.

As an example, the proposed algorithm is applied to the A_2BC alloy on the bcc lattice. The possible ranges of the order parameters are presented in Fig. 1 for (a) $\alpha_{AB}^{(i)}$, (b) $\alpha_{AC}^{(i)}$ and (c) $\alpha_{BC}^{(i)}$.

3. Conclusions

A simplified algorithm based on the method of concentration waves is proposed, which makes it possible to determine the ranges of possible changes in the short-range order parameters α_i for alloys with the composition A_2BC, given on an arbitrary lattice of arbitrary dimension N. The parameters α_i, together with the coordination numbers N_i and the ordering energies ω_i, determine the potential energy of the structures. The described algorithm is based on several essential assumptions. In particular, it does not take into account the deformation of the lattice upon permutation of atoms of sorts A, B and C, in other words, it is assumed that atoms of different sorts have the same size. The proposed procedure works much faster than full enumeration algorithm, and, for structures on three-dimensional lattices, the gain in time will be even more significant than for two-dimensional lattices.

Methodology developed in this study can be applied to the lattices of any dimension, in particular, to the novel 2D materials [11-15].

Acknowledgments

V.A. Valitov expresses gratitude for the financial support from the Russian Science Foundation, Grant No. 18-19-00685. This work was supported by the State Assignment of IMSP RAS No. AAAA-A17-117041310220-8.

References

[1] Khachaturyan A G 1974 *The Theory of Phase Transformations and the Structure of Solid Solutions* (Moscow: Nauka) 384 p [in Russian]
[2] Starenchenko S V, Kozlov E V and Starenchenko V A 2007 *Regularities of the Order-Disorder Thermal Phase Transition in Alloys with Superstructures* (Tomsk: Publishing House of Scientific and Technical Literature) 268 p. [in Russian]
[3] Khalikov A R 2011 *Fundamental'nye Problemy Sovremennogo Materialovedenia* **8** 109 [in Russian]
[4] Khalikov A R, Sharapov E A, Korznikova E A, Valitov V A and Dmitriev S V 2018 *Fundamental'nye problemy sovremennogo materialovedenia* **15** 482 [in Russian]
[5] Lobzenko I P, Khalikov A R, Sharapov E A, Korznikova E A, Semenov A S and Dmitriev S V 2019 *Fundamental'nye problemy sovremennogo materialovedenia* **16** 47 [in Russian]
[6] Khalikov A R, Sharapov E A, Korznikova E A, Potekaev A I, Starostenkov M D, Galieva E V and Dmitriev S V 2019 *Russ. Phys. J.* **62** 691

IOP Conf. Series: Materials Science and Engineering 1008 (2020) 012072 doi:10.1088/1757-899X/1008/1/012072

[7] Khalikov A R and Iskandarov A M 2013 *Rus. Phys. J.* **55** 1470
[8] Pavlukhina O O, Sokolovskiy V V, Buchelnikov V D and Zagrebin M A 2019 *Lett. Mater.* **9** 395
[9] Zagrebin M, Derevyanko S, Sokolovskiy V and Buchelnikov V 2018 *Lett. Mater.* **8** 21
[10] Cowley J M 1950 *Phys. Rev.* **77** 669
[11] Barani E, Korznikova E A, Chetverikov A P, Zhou K, Dmitriev S V 2017 *Phys. Lett.* A **381** 3553
[12] Kistanov A A, Khadiullin S K, Dmitriev S V and Korznikova E A 2019 *ChemPhysChem* **20** 575
[13] Babicheva R I, Dahanayaka M, Liu B, Korznikova E A, Dmitriev S V, Wu M S and Zhou K 2020 *Mater. Sci. Eng.* B **259** 114569
[14] Yoon T, Kang S, Yeob Kang T and Kim T-S 2019 *Comp. Model. Eng. Sci.* **120** 351
[15] Shcherbinin S A, Zhou K, Dmitriev S V, Korznikova E A, Davletshin A R and Kistanov A A 2020 *J. Phys. Chem.* C **124** 10235

IOP Conf. Series: Materials Science and Engineering 1008 (2020) 012073 doi:10.1088/1757-899X/1008/1/012073

Properties of one-dimensional nonlinear vibrational modes in triangular lattice with Lennard-Jones interactions

I R Sunagatova[1], A M Subkhangulova[1], M N Semenova[2], D I Borisov[3,4,5], A S Semenov[2] and S V Dmitriev[6,7]

[1] Bashkir State University, Zaki Validi Str. 32, Ufa 450076, Russia
[2] Mirny Polytechnic Institute, the branch of Ammosov North-Eastern Federal University, Mirny, Sakha (Yakutia), Russia
[3] Institute of Mathematics, Ufa Federal Research Center, Russian Academy of Sciences, Chernyshevsky Str. 112, Ufa 450008, Russia
[4] Bashkir State Pedagogical University named after M. Akhmulla, October Rev. Str. 3a, Ufa 450000, Russia
[5] University of Hradec Králové, 62, Rokitanského, Hradec Králové 50003, Czech Republic
[6] Institute for Metals Superplasticity Problems of RAS, 450001, Khalturin St. 39, Ufa, Russia
[7] Institute of Molecule and Crystal Physics, Ufa Federal Research Centre of the Russian Academy of Sciences, 71 October Avenue, Ufa 450054, Russia

E-mail: sunagatova66@gmail.com

Abstract. Crystal lattices support delocalized nonlinear vibrational modes (DNVMs), which are determined solely by the lattice point symmetry, and are exact solutions of the equations of atomic motion for any interatomic potential. DNVMs can be used for setting initial conditions to excite spatially localized vibrational modes called discrete breathers (DBs). In this study, DNVMs derived for instability can result in the formation of localized vibrational modes concentrating a significant part of the lattice energy. In some cases, localized vibrational modes can be obtained by imposing localizing functions upon DNVM.

1. Introduction

Well-known spatially localized vibrational modes in nonlinear lattices are called discrete breathers (DBs) [1–3]. It has been shown theoretically and experimentally that DBs exist in a number of discrete physical systems of different nature [4] including crystals [5] such as composed of carbon and hydrocarbon [6–14], *h*-BN [15], crystals with ionic [16–18] and covalent [19, 20] bonding, ordered alloys and intermetallics [21–23], various metals [24–30] and metal hydrides [31]. It is important to investigate properties of DBs because they can take part in defect structure evolution of crystalline solids [28] and affect macroscopic properties, for example, heat capacity [32], thermal expansion [33], thermal conductivity [34, 35] and stiffness [36]. DBs can be important for tribology [37].

DBs can be defined as spatially localized vibrational modes in nonlinear lattices. In most of the works on DBs it is assumed that they are localized in all spatial dimensions. However, it is clear that in a n-dimensional lattice DB can be localized in $k<n$ dimensions and delocalized in the remaining $n-k$ dimensions. For the first time, one-dimensional DBs ($k=1$) were analysed in two-dimensional lattice of graphene ($n=2$) by Baimova in the work [38]. Later, Bachurina has analysed linear ($k=1$) and planar ($k=2$) DBs in different metals with fcc lattice ($n=3$) [25, 26]. In those works, the delocalized nonlinear vibrational modes (DNVMs) were used for excitation of DBs.

Recall that DNVMs (originally called as the bushes of normal nonlinear modes [39–41]) are exact solutions of nonlinear dynamic equations, determined solely by the point symmetry of the crystal. Dynamics of N-component DNVM can be described by N coupled equations of motion. DNVMs can be used for excitation of DBs, for example, in fcc metals [25, 26], in triangular Morse lattice [42, 43] and in graphene [44].

In the present study we reconsider the one-dimensional DBs in triangular Lennard-Jones lattice [45] excited with the help of one-component DNVMs reported in [40] for a chain of coupled particles. It is important to note that DNVMs are unstable for amplitudes above a threshold value [46–51]. In this study, parameters of the unstable DNVMs are determined before the instability develops.

2. Simulation setup

A two-dimensional triangular lattice with an interatomic distance d is analyzed. Cartesian coordinate system with the x axis aligned with close-packed atomic rows is used. Interatomic interactions are described by the Lennard-Jones potential

$$U_{LJ}(r) = L\left[\left(\frac{r_l}{r}\right)^{12} - 2\left(\frac{r_l}{r}\right)^{6}\right], \tag{1}$$

where L and r_l are the parameters. Parameter L defines the binding energy and r_l is the equilibrium distance for a pair of atoms. We take $L=1$ and $r_l=1$, using them as the units of energy and distance, respectively. By a proper choice of the unit of time we also set the atom mass $M=1$. The cut-off radius is taken to be $5r_l$. Relaxation of the triangular lattice with these parameters give the equilibrium interatomic distance $d = 0.99027777$. For solving the equations of motion the sixth-order Stormer method is employed with the time step $\tau = 10^{-3}$. Periodic boundary conditions are used. The computational cell size includes 12×200 atoms along the x and the y-axis, respectively.

The following initial conditions are used (see figure 1). There exist four one-component DNVMs in a chain of particles. In a close-packed atomic row parallel to the x-axis one of them are excited by applying initial displacements along the x-axis as shown in figure 1(a-d). Atoms vibrating in the excited row produce kind of local thermal expansion. In order to take this thermal expansion into account, the atomic half-planes above and below the excited row are shifter as the rigid bodies by Y and $-Y$, respectively. Initial velocities of all atoms in the computational cell are equal to zero. Displacement patterns in DNVMs have period p. Modes from 1 to 4 have the periods 2, 3, 4 and 4, respectively.

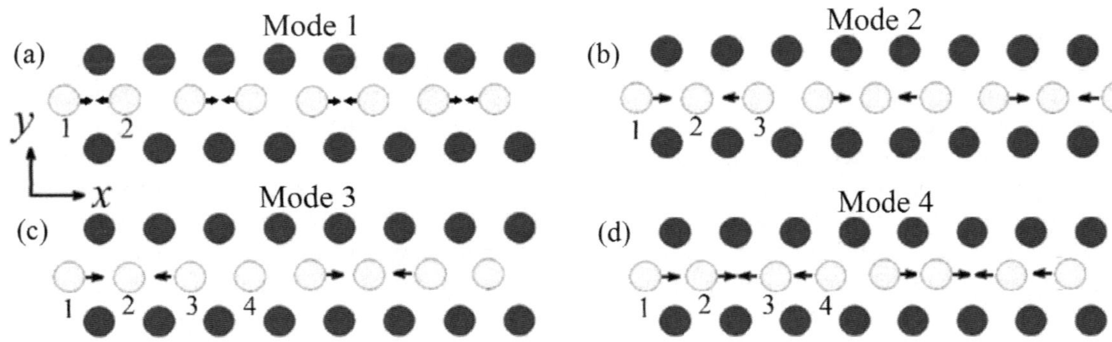

Figure 1. (a-d) Vibrational patterns used for excitation of one-dimensional DBs, 1 to 4, respectively, in triangular lattice. The x axis is directed along a close-packed atomic row, and the y axis is normal to the x axis. Only atoms in one close-packed row (shown in light colour) are initially excited. In (a) and (d) all atoms vibrate, in (b) and (c) some atoms are at rest. Period of the vibrational mode in (a) is equal to 2, in (b) to 3, and in (c) and (d) to 4. The half-planes above and below the excited atomic row are initially shifted as the rigid bodies by Y and $-Y$, respectively. DNVMs for the chain have been obtained and analyzed in [40,52,53].

3. Simulation results

In our simulations for each of the studied DNVMs we analyse the properties of the corresponding one-dimensional DBs. In figure 2 the following results are presented for each of the four studied DBs: (a) DB amplitude as the function of time, (b) DB frequency as the function of amplitude, (c) energy per atoms as the function of amplitude and (c) shift Y of the atomic half-spaces away from the excited atomic row as the function of amplitude. Curves for the modes from 1 to 4 are numbered.

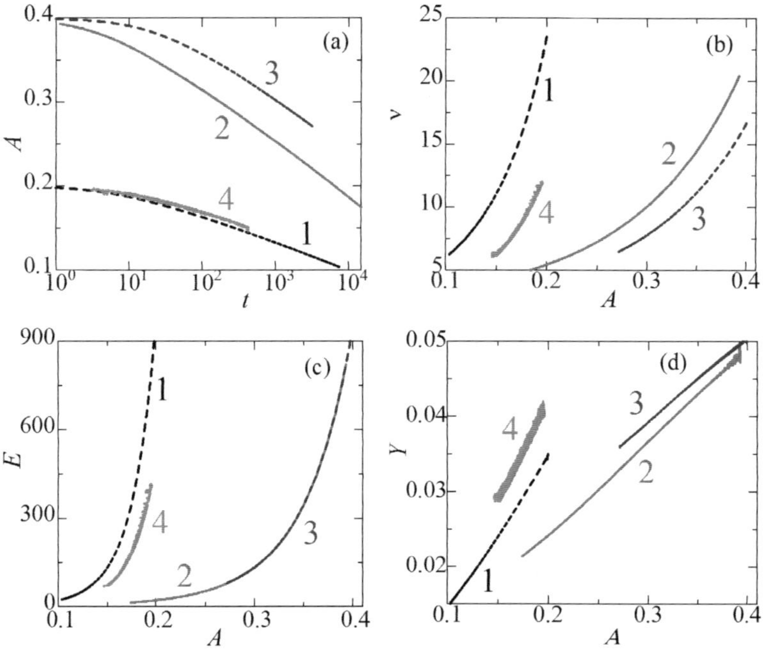

Figure 2. (a) DB amplitude as the function of time. (b) DB frequency as the function of amplitude. (c) Energy per atoms as the function of amplitude. (c) Shift of the atomic half-spaces away from the excited atomic row as the function of amplitude.

As can be seen from figure 2(a), DB amplitude decrease with time because they constantly radiate energy in the form of small-amplitude phonons. Recall that the DNVMs are the exact solutions for the chain of particles, but in our simulations the chain of excited atoms interacts with the rest of the two-dimensional lattice and DBs are not the exact solutions for triangular lattice and some energy radiation takes place. In figure 2(b) one can see that all four DBs demonstrate hard-type nonlinearity with frequencies increasing with the amplitude. DB 1 has the highest frequency for given amplitude. DB 4 has the second highest frequency. DB 3 has the lowest frequency and DB 2 the second lowest frequency. DBs 1 and 4 have relatively high frequencies because in the corresponding DNVMs all atoms move, while in DB 2 one third of atoms is at rest and in the lowest frequency DB3 a half of atoms is at rest (see figure 1). From figure 2(c) we deduce that, as expected, DB energy decreases with decreasing amplitude. Finally, looking at figure 2(d), we can see that the shift of the half-planes of the crystal away from the excited row depends almost linearly on the DB amplitude. DBs 4 produces greater shift than DB 1. DB 2 produces the smallest shift.

4. Conclusions

Molecular dynamics simulations were used to analyze properties of four one-dimensional DBs in triangular Lennard-Jones lattice. DBs were excited with the help of DNVMs derived for the chain of particles by Chechin with co-authors [40]. DB amplitudes as the functions of time were calculated. DB amplitude decreases with time for all studied DBs indicating that they are not the exact solutions of the equations of motion.

A natural continuation of the presented study could be the analysis of one-dimensional DBs in 3D crystals, e.g., in fcc and bcc metals.

IOP Conf. Series: Materials Science and Engineering 1008 (2020) 012073 doi:10.1088/1757-899X/1008/1/012073

Acknowledgments
This work was supported by the State Assignment of IMSP RAS No. AAAA-A17-117041310220-8.

References

[1] Dolgov A S 1986 *Sov. Phys. Solid State* **28** 907
[2] Sievers A J and Takeno S 1988 *Phys. Rev. Lett.* **61** 970
[3] Page J B 1990 *Phys. Rev.* B **41** 7835
[4] Flach S and Willis C R 1998 *Phys. Rep.* **295** 181
[5] Dmitriev S V, Korznikova E A, Baimova J A and Velarde M G 2016 *Phys. Usp.* **59** 446
[6] Shimada T, Shirasaki D and Kitamura T 2010 *Phys. Rev.* B **81** 035401
[7] Yamayose Y, Kinoshita Y, Doi Y, Nakatani A and Kitamura T 2007 *Europhys. Lett.* **80** 40008
[8] Kinoshita Y, Yamayose Y, Doi Y, Nakatani A and Kitamura T 2008 *Phys. Rev.* B **77** 024307
[9] Liu B, Baimova J A, Dmitriev S V, Wang X, Zhu H and Zhou K J 2013 *Physica* D **46** 305302
[10] Chechin G M, Dmitriev S V, Lobzenko I P and Ryabov D S 2014 *Phys. Rev.* B **90** 045432
[11] Khadeeva L Z, Dmitriev S V and Kivshar Y S 2011 *JETP Lett.* **94** 539
[12] Baimova J A, Dmitriev S V and Zhou K 2012 *Europhys. Lett.* **100** 36005
[13] Lobzenko I P, Chechin G M, Bezuglova G S, Baimova Yu A, Korznikova E A and Dmitriev S V 2016 *Phys. Solid State.* **58** 633
[14] Evazzade I, Lobzenko I P, Korznikova E A, Ovid'ko I A, Roknabadi M R and Dmitriev S V 2017 *Phys. Rev.* B **95** 035423
[15] Barani E, Korznikova E A, Chetverikov A P, Zhou K and Dmitriev S V 2017 *Phys. Lett.* A **381** 3553
[16] Kiselev S A and Sievers A J 1997 *Phys. Rev.* B **55** 5755
[17] Khadeeva L Z and Dmitriev S V 2010 *Phys. Rev.* B **81** 214306
[18] Riviere A, Lepri S, Colognesi D and Piazza F 2019 *Phys. Rev.* B **99** 024307
[19] Voulgarakis N K, Hadjisavvas G, Kelires P C and Tsironis G P 2004 *Phys. Rev.* B **69** 113201
[20] Murzaev R T, Bachurin D V, Korznikova E A and Dmitriev S V 2017 *Phys. Lett.* A **381** 1003
[21] Medvedev N N, Starostenkov M D and Manley M E 2013 *J. Appl. Phys.* **114** 213506
[22] Starostenkov M D, Potekaev A I, Dmitriev S V, Zakharov P V, Eremin A M and Kulagina V V 2016 *Russ. Phys. J.* **58** 1353
[23] Zakharov P V, Korznikova E A, Dmitriev S V, Ekomasov E G and Zhou K 2019 *Surf. Sci.* **679** 1
[24] Haas M, Hizhnyakov V, Shelkan A, Klopov M and Sievers A J 2011 *Phys. Rev.* B **84** 144303
[25] Bachurina O V 2019 *Comp. Mater. Sci.* **160** 217
[26] Bachurina O V 2019 *Model. Simul. Mater. Sci. Eng.* **27** 055001
[27] Murzaev R T, Kistanov A A, Dubinko V I, Terentyev D A and Dmitriev S V 2015 *Comp. Mater. Sci.* **98** 88
[28] Terentyev D A, Dubinko A V, Dubinko V I, Dmitriev S V, Zhurkin E E and Sorokin M V 2015 *Model. Simul. Mater. Sci. Eng.* **23** 085007
[29] Murzaev R T, Babicheva R I, Zhou K, Korznikova E A, Fomin S Y, Dubinko V I and Dmitriev S V 2016 *Eur. Phys. J* B **89** 168
[30] Bachurina O V, Murzaev R T, Semenov A S, Korznikova E A and Dmitriev S V 2018 *Phys. Solid State* **60** 989
[31] Dubinko V, Laptev D, Terentyev D, Dmitriev S V and Irwin K 2019 *Comp. Mater. Sci.* **158**, 389
[32] Mihaila B et al 2006 *Phys. Rev. Lett.* **96** 076401
[33] Manley M E 2010 *Acta Mater.* **58** 2926
[34] Xiong D, Saadatmand D, Dmitriev S V 2017 *Phys. Rev.* E **96** 042109
[35] Wang J, Dmitriev S V and Xiong D 2020 *Phys. Rev. Res.* **2** 013179
[36] Korznikova E A, Morkina A Y, Singh M, Krivtsov A M, Kuzkin V A, Gani V A, Bebikhov Y V and Dmitriev S V 2020 *Eur. Phys. J.* B **93** 123
[37] Popov V 2019 *Facta Universitatis, Ser.: Mech. Eng.* **17** 39
[38] Baimova J 2016 *Lett. Mater.* **6** 31

[39] Chechin G M and Sakhnenko V P 1998 *Physica* D **117** 43
[40] Chechin G M, Ryabov D S and Zhukov K G 2005 *Physica* D **203** 121
[41] Chechin G M, Ryabov D S and Shcherbinin S A 2017 *Lett. Mater.* **6** 9
[42] Fomin S Yu and Korznikova E A 2016 *Lett. Mater.* **6** 57
[43] Kistanov A A, Semenov A S and Dmitriev S V 2014 *J. Exp. Theor. Phys.* **119** 766
[44] Barani E, Lobzenko I P, Korznikova E A, Soboleva E G, Dmitriev S V, Zhou K and Marjaneh A M 2017 *Eur. Phys. J* B **90** 38
[45] Semenov A S, Murzaev R T, Bebikhov Y V, Kudreyko A A and Dmitriev S V 2020 *Lett. Mater.* **10** 185
[46] Abdullina D U, Semenova M N, Semenov A S, Korznikova E A and Dmitriev S V 2019 *Eur. Phys. J.* B **92** 249
[47] Burlakov V M and Kiselev S 1991 *Sov. Phys. JETP* **72** 854
[48] Ikeda K, Doi Y, Feng B F and Kawahara T 2007 *Physica* D **225** 184
[49] Korznikova E A, Bachurin D V, Fomin S Yu, Chetverikov A P and Dmitriev S V 2017 *Eur. Phys. J.* B **90** 23
[50] Shcherbinin S A, Semenova M N, Semenov A S, Korznikova E A, Chechin G M, Dmitriev S V 2019 *Phys. Solid State* **61** 2139
[51] Bachurina O V, Murzaev R T and Bachurin D V 2019 *J. Micromech. Mol. Phys.* **04** 1950001
[52] Poggi P and Ruffo S 1997 *Physica* D **103** 251
[53] Chechin G M and Shcherbinin S A 2015 *Commun. Nonlinear Sci.* **22** 244

AUTHOR INDEX